EMERGENT COMPUTATION

SPECIAL ISSUES OF *PHYSICA D*

The titles in this series are paperback, readily accessible editions of the Special Volumes of *Physica D*, produced by special agreement with Elsevier Science Publishers B.V.

Emergent Computation: Self-Organizing, Collective, and Cooperative Phenomena in Natural and Artificial Computing Networks, edited by Stephanie Forrest

EMERGENT COMPUTATION

SELF-ORGANIZING, COLLECTIVE, AND COOPERATIVE PHENOMENA IN NATURAL AND ARTIFICIAL COMPUTING NETWORKS

edited by
Stephanie FORREST

A Bradford Book
The MIT Press
Cambridge, Massachusetts
London, England

First MIT Press edition, 1991

©1991 Elsevier Science Publishers B.V.

All rights reserved. No part of this book may be reproduced in any form by any electronic or mechanical means (including photocopying, recording, or information storage and retrieval) without permission in writing from the publisher.

This book was printed and bound in the Netherlands.

Library of Congress Cataloging-in-Publication Data

Emergent computation: self-organizing, collective, and cooperative phenomena in natural and artificial computing networks / edited by Stephanie Forrest.
    p.   cm.---(Special issues of physica D)
"A Bradford book."
Includes bibliographical references and index.
ISBN 0-262-56057-7
  1. Parallel processing(Electronic computers)--Congresses. 2. Computer networks--Congresses.
I. Forrest, Stephanie. II. Series.
QA76.58.E44 1991
004'.35--dc20
                                                    90-46186
                                                      CIP

PREFACE TO THE MIT PRESS EDITION

The conference on Emergent Computation was the ninth in a series of annual interdisciplinary conferences organized by the Center for Nonlinear Studies (CNLS) and held at Los Alamos National Laboratory. The conference consisted of twenty-four invited talks, an extensive poster session, and lots of discussion. Papers for the Proceedings were selected from both the invited talks and the poster session. All of the papers appearing in this volume have been refereed according to the standards of Physica D. In selecting papers to appear in the Proceedings we required both that the paper represent a novel contribution in its own field and that it say something relevant about emergent computation. The Proceedings are published by Elsevier Science Publishers B.V. as a special issue of Physica D. The MIT Press is publishing the Proceedings in a paperback edition for a worldwide audience.

Our goal has been to define the area of emergent computation and to bring the principles of computation within the scope of nonlinear studies. Descriptive models of real physical systems demonstrate how natural chaos can arise. People, however, create computational systems to achieve specific goals such as learning to solve a problem, building a robust model of an environment, or maintaining a computer network. Evolution also provides examples of living emergent systems. In all of these systems, the major question from an emergent computation perspective is how an architecture (whether hardware, software, or biological) with many interactions and often unpredictable effects can come to be used effectively.

It is a great pleasure to thank my many friends at Los Alamos for their generous help with the conference. The organizing committee, Chris Barrett, John George, Alan Lapedes, George Papcun, and Bryan Travis, played a major role in shaping the conference, particularly in selecting invited speakers. I am also grateful to the following scientists who chaired the sessions: Randy Michelson, George Papcun, Steen Rasmussen, Bryan Travis, and Alan Lapedes. The anonymous reviewers did an outstanding job and made a significant contribution to the overall quality of the proceedings. Marian Martinez and the rest of the CNLS office staff (Barbara, Dorothy, Frankie, Lisa, and Tanya) did a tremendous amount of administrative work to make the conference a success. They were the true organizers of the conference. Leeroy Herrera coordinated our use of the conference facilities. Susan Carlson, Jim Cruz, and Marjorie Mascheroni designed the conference poster. Fred Carey, Gary Doolen, Doyne Farmer, Mac Hyman, Erica Jen, Chris Langton, Tim Thomas, and many others all made useful suggestions and helped me in their own ways. David Campbell, Director of CNLS, was a tremendous resource, providing personal support, resolving many administrative details, and delivering an excellent talk at the conference banquet. Sig Hecker, Director of the Los Alamos National Laboratory, made the welcoming remarks. Finally, I am grateful to Don Austin of the Scientific Computing Staff of the Office of Energy Research, the Department of Energy, for his continuing financial support of the CNLS conferences.

Stephanie Forrest
Los Alamos, 1990

CONTENTS

Preface

Emergent computation: self-organizing, collective, and cooperative phenomena in natural and artificial computing networks. Introduction to the Proceedings of the Ninth Annual CNLS Conference
S. Forrest ... 1

Computation at the edge of chaos: phase transitions and emergent computation
C.G. Langton ... 12

The performance of cooperative processes
B.A. Huberman ... 38

Collective behavior of predictive agents
J.O. Kephart, T. Hogg and B.A. Huberman ... 48

Toward diagnosis as an emergent behavior in a network ecosystem
R.A. Maxion ... 66

The complex behavior of simple machines
R. Machlin and Q.F. Stout ... 85

Computer arithmetic, chaos and fractals
J. Palmore and Ch. Herring ... 99

The Coreworld: emergence and evolution of cooperative structures in a computational chemistry
S. Rasmussen, C. Knudsen, R. Feldberg and M. Hindsholm ... 111

Requirements for evolvability in complex systems: orderly dynamics and frozen components
S.A. Kauffman ... 135

A Rosetta Stone for connectionism
J.D. Farmer ... 153

Concerning the emergence of tag-mediated lookahead in classifier systems
J.H. Holland ... 188

Learning and bucket brigade dynamics in classifier systems
M. Compiani, D. Montanari and R. Serra ... 202

Emergent behavior in classifier systems
S. Forrest and J.H. Miller ... 213

Co-evolving parasites improve simulated evolution as an optimization procedure
W.D. Hillis ... 228

Computer symbiosis – emergence of symbiotic behavior through evolution
T. Ikegami and K. Kaneko ... 235

Using genetic search to exploit the emergent behavior of neural networks
J.D. Schaffer, R.A. Caruana and L.J. Eshelman ... 244

Perceptron redux: emergence of structure
S.W. Wilson ... 249

Contents

An energy function for specialization
  W. Banzhaf and H. Haken — 257

A stochastic version of the delta rule
  S.J. Hanson — 265

Synchronous or asynchronous parallel dynamics. Which is more efficient?
  I. Kanter — 273

On the nature of explanation: a PDP approach
  P.M. Churchland — 281

Parallel simulated annealing techniques
  D.R. Greening — 293

Geometric learning algorithms
  S.M. Omohundro — 307

The emergence of understanding in a computer model of concepts and analogy-making
  M. Mitchell and D.R. Hofstadter — 322

The symbol grounding problem
  S. Harnad — 335

Selectionist models of perceptual and motor systems and implications for functionalist theories of brain function
  G.N. Reeke Jr. and O. Sporns — 347

Bifurcation and category learning in network models of oscillating cortex
  B. Baird — 365

Non-linear dynamical system theory and primary visual cortical processing
  R.M. Siegel — 385

A dynamical system view of cerebellar function
  J.D. Keeler — 396

Neuromagnetic studies of human vision: noninvasive characterization of functional neural architecture
  J.S. George, C.J. Aine and E.R. Flynn — 411

Computational connectionism within neurons: a model of cytoskeletal automata subserving neural networks
  S. Rasmussen, H. Karampurwala, R. Vaidyanath, K.S. Jensen and S. Hameroff — 428

List of contributors — 450
Analytic subject index — 451

EMERGENT COMPUTATION:
SELF-ORGANIZING, COLLECTIVE, AND COOPERATIVE PHENOMENA IN NATURAL AND ARTIFICIAL COMPUTING NETWORKS

INTRODUCTION TO THE PROCEEDINGS OF THE NINTH ANNUAL CNLS CONFERENCE

Stephanie FORREST

Center for Nonlinear Studies and Computing Division, MS-B258, Los Alamos National Laboratory, Los Alamos, NM 87545, USA

Parallel computing has typically emphasized systems that can be explicitly decomposed into independent subunits with minimal interactions. For example, most parallel processing systems achieve speedups by identifying code and data segments that can be executed simultaneously. Under this approach, interactions among the various segments are managed directly, through synchronization, and communication among components is viewed as an inherent cost of computation. As a result, most extant parallel systems require substantial amounts of overhead to manage and coordinate the activities of the various processors, and they obtain speedups that are considerably less than a linear function of the number of processors.

An alternative approach exploits the interactions among simultaneous computations to improve efficiency, increase flexibility, or provide a more natural representation. Researchers in several fields have begun to explore computational models in which the behavior of the entire system is in some sense more than the sum of its parts. These include connectionist models [46], classifier systems [22], cellular automata [5, 7, 56], biological models [11], artificial-life models [33], and the study of cooperation in social systems with no central authority [2]. In these systems interesting global behavior *emerges* from many local interactions. When the emergent behavior is also a computation, we refer to the system as an *emergent computation*.

The distinction between standard and emergent computations is analogous to the difference between linear and nonlinear systems. Emergent computations arise from nonlinear systems while standard computing practices focus on linear behaviors (see section 2.2). The idea that interactions among simple deterministic elements can produce interesting and complex global behaviors is well accepted in the physical sciences. However, the field of computing is oriented towards building systems that accomplish specific tasks, and emergent properties of complex systems are inherently difficult to predict. Thus, it is not immediately obvious how architectures (either hardware or software) that have many interactions with often unpredictable effects can be used effectively, and it is for this reason that I have chosen to use the term "emergent computation" instead of referring more broadly to nonlinear properties of computational systems. The premise of emergent computation is that interesting and useful computational systems can be constructed by exploiting interactions among primitive components, and further, that for some kinds of problems (e.g. modeling intelligent behavior) it may be the only feasible method.

To date, there has been no unified attempt to specify carefully what constitutes an emergent computation or to determine what properties are required of the supporting architectures that generate them. This volume contains the Proceedings of a recent Conference held at Los Alamos National Laboratory devoted to these questions. Emergent computation is potentially relevant to several areas, including adaptive systems, parallel processing, and cognitive and biological modeling, and the Proceedings reflects this diversity, with contributions from physicists, computer scientists, biologists, psychologists, and philosophers. The Proceedings are thus intended for an interdisciplinary audience. Each paper addresses this by providing introductory explanations beyond that required for a field-specific publication. In this introduction I hope to stimulate discussion of emergent computation beyond the scope of the actual conference. First, a definition is proposed and several detailed examples are presented. Then several common themes are highlighted, and the contents is briefly reviewed.

1. What is emergent computation?

It is increasingly common to describe physical phenomena in terms of their information processing properties [57, 58]. However, we wish to distinguish emergent computation from the general emergent properties of complex phenomena. We do this by requiring that both the explicit and the emergent levels of a system be computations. For example, a Rayleigh–Bénard convecting flow, in which the dynamics of the fluid particles follow a chaotic path, would not necessarily be considered a form of emergent computation.

The requirements for emergent computation are quite similar to those proposed by Hofstadter in his paper on subcognition [19]. He stresses that information which is absent at lower levels can exist at the level of collective activities. This is the essence of the following constituents of emergent computation:

(i) A collection of agents, each following explicit instructions;

(ii) Interactions among the agents (according to the instructions), which form implicit global patterns at the macroscopic level i.e. epiphenomena;

(iii) A natural interpretation of the epiphenomena as computations.

The term "explicit instructions" refers to a primitive level of computation, also called "micro-structure", "low-level instructions", "local programs", "concrete structure", and "component subsystems". In a typical case, such as cellular automaton, each cell acts as an agent executing the instructions in its state-transition table. However, in some cases the coding for an instruction may not be distinguished from the agent that executes it. The important point is that the explicit instructions are at a different (and lower) level than the phenomena of interest. The level of an instruction is determined by the entity that processes it. For example, if the low-level instructions were machine code, they would be executed by hardware, while higher-level instructions would be interpreted by "virtual machines" simulated by the lower-level machine code instructions. The higher-level instructions would be implicit, although not necessarily the product of interactions (see section 2.2 on superposition).

There is a tension between low-level explicit computations and the patterns of their interaction, and the interaction among the levels is important. Global patterns may influence the behavior of the lower-level local instructions, that is, there may be feedback between the levels. Patterns that are interpretable as computations process information, which distinguishes emergent computation from the interesting global properties of many complex systems such as the Rayleigh–Bénard experiment mentioned earlier.

Central to the definition is the question of to what extent the patterns are "in the eye of the beholder", or interpreted, and to what extent they are inherent in the phenomena itself. This issue arises because the phenomena of interest are implicit rather than explicit. Note that to a lesser extent the interpretation problem also exists in standard computation. The difference is that in emergent computations there is no one address (or set of addresses) where one can read out an answer. Thus, the time for interpretation is likely to be much lower for standard computations than emergent ones. Currently, many emergent computations are interpreted by the perceptual system of the person running the experiment. Thus, when conducting a cellular automaton experiment, researchers typically rely on graphics-based simulations to reveal the phenomena of interest. While quantitative measures can be developed in some cases to interpret the results, scientific visualization techniques are an integral part of most current emergent computations.

According to the Church–Turing thesis, a Turing machine can both implement any definable computation and simulate any set of explicit instructions we might choose as the basis of an emergent computation [23]. Thus, the concept of emergent computation cannot contribute magical computational properties. Rather, we are advocating a way of thinking about the design of computational systems that could potentially lead to radically different architectures which are more robust and efficient than current designs[#1].

A related question is whether or not emergent computations can be implemented in more traditional ways (i.e. can the emergent patterns be encoded as a set of explicit instructions instead of indirectly as implicit patterns?). While in some cases it may be possible to encode the emergent patterns directly in some language or machine, there are several advantages to an emergent-computation approach, including efficiency, flexibility, representation, and grounding. First, implementing computations indirectly as emergent patterns may provide implementation efficiencies because of the need for less control over the different components (e.g. processes). As mentioned earlier, a high proportion of computing time is devoted to managing interactions among processes. Other kinds of efficiencies may also be realized, including efficiencies of cost through the use of multiple cheap components, efficient uses of programmer time, and raw computational speed through the use of massive parallelism. Second, flexibility is important for systems that must interact with complex and dynamic environments, e.g. intelligent systems. For these systems, it is impossible to get enough flexibility from explicit instructions; for realistic environments, it is just not possible to program in all contingencies ahead of time. Therefore, the flexibility must appear at the emergent level. The interaction between the instructions and the environment (or between emergent properties of the instructions and the environment) is important, and there are global patterns (symbols, etc.) associated with this instruction–environment interaction. Third, the advantage in representation arises in systems for which it is difficult to articulate a formal description of the emergent level. Several authors have argued for the impossibility of such an undertaking for systems of sufficient complexity such as weather patterns and living systems [19, 33, 35, 48]. In these circumstances, emergent systems may provide the most natural model. Finally, the grounding issue arises if the emergent patterns are intended as real phenomena or models of real phenomena (as in cognitive modeling). In this circumstance, the intended interpretation of a purely formal model (e.g. symbolic models of artificial intelligence) becomes problematic since the model is not connected to (grounded in) the domain of interest (by e.g. a sensory interface). Emergent-computation models can address this problem by using low-level explicit instructions that are directly connected to the domain. Harnad's paper discusses the grounding problem in detail [17].

At the architectural level, there are two criteria that capture the spirit of emergent computation: efficacy and efficiency. The criterion of computational efficacy is met by systems in which each

---

[#1] Even if in principle emergent computations can be simulated by a Turing machine, interpreting the resulting patterns as computations is likely to be so difficult as to be infeasible.

computational unit has limited processing power (e.g. a finite state machine) and in which the collective system is computationally more powerful (e.g. a Turing machine). The criterion of computational efficiency can be met by parallel models that are capable of linear or better than linear speedups relative to the number of processors used to solve the problem. (Other criteria can be imagined that are less strict. For example, it may be reasonable to construct systems in which not all of the processors are used all of the time. In these cases, the dimensions of time and number of processors might be combined to obtain a reasonable definition somewhat different from that mentioned above.) Previous work on computational systems with interesting collective/emergent properties has generally focused on some variant of the first of these criteria and ignored the second.

2. Example problems

In this section, three concrete examples are presented to illustrate what sorts of problems emergent computation can address and provide a framework for interpreting the definition. The parallel processing example shows how emergent computation can lead to efficiency improvements. The section on programming languages establishes the connection between emergent computation and nonlinear systems. Finally, two search techniques are compared to show how the emergent-computation approach to a problem differs from other more conventional approaches.

2.1. Parallel processing

Consider the problem of designing a large complex computational system to perform reliably and efficiently. By "large complex" we mean that there are many components and many interactions among the components. The computational system could be a complicated algorithm that we would like to run in parallel, it could be distributed over many machines with possibly heterogeneous operating systems, or it could simply be a large, evolving software package being modified simultaneously by several different programmers. In the following, we will focus on the parallelization example, but similar arguments can be made for the distributed systems and software engineering aspects of the problem.

The conventional approach to such a problem looks for code and/or data segments that can be executed independently, and hence simultaneously. With this view, it is important to minimize the interactions among the various components. Synchronization strategies are defined to manage communication among the independent components. The controlling program *knows* about all possible interactions and manages them directly. Thus, interactions are viewed as costs to be minimized. As mentioned earlier, most extant parallel systems consume substantial amounts of overhead managing and coordinating the activities of their processors. This is because the flow of data and control rarely match exactly the interconnection topology of the parallel machine. Potential speedups are also limited by inherently sequential components within a computation, a phenomenon quantified by Amdahl's law [1]. For these reasons, the speedups that are achieved by most parallel systems are considerably less than a linear function of the number of processors.

Emergent computation suggests a view of parallelism in which the interactions among components lead to problem solutions with potentially better than linear performance. For example, a system that performs explicit search at the concrete architecture level, but is implicitly searching a much larger space (see section 2.3) meets this criterion.

The claim that superlinear speedups are in principle possible is controversial. The standard counterargument is as follows: if a process P runs in t time on n processors, then there exists a sequential machine that can simulate P in at most $(k_1 nt) + k_2$ time steps, where k_1 and k_2 are constants, so the speedup is only by a factor of n. This counterargument ignores the time required to interpret the results. If the result of the computation is a global pattern (e.g. a pattern of states in a cellular automaton distributed across several time steps), then the procedure for recognizing that same pattern on a sequential machine might require as much computation as the original computation.

Even allowing for the theoretical possibility of superlinear speedups, one might question whether or not it is feasible to actually construct such a system. The following simple example illustrates how interactions among components can provably help the efficiency of a computation. In formal models of parallel computation, there are various assumptions about what happens when two independent processors try to write to the same location in global memory simultaneously. Some of these assumptions forbid any interaction between the two simultaneous writes: for example, one processor is allowed to dominate and write successfully while the other processor is forced to wait (an "Exclusive Write"). Others exploit the interaction: for example, one version of the "Concurrent Write" model prevents both processors from writing but records the collision as a "?" in the memory cell, destroying the previous contents. The Concurrent Write model turns out to be provably stronger than the Exclusive Write in the sense that certain parallel algorithms can be implemented more efficiently with Concurrent Write than they can with Exclusive Write (for example, computing certain kinds of disjunction [24]). The trick is that if the interaction is recorded as a "?" then both processors that tried to write can inspect that memory cell and determine that there was a collision. This information can be exploited in certain circumstances to produce more efficient algorithms. Thus, in the Concurrent Write model, write collisions (interactions) are shown to be a useful form of computation, leading to performance improvements, even if the collisions themselves do not result in transmitted values reaching their destination. This small example meets the criterion of "computing by interaction", although the interactions are recorded explicitly rather than implicitly, as we would expect in a truly emergent computation.

Generally, we expect the emergent-computation approach to parallelism to have the following features: (1) no central authority to control the overall flow of computation, (2) autonomous agents that can communicate with some subset of the other agents directly, (3) global cooperation (see section 3) that emerges as the result of many local interactions, (4) learning and adaptation replacing direct programmed control, and (5) the dynamic behavior of the system taking precedence over static data structures.

2.2. Programming languages and the superposition principle

Emergent computation arises from interaction among separate components. There are several ways in which the standard approach to programming-language design minimizes the potential for emergent computation. This example explores the connection between emergent computation and nonlinear systems.

The notation, or syntax, used to express computer programs is for the most part *context free*. Roughly, this means that legal programs are required to be written in such a way that the legality (whether or not the program is syntactically correct) of any one part of the program can be determined independently of the other parts. While this is a very powerful property (among other things, it makes it possible to build efficient compilers), emergent computations are almost certainly not context free since they arise from interactions among components. However, the low-level instructions that generate emergent computations may well be context free.

The semantics of programming languages can be described by any of several different standard mathematical models [52]. These models describe how the syntax should be interpreted, that is, what a program *means* (more specifically, what function it computes). The meaning of a program helps determine the set of low-level machine instructions that are executed when a program *runs*. The standard approaches to programming-language semantics discourage emergent computation. For example, in the denotational semantics approach [52], the meaning of a program is determined by composing the meanings of its constituents. The meaning of an arithmetic expression $A + B$ might be written as follows: (read ⟦*expression*⟧ as "the meaning of expression")[#2]:

$$⟦A \oplus B⟧ = ⟦A⟧ + ⟦B⟧.$$

Thus, the meaning of A is isolated from B, and can be computed independently of it. Similar expressions can be written for all the common programming constructs, including assignment statements, conditional statements, and loops. By contrast, we expect that in an emergent computation there would be interactions between components that would not interact in standard computation.

This compositional approach to programming-language semantics is analogous to the superposition principle in physics, which states that for homogeneous linear differential equations, the sum of any two solutions is itself a solution. Systems that obey the superposition principle are linear and thus incapable of generating complicated behaviors associated with nonlinear systems such as chaos, solitons, and self-organization. Similarly in the domain of programming languages, the ability to define the meaning of context-free programs in terms of their constituent parts indicates that there are few if any interactions between the meaning of one part and the meaning of another. In these sorts of languages and models, the goal is to minimize side effects that could lead to inadvertent interactions (e.g. changing the value of a global variable) – once again, emergent computation is primarily computation by side effect. The analogy between the superposition principle in physics and the compositional approach of denotational semantics suggests that something like the

---

[#2] Two different plus symbols are used to distinguish between the symbol plus ($\oplus$) and the operation that implements it (+).

distinction between linear and nonlinear models in physics exists in computational systems. Note, however, that it is possible to write programs in context-free languages that have nonlinear behaviors when executed, e.g. a simple logistic map, just as it may be possible to write the low-level instructions for an emergent-computation system in a context-free language.

While nonlinear computational systems are more difficult to engineer than linear ones, they are capable of much richer behavior. The role of enzymes in catalysis provides a nice example of how nonlinear effects can arise from simple recombinations of compounds [10]. More generally, consider the problem of recombination in adaptive systems. If one can detect combinations that yield effects not anticipated by superposition, then those combinations can be exploited in various ways that are not available in a model based on principles of superposition.

A final example of how the principle of superposition pervades standard programming languages is provided by the Church–Rosser theorem [9]. The λ calculus defines a formal representation for functions and is closely related to the Lisp programming language. In the λ calculus, various substitution and conversion rules are defined for reducing λ expressions to normal form. The Church–Rosser theorem (technically, one of its corollaries) shows that no λ expression can be converted to two different normal forms (for example, by applying reductions in different order). This is another example of how computer science gets a lot of leverage out of systems that have something like the principle of superposition. Since nonlinear systems often have the property that operations applied in different orders have different effects, emergent computations will not in general have nice simplification rules like the Church–Rosser theorem.

2.3. Search

The problem of searching a large space of possibilities for an acceptable solution, a particular datum, or an optimal value is one of the most basic operations performed by a computer. Intelligent systems are often described in terms of their capabilities for "intelligent" search – that is the ability to search an intractably large space for an acceptable solution, using knowledge-based heuristics, previous experience, etc. The various techniques of intelligent search provide a sharp contrast between emergent computation and traditional approaches to computation. A classical approach to the problem of search is that of an early artificial intelligence program, the general problem solver (GPS) [39], while the emergent-computation approach is illustrated by the genetic algorithm [20].

GPS uses means–ends analysis to search a state space to find some predetermined goal state. GPS works by defining subgoals part way between the start state and the goal state, and then solving each of the subgoals independently (and recursively). Under this approach, the domain of problem solving is viewed as "nearly decomposable" [50], meaning that for the most part each subgoal can be solved without knowledge of the other subgoals in the system. The overall approach taken by GPS is still prevalent in artificial intelligence, the recent work on SOAR [31] being a good example.

In contrast, genetic algorithms [14, 20] show how emergent computation can be used to search large spaces. There are two levels of the algorithm, explicit and implicit. At the mechanistic or explicit level, a genetic algorithm consists of:

(i) A population of randomly chosen bit strings: $P \subseteq \{0, 1\}^l$, representing an initial set of guesses, where l is a fixed positive integer denoting the length in bits of a guess;

(ii) A fitness function: F: guesses $\rightarrow \mathbb{R}$, where $\mathbb{R}$ denotes the real numbers;

(iii) A scheme for differentially reproducing the population based on fitness, such that more copies are made of more fit individuals and fewer or no copies of less fit ones;

(iv) A set of "genetic" operators (e.g. mutation, crossover, and inversion) that modify individuals to produce new guesses;

(v) Iteration for many generations of the cycle: evaluation of fitness, differential reproduction, and application of operators. Over time, the population will become more like the successful individuals of previous generations and less like the unsuccessful ones.

At the virtual, or implicit, level, we can interpret the genetic algorithm as searching a higher-order space of patterns, the space of hyperplanes in $\{0, 1\}^l$. When one individual is evaluated by the fitness function, many different hyperplanes are being sampled simultaneously – the so-called implicit parallelism of the genetic algorithm. For example, evaluating the string 000 provides information about the following hyperplanes[#3]:

$$000, 00\#, 0\#0, \#00, 0\#\#, \#0\#, \#\#0, \#\#\#.$$

Populations undergoing reproduction and cross-over (with some other special conditions) are guaranteed exponentially increasing samples of the observed best schemata (a property described in refs. [14, 20]). Thus, performance improvements provably arise from the collective properties of the individuals in the population over time. The population serves as a distributed database that implicitly contains recoverable information about the multitudes of hyperplanes (because each individual serves in the sample set of many hyperplanes). Put another way, the population reflects the ongoing statistics of the search over time.

Several aspects of emergent computation are illustrated by this example. The algorithm is very flexible, allowing it to track changes in the environment. Since the statistical record of the search is distributed across the population of individuals, interpretation is an issue if there is a need to recover the statistics explicitly. Normally it is sufficient to look at a few typical individuals or to

---

[#3] The # symbol means "don't care". Thus, #00 denotes the pattern, or *schema*, which requires that the second 2 bits be set to 0 and will accept a 0 or a 1 in the first bit position. The space of possible schemata is the space of hyperplanes in $\{0, 1\}^l$. (See ref. [14] for an introduction to both the mechanism and theory of the genetic algorithm.)

treat the best individual seen as the "answer" to the problem. The potential efficiency of emergent computation is also demonstrated through the use of implicit parallelism. There is a price, however. While the algorithm is highly efficient, it achieves its efficiency through sampling. This means that there is some loss of accuracy (see Greening's paper in these Proceedings [15] for a careful treatment of this issue).

3. Themes of emergent computation

Three important and overlapping themes of systems that exhibit emergent computation are self-organization, collective phenomena, and cooperative behavior. Here, we use the term self-organization to mean the spontaneous emergence of order from an initially random system, but see ref. [40] for a detailed formulation of self-organization. Collective phenomena are those in which there are many agents, many interactions among the agents, and an emphasis on global patterns. A third component of emergent computation is the notion of cooperative behavior, i.e. that the whole is somehow more than the sum of the parts. In this section, these three themes are illustrated in the context of several examples.

One of the most compelling examples comes from nature in the form of ant colonies. The actions of any individual ant are quite limited and apparently random, but the collective organization and behavior of the colony is highly sophisticated, including such activities as mass communication and nest building [53, 54]. In the absence of any centralized control, the collective entity (the colony) can "decide" (the decision itself is emergent) when, where, and how to build a nest – self-organizing, collective, and cooperative behavior in the extreme. Clearly, many of the activities in an ant colony involve information-processing, such as laying trails from the nest to potential food sites, communicating the quality and quantity of food at a particular site, etc. By making an analogy between the cells in a cellular automaton and individual ants, Langton has described computational models that emulate some of the important information-processing aspects of ant colonies [32].

Kauffman's article in these Proceedings [28] explores self-organizing behavior in simple randomly connected networks of Boolean function. These networks spontaneously organize themselves into regular structures of "frozen components" that are impervious to fluctuating states in the rest of the network. The tendency of a network to exhibit this and other self-organizing behaviors is related to various structural properties of the network and more generally to the problem of adaptation.

Not all examples of emergent computation are beneficial. The Internet (a nationwide network for exchanging electronic mail) was designed so that messages would be routed somewhat randomly (there are usually many different routes that a message may take between two Internet hosts). The intent is for message traffic to be evenly distributed across the various hosts. However, in some circumstances the messages have been found to self-organize into a higher-level structure, called a token-passing ring, so that all of the messages collect at one node, and then are passed along to the next node in the ring [26]. In this case, the self-organization is highly detrimental to the overall performance of the network. The behavior raises the question of what, if any, low-level protocols could reliably prevent harmful self-organizing behavior in a system like the Internet.

In a computational setting, there are at least two quite different types of cooperation: (1) program correctness, and (2) resource allocation. In this context, program correctness means that a collection of independent instructions evolves (more accurately, coevolves) over time in such a way that their interactions result in the desired global behavior. That is, the adaption takes place at the instruction level, but the behavior of interest is at the collective level. If the collective instructions (a program) learn the correct behavior, we say that they are cooperating. Holland's classifier systems (see papers in these Proceedings) are a good example of this sense of cooperation. The second meaning of cooperation occurs when some shared resource on a local area network (e.g., CPU time, printers, network access, etc.) is allocated efficiently among a set of distributed processes. The Huberman and Kephart et al. papers in this volume [24, 30] discuss how robust resource-allocation strategies can emerge in distributed systems.

4. Review of contents

This introduction has described one view of emergent computation. The conference produced several themes and topics of its own. In particular, the themes of design (how to construct such systems), learning and the importance of preexisting structure, the role of parallelism, and the tension between cooperative and competitive models of interaction are central to many of the papers in the Proceedings. Emergent-computation systems can be constructed either by adapting each individual component independently or by tinkering with all of the components as a group. Wilson's paper [55] addresses this issue of local- versus system-level design. Learning is clearly central to emergent computation, since it provides the most natural way to control such a system. Several papers in the Proceedings (Baird, Banzhaf and Haken, Hansen, Omohundro, Schaffer et al. [3, 4, 16, 41, 47]) focus on specific learning issues, and many others use learning as an integral part of their system. The role of parallelism in emergent computation is often assumed. However, Machlin and Stout's paper [36] challenges that assumption, and Greening's paper [15] explores the consequences of using parallelism efficiently.

The papers have been grouped roughly into the following subject areas: (1) artificial networks, (2) learning and adaptation, and (3) biological networks. Thus, all of the papers on biological networks are grouped together, although they emphasize different aspects of problems of emergent computation.

There is a wide range of papers concerned with emergent behavior and computing. Langton's paper [34] illustrates the importance of phase transitions to emergent computation. Huberman [24], Kephart et al. [30], and Maxion [37] discuss emergent behaviors in computing networks. Machlin and Stout's paper [36] illustrates how very simple Turing machines can exhibit interesting and complex behavior. Palmore and Herring's paper [42] provides an example of the connection between emergent computation and real computing procedures (computer arithmetic). Rasmussen's paper [44] uses a simple model of computer memory to show how cooperative "life-like" structures can emerge under various conditions. Finally, Kauffman's paper [28] explores the self-organizing properties of simple Boolean networks.

The adaptive systems aspect of emergent computation is a dominant theme in the Proceedings, and Farmer's paper [10] relates various models of learning through the common thread of adaptive dynamics. Papers on classifier systems and genetic algorithms range from proposals for new mechanisms (Holland [21]) to methods for analyzing classifier system behavior (Compiani et al. [8] and Forrest and Miller [12]), to bridges between genetic algorithms and neural networks (Schaffer et al. [47] and Wilson [55]). Two papers (Hillis [18], Ikegami and Kaneko [25]) explore how interactions between hosts and parasites can improve the global behavior of an evolutionary system. Banzhaf and Haken's [4], Hanson's [16], Kanter's [27], and Churchland's [6] papers describe connectionist models of learning; Greening's paper [15] discusses parallel simulated annealing techniques; Omohundro [41] examines geometric learning algorithms. Papers on the emergence of symbolic reasoning systems from subsymbolic components include Mitchell and Hofstadter [38] (models of analogy-making) and Harnad [17] (connectionism and the symbol-grounding problem).

Several papers describe emergent computations in different biological systems, ranging from the cortex to the cytoskeleton. Reeke and Sporns [45] discuss perceptual and motor systems. Two papers

(Baird [3] and Siegel [49]) focus on the cortex, Keeler's paper [29] examines cerebellar function, and George et al. [13] consider vision. Finally, Rasmussen et al. [43] present a connectionist model of the cytoskeleton.

Acknowledgements

I am grateful to Doyne Farmer, John Holland, Melanie Mitchell, and Quentin Stout for their careful reading of the manuscript and many helpful suggestions. Chris Langton and I have had many productive discussions of these ideas over the years.

References

[1] G.M. Amdahl, Validity of the single processor approach to achieving large-scale computing capabilities, AFIPS Conf. Proc. (1967) 483–485.

[2] R. Axelrod, An evolutionary approach to norms, Am. Political Sci. Rev. (1986) 80.

[3] B. Baird, Bifurcation and learning in oscillating neutral network models of cortex, Physica D 42 (1990) 365–384, these Proceedings.

[4] W. Banzhaf and H. Haken, An energy function for specialization, Physica D 42 (1990) 257–264, these Proceedings.

[5] A.W. Burks, ed., Essays on Cellular Automata (University of Illinois Press, Urbana, IL, 1970).

[6] P.M. Churchland, On the nature of explanation: a PDP approach, Physica D 42 (1990) 281–292, these Proceedings.

[7] E.F. Codd, Cellular Automata (Academic Press, New York, 1968).

[8] M. Compiani, D. Montanari and R. Serra, Learning and bucket brigade dynamics in classifier systems, Physica D 42 (1990) 202–212, these Proceedings.

[9] H.B. Curry and R. Feys, Combinatory Logic, Vol. I (North-Holland, Amsterdam, 1968).

[10] J.D. Farmer, A Rosetta Stone for connectionism, Physica D 42 (1990) 153–187, these Proceedings.

[11] J.D. Farmer, N.H. Packard and A.S. Perelson, The immune system, adaption, and machine learning, Physica D 22 (1986) 187–204.

[12] S. Forrest and J. Miller, Emergent behaviors of classifier systems, Physica D 42 (1990) 213–227, these Proceedings.

[13] J.S. George, C.J. Aine and E.R. Flynn, Neuromagnetic studies of human vision: noninvasive characterization of functional architecture, Physica D 42 (1990) 411–427, these Proceedings.

[14] D.E. Goldberg, Genetic Algorithms in Search Optimization, and Machine Learning (Addison–Wesley, Reading, MA, 1989).

[15] D.R. Greening, Parallel simulated annealing techniques, Physica D 42 (1990) 293–306, these Proceedings.

[16] S.J. Hanson, A stochastic version of the delta rule, Physica D 42 (1990) 265–272, these Proceedings.

[17] S. Harnad, The symbol grounding problem, Physica D 42 (1990) 335–346, these Proceedings.

[18] W.D. Hillis, Co-evolving parasites improve simulated evolution as an optimization procedure, Physica D 42 (1990) 228–234, these Proceedings.

[19] D.R. Hofstadter, Artificial intelligence: subcognition as computation, Technical Report 132, Indiana University, Bloomington, IN (1982).

[20] J.H. Holland, Adaption in Natural and Artificial Systems (University of Michigan Press, Ann Arbor, MI, 1975).

[21] J.H. Holland, Concerning the emergence of tag-mediated lookahead in classifier systems, Physica D 42 (1990) 188–201, these Proceedings.

[22] J.H. Holland, K.J. Holyoak, R.E. Nisbett and P. Thagard, Induction: Processes of Inference, Learning, and Discovery (MIT Press, Cambridge, MA, 1986).

[23] J.E. Hopcroft and J.D. Ullman, Introduction to Automata. Theory, Languages, and Computation (Addison–Wesley, Reading, MA, 1979).

[24] B.A. Huberman, The performance of cooperative processes, Physica D 42 (1990) 38–47, these Proceedings.

[25] T. Ikegami and K. Kaneko, Computer symbiosis – emergence of symbiotic behavior through evolution, Physica D 42 (1990) 235–243, these Proceedings.

[26] V. Jacobson, personal communication.

[27] I. Kanter, Synchronous or asynchronous parallel dynamics – Which is more different, Physica D 42 (199) 273–280, these Proceedings.

[28] S.A. Kauffman, Requirements for evolvability in complex systems: orderly dynamics and frozen components, Physica D 42 (1990) 135–152, these Proceedings.

[29] J.D. Keeler, A dynamical systems view of cerebellar function, Physica D 42 (1990) 396–410, these Proceedings.

[30] J.O. Kephart, T. Hogg and B.A. Huberman, Collective behavior of predictive agents, Physica D 42 (1990) 48–65, these Proceedings.

[31] J.E. Laird, A. Newell and P.S. Rosenbloom, Soar: an architecture for general intelligence, Artificial Intelligence 33 (1987) 64.

[32] C.G. Langton, Studying artificial life with cellular automata, Physica D 22 (1986) 120–149.

[33] C.G. Langton, ed., Artificial Life, Santa Fe Institute Studies in the Sciences of Complexity (Addison–Wesley, Reading, MA, 1989).

[34] C.G. Langton, Computation at the edge of chaos: phase transitions and emergent computation, Physica D 42 (1990) 12–37, these Proceedings.

[35] E.N. Lorenz, Deterministic nonperiodic flow, J. Atmos. Sci. 20 (1963) 130–141.

[36] R. Machlin and Q.F. Stout, The complex behavior of simple machines, Physica D 42 (1990) 85–98, these Proceedings.

[37] R.A. Maxion, Toward diagnosis as an emergent behavior in a network ecosystem, Physica D 42 (1990) 66–84, these Proceedings.

[38] M. Mitchell and D.R. Hofstadter, The emergence of understanding in a computer model of concepts and analogy-making, Physica D 42 (1990) 322–334, these Proceedings.

[39] A. Newell and H.A. Simon, A program that simulates human thought, in: Computers and Thought, eds. E.A. Feigenbaum and J. Feldman (McGraw-Hill, New York, 1963) 279–296.

[40] G. Nicolis and I. Prigogine, Self-Organization in Nonequilibrium Systems (Wiley, New York, 1977).

[41] S.M. Omohundro, Geometric learning algorithms, Physica D 42 (1990) 307–321, these Proceedings.

[42] J. Palmore and C. Herring, Computer arithmetic, chaos and fractals, Physica D 42 (1990) 99–110, these Proceedings.

[43] S. Rasmussen, H. Karampurwala, R. Vaidyanath, K.S. Jensen and S. Hameroff, Computational connectionism with neutrons: a model of cytoskeletal automata subserving neural networks, Physica D 42 (1990) 428–449, these Proceedings.

[44] S. Rasmussen, C. Knudsen, R. Feldberg and M. Hindsholm, The Coreworld: emergence and evolution of cooperative structures in a computational chemistry, Physica D 42 (1990) 111–134, these Proceedings.

[45] G.N. Reeke Jr. and O. Sporns, Selectionist models of perceptual and motor systems and implications for functionalist theories of brain function, Physica D 42 (1990) 347–364, these Proceedings.

[46] D.E. Rumelhard, J. L. McClelland and the PDP Research Group, Parallel Distributed Processing: Explorations in the Microstructure of Cognition (MIT Press, Cambridge, MA, 1986).

[47] J.D. Schaffer, R.A. Caruana and L.J. Eshelman, Using genetic search to exploit the emergent behavior of neural networks, Physica D 42 (1990) 244–248, these Proceedings.

[48] R. Shaw, Strange attractors, chaotic behavior, and information flow, Z. Naturforsch. 36a (1981) 80–112.

[49] R.M. Siegel, Non-linear dynamical system theory and primary visual cortical processing, Physica D 42 (1990) 385–395, these Proceedings.

[50] H.A. Simon, The Sciences of the Artificial (MIT Press, Cambridge, MA, 1969).

[51] Q. Stout, personal communication.

[52] J.E. Stoy, Denitational Semantics: The Scott–Strachey Approach to Programming Language Theory (MIT Press, Cambridge, MA, 1977).

[53] E.O. Wilson, The Social Insects (Belknap/Harvard Univ. Press, Cambridge, MA, 1971).

[54] E.O. Wilson, Sociobiology (Belknap/Harvard Univ. Press, Cambridge, MA, 1975).

[55] S.W. Wilson, Perceptron redux: emergence of structure, Physica D 42 (1990) 249–256, these Proceedings.

[56] S. Wolfram, Universality and complexity in cellular automata, Physica D 10 (1984) 1–35.

[57] W.H. Zurek, Algorithmic randomness and physical entropy, Phys. Rev. A 40 (1989) 4731–4751.

[58] W.H. Zurek, Thermodynamic cost of computation, algorithm complexity and the information metric, Nature 341 (1989) 119–124.

COMPUTATION AT THE EDGE OF CHAOS:
PHASE TRANSITIONS AND EMERGENT COMPUTATION

Chris G. LANGTON

Complex Systems Group, Theoretical Division, Los Alamos National Laboratory, Los Alamos, NM 87454, USA

In order for computation to emerge spontaneously and become an important factor in the dynamics of a system, the material substrate must support the primitive functions required for computation: the transmission, storage, and modification of information. Under what conditions might we expect physical systems to support such computational primitives?

This paper presents research on cellular automata which suggests that the optimal conditions for the support of information transmission, storage, and modification, are achieved in the vicinity of a phase transition. We observe surprising similarities between the behaviors of computations and systems near phase transitions, finding analogs of computational complexity classes and the halting problem within the phenomenology of phase transitions.

We conclude that there is a fundamental connection between computation and phase transitions, especially second-order or "critical" transitions, and discuss some of the implications for our understanding of nature if such a connection is borne out.

1. Introduction

Most of the papers in these Proceedings assume the existence of a physical system with the capacity to support computation, and inquire after the manner in which processes making use of this capacity might emerge spontaneously.

In this paper, we will focus on the conditions under which this *capacity to support computation* itself might emerge in physical systems, rather than on how this capacity might ultimately come to be utilized.

Therefore, the fundamental question addressed in this paper is the following:

Under what conditions will physical systems support the basic operations of information transmission, storage, and modification constituting the capacity to support computation?

This question is difficult to address directly. Instead, we will reformulate the question in the context of a class of formal abstractions of physical systems: cellular automata (CAs). Our question, thus, becomes:

Under what conditions will cellular automata support the basic operations of information transmission, storage, and modification?

This turns out to be a tractable problem, with a somewhat surprising answer; one which leads directly to a hypothesis about the conditions under which computations might emerge spontaneously in nature.

1.1. Overview

First, we introduce cellular automata and a simple scheme for parameterizing the space of all possible CA rules. We then apply this parameterization scheme to the space of possible one-dimensional CAs in a *qualitative* survey of the different dynamical regimes existing in CA rule space and their relationship to one another. Next, we present a *quantitative* picture of these structural relationships, using data from an extensive survey of two-dimensional CAs. Finally, we review the observed relationships among dynamical regimes, and discuss their implications for the more general question raised in the introduction.

1.2. Results

We find that by selecting an appropriate parameterization of the space of CAs, one observes a *phase transition* between highly *ordered* and highly *disordered* dynamics, analogous to the phase transition between the *solid* and *fluid* states of matter. Furthermore, we observe that CAs exhibiting the most complex behavior – both qualitatively and quantitatively – are found *generically* in the vicinity of this phase transition. Most importantly, we observe that CAs in the transition region have the greatest potential for the support of information storage, transmission, and modification, and therefore for the emergence of computation.

These observations suggest that there is a fundamental connection between phase transitions and computation, leading to the following hypothesis concerning the emergence of computation in physical systems:

Computation may emerge spontaneously and come to dominate the dynamics of physical systems when those systems are at or near a transition between their *solid* and *fluid* phases, especially in the vicinity of a second-order or "critical" transition.

This hypothesis, if borne out, has many implications for understanding the role of information in nature.

Perhaps the most exciting implication is the possibility that life had its origin in the vicinity of a phase transition, and that evolution reflects the process by which life has gained local control over a successively greater number of environmental parameters affecting its ability to maintain itself at a critical balance point between order and chaos.

1.3. Cellular automata

In this section, we review cellular automata, introduce a parameterization of the space of possible CA rules, and discuss computation in CAs.

Cellular automata are discrete space/time logical universes, obeying their own local physics [26, 3, 5, 27, 28].

Space in CAs is partitioned into discrete volume elements called "cells" and time progresses in discrete steps. Each cell of space is in one of a finite number of states at any one time. The physics of this logical universe is a deterministic, local physics. "Local" means that the state of a cell at time $t + 1$ is a function only of its own state and the states of its immediate neighbors at time t. "Deterministic" means that once a local physics and an initial state of a ČA has been chosen, its future evolution is uniquely determined.

1.4. Formal definition of cellular automata

Formally, a cellular automaton is a D-dimensional lattice with a finite automaton residing at each lattice site. Each automaton takes as input the states of the automata within some *finite* local region of the lattice, defined by a neighborhood template $\mathcal{N}$, where the dimension of $\mathcal{N} \leq D$. The size of the neighborhood template, $|\mathcal{N}|$, is just the number of lattice points covered by $\mathcal{N}$. By convention, an automaton is considered to be a member of its own neighborhood. Two typical two-dimensional neighborhood templates are:

five cell neighborhood   nine cell neighborhood

Each finite automaton consists of a finite set of *cell states* Σ, a finite *input alphabet* α, and a *transition function* Δ, which is a mapping from the set of neighborhood states to the set of cell states. Letting $N = |\mathcal{N}|$:

$$\Delta: \Sigma^N \to \Sigma.$$

The *state* of a neighborhood is the cross product of the states of the automata covered by the neighborhood template. Thus, the input alphabet

α for each automaton consists of the set of possible neighborhood states: $\alpha = \Sigma^N$. Letting $K = |\Sigma|$ (the number of cell states) the size of α is equal to the number of possible neighborhood states

$$|\alpha| = |\Delta| = |\Sigma^N| = K^N.$$

To define a transition function Δ, one must associate a unique next state in Σ with each possible neighborhood state. Since there are $K = |\Sigma|$ choices of state to assign as the next state for each of the $|\Sigma^N|$ possible neighborhood states, there are $K^{(K^N)}$ possible transition functions Δ that can be defined. We use the notation $\mathscr{D}_N^K$ to refer to the set of all possible transition functions Δ which can be defined using N neighbors and K states.

1.5. Example

Consider a two-dimensional cellular automaton using 8 states per cell, a rectangular lattice, and the five-cell neighborhood template shown above. Here $K = 8$ and $N = 5$, so $|\Delta| = K^N = 8^5 = 32\,768$ and there are thus 32 768 possible neighborhood states. For each of these, there is a choice of 8 states as the next cell state under Δ, so there are $K^{(K^N)} = |\mathscr{D}_N^K| = 8^{(8^5)} \approx 10^{30\,000}$ possible transition functions using the 5-cell neighborhood template with 8 states per cell, an exceedingly large number.

2. Parameterizing the space of CA rules

$\mathscr{D}_N^K$, the set of possible transition functions Δ for a CA of K states and N neighbors, is fixed once we have chosen the number of states per cell and the neighborhood template. However, there is no intrinsic order within $\mathscr{D}_N^K$; it is a large, undifferentiated space of CA rules.

Imposing a *structure* on this undifferentiated space of CA rules allows us to define a natural ordering on the rules, and provides us with an index into the rule space. The ideal ordering scheme would partition the space of CA rules in such a manner that rules from the same partition would support similar dynamics. Such an ordering on $\mathscr{D}_N^K$ would allow us to observe the way in which the dynamical behaviors of CAs vary from partition to partition.

The location in this space of the partitions supporting the transmission, modification, and storage of information, relative to the location of partitions supporting *other* possible dynamical behaviors should provide us with insight into the conditions under which we should expect computation to emerge in CAs.

2.1. The λ parameter

We will consider only a subspace of $\mathscr{D}_N^K$, characterized by the parameter λ [18, 17].

The λ parameter is defined as follows. We pick an arbitrary state $s \in \Sigma$, and call it the *quiescent* state s_q. Let there be n transitions to this special quiescent state in a transition function Δ. Let the remaining $K^N - n$ transitions in Δ be filled by picking randomly and uniformly over the other $K - 1$ states in $\Sigma - s_q$. Then

$$\lambda = \frac{K^N - n}{K^N}. \qquad (1)$$

If $n = K^N$, then *all* of the transitions in the rule table will be to the quiescent state s_q and $\lambda = 0.0$. If $n = 0$, then there will be *no* transitions to s_q and $\lambda = 1.0$. When all states are represented equally in the rule table, then $\lambda = 1.0 - 1/K$.

The parameter values $\lambda = 0.0$ and $\lambda = 1.0 - 1/K$ represent the most homogeneous and the most heterogeneous rule tables, respectively. The behavior in which we will be interested is captured between these two parameter values. Therefore, we experiment primarily with λ in this range.

2.2. Searching CA space with the λ parameter

In the following, we use the λ parameter as a means of sampling $\mathscr{D}_N^K$ in an ordered manner. We do this by stepping through the range $0.0 \leq \lambda \leq$

$1.0 - 1/K$ in discrete steps, randomly constructing Δ functions for each λ point. Then we run CAs under these randomly constructed Δ functions, collecting data on various measures of their dynamical behavior. Finally, we examine the behavior of these measures as a function of λ.

Δ functions are constructed in two ways using λ. In the "random-table method", λ is interpreted as a bias on the random selection of states from Σ as we sequentially fill in the transitions that make up a Δ function. To do this, we step through the table, flipping a λ-biased coin for each neighborhood state. If the coin comes up tails, with probability $1.0 - \lambda$, we assign the state s_q as the next cell state for that neighborhood state. If the coin comes up heads, with probability λ, we pick one of the $K - 1$ states in $\Sigma - s_q$ at uniform random as the next cell state.

In the "table-walk-through" method, we start with a Δ function consisting entirely of transitions to s_q, so that $\lambda = 0.0$ (but note restrictions below). New transition tables with higher λ values are generated by randomly replacing a few of the transitions to s_q in the current function with transitions to other states, selected randomly from $\Sigma - s_q$. Tables with *lower* λ values are generated by randomly replacing a few transitions that are *not* to s_q in the current table by transitions to s_q.

Thus, under the table-walk-through method, we progressively perturb "the same table", whereas under the random-table method, each new table is generated from scratch.

2.3. Further restrictions on CAs

In order to make our studies more tractable, we impose two further conditions on the rule space. First, a strong *quiescence condition*: all neighborhood states uniform in cell state s_i will map to state s_i. Second, an *isotropy condition*: all planar rotations of a neighborhood state will map to the same cell state. These restrictions mean that arrays uniform in any single state will remain so, and that the physics cannot tell which way is up, so to speak.

2.4. Discussion

λ is not necessarily the best parameter. One can improve on λ in a number of ways. For instance, Gutowitz [12, 11] has defined a hierarchy of parameterization schemes in which λ is the simplest scheme, mean field theory constitutes the next simplest scheme, and so on.

However, λ suffices to reveal a great deal about the overall structural relationships between the various dynamical regimes in CA rule space, and it is very useful to get a feel for the "lay of the CA landscape" at this low-resolution level before increasing the resolution and surveying finer details. For one thing, λ helps restrict the area of search to a particularly promising "spot", which is useful because higher-order parameterizations map CA rule space onto *many* dimensions, whereas λ is a one-dimensional parameter.

λ discriminates well between dynamical regimes for "large" values of K and N, whereas λ discriminates poorly for small values of K and N. For example, for a 1D CA with $K = 2$, and $N = 3$, λ is only roughly correlated with dynamical behavior. This may explain why the relationships reported here were not observed in earlier work on classifying CA dynamics [29, 28], as these investigations were carried out using CAs with minimal values of K and N.

For these reasons, we employ CAs for which $K \geq 4$ and $N \geq 5$, which results in transition tables of size $4^5 = 1024$ or larger.

2.5. Computation in CAs

Cellular automata can be viewed either as computers themselves or as *logical universes* within which computers may be embedded.

On the first view, an initial configuration constitutes the data that the physical computer is working on, and the transition function implements the algorithm that is to be applied to the data. This is the approach taken in most current applications of cellular automata, such as image processing.

On the second view, the initial configuration itself constitutes a computer, and the transition function is seen as the "physics" obeyed by the parts of this embedded computer. The algorithm being run and the data being manipulated are functions of the precise state of the initial configuration of the embedded computer. In the most general case, the initial configuration will constitute a universal computer.

We can always take the first point of view, but what we are interested in here is the question: when is it possible – even necessary – to adopt the second point of view to understand the dynamics of a CA?

That CAs are capable of supporting universal computation has been known since their invention by Ulam and von Neumann in the late 40's. Von Neumann's proof of the possibility of machine self-reproduction involves the demonstration of the existence of a universal computer/constructor in a 29-state CA [26]. Since then, Codd [5], Smith [24], Conway and co-workers [2], Fredkin and Toffoli [7] – to name but a few – have found much simpler CA rules supporting universal computation.

All of these proofs involve the embedding of a computer within the CA, or at least they show that all of the important parts of such a computer could be implemented and that those parts are sufficient to construct a computer. Some of these proofs involve the construction of Turing machines, others involve the construction of stored-program computers.

All of these constructs rely on three fundamental features of the dynamics supported by the underlying transition function physics. First, the physics must support the *storage* of information, which means that the dynamics must preserve local state information for arbitrarily long times. Second, the physics must support the *transmission* of information, which means that the dynamics must provide for the propagation of information in the form of *signals* over arbitrarily long distances. Third, stored and transmitted information must be able to interact with one another, resulting in a possible modification of one or the other.

These fundamental properties must be provided by *any* dynamical system if it is to support computation. Taken together, they require that any dynamical system supporting computation *must exhibit arbitrarily large correlation lengths in space and time*. These correlation lengths must be *potentially* infinite, but not *necessarily* so. Codd [5] refers to this situation as one in which the propagation of information must be *unbounded* in principle but *boundable* in practice.

2.6. Wolfram's qualitative CA classes

Wolfram [29] has proposed the following four qualitative classes of CA behavior:

Class I evolves to a homogeneous state.
Class II evolves to simple separated periodic structures.
Class III yields chaotic aperiodic patterns.
Class IV yields complex patterns of localized structures.

Wolfram finds the following analogs for his classes of cellular automaton behaviors in the field of dynamical systems.

Class I cellular automata evolve to *limit points*.
Class II cellular automata evolve to *limit cycles*.
Class III cellular automata evolve to *chaotic* behavior of the kind associated with *strange attractors*.
Class IV cellular automata "effectively have very long *transients*".

This association of class IV CAs with "very long transients" will figure "critically" in what follows.

Wolfram suggests that class IV CAs are capable of supporting computation, even universal computation, and that it is this capacity that makes their behavior so complex. This paper supports Wolfram's hypothesis, and offers an explanation for both the existence of these classes and their relationship to one another.

In their surveys of 1D and 2D CAs, Packard and Wolfram [23] hypothesized that class IV CAs constitute a set of measure 0. This means that class IV behaviors should be infinitely hard to find in the "thermodynamic limit" of an infinitely large CA rule space. However, it turns out that they are *not* hard to find in rule spaces that are far from the thermodynamic limit. By locating class IV behaviors in these non-limiting rule spaces and tracking the manner in which they become vanishingly rare as one goes to larger rule spaces, we can derive a general theory about where to locate rules likely to support computation in *any* CA rule space.

3. Qualitative overview of CA dynamics

In this section, we present a series of examples illustrating the changes observed in the dynamical behavior of one-dimensional CAs as we alter the λ parameter throughout its range using the table-walk-through method. For these CAs, $K = 4$, $N = 5$ (i.e. *two* cells on the left and *two* cells on the right are included in the neighborhood template). The arrays consist of 128 sites connected in a circle, resulting in periodic boundary conditions. Each array is started from a random initial configuration on the top line, and successive lines show successive time steps in the evolution.

For each value of λ, we show two evolutions. The arrays in fig. 1 are started from a uniform random initial configuration over all 128 sites, while those in fig. 2 are started from configurations whose sites are all 0, with the exception of a patch of 20 randomized sites in the middle.

Fig. 1 illustrates the kinds of structures that develop, as well as the typical transient times before these structures are achieved. Fig. 2 illustrates the relative spread or collapse of the area of dynamical activity with time. For those values of λ exhibiting long transients, we have reduced the scale of the arrays in order to display longer evolutions.

We start with $\lambda \approx 0.00$. Note that under the strong quiescence condition mentioned above we cannot have $\lambda = 0.00$ exactly. The primary features observed as we vary λ throughout its range are itemized below.

$\lambda \approx 0.00$ All dynamical activity dies out after a single time step, leaving the arrays uniform in state s_q. The area of dynamical activity has collapsed to zero.

$\lambda \approx 0.05$ The dynamics reaches the uniform s_q fixed point after approximately 2 time steps.

$\lambda \approx 0.10$ The homogeneous fixed point is reached after 3 or 4 time steps.

$\lambda \approx 0.15$ The homogeneous fixed point is reached after 4 or 5 time steps.

$\lambda \approx 0.20$ The dynamics reaches a periodic structure which will persist forever (fig. 1, $\lambda = 0.20$). Transients have increased to 7 to 10 time steps as well. Note that the evolution does not necessarily lead to periodic dynamics (fig. 2, $\lambda = 0.20$).

$\lambda \approx 0.25$ Structures of period 1 appear. Thus, there are now three different possible outcomes for the ultimate dynamics of the system, depending on the initial state. The dynamics may reach a *homogeneous* fixed point consisting entirely of state s_q, or it may reach a *heterogeneous* fixed point consisting mostly of cells in state s_q with a sprinkling of cells stuck in one of the other states, or it may settle down to periodic behavior. Notice that the transients have lengthened even more.

$\lambda \approx 0.30$ Transients have lengthened again.

$\lambda \approx 0.35$ Transient length has grown significantly, and a new kind of periodic structure with a longer period has appeared (fig. 1, $\lambda = 0.35$). Most of the previous structures are still possible, hence the spectrum of dynamical possibilities is broadening.

$\lambda \approx 0.40$ Transient length has increased to about 60 time steps, and a structure has appeared with a period of about 40 time steps. The area of dynamical activity is

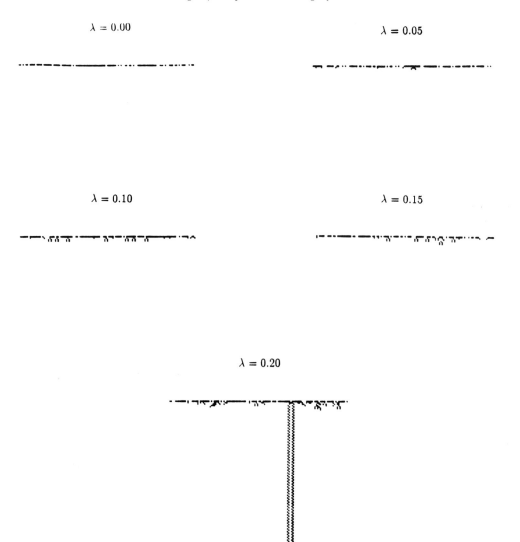

Fig. 1. Evolutions of one-dimensional, $K = 4$, $N = 5$ CAs from *fully* random initial configurations over $0.0 < \lambda \leq 0.75$. As λ is increased the structures become more complicated, and the transients grow in length until they become arbitrarily long at $\lambda \approx 0.50$. For $0.50 < \lambda \leq 0.75$, the transient lengths *decrease* with increasing λ, as indicated by the arrows to the right of the evolutions.

still collapsing down onto isolated periodic configurations.

$\lambda \approx 0.45$ Transient length has increased to almost 1000 time steps (fig. 1, $\lambda = 0.45$). Here, the structure on the right appears to be periodic, with a period of about 100 time steps. However, after viewing several cycles of its period, it is apparent that the whole structure is moving to the left, and so this pattern will not recur precisely in its same position until it has cycled at least once around the array. Furthermore, as it propagates to the left, this structure eventually annihi-

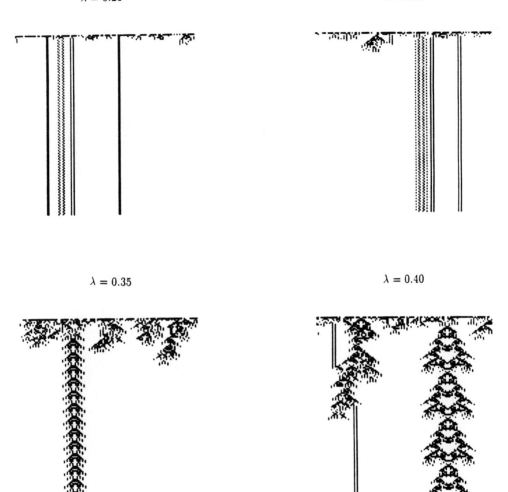

Fig. 1. Continued

lates a period-1 structure after about 800 time steps. Thus, the transient length before a periodic structure is reached has grown enormously. It turns out that even after one orbit around the array, the periodic structure does not return exactly to its previous position. It must orbit the array 3 times before it repeats itself exactly. As it has shifted over only 3 sites after its quasi-period of 116 time steps, the true period of this structure is 14 848 time steps. Here, the area of dynamical activity is at a balance point between collapse and expansion.

$\lambda \approx 0.50$ Typical transient length is on the order of 12 000 time steps. After the transient, the dynamical activity settles down to periodic behavior, possibly of period one as shown in the figure. Although, the dynamics eventually becomes simple, the transient time has increased dramatically. Note in fig. 2 that the general tendency now is that the area of

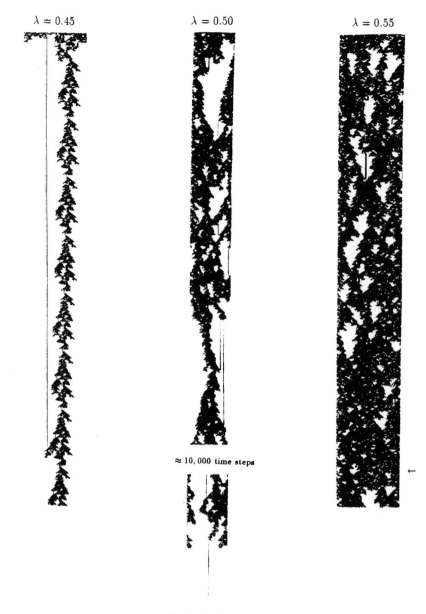

Fig. 1. Continued

dynamical activity *expands* rather than contracts with time. There are, however, large fluctuations in the area covered by dynamical activity, and it is these fluctuations which lead to the eventual collapse of the dynamics.

$\lambda \approx 0.55$ We have entered a new dynamical regime in which the transients have become so long that — for all practical purposes — they *are* the steady state behavior of the system over any period of time for which we can observe them. Whereas before, the dynamics *eventually* settled down to periodic behavior, we are now in a regime in which the dynamics typically settles down to ef-

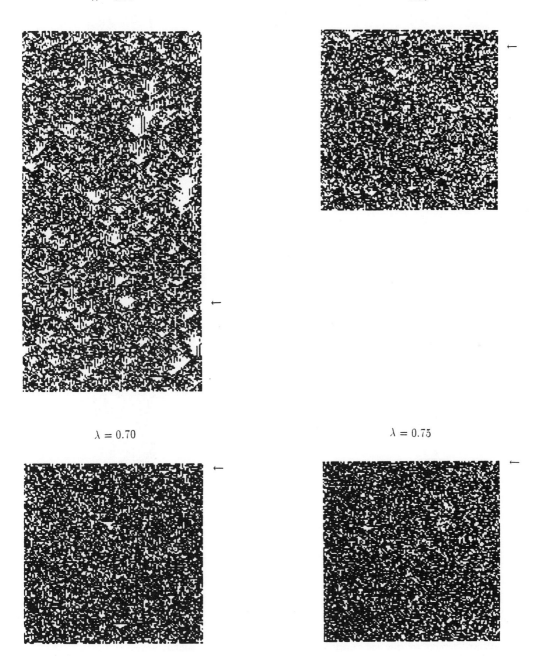

Fig. 1. Continued

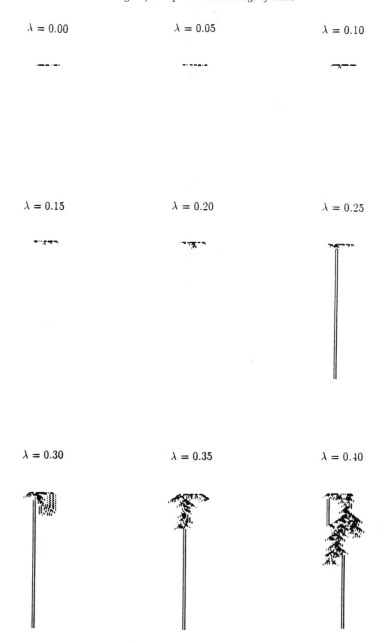

Fig. 2. Evolutions of one-dimensional, $K=4$, $N=5$ CAs from *partially* random initial configurations over $0.0 < \lambda \leq 0.75$. This series illustrates the change in the rate of spread of the dynamics from negative for $\lambda < 0.45$, to positive for $\lambda > 0.45$. For $\lambda \approx 0.45$, the dynamics is balanced between collapse and expansion, giving rise to particle-like solitary waves.

fectively *chaotic* behavior. Furthermore, the previous trend of transient length *increasing* with increasing λ is reversed. The arrow to the right of the evolutions of figs. 1, $\lambda = 0.55$–0.75 indicates the approximate time by which the site-occupation density has settled down to within 1% of its long-time average. Note that the area of dynamical activity expands more rapidly with time.

$\lambda \approx 0.60$ The dynamics are quite chaotic, and the transient length to "typical" chaotic be-

$\lambda = 0.45$ $\lambda = 0.50$ $\lambda = 0.55$

Fig. 2. Continued

havior has decreased significantly. The area of dynamical activity expands more rapidly with time.

$\lambda \approx 0.65$ Typical chaotic behavior is achieved in only 10 time steps or so. The area of dynamical activity is expanding at about one cell per time step in each direction, approximately half of the maximum possible rate for this neighborhood template.

$\lambda \approx 0.70$ Fully developed chaotic behavior is reached in only 2 time steps. The area

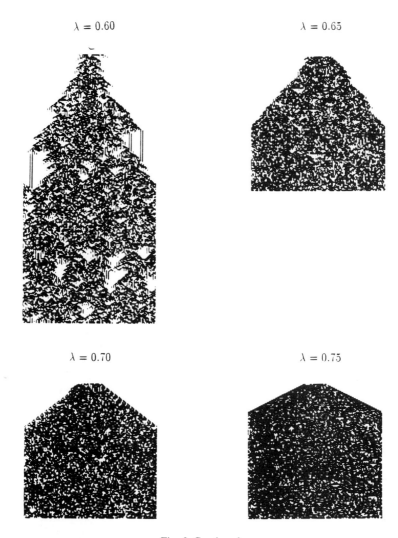

Fig. 2. Continued

of dynamical activity is expanding even more rapidly.

$\lambda \approx 0.75$ After only a single time step, the array is essentially random and remains so thereafter. The area of dynamical activity spreads at the maximum possible rate.

Therefore, by varying the λ parameter throughout $0.0 < \lambda \leq 0.75$ over the space of possible $K = 4$, $N = 5$, 1D cellular automata, we progress from CAs exhibiting the maximal possible order to CAs exhibiting the maximal possible disorder. At intermediate values of λ, we encounter a *phase transition* between periodic and chaotic dynamics, and while the behavior at either end of the λ spectrum seems "simple" and easily predictable, the behavior in the vicinity of this phase transition seems "complex" and unpredictable.

4. Comments on qualitative dynamics

There are several observations to be made about the 1D examples of section 3.

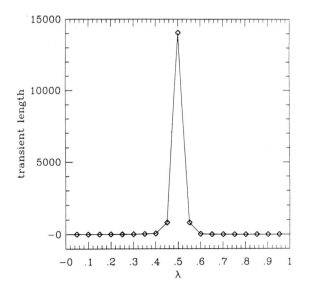
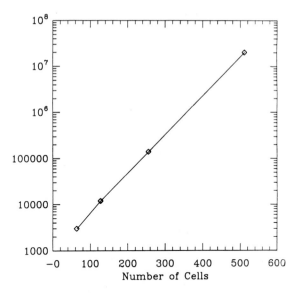

Fig. 3. Average transient length as a function of λ in an array of 128 cells.

Fig. 4. Growth of average transients as a function of array size for λ = 0.50.

First, transients grow rapidly in the vicinity of the transition between ordered and disordered dynamics, a phenomenon known in the study of phase transitions as *critical slowing down*. The relationship between transient length and λ is plotted in fig. 3.

Second, the size of the array has an effect on the dynamics only for intermediate values of λ. For low values of λ, array size has *no* discernible effect on transient length. Not until λ ≈ 0.45 do we begin to see a small difference in the transient length as the size of the array is increased. For λ = 0.50, however, array size has a significant effect on the transient length. The growth of transient length as a function of array size for λ = 0.50 is plotted in fig. 4. The essentially linear relationship on this log-normal plot suggests that transient length depends *exponentially* on array size at λ = 0.50. As we continue to raise λ beyond 0.50, although the dynamics is now settling down to effectively chaotic behavior instead of periodic behavior, the transient lengths are getting *shorter* with increasing λ, rather than longer. A number of statistical measures (see ref. [17]) reveal that the time it takes to reach "typical" behavior decreases as λ increases past the transition point. Further-more, transient times exhibit decreasing dependence on array size as λ is increased past the transition point. By the time all states are represented uniformly in the transition table – at λ = 0.75 in this case – the transient lengths exhibit *no* dependence on array size – just as was the case for low values of λ.

Third, the overall evolutionary pattern in time appears more random as λ → 0.75. This observation is borne out by various entropy and correlation measures (see section 5). λ = 0.75 represents the state of maximal dynamical disorder.

Fourth, the transition region supports both static and propagating structures (fig. 1, λ = 0.45.) These particle-like structures are essentially *solitary waves*, quasi-periodic patterns of state change, which – like the "gliders" in Conway's Game of Life [8] – propagate through the array, constantly moving with respect to the fixed background of the lattice. The λ value for the Game of Life (λ_{Life} = 0.273) lies within the transition region for $K = 2$, $N = 9$ 2D CAs. Fig. 5 traces the time evolution of an array of 512 sites, and shows that the rule governing the behavior of fig. 1, λ = 0.45 supports several different kinds of particles, which interact with each other and with static periodic

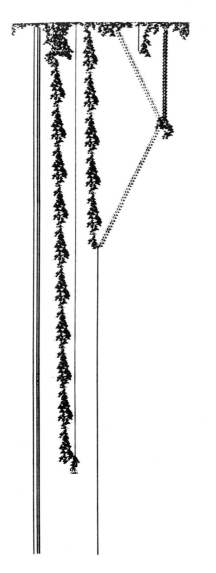

Fig. 5. Propagating structures and their interactions in an array of 512 cells with $\lambda = 0.45$.

storage elements in the construction of a general purpose computer [2].

4.1. Complications

Finally, it must be pointed out that although the examples presented illustrate the general behavior of the dynamics as a function of λ, the story is not quite as simple as we have presented it here. The story is complicated by two factors, which will be detailed in the next section.

First, different traversals of λ space using the table-walk-through method make the transition to chaotic behavior at different λ values, although there is a well defined distribution around a mean value. Second, one does not always capture a second-order phase transition as neatly as in this example. Often, the dynamics jumps directly from fairly ordered to fairly disordered behavior, suggesting that both first- and second-order transitions are possible.

Despite these complications, the overall picture is clear: as we survey CA rule spaces using the λ parameter, we encounter a phase transition between periodic and chaotic behavior, and the most complex behavior is found in the vicinity of this transition, both qualitatively and quantitatively.

structures in complicated ways. Note that the collision of a particle with a static periodic structure produces a particle traveling in the opposite direction. These propagating and static structures can form the basis for signals and storage, and interactions between them can modify either stored or transmitted information in the support of an overall computation. The proof that the Game of Life is computation-universal employs propagating "gliders" as signals and the period-2 "blinkers" as

5. Quantitative overview of CA dynamics

In this section, we present a brief quantitative overview of the structural relations among the dynamical regimes in CA rule spaces as revealed by the λ parameter[#1].

The results of this section are based on experiments using 2D CAs with $K = 8$ and $N = 5$. Arrays are typically of size 64×64, and again, periodic boundary conditions are employed.

[#1] The results presented here summarize my Thesis research [17]. The reader is referred to that work for a more detailed presentation of the results in this section.

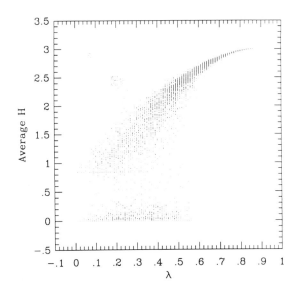

Fig. 6. Average single cell entropy $\overline{H}$ over λ space for approximately 10 000 CA runs. Each point represents a different transition function.

5.1. Measures of complexity

The measures employed were chosen for their collective ability to reveal the presence of information in its various forms within CA dynamics.

5.1.1. Shannon entropy

We use Shannon's entropy H to measure basic information capacity. For a discrete process A of K states[#2]:

$$H(A) = - \sum_{i=1}^{K} p_i \log p_i. \quad (2)$$

Fig. 6 shows the average entropy per cell, $\overline{H}$, as a function of λ for approximately 10 000 CA runs. The random-table method was employed, so each point represents a distinct random transition table.

First, note the overall envelope of the data and the large variance at most λ points. Second, note the sparsely populated gap over $0.0 \leq \lambda \leq 0.6$ and between $0.0 \leq \overline{H} \leq 0.84$. This distribution appears to be bimodal, suggesting the presence of a phase transition. Third, note the rapid decrease in variability as λ is raised from ~ 0.6 to its maximum value of 0.875.

Two other features of this plot deserve special mention. First, the abrupt cutoff of low $\overline{H}$ values at $\lambda \approx 0.6$ corresponds to the *site–percolation* threshold $P_c \approx 0.59$ for this neighborhood template. Thus, we may suppose that, since λ is a dynamical analog of the site occupation probability P, the *dynamical* percolation threshold for a particular neighborhood template is bounded above by the *static* percolation threshold P_c. This is borne out by experiments with other neighborhood templates. For instance, the 9-neighbor template exhibits a sharp cutoff at $\lambda \approx 0.4$, which corresponds well with the site percolation threshold $P_c \approx 0.402$ for this lattice.

The second feature is the "ceiling" of the gap at $\overline{H} \approx 0.84$. This turns out to be the average entropy value for one of the most commonly occurring chaotic rules. In such rules the dynamics has collapsed onto only two states – s_q and one other – and the rule is such that a mostly quiescent neighborhood containing one non-quiescent state maps to that non-quiescent state. In 1D CAs, such rules give rise to the familiar triangular fractal pattern known as the Sierpiński gasket. There are many ways to achieve such rules, and they can be achieved at very low λ values. Most of the low-λ chaotic rules are of this type.

The entropy data of fig. 6 suggest an anomaly at intermediate parameter values, possibly a phase transition between two kinds of dynamics. Since there seems to be a discrete jump between low and high entropy values, the evidence points to a first-order transition, similar to that observed between the solid and fluid phases of matter. However, the fact that the gap is not completely empty suggests the possibility of second-order transitions as well.

The table-walk-through method of varying λ reveals more details of the structure of the entropy data. Fig. 7 shows four superimposed examples of the change in the average cell entropy as we vary the λ value of a table. Notice that in each of the four cases the entropy remains fairly close to zero

---

[#2] Throughout, log is taken to the base 2, thus the units are bits.

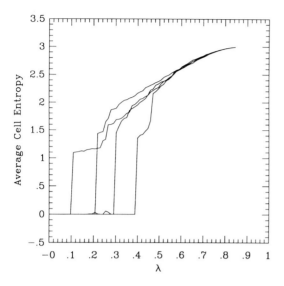

Fig. 7. Superposition of 4 transition events. Note the different λ values at which the transitions take place.

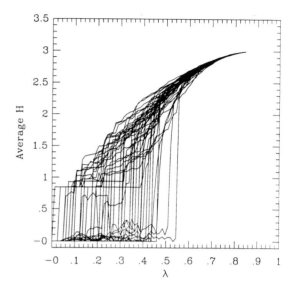

Fig. 8. Superposition of 50 transition events, showing the internal structure of fig. 6.

until – at some critical λ value – the entropy jumps to a higher value, and proceeds fairly smoothly towards its maximum possible value as λ is increased further. Such a discontinuity is a classic signature of a first-order phase transition. Most of our complexity measures exhibit similar discontinuities at the same λ value *within a particular table*.

Notice also that the λ value at which the transition occurs is different for each of the four examples. Obviously, the same thing – a jump – is happening as we vary λ in each of these examples, but it happens at different values of λ. When we superimpose 50 runs, as in fig. 8. we see the internal structure of the entropy data envelope plotted in fig. 6.

Since we have located the transition events, we may line up these plots by the events themselves, rather than by λ, in order to get a clearer picture of what is going on before, during, and after the transition. This is illustrated in fig. 9. The abcissa is now measured in terms of Δλ: the distance from the transition event. Fig. 10 shows the same data as fig. 8 but lined up by Δλ.

5.1.2. Mutual information

In order for two distinct cells to cooperate in the support of a computation, they must be able

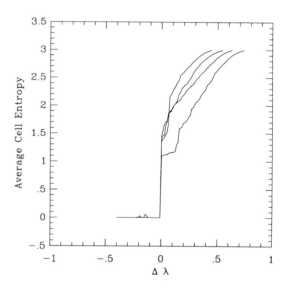

Fig. 9. Plots lined up by the transition event, rather than by λ. Δλ is the distance from the transition event.

to affect one another's behavior. Therefore, we should be able to find correlations between events taking place at the two cells.

The mutual information $I(A; B)$ between two cells A and B can be used to study correlations in systems when the values at the sites to be measured cannot be ordered, as is the case for the states of the cells in cellular automata [19].

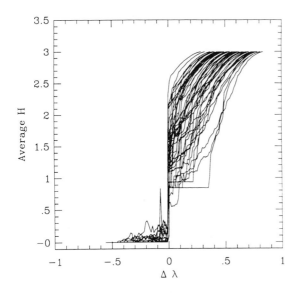
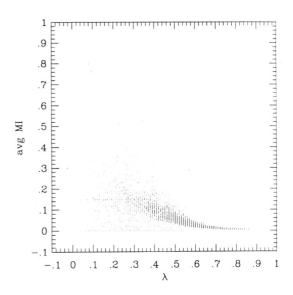

Fig. 10. Superposition of 50 transition events lined up by Δλ. Compare with fig. 8.

Fig. 11. Average mutual information between a cell and itself at the next time step.

The mutual information is a simple function of the individual cell entropies, $H(A)$ and $H(B)$, and the entropy of the two cells considered as a joint process, $H(A, B)$:

$$I(A; B) = H(A) + H(B) - H(A, B). \qquad (3)$$

This is a measure of the degree to which the state of cell A is correlated with the state of cell B, and vice versa.

Fig. 11 shows the average mutual information between a cell and itself at the next time step. Note the tight convergence to low values of the mutual information for high λ and the location of the highest values.

The increase of the mutual information in a particular region is evidence that the correlation length is growing in that region, further evidence for a phase transition.

Fig. 12 shows the behavior of the average mutual information as λ is varied, both against λ and $\Delta\lambda$. The average mutual information is essentially zero below the transition point, it jumps to a moderate value at the transition, and then decays slowly with increasing λ. The jump in the mutual information clearly indicates the onset of the chaotic regime, and the decaying tail indicates the approach to effectively random dynamics. The lack of correlation between even adjacent cells at high λ means that cells are *acting* as if they were independent of each other, even though they are causally connected. The resulting global dynamics is the same as if each cell picked its next state at uniform random from among the K states, with no consideration of the states of its neighbors. This kind of global dynamics is predictable in the same statistical sense that an ideal gas is globally predictable. In fact it is appropriate to view this dynamical regime as a hot gas of randomly flipping cells.

Fig. 13 shows the average mutual information curves for several different temporal and spatial separations. Note that the decay in both time and space is slowest in the middle region.

At intermediate λ values, the dynamics support the preservation of information locally, as indicated in the peak in correlations between distinct cells. If cells are cooperatively engaged in the support of a computation, they must exhibit some – but not *too* much – correlation in their behaviors. If the correlations are too strong, then the cells are overly dependent, with one mimicing the other – not a cooperative computational enterprise.

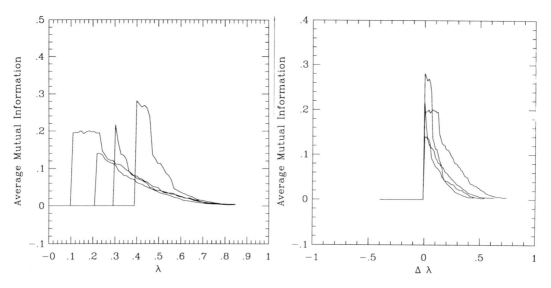

Fig. 12. Average mutual information versus λ and $\Delta\lambda$. The mutual information in this case is for a single time step at a single cell.

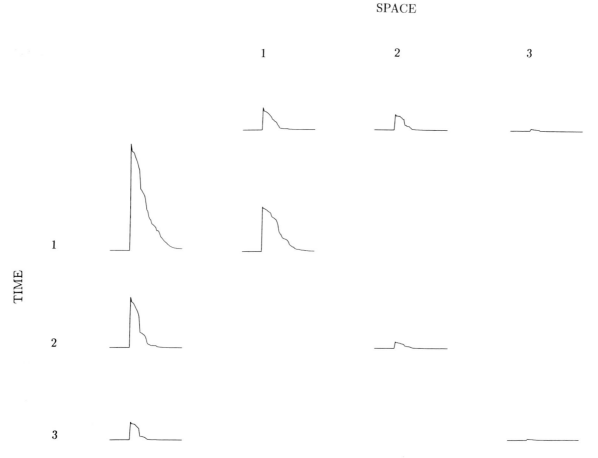

Fig. 13. Decay of average mutual information in space and time.

On the other hand, if the correlations are too small, then the cells are overly *independent*, and again, they cannot cooperate in a computational enterprise, as each cell does something totally unpredictable in response to the state of the other. Correlations in behavior imply a kind of common code, or protocol, by which changes of state in one cell can be recognized and understood by the other as a *meaningful signal*. With no correlations in behavior, there can be no common code with which to communicate information.

6. Mutual information and entropy

It is often useful to examine the way in which observed measures behave when plotted against one another, effectively removing the (possibly unnatural) ordering imposed by the control parameter.

Of the measures we have looked at, the most informative pair when plotted against each other are the mutual information and the average single cell entropy. The relationship between these two measures is plotted in fig. 14. Again, we see clear evidence of a phase transition.

The envelope of the relationship is bounded below the transition by the linear bound that H places on the mutual information. All of the points on this line are for periodic CAs. This line intersects the curve bounding the envelope *above* the transition at an entropy value $H_c \approx 0.32$ on the normalized entropy scale.

This is a *very* informative plot. There is a clear, sharply defined maximum value of mutual information at a specific value of the entropy, and the mutual information falls off rapidly on either side. This seems to imply that there is an *optimal working entropy* at which CAs exhibit large spatial and temporal correlations. Why should this be the case?

Briefly, information storage involves *lowering entropy* while information transmission involves *raising entropy* [10]. In order to compute, a system must do both, and therefore must effect a trade-off between high and low operating entropy. It would seem from the work reported here that this trade-off is optimized in the vicinity of a phase transition.

A similar relationship has been observed by Crutchfield at Berkeley in his work on the transition to chaos in continuous dynamical systems [6]. This relationship is illustrated in fig. 15. Briefly, the ordinate of this plot – C – is a measure of the size of the minimal finite state machine required to recognize strings of 1's and 0's generated by a dynamical system (the logistic map, in this case) when these strings are characterized by the normalized per-symbol entropy listed on the abcissa. The observance of this same fundamental entropy/complexity relationship in these different classes of dynamical systems is very exciting.

These relationships support the view that, rather than increasing monotonically with randomness – as is the case for the usual measures of complexity, such as that of Chaitin and Kolmogorov [4, 16] – complexity increases with randomness only up to a point – *a phase transition* – after which complexity *decreases* with further in-

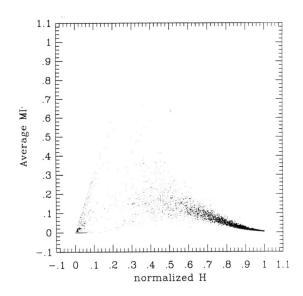

Fig. 14. Average mutual information versus average single cell entropy $\overline{H}$. The mutual information in this case is computed between a cell and itself at the next time step. The entropy is normalized to 1.0.

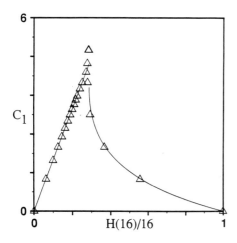

Fig. 15. Crutchfield's plot of machine complexity versus normalized per-symbol entropy for the logistic map. Compare with fig. 14.

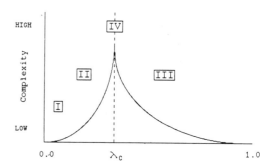

Fig. 16. Location of the Wolfram classes in λ space.

creases in randomness, so that total disorder is just as "simple", in a sense, as total order. Complex behavior involves a mix of order and disorder.

7. Phase transitions and computation

What does all of this tell us about emergent computation? The answer is that *information* becomes an important factor in the dynamics of CAs in the vicinity of the phase transition between periodic and chaotic behavior. Only in the vicinity of this phase transition can information propagate over long distances without decaying appreciably. This allows for the long-range correlations in behavior, sensitivity to "size", extended transients, etc., which are necessary for the support of computation. By contrast, the ordered regime does not allow information to propagate at all, whereas the disordered regime propagates effects too well, causing information to decay rapidly into random noise.

If it is true that these phase-transition dynamics – especially "critical" or second-order dynamics – support the possibility of emergent computation, then we should be able to find analogs for various well-known features of computation in the phenomenology of phase transitions, and vice versa. In the following sections, we point out several possible analogs, and offer an interpretation which suggests that computation as we know it is really just a special case of a more universal physical phenomenon.

7.1. Locating the Wolfram classes

First, there is an obvious mapping of the Wolfram classes onto the spectrum of dynamical possibilities over the λ space: classes I and II constitute the *ordered* phase, while class III constitutes the *disordered* phase. Because of their long transients, propagating structures, large correlation lengths, and other statistical properties, the only logical choice for the location of class IV CAs is *at* the transition between these two phases of dynamical behavior. Fig. 16 shows how the Wolfram classes fit into the λ spectrum.

This also explains why one expects class IV CAs to constitute a set of measure 0. In the thermodynamic limit, the phase transition is located along a $(K-2)$-dimensional hyperplane in the rule space for K-state CAs (see ref. [17]). Hyperplanes embedded in higher-dimensional spaces constitute sets of measure 0. However, if we know where to look for a set of measure 0, we can find many instances. As we go to the thermodynamic limit, we can locate the phase transition more and more precisely, and hence we should be able to locate class IV CAs in arbitrarily large rule spaces even though they constitute a set of measure 0.

If Wolfram is correct in attributing the capacity for universal computation to class IV CAs, then when we locate class IV CAs at a phase transition, we are also locating universal computation at a phase transition.

7.2. Complexity classes

One obvious property of computations for which we would like to find an analog in phase-transition phenomena is the existence of the various complexity classes. Some computations may be performed using an amount of time or space which is only a linear – or even a constant – function of the "size" of the input, while other computations exhibit polynomial, or even exponential dependence [9]. Where can we find a natural analog of these complexity classes within the phenomenology of phase transitions?

The obvious answer is in the divergence of transient times as one approaches the phase transition. As illustrated in the qualitative dynamics of 1D CAs, for λ values far from the transition point, transients die out in time which is independent of the size of the array. As λ approaches the transition point, transients begin to show more and more dependence on array size. For values of λ very near a "critical" transition, this size dependence appears to be exponential or worse. This is true whether we approach the transition from the ordered regime or the disordered regime, which suggests that in addition to the familiar complexity-class hierarchy for halting computations, there should be a similar complexity-class hierarchy for *non*-halting computations.

7.3. The Halting problem

This last point brings up another property of computation which should be reflected in phase-transition dynamics.

Some computations halt, and some do not. For some computations, we can decide whether or not they will halt. However, Turing demonstrated that for certain classes of machines this "halting problem" is undecidable: there exist computations for which it is not possible to decide whether or not they will halt.

Thus, with respect to our ability to decide the ultimate outcome of computations, there are essentially three possibilities: we can determine that they will halt, we can determine that they will not halt, or we cannot determine whether or not they will halt.

As we have seen, there are three similar possibilities for the ultimate outcome of the evolutions of CAs. CAs below the transition point rapidly "freeze up" into short-period behavior from any possible initial configuration. On the other hand, CAs above the transition point will never freeze into periodic behavior, settling down rapidly instead to chaotic behavior. Thus, we can predict the ultimate dynamics of CAs away from the transition point with a high degree of certainty.

For CAs in the vicinity of the transition, however, both of these ultimate dynamical outcomes are possible, and because of the extended transients, it will be "effectively" undecidable whether a particular rule operating on a particular initial configuration will ultimately lead to a frozen state or not for this range of λ.

Thus, we can identify a natural analog of Turing's Halting problem in what we call the *Freezing problem*: for an arbitrary CA in the vicinity of the transition point, will the dynamics ultimately "freeze up" into short-period behavior or not? It is quite likely that the freezing problem is undecidable.

8. The natural domain of information

Let us now lay out in general outline an interpretation that will tie together all of these disparate phenomena into a coherent picture of the nature of computation. The reader should bear in mind that this interpretation, although strongly supported by evidence, is only a conjecture at this point; many details remain to be worked out.

8.1. Solids, fluids, and dynamics

We propose that the *solid* and *fluid* phases of matter, with which we are so familiar from everyday experience, are much more fundamental aspects of nature than we have supposed them to be. Rather than merely being possible states of matter, they constitute *two fundamental universality classes of dynamical behavior*.

We know solids and fluids primarily as states of matter because up until quite recently, everything that exhibited dynamical behavior was made up of some kind of material. Now, however, with the availability of computers, we are able to experiment with dynamics abstracted from any particular material substrate. The findings reported in this paper suggest that for dynamical systems in general – whether purely formal or manifestly material – there are primarily only two ultimate dynamical possibilities.

However, these two universality classes are separated by a *phase transition*. The dynamics of systems within this transition region – especially the "critical" systems – appear to support the basic mechanisms necessary for information transmission, storage, and modification, and therefore provide the capacity for emergent computation. Thus, a third possibility is that systems can be constructed in such a way that they manage to avoid either of the two primary dynamical outcomes by maintaining themselves on indefinitely extended transients.

It is a system's capacity for supporting a dynamics of information that allows complex behavior in the vicinity of a phase transition. This in turn allows for the possibility of the freezing problem. Since computers and computations are specific instances of material and formal systems respectively, they are also ultimately bound by these universality classes. Therefore, if this interpretation is correct, the halting problem can be seen as a specific instance of the more general freezing problem for dynamical systems. We can therefore view computations as special instances of the kinds of processes that occur in a physical system in the vicinity of a solid/liquid or a liquid/vapor transition.

8.2. Related work

Others have been working on the problem of finding structure in the rule spaces of cellular automata and other, similar spatially distributed dynamical systems.

In my initial investigations with the λ parameter [18], I suggested that Wolfram's class IV CAs constituted a transition between class II and class III, that is, between periodic and chaotic dynamics.

Kauffman [14, 13] has investigated a class of related dynamical systems known as *Boolean nets*, in which he finds a similar phase transition between ordered and disordered dynamics.

Vichniac, Tamayo and Hartman [25] discovered that the Wolfram classes could be recovered by varying the frequency of two simple rules in an inhomogeneous cellular automaton. They also suggested a relation between critical slowing down and the halting problem.

Packard and Li [20] have mapped out the space of "elementary" $K=2$, $N=3$, 1D CAs fairly completely, using a parameterization scheme similar to λ.

Packard [22] has also performed an interesting series of experiments in which he "adapts" CA rules by selecting for certain behaviors. He finds an initially random population of rules will drift towards the phase-transition region. His interpretation of this phenomenon is that it is easier to find rules which will *compute* the desired behavior – by making use of a general computational capacity – than it is to find rules that are "hard-wired" to produce *only* the desired behavior.

McIntosh [21] has applied the mean-field approach of Gutowitz [12, 11] and suggests that the Wolfram classes can be distinguished on the basis of simple features of the mean field theory curves.

Wootters [30] has applied mean-field theory to explain the results from the λ parameter, and has been able to reproduce many of the features of fig. 6.

Together with Crutchfield's work mentioned earlier, these results collectively point to the existence of a phase transition in the spectrum of dynamical systems, and also suggest that the complex dynamics of systems in the vicinity of a phase transition rest on a fundamental capacity for processing information.

8.3. Questions

There are many questions that need to be addressed. For instance, can the "fluid" dynamical systems be further divided up into "gases" and "liquids"? There is some evidence for both solid/liquid and liquid/gas transitions in the space of CAs [17].

How might these issues be addressed by statistical mechanics, which has been very effective in treating phase transitions in general? Can analogs for temperature, pressure, volume, and energy be found? There is some evidence that equivalent measures can be defined [6, 17]. On the other hand, it is possible that statistical mechanics alone will not be able to fully treat phase-transition phenomena without being augmented by ideas from the theory of computation.

What are the implications for optimization techniques such as simulated annealing [15], which call for extended stays in the vicinity of the freezing point? It is interesting that this is the very point at which we would expect information processing to emerge spontaneously within the system being annealed – suggesting that the real reason for hovering in the vicinity of the freezing point is to allow the system to compute its own solution via an emergent computation.

How are the notions reported here related to Bak's self-organized criticality [1]? In many ways, it seems that Bak has discovered that dynamical systems can be made to *boil* when driven in the right way, which is a phenomenon we would expect at a phase transition. In fact, Bak has suggested that Conway's game of Life is a self-organized critical system, although he does not bring Life's computational capacity into the discussion.

Finally, what are the implications for understanding the origin and evolution of life? One of the most exciting implications of this point of view is that life had its origin in just these kinds of extended transient dynamics. Looking at a living cell, one finds phase-transition phenomena everywhere. The point of view advocated here would suggest that we ourselves are examples of the kind of "computation" that can emerge in the vicinity of a phase transition given enough time.

Now nature is not so beneficent as to maintain conditions at or near a phase transition forever. Therefore, in order to survive, the early extended transient systems that were the precursors of life as we now know it had to gain control over their own dynamical state. They had to learn to maintain themselves on these extended transients in the face of fluctuating environmental parameters, and to steer a delicate course between too much order and too much chaos, the Scylla and Charybdis of dynamical systems. Such transient systems must have "discovered" how to make use of their intrinsic information processing capability in order to sense and respond to their local environment. Evolution has been the process by which such systems have managed to gain local control over more and more of the environmental variables affecting their ability to maintain themselves on extended transients with essentially open futures.

9. Conclusion

Von Neumann observed that[#3]:

"There is thus this completely decisive property of complexity, that there exists a critical size below which the process of synthesis is degenerative, but above which the phenomenon of synthesis, if

---

[#3] John von Neumann, in his 1949 University of Illinois lectures on the Theory and Organization of Complicated Automata [26].

properly arranged, can become explosive, in other words, where syntheses of automata can proceed in such a manner that each automaton will produce other automata which are more complex and of higher potentialities than itself".

Although we are using a slightly different sense of "complexity" than von Neumann, the results of this paper support his observation. More importantly, however, we suggest that a similar observation can be made in the case of *too much* "complexity": *above* a certain level of "complexity", the process of synthesis is also degenerative.

In other words, we find that there exist an *upper* limit as well as a *lower* limit on the "complexity" of a system if the process of synthesis is to be non-degenerative, constructive, or open ended. We also find that these upper and lower bounds seem to be fairly close together and are located in the vicinity of a phase transition.

As the systems near the phase transition exhibit a range of behaviors which reflects the phenomenology of computations surprisingly well, we suggest that we can locate computation within the spectrum of dynamical behaviors at a phase transition here at the "edge of chaos".

Acknowledgements

Many people have contributed to the ideas presented here. I have benefitted greatly from conversations with Richard Bagley, Jim Crutchfield, Doyne Farmer, Howard Gutowitz, Hyman Hartman, Stuart Kauffman, Wentian Li, Norman Packard, Steen Rasmussen, Rob Shaw and Bill Wooters. Stephanie Forrest has been a long time intellectual companion and critic, and served as midwife in the delivery of this paper.

References

[1] P. Bak, C. Tang and K. Wiesenfeld, Self-organized criticality, Phys. Rev. A 38 (1988) 364–374.

[2] E. Berlekamp, J.H. Conway and R. Guy, Winning Ways for Your Mathematical Plays (Academic Press, New York, 1982).

[3] A.W. Burks, Essays on Cellular Automata (University of Illinois Press, Urbana, IL, 1970).

[4] G. Chaitin, J. Assoc. Comput. Mach. 13 (1966) 145.

[5] E.F. Codd, Cellular Automata (Academic Press, New York, 1968).

[6] J.P. Crutchfield and K. Young, Computation at the onset of chaos, in: Complexity, Entropy, and Physics of Information, ed. W. Zurek (Addison–Wesley, Reading, MA, 1990).

[7] E. Fredkin and T. Toffoli, Conservative logic, Int. J. Theor. Phys. 21 (1982) 219–253.

[8] M. Gardner, Mathematical games: The fantastic combinations of John Conway's new solitaire game 'Life', Sci. Am. 223(4) (October 1979) 120–123.

[9] M.R. Garey and D.S. Johnson, Computers and Intractability (Freeman, San Fransisco, 1979).

[10] L.L. Gatlin, Information Theory and the Living System (Columbia Univ. Press, New York, 1972).

[11] H.A. Gutowitz, A hierarchical classification of cellular automata, in: Proceedings of the 1989 Cellular Automata Workshop, ed. H.A. Gutowitz (North-Holland, Amsterdam, 1990), Physica D, to be published.

[12] H.A. Gutowitz, J.D. Victor and B.W. Knight, Local structure theory for cellular automata, Physica D 28 (1987) 18–48.

[13] S.A. Kauffman, Emergent properties in random complex automata, Physica D 10 (1984) 145–156.

[14] S.A. Kauffman, Metabolic stability and epigenesis in randomly constructed genetic nets, J. Theor. Biol. 22 (1969) 437–467.

[15] S. Kirkpatrick, C.D. Gelatt and M.P. Vecchi, Optimization by simulated annealing, Science 220 (1983) 671–680.

[16] A.N. Kolmogorov, Prob. Inf. Transm. 1 (1965) 1.

[17] C.G. Langton, Computation at the edge of chaos, Ph.D. Thesis, University of Michigan (1990).

[18] C.G. Langton, Studying artificial life with cellular automata, Physica D 22 (1986) 120–149.

[19] W. Li, Analyzing Complex Systems, Ph.D. Thesis, Columbia University (1989).

[20] W. Li and N.H. Packard, Structure of elementary cellular automata rule-space, Complex Systems, submitted for publication (1990).

[21] H.V. McIntosh, in: Proceedings of the 1989 Cellular Automata Workshop, ed. H.A. Gutowitz (North-Holland, Amsterdam), Physica D, to be published.

[22] N.H. Packard, Adaptation toward the edge of chaos, Technical Report, Center for Complex Systems Research, University of Illinois, CCSR-88-5 (1988).

[23] N.H. Packard and S. Wolfram, Two-dimensional cellular automata, J. Stat. Phys. 38 (1985) 901.

[24] A.R. Smith III, Simple computation-universal cellular spaces, J. Assoc. Comput. Mach. 18 (1971) 339–353.

[25] G.Y. Vichniac, P. Tamayo and H. Hartman, Annealed and quenched inhomogeneous cellular automata, J. Stat. Phys. 45 (1986) 875–883.

[26] J. von Neumann, Theory of self-reproducing automata, 1949 University of Illinois Lectures on the Theory and Organization of Complicated Automata, ed. A.W. Burks (University of Illinois Press, Urbana, IL, 1966).

[27] S. Wolfram, Statistical mechanics of cellular automata, Rev. Mod. Phys. 55 (1983) 601–644.

[28] S. Wolfram, ed., Theory and Applications of Cellular Automata (World Scientific, Singapore, 1986).

[29] S. Wolfram, Universality and complexity in cellular automata, Physica D 10 (1984) 1–35.

[30] W.T. Wootters and C.G. Langton, Is there a sharp phase transition for deterministic cellular automata?, in: Proceedings of the 1989 Cellular Automata Workshop, ed. H.A. Gutowitz, to appear in Physica D (1990).

THE PERFORMANCE OF COOPERATIVE PROCESSES

Bernardo A. HUBERMAN

Dynamics of Computation Group, Xerox Palo Alto Research Center, Palo Alto, CA 94304, USA

Computational processes in distributed networks without global controls resemble a community of concurrent agents which, in their interactions, strategies, and competition, for resources behave like whole ecologies. This brings to mind the spontaneous appearance of organized behavior in biological and social systems, where agents can engage in cooperative strategies while working on the solution of particular problems. This paper analyzes the performance characteristics of interacting processes engaged in cooperative problem solving. It shows that for a wide class of problems, there is a highly nonlinear and universal increase in performance due to the interactions between agents. In some cases this is further enhanced by sharp phase transitions in the topological structure of the problem. These results are illustrated in the context of three prototypical search examples.

1. Introduction

Propelled by advances in software, hardware and interconnectivity, computational systems are starting to spread throughout offices, laboratories, countries and continents. Unlike stand-alone computers, these growing networks seldom offer centralized scheduling and resources allocation. Instead, computational processes consisting in the active execution of programs, migrate from workstations to printers, servers and other machines of the network as the need arises, without knowledge of the state of remote computers or their availability at run time. They thus become a community of concurrent processes which, in their interactions, strategies, and competition for resources, behave like whole ecologies.

This analogy between distributed computation and natural ecologies brings to mind the spontaneous appearance of organized behavior in biological and social systems, where agents can engage in cooperative strategies while working on the solution of particular problems. As the examples of the scientific community, social organizations, the economy, and biological ecosystems show, a collection of interacting agents individually trying to solve a problem using different techniques can significantly enhance the performance of the system as a whole.

The essential characteristic of distributed cooperative problem solving is that locally programmed agents exchange messages reporting on their partial success towards completion of a goal. Such tasks generally require adaptability to unexpected events, dealing with imperfect and conflicting information from many sources, and acting before all relevant information is available. In particular, incorrect information can arise not only from hardware limitations but also from computations using probabilistic methods, heuristics, rules with many exceptions, or learning procedures that result in overgeneralization. Similarly, delays in receiving needed information can be due to the time required to fully interpret signals in addition to physical communication limitations.

Recently, we showed that the dynamical behavior of highly interacting agents in a computational ecology exhibits a wide range of behaviors [5]. By allowing for cooperation between agents engaged in different tasks, one can obtain improved performance of the system as a whole. In many such cases, those agents making the most progress per unit time are the ones that set the overall performance. For example, consider a concurrent constrained search in a large database in which a number of agents are looking for an item which satisfies the constraints. The overall search time is determined by the agent which arrives at the answer first, thereby terminating all related processes. Thus an interesting question concerns the distribution of performance among the agents, particularly the nature of those performing exceedingly well. This emphasis on performance distribution highlights the need to study more than just the average behavior of highly interacting systems.

To determine the distribution of performance, one might expect it necessary to know the details of the cooperating processes. Fortunately, however, highly cooperative systems, when sufficiently large, display a universal distribution of individual performance, largely independent of the detailed nature of either the individual processes or the particular problem being tackled. In particular, this predicts an extended tail of high performance and can be expected to apply when such performance requires successful completion of a number of nearly independent steps or subtasks. For instance, this distribution has been observed to describe a wide range of systems [1, 3] including scientific productivity [9], species diversity in ecosystems [6], and income distributions in national economies [7]. We therefore conjectured that distributed systems operating as computational ecosystems will display the same quantitative characteristics [5].

In this paper, I will discuss the performance characteristics of interacting processes engaged in cooperative problem solving. For a wide class of problems, there is a highly nonlinear increase in performance due to the interactions between agents. In some cases this is further enhanced by sharp phase transitions in the topological structure of the problem. These results will be illustrated in the context of three prototypical search examples.

The first considers a general search for a particular goal among a number of states. The second describes the further enhancement of performance due to phase transitions in a hierarchical search problem. The final example concerns a search for a good, but not necessary optimal, state. Throughout these examples we show how the existence of a diverse society of processes is required to achieve this performance enhancement.

2. Concurrent search

Consider the case of heuristically guided search, which applies to a wide range of problems [8]. A search procedure can be thought of as a process which examines a series of states until a particular goal state is obtained. These states typically represent various potential solutions of a problem, usually obtained through a series of choices. Various constraints on the choices can be employed to exclude undesirable states. Examples range from well-defined problem spaces, as in chess, to problems in the physical world such as robot navigation.

As a specific example, consider the case of a d-dimensional vector, each of whose components can take b different values. The search consists of attempting to find a particular suitable value (or goal) among the b^d possible states. It is thus a simple instance of constrained search involving the assignment of values to components of a vector subject to a number of constraints. A random search through the space will, on average, find the goal only after examining one half of the possibilities, an extremely slow process for large problems (i.e., the required time is exponential in d, the number of components to be selected). Other specific approaches can be thought of as defining an order in which the possible states are examined, with the ensuing performance characterized by where in this sequence of states the goal appears. We now suppose that n agents or processes are cooperating on the solution of this problem, using a variety of heuristics, and that the problem is completed by the first agent to find the solution. The heuristic used by agent i can be simply characterized by the fraction f_i, between 0 and 1, of unproductive states that it examines before reaching the goal. A perfect heuristic will thus correspond to $f_i = 0$ and one which chooses at random has $f_i = 1/2$.

In addition to their own search effort, agents can exchange information regarding the likely location of the goal state within the space. In terms of the sequence of states examined by a particular agent, the effect of good hints is to move the goal toward the beginning of the sequence by eliminating from consideration states that would otherwise have to be examined. A simple way to characterize a hint is by the fraction of unproductive nodes (and that would have otherwise been examined before reaching the goal) that the hint removes from the search. Since hints need not always be correctly interpreted, they can also lead to an increase in the actual number of nodes examined before the answer is found. For such cases, we suppose that the increase, on average, is still proportional to the amount of remaining work, i.e. bad hints will not cause the agent to nearly start over when it is already near the goal but will instead only cause it to reintroduce a small number of additional possibilities. Note that the effectiveness of hints depends not only on the validity of their information, but also on the ability of recipients to interpret and use them effectively. In particular, the effect of the same hint sent to two different agents can be very different.

A simple example of this characterization of hint effectiveness is given by a concurrent search by many processes. Suppose there are a number of characteristics of states that are important (such as gender, citations, and subfield in a database). Then a particular hint specifying gender, say, would eliminate one half of all remaining states in a process that is not explicitly examining gender.

To the extent that the fractions of unproductive nodes pruned by the various hints are independent, the fraction of nodes that an agent i will have to consider is given by

$$f_i = f_i^{\text{initial}} \prod_{j \neq i} f_{j \to i}^{\text{hint}}, \tag{1}$$

where $f_{j \to i}^{\text{hint}}$ is the fraction of unproductive nodes eliminated by the hint the agent i receives from agent j, and f_i^{initial} characterizes the performance of the agent's initial heuristic. Note that hints that are very noisy or uninterpretable by the agent correspond to a fraction equal to one because they do not lead to any pruning. Conversely, a perfect hint would directly specify the goal and make f_i equal to zero. Furthermore, we should note that since hints will generally arrive over time during the search, the fractions characterizing the hints are interpreted as effective values for each agent, i.e. a good hint received late, or not utilized, will have a small effect and a corresponding hint fraction near one.

The assumption of independence relies on the fact that the agents broadcast hints that are not overlapping, i.e. the pruning of two hints will not be correlated. This will happen whenever the agents are diverse enough so as to have different procedures for their own searches. If the agents were all similar, i.e. the pruning was the same for all of them, the product in eq. (1) would effectively only have one factor. For immediate cases, the product would only include those agents which differ from each other in the whole population. As additional consideration, the overall heuristic effectiveness f_i must not exceed one, so there is a limit to the number and placement of independent hint fractions larger than one that can appear in eq. (1). We therefore define n_{eff} to be the effective number of diverse agents, which in turn defines the actual number of terms in the product of eq. (1). This leads to a direct dependence of the pruning effectiveness of the diversity of the system. Although the hints that individual agents find useful need not come from the same sources, for simplicity we suppose the number of diverse hints received by each agent is the same.

We now derive the law that regulates the pruning effectiveness among agents. By taking logarithms in eq. (1), one obtains

$$\log f_i = \log f_i^{\text{initial}} + \log f_{1 \to i}^{\text{hint}} + \cdots + \log f_{n_{\text{eff}} \to i}^{\text{hint}}, \tag{2}$$

where we have included only terms arising from diverse hints. If the individual distributions of the logarithms of the fractions satisfy the weak condition of having a finite variance, and if the number of hints is large, then the central limit theorem applies. Therefore, the values of $\log f_i$ for the various agents will be normally distributed around its mean, μ, with standard deviation σ, i.e. according to $N(\mu, \sigma, \log f_i)$. Here μ and σ^2 are the mean and variance of the $\log f_i$ of the various agents, which are given by the sum of the corresponding moments of the individual terms in the sum. In other words, f itself is distributed according to the log-normal distribution [1]

$$\text{Prob}(f) = \frac{1}{\sigma f \sqrt{2\pi}} \exp\left[-(\log f - \mu)^2 / 2\sigma^2\right]$$

$$= \frac{N(\mu, \sigma, \log f)}{f}, \tag{3}$$

which gives the probability density for a given agent to have various values of f. The mean value of f is $m = \exp(\mu + \frac{1}{2}\sigma^2)$ and its variance is given by $m^2[\exp(\sigma^2) - 1]$. This distribution is highly asymmetric with a long tail, signifying an enormous range of performance among the individual agents.

To examine the effect of hints, we measure performance for the agents in terms of the speed at which they solve the problem. This is given by

$$S \equiv \frac{\text{size of search space}}{\text{time to reach goal}}$$

$$= \frac{b^d}{f(b^d - 1) + 1}, \tag{4}$$

where the time required to find the goal is just the number of states that were actually examined during the search. For the large search spaces of interest here, this will be approximately given by $1/f$ except for very small values of f. When a variable such as f is log-normally distributed, so is any power of it, in particular $1/f$. Hence the log-normal distribution of f derived above will produce a similar distribution for corresponding values of S. In practice, of course, there is a finite upper limit on performance (given, in this case, by $S_{\max} = b^d$) even though f can be arbitrarily small. This implies an eventual cutoff in the distribution at extremely high performance levels. Nevertheless, the extended tail of the log-normal distribution can be expected to adequately describe the enhancement in performance due to exchange of hints for values well below this maximum.

As a concrete case, we consider the situation in which hints, on the average, neither advance nor hinder performance of the system as a whole, i.e. the mean value of the hint fractions is one, which can be considered a worst case scenario. Thus, any improvement of the overall performance of the system will come from the tail of the distribution. Specifically, we take the $f_{j \to i}^{\text{hints}}$ values to be normally distributed according to $N(1, 0.02, f)$. We also take the initial performance of the agents (i.e. speed S without hints) to be normally distributed according to $N(4, 0.05, S)$ which corresponds to somewhat better than random search. These choices ensure that there is a negligible chance for S to reach its maximum, so the log-normal distribution will accurately describe the high performance tail in the range of interest. The resulting distributions are compared in fig. 1.

Because of the enhanced performance tail, a collection of cooperating agents is far more likely to have a few high performers than the noninteracting case. This can be seen by examining the tail of the distributions, particularly the top percentiles of performance. In particular, for a system with n agents the expected top performer will be in the top $100/n$ percentile. This can be quantified by specifying the speed reached or exceeded by the top performers. With no hints, the top 0.1

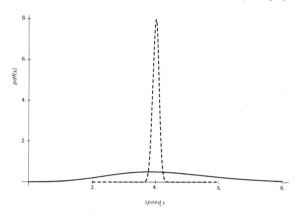

Fig. 1. Distribution of agents according to their performance S in a search with $b = 5$ and $d = 20$. The dashed curve corresponds to the noninteracting case of no hints exchanged during the search. The solid gray curve corresponds to $n_{eff} = 10$, and the solid black one to $n_{eff} = 100$. Notice the appearance of a long tail in the interacting cases, which results in an improvement in performance. The area under each of the curves is one. The nature of the hints is such that, on the *average*, they neither enhance nor retard the search procedure.

percentile is located at a speed of 4.15. On the other hand, this percentile moves up to 4.89 and 7.58 when $n_{eff} = 10$ and $n_{eff} = 100$ respectively, where n_{eff} is the effective number of cooperating agents. Note that the top 0.1 percentile characterizes the best performance to be expected in a collection of 1000 cooperating agents. The enhancement of the top performers increases as higher percentiles or larger diversity are considered, and shows the highly nonlinear multiplicative effect of cooperative interactions.

A collection of agents could manage to exchange hints such that on average they *increase* the performance of a system. In this case, in addition to the enhanced tail there will also be a shift of the overall performance curve toward higher values due to the multiplication of a number of factors greater than one. Notice that in such a case high diversity is less important in setting the improved performance than was in the example. We should remark, however, that this scenario breaks down for very high performance agents due to the upper bound on the maximum speed $S_{max} = b^d$.

3. Hierarchical search and phase transitions

A very important class of problem solving involves heuristic searches in tree structures. Thus it is important to elucidate how the above considerations of cooperation apply to this case. In particular, suppose the search takes place in a tree with branching ratio b and depth d, so the total number of nodes in the tree is given by

$$N_{total} = \frac{b^{d+1} - 1}{b - 1}. \tag{5}$$

This can be viewed as an extension of the previous example in that successive levels of the tree correspond to choices for successive components of the desired vector, with the leaves of the tree corresponding to fully specified vectors. The additional tree structure becomes relevant when the heuristic can evaluate choices based on vectors with some components unspecified. These evaluations offer the possibility of eliminating large groups of nodes at once.

The search proceeds by starting at the root and recursively choosing which nodes to examine at successively deeper levels of the tree. At each node of the tree there is one correct choice, in which the search gets one step closer to the goal. All other choices lead away from the goal. The heuristic used by each agent can then be characterized by how many choices are made at a particular node before the correct one is reached. The perfect heuristic would choose correctly the first time, and would find the goal in d time steps, whereas the worst one would choose the correct choice last, and hence be worse than random selection. To characterize an agent's heuristic, we assume that each incorrect choice has a probability p of being chosen by the heuristic before the correct one. Thus, when the branching ratio is 2, the perfect heuristic corresponds to $p = 0$, random to $p = 0.5$, and worst to $p = 1$. For simplicity, we suppose the heuristic effectiveness, as measured by p, is uniform throughout the tree. Alternatively, p can be thought of as the value of the effectiveness averaged over all nodes in the tree. In the latter case, any particular correlations between nodes are ignored, in the spirit of a mean-field theory, which can be expected to apply quite well in large-scale problems. Note that while p specifies the fraction of incorrect choices made before the correct one on average throughout the tree, this probabilistic description allows for variation among the nodes.

The a posteriori effect of hints received from other agents can be described as a modification to an agent's value of p. Assuming independence among the hints received, this probability is given by

$$p_i = p_i^{\text{initial}} \prod_{j=1}^{n_{\text{eff}}} f_{j \to i}^{\text{hint}}, \quad (6)$$

where p_i^{initial} characterizes the agent's initial heuristic and the hint fractions are the same as introduced in section 2, but now averaged over the entire tree. By supposing the various quantities appearing in eq. (6) are random variables, we again obtain the universal log-normal distribution (over the set of agents) of heuristic effectiveness when there are a large number of agents exchanging hints.

Given this distribution in *local* decision effectiveness, we now need the distribution of performance in the full search problem, i.e. the rate at which the search for the goal is completed. This relationship is more complex than in the unstructured example considered above, and in particular it produces a phase transition in over-all agent performance at a critical value of p [4]. This sharp transition leads to the possibility of an additional enhancement in performance.

Specifically, the overall performance is related to the time T, or number of steps, required to reach the goal from the root of the tree. To quantify the search performance, we consider the search speed given by

$$S \equiv \frac{\text{number of nodes in the tree}}{\text{number of steps to the goal}} = \frac{N_{\text{total}}}{T}. \quad (7)$$

To compare trees of different depths, it is convenient to normalize this to the maximum possible speed, namely $S_{\text{max}} = N_{\text{total}}/d$, giving the normalized speed $s \equiv S/S_{\text{max}} = d/T$.

Because of the probabilistic characterization of the heuristic for each agent, T is a random variable. It is determined by two contributions: the length of the correct path to the goal (equal to the depth of the search tree, d), plus the number of nodes visited in every incorrectly chosen subtree along the way to the goal, in itself a random variable. While the actual probability distribution of T values for a given value of p is complicated, one can show that the average number of steps required to reach the goal is given by ref. [4] as

$$\langle T \rangle = d + \frac{(\mu - p)(d - \mu - d\mu + \mu^{1+d})}{(\mu - 1)^2}, \quad (8)$$

where $\mu = bp$. As the depth of the tree increases, this becomes increasingly singular around the value $\mu = 1$, indicating a sudden transition from linear to exponential search. This is illustrated in fig. 2, which shows the behavior of the average normalized search speed $\tilde{s} \equiv d/\langle T \rangle$ as a function of the local decision effectiveness characterized by p. Near the transition, a small change in the local effectiveness of the heuristic has a major impact

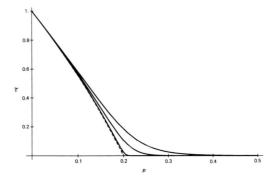

Fig. 2. Plot of $\tilde{s}$ versus local decision effectiveness for trees with branching ratio 5 and depths 10, 20 and 100. The distinction between the linear regime ($p < 0.2$) and the exponential one becomes increasingly sharp as the depth increases. The dashed curve is the limit for an infinitely deep tree and shows the abrupt change at $p = 0.2$ from linear to exponential search.

on the global behavior of large-scale search problems. The existence of such a phase transition implies that, in spite of the fact that the average behavior of cooperative algorithms may be far into the exponential regime, the appearance of an extended tail in performance makes it possible for a few agents to solve the problem in polynomial time. In such cases, one obtains a dramatic improvement in overall system performance by combining these two effects. We should note that other search topologies such as general graphs also exhibit these phase transitions [2], so these results can apply to a wide range of topologies found in large-scale search problems.

Finally, to illustrate the result of combining diverse hints with the phase transition in tree searches, we evaluate the distribution of relative global speed s for the agents searching in a tree with a branching ratio $b = 5$ and depth $d = 20$. This combines the distribution of local decision effectiveness with its relation to global speed. As in the previous example, we suppose hints on average neither help nor hinder the agents. In particular, we take the $f_{j \to i}^{\text{hints}}$ values to be normally distributed according to $N(1, 0.015, f)$. We also take the initial performance of the agents (i.e. p_i^{initial}) to be normally distributed according to $N(0.33, 0.0056, p)$, which corresponds to a bit better than random search. The resulting distributions were evaluated through simulations of the search process and are compared in fig. 3, on a logarithmic scale to emphasize the extended tails.

In this case, the enhancement of the global performance of the system is most dramatic at the higher end of the distribution, not all of which is shown in the figure. In this example, the top 0.1, percentile agents will have an enhancement of global speed over the case of no hints by factors of 2 and 41 for 10 and a 100 hints respectively. This illustrates the nonlinear relation between performance, number of agents and diversity of hints.

4. Satisficing searches

In many heuristic search problems, the exponential growth of the search time with problem size forces one to accept a satisfactory answer rather than an optimal one. In such a case, the search returns the best result found in a fixed amount of time rather than continuing until the optimal value is found. If the search returns have a good value, they can provide acceptable solutions to the problem without the cost involved in obtaining the true optimum. A well-known instance is the traveling salesman problem, consisting of a collection of cities and distances between them and an attempt to find the shortest path which visits each of them. The time required to find this path grows exponentially with the number of cities if $P \neq NP$.

For large instances of the problem, one must settle instead for paths that are reasonably short but not optimal.

In these cases of limited search time, the extended tails of the cooperative distributions discussed above result in a better value returned compared to cases in which hints are not used. To see this we consider an unstructured search problem where the states have various values v, which we take to be integers between 0 and some maximum V. In the previous examples, one could view the single goal as having the maximum value while

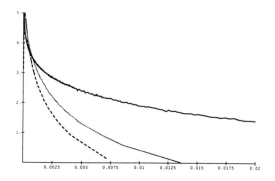

Fig. 3. Distribution of agents (on a log scale) as a function of relative global speed s for a concurrent search in a tree with $b = 5$, and $d = 20$. The dashed line corresponds to the case of no hints being exchanged during the search. The solid gray curve corresponds to $n_{\text{eff}} = 10$, and the solid black one to $n_{\text{eff}} = 100$. Note the enhancement of the high-performance tail as the diversity of exchanged hints is increased.

all other states have a value of 0. To allow for the possible usefulness of nonoptimal states, we suppose that their values are distributed throughout the range. In order that a simple random search is unlikely to be effective, we need relatively few states with good values. A simple distribution of values satisfying these requirements is given by the binomal distribution:

$$m_v = \binom{V}{v} 3^{V-v} \qquad (9)$$

where m_v is the number of states with value v. Note that this has exactly one state with the maximum value and most states have smaller values clustered around the average $V/4$.

For problems of this kind, the effectiveness of a heuristic is determined by how well it can discriminate between states of high and low value. When faced with selecting among states with a range of values, a good heuristic will tend to pick those states with high value. That is, the likelihood of selecting a state will increase with its value. Moreover, this increase will become more rapid as the heuristic improves. As a concrete example, we suppose that the heuristics used by the various agents in the search are characterized by a discrimination parameter α such that states with value v are selected by the heuristic with relative probability α^v. Large values of α provide excellent discrimination while $\alpha = 1$ corresponds to random selections. In terms of our previous examples, in which only the goal had a nonzero value, the relative selection probabilities were 1 for the goal and p for all other states. Thus we see that this characterization of heuristic discrimination identifies α^V with $1/p$ in the case of only two distinct values.

As in the previous examples, cooperation among diverse agents leads to a log-normal distribution of selection probability values among the agents. Here this means that the α values will themselves be log-normally distributed. Instead of focusing on the time required to find the answer, we can examine the distribution of values returned by the various agents in a given interval of time. As an extreme contrast with the previous examples, which continued until the goal was found, we allow each agent to examine only one state, selected using the heuristic. The value returned by the agent will then correspond to this state. (If additional time were available, the agents would continue to select according to their heuristic and return the maximum value found.) This simplifications can be used to obtain the distribution of returned values resulting from interactions among the agents as a function of the number of diverse agents, n_{eff}.

Since all points are available to be selected, the probability that an agent operating with a heuristic discrimination level of α will select a state with value v is

$$p(\alpha, v) = \frac{m_v \alpha^v}{\sum_{u=0}^{V} m_u \alpha^u} = \binom{V}{v} \frac{\left(\frac{1}{3}\alpha\right)^v}{\left(1 + \frac{1}{3}\alpha\right)^V}. \qquad (10)$$

To finally obtain the distribution of values returned by the agents, this must be integrated over the distribution of α values. When hints are exchanged, this parameter will be distributed log-normally with a mean μ and standard deviation σ depending on the corresponding values for the hint fractions. The result can be written as

$$P(v) = \frac{1}{\sqrt{2\pi}} \binom{V}{v} \exp\left[v\tilde{\mu} + \tfrac{1}{2}(v\sigma)^2\right]$$
$$\times \int_{-\infty}^{\infty} dt \exp\left(-\tfrac{1}{2}t^2\right)$$
$$\times \left[1 + \exp(\tilde{\mu} + v\sigma^2 + \sigma t)\right]^{-V}, \qquad (11)$$

where $\tilde{\mu} \equiv \mu - \ln 3$.

The distributions are compared in fig. 4 for the case in which the initial agents' heuristic has $\alpha = 1.5$ (i.e. a bit better than random value discrimination) and the hint fractions are distributed according to $N(1, 0.05)$, again giving a case in which the hints, on average, neither help nor hinder the search. In this case, the top 0.1 percentile level is at a value $v = 52$ when $n_{\text{eff}} = 10$ and $v = 70$ when

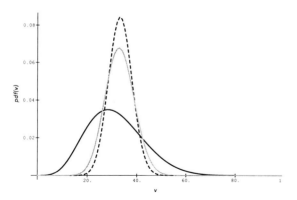

Fig. 4. Distribution of values returned in a satisfying search with $V = 100$. The dashed curve shows the distribution for the non-interacting case. The solid gray curve corresponds to $n_{\text{eff}} = 10$, and the solid black one is for $n_{\text{eff}} = 100$. The area under each curve is one.

$n_{\text{eff}} = 100$. This compares with the non-interacting case in which this performance level is at $v = 48$.

5. Diversity and cooperation

We have shown that the effectiveness of hints exchanged among agents depends critically on how independently they are able to prune the search space. At one extreme, when all the agents use the same technique, the hints do not provide any additional pruning. Similarly, if the various agents randomly search through the space and only report the nodes they have already examined, this will not significantly help the other agents.

Thus we are naturally led to consider the diversity of interactions as a fundamental characterization of these systems. By the above arguments, those systems with the highest diversity can be expected to most effectively utilize the information gained from the various cooperative agents, and have the highest overall performance. Since the emergence of complex behavior, i.e. high performance on difficult problems requiring adaptability and dealing with imperfect information, is an important issue, these results suggest that diversity is a key ingredient in the successful completion of distributed problem solving.

Although the assumption of step by step completion independence and the total neglect of cost involved in exchanging information may seem overly restrictive, there is ample evidence that the enhanced tail of the log-normal distribution in performance applies to many different systems. For instance, in a study of the statistics of individual variations of productivity in research laboratories, Shockley discovered a log-normal distribution in the individual productivity, regardless of field or laboratory location [9]. In a totally unrelated endeavor, that of animal ecology, several investigators have discovered a log-normal distribution of abundances in species as different as nesting birds, butterflies and moths captured in light traps [6]. Since this is related to the ability to survive through a number of events in the ecosystem, it mimics the process whereby agents successfully solve given problems, and can be described by the model given above. Moreover, a study of the income distribution in families and single individuals in the US and many other economies shows a log-normal distribution [7], which can be ascribed to the many necessary steps that lead to the accumulation of wealth. These examples suggest that there is an underlying universal law describing the behavior of large systems of interdependent entities. Therefore, if resources are allocated to individual agents in computational ecologies based on their perceived progress, one expects a consequent wide distribution in their utilization.

In summary, we have analyzed the performance characteristics of interacting processes engaged in cooperative problem solving. We showed that for a wide class of problems, there can be a highly nonlinear and universal increase in performance due to the interactions between the agents. In some cases this is further enhanced by sharp phase transitions in the topological structure of the problem. In spite of its simplicity, the model produces universal results which have been observed in a number of other systems. This makes us believe that it might be applicable to cooperative processes in distributed computation. Last but not least, these results provide a quantitative indication of the power of collaboration among diverse processes when engaged in distributed problem solving.

Acknowledgements

This work was done in collaboration with Tad Hogg. It was partially supported by ONR contract N00014-82-0699.

References

[1] J. Aitchison and J.A.C. Brown, The Log-normal Distribution (Cambridge University Press, Cambridge, 1957).
[2] B. Bollobas, Random Graphs (Academic Press, New York, 1985).
[3] E.L. Crow and K. Shimizu, eds., Lognormal Distributions: Theory and Applications (Dekker, New York, 1988).
[4] B.A. Huberman and T. Hogg, Phase transitions in artificial intelligence systems, Artificial Intelligence 33 (1987) 155–171.
[5] B.A. Huberman and T. Hogg, The behavior of computational ecologies, in: The Ecology of Computation, ed. B.A. Huberman (North-Holland, Amsterdam, 1988) pp. 71–115.
[6] C.J. Krebs, Ecology (Harper and Row, New York, 1972).
[7] E.W. Montroll and M.F. Shlesinger, On $1/f$ noise and other distributions with long tails, Proc. Natl. Acad. Sci. US 79 (1982) 3380–3383.
[8] J. Pearl, Heuristics: Intelligent Search Strategies for Computer Problem Solving (Addison–Wesley, Reading, MA, 1984).
[9] W. Schockley, On the statistics of individual variations of productivity in research laboratories, Proc. IRE 45 (1957) 279–290.

COLLECTIVE BEHAVIOR OF PREDICTIVE AGENTS

Jeffrey O. KEPHART, Tad HOGG and Bernardo A. HUBERMAN
Dynamics of Computation Group, Xerox Palo Alto Research Center, Palo Alto, CA 94304, USA

We investigate the effect of predictions upon a model of coevolutionary systems which was originally inspired by computational ecosystems. The model incorporates many of the features of distributed resource allocation in systems comprised of many individual agents, including asynchrony, resource contention, and decision-making based upon incomplete knowledge and delayed information. Previous analyses of a similar model of non-predictive agents have demonstrated that periodic or chaotic oscillations in resource allocation can occur under certain conditions, and that these oscillations can affect the performance of the system adversely. In this work, we show that the system performance can be improved if the agents do an adequate job of predicting the current state of the system. We explore two plausible methods for prediction – technical analysis and system analysis. Technical analysts are responsive to the behavior of the system, but suffer from an inability to take their own behavior into account. System analysts perform extremely well when they have very accurate information about the other agents in the system, but can perform very poorly when their information is even slightly inaccurate. By combining the strengths of both methods, we obtain a successful hybrid of the two prediction methods which adapts its model of other agents in response to the observed behavior of the system.

1. Introduction

In many economic and social systems, perceptions and expectations about the current or future behavior of the system play an important role in determining an individual's action. Bullish and bearish behavior on the stock and futures exchanges and the unpredictability of market share among competing technologies [1] are familiar examples of how uncertainty and expectations can foment highly volatile dynamics and self-reinforcing behavior in economic systems. In the social arena, self-reinforcing perceptions about what is "in" can greatly influence tastes and behavior, resulting in fads which appear and disappear with surprising suddenness. In such systems, individuals continually adapt their expectations so as to improve their estimates or predictions of the behavior of the rest of the system and then modify their behavior accordingly. The result is a "coevolutionary" [2] system in which all of the individuals are simultaneously trying to adapt to one another.

In this work, we gain some insight into how predictions influence the dynamics of coevolutionary systems by focusing upon a particular model in which the issue of predictions arises very naturally. This probabilistic, non-linear dynamical model, originally developed in order to understand the behavior of computational ecosystems, describes a system of many individual agents which make independent, asynchronous decisions about resource allocation based upon imperfect and delayed information. Previous studies of *non-predictive* agents in such systems have revealed that periodic or chaotic oscillations in resource allocation can occur under certain conditions [3–5], and that these oscillations have an adverse effect upon the performance of the system. These undesirable phenomena can be traced to the fact that the agents do not take the information delay into account. However, if the agents which comprise the system were able to make accurate predictions of its current state, the information delay could be overcome, and the system would perform well.

The task of predicting the future in a coevolutionary system is more difficult than one might imagine. We shall explore the strengths and weaknesses of two basic methods of prediction – one based upon technical analysis and the other upon system analysis. As well shall see, prediction strategies based upon technical analysis which appear to be reasonable from the perspective of a single agent can be disastrous when adopted by a substantial fraction of the agents. Prediction strategies based upon system analysis yield self-consistent results when adopted by all of the agents, but are overly sensitive to the accuracy of the information that agents have about one another. Fortunately, by combining the strengths of both techniques, we can obtain a hybrid agent – the "adaptive system analyst" – which makes predictions which are sufficiently accurate to ensure good system performance.

After a brief summary of our model in section 2, we analyze systems of predictive agents and derive a quantitative relationship between prediction accuracy and overall system performance in section 3. Then, in section 4, we review the behavior of non-predictive agents. The rest of the paper is devoted to a search for methods which yield accurate predictions and result in good overall performance. In section 5, we present two examples of technical analysts: linear extrapolaters and cyclical trend analysts. We illustrate the behavior of system analysts with three examples in section 6. Finally, in section 7, we introduce adaptive system analysts, which observe the behavior of the system in order to adapt their assumptions about other agents. Section 8 provides a summary of our results and a brief discussion of the many unresolved issues which remain.

2. Model

In this paper, we shall analyze a model of a coevolutionary system which was first introduced in the context of computational ecosystems [3–5]. In the model, A agents are free to choose (independently and asynchronously from one another) from among R resources according to their perceived (not necessarily correct) payoffs. The resources might represent various types of hardware or software, such as memory, time slices on a processor, a database, or a proprietary algorithm. In general, the payoff G_r for using a particular resource r depends upon the number of agents already using r. In strictly competitive situations, in which the agents are completely independent of one another and thus prefer resources solely on the basis of their capacity, the payoff for using a resource decreases monotonically with the number of agents using it. However, in more cooperative situations, a resource's payoff could be maximal when a number of agents are using it. For example, this could occur if agents communicate results to one another and there is a significant communication cost between agents using different resources. Other important features included in the model are uncertainties and delays in the agents' information about resource usage.

Fig. 1 illustrates in some detail our model of how an individual agent chooses a resource. At a particular time t, it predicts $f^*(t)$, the current fraction of agents using each resource, from $f(t-\tau)$, the fraction of agents which were using each resource at time $t-\tau$, where τ is the time required for information to reach and be processed by each agent. This prediction step is the main focus of this paper. After the agent has somehow generated a prediction of the current state of the system, it evaluates its perceived instantaneous payoff (i.e. the award it expects to receive at time t) for using each resource. The use of an instantaneous payoff, which simplifies the analysis, can be interpreted as meaning that the agent is completely myopic in its evaluation of the payoff, i.e. it considers only its estimate of the current state of the system, without regard for the future. By adding Gaussian noise with zero mean and standard deviation σ to each payoff, we are able to model several different effects: (i) intentional inclusion of randomness in the decision procedure,

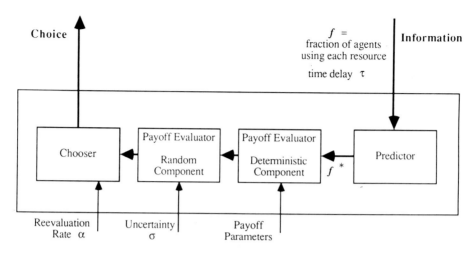

Fig. 1. Model of the decision-making procedure of the agents.

(ii) heterogeneity in the payoffs G_r among the population of agents, or (iii) imperfections or uncertainties in the information $f(t-\tau)$. In previous work we have established that uncertainty can have an important influence upon the stability of the system. After the deterministic and the stochastic portions of the payoffs are evaluated for each resource, the agent chooses the resource associated with the highest payoff. To model asynchronicity, we assume that each agent reevaluates its choice at random intervals which are chosen according to an exponential distribution with mean $1/\alpha$.

Using the complementary techniques of theoretical analysis and simulation, we can measure the fraction of agents $f(t)$ using each resource as a function of time, from which we can calculate a number of quantities which reflect different aspects of the collective behavior of the system. We can then explore how this behavior depends upon the various parameters of the model: the reevaluation rate α, the time delay τ, the uncertainty σ, the payoffs $G_r(f_r)$, and the prediction strategy. The theoretical analysis, described more fully in section 3, starts with rate equations for the movement of agents between resources. When there are many agents in the system, the rate equations are well-approximated by a deterministic differential-delay equation for $f(t)$. The simulation, a straightforward event-driven implementation of the model, helps us to assess both the validity and the approximations in the theoretical analysis and the effects of finite system size.

3. Analysis of predictive agents

In this section, we analyze the behavior of a system of predictive agents without making any assumptions about *how* the prediction is made. Then, we derive a relationship between overall system performance and prediction accuracy. For simplicity, we shall confine our analysis to systems with two resources and homogeneous agents (i.e. all agents share the same payoffs $G_1(f_1)$ and $G_2(f_2)$). It suffices to concentrate on the dynamics of $f \equiv f_1$, the fraction of the agents using resource 1, since $f_2 = 1 - f_1$.

In physical systems, the mean-field approximation is appropriate for analyzing systems of many particles with long-range interactions. In our model, many agents interact with one another via the global quantity f. Therefore, it is not surprising that analysis based upon this approximation provides an accurate description of the behavior of our model, as we have established previously [4, 5]. Starting with Huberman and Hogg's rate equations for the probability distribution $P_i(t)$ for i agents to be using resource 1 at time t [3, 5], one can first use the mean-field approximation to derive an equation for the ensemble average $\langle f(t) \rangle$. The result is then modified phenomenologically to account for the fact that the agents do not know the current state of the system, but can only predict it. The resultant dynamical equation is [3, 5]:

$$\frac{d\langle f(t) \rangle}{dt} = \alpha [\rho(\langle f^*(t) \rangle) - \langle f(t) \rangle], \quad (1)$$

where f^* is the predicted fraction of agents using resource 1 and $\rho(f^*)$ is the probability that a particular agent will choose resource 1, given that it believes the current state of the system is f^*. For the sake of notational simplicity, we henceforth omit the ensemble averages in eq. (1), interpreting the result as a fully deterministic equation for the time-evolution of $f(t)$, the fraction of agents using resource 1.

The decision-probability function $\rho(f^*)$ summarizes the deterministic and stochastic components of the payoffs and is related to them via an equation given in ref. [5]. Figs. 8 and 11, which appear later in this paper, exemplify the typical form of $\rho(f^*)$. When the uncertainty parameter σ is zero, $\rho(f^*)$ equals either 1 or 0, depending upon whether the higher payoff is obtained for resource 1 or resource 2 for the given value of f^*. To get some insight into the nature of eq. (1) in this case, suppose for the moment that each agent predicts the correct value of f^*, i.e. $f^*(t) = f(t)$. (This would certainly be the case if there were no time delays.) Then, according to eq. (1), f will increase if the payoff for using resource 1 is greater and decrease if the payoff for using resource 2 is greater. As a result, the system stabilizes at a value of f for which the two payoffs are equal – a Nash equilibrium point. Now suppose that the uncertainty parameter σ has some non-zero value. In this case, $\rho(f^*)$ acts as a fuzzy decision boundary, the width of which is determined by σ. The position of the boundary also depends somewhat upon σ, shifting to a value $f_N(\sigma)$ obtained by setting the left-hand side of eq. (1) to zero:

$$f_N = \rho(f_N). \quad (2)$$

In this Nash equilibrium, all of the agents choose resource 1 with probability f_N and resource 2 with probability $1 - f_N$. In the more general case in which the prediction $f^*(t)$ is not necessarily equal to $f(t)$, the equilibrium f_{eq} is given by

$$f_{eq} = \rho(f_{eq}^*(f_{eq})), \quad (3)$$

where f_{eq}^* is the predicted state of the system given that it is in the equilibrium f_{eq}. As we shall see in section 4, this equilibrium may or may not be stable. (Henceforth, f_N shall denote a special case of f_{eq} for which $f_{eq} = f_{eq}^*(f_{eq})$. Thus, in an equilibrium f_N, the agents correctly predict that equilibrium.)

One can also derive from the rate equations an equation of motion for the variance $s^2(t) \equiv \langle f^2(t) \rangle - \langle f(t) \rangle^2$:

$$\frac{d\langle s^2(t) \rangle}{dt}$$
$$= \alpha \{ A^{-1}[\langle f \rangle + \rho(\langle f^* \rangle) - 2\langle f \rho(f^*) \rangle]$$
$$+ 2[\langle f \rho(f^*) \rangle - \langle f \rangle \rho(\langle f^* \rangle) - s^2] \}, \quad (4)$$

where the dependence of f and f^* upon t have been suppressed for the sake of brevity and A is the number of agents. If the system is able to reach a stationary equilibrium f_{eq}, we can set the left-hand side of eq. (4) to zero and substitute eq. (3) to obtain the time-independent variance:

$$s_{eq}^2 = A^{-1} \{ f_{eq}(1 - f_{eq}) + (A - 1) \times [\langle f \rho(f^*) \rangle - f_{eq}^2] \}. \quad (5)$$

Assuming ergodicity, we can reinterpret eq. (5) as the time average of the variance of f for a typical member of the ensemble, which is the quantity in which we are interested. In general, if the fluctuations s_{eq} are sufficiently small, the term in square brackets in eq. (5) reduces to $\rho'(f_{eq}^*)s_{eq}^2 R$, where $R = \langle \delta f \delta f^* \rangle / \langle \delta f \delta f \rangle$ is the correlation between the predicted and actual behavior, $\delta f = f - f_{eq}$ and $\delta f^* = f^* - f_{eq}^*$. Then eq. (5) reduces to

$$s_{eq}^2 = \frac{f_{eq}(1-f_{eq})}{A-(A-1)\rho'(f_{eq}^*)R}. \qquad (6)$$

Eq. (6) has some important ramifications for systems in equilibrium. If the correlation $R > 1/\rho'(f_{eq}^*)$ (i.e. the predicted and actual values of f are not too *anti*-correlated[1]) and the number of agents A is at least moderately large, the fluctuations s_{eq} about the equilibrium f_{eq} scale as $1/\sqrt{A}$ regardless of the estimate $f^*(t)$. However, if $R \leq 1/\rho'(f_{eq}^*)$, the fluctuations become macroscopic (i.e. they do not approach 0 in the limit of an infinitely large system). In all of the systems we have observed, this corresponds to oscillatory behavior, in which case none of the above derivations for equilibrium systems are applicable.

One very important aspect of a system's behavior is its performance, which we define to be the time average of the total payoff to all of the agents in the system. This quantity is maximized if the system is stabilized at the value f_{opt}, given by

$$\frac{d}{df}[fG_1(f)+(1-f)G_2(1-f)]|_{f=f_{opt}} = 0. \qquad (7)$$

If the time delay and uncertainty are both zero, the equilibrium is given by

$$G_1(f_N) = G_2(1-f_N). \qquad (8)$$

Thus, in general, f_{opt} has no relationship to the equilibrium f_N – a dilemma which comes under the rubric of the "tragedy of the commons problem" in economics [6]. However, in order to have a convenient landmark to gauge system performance and to sidestep the public goods issue, we

---

[1] As illustrated later in figs. 8 and 11, the slope $\rho'(f_{eq}^*)$ is generally negative.

shall deliberately choose payoffs for which f_{opt} and $f_N(\sigma = 0)$ (the equilibrium in the limit of zero time delay and zero uncertainty) are equal. (To facilitate comparison, we shall further restrict the payoffs such that $f_{opt} = 0.75$ for all of the examples presented in this paper.)

Thus, in order for a system to perform well, the following two conditions must be satisfied. First, the equilibrium f_{eq} as given by eq. (3) must be close to f_{opt} even when the time delay and the uncertainty are greater than zero. Second, the correlation R between the actual and predicted behavior must be positive (or at least not too negative). The greater the correlation, the smaller will be the fluctuations about f_{eq}.

We shall call upon these analytic results to help interpret our observations throughout the remainder of this paper, which shall be devoted to an investigation of three different prediction strategies: naive prediction, technical analysis, and system analysis.

4. Non-predictive agents

We refer to agents which use the prediction strategy $f^*(t) = f(t-\tau)$ as non-predictive, or "naive", agents. Their typical behavior is exemplified by the phase diagram of fig. 2, which relates the behavior of a chosen homogeneous, two-resource system to the product of the reevaluation rate and the time delay, $\beta \equiv \alpha \tau$, and the uncertainty, σ. For a fixed value of the uncertainty, one can pass through a number of different behavioral regimes as the delay is increased. For sufficiently small delays, the system relaxes to $f_N(\sigma)$ without any oscillation (unshaded region of fig. 2). When the normalized delay β is increased above the critical value β_1, the system exhibits damped oscillations about f_N. Above the critical threshold β_2, the equilibrium becomes unstable, and the system exhibits persistent oscillations. If the difference between the payoffs is sufficiently nonlinear, as it is in this example, other forms of behavior, including period-doubling and chaos, can be observed as

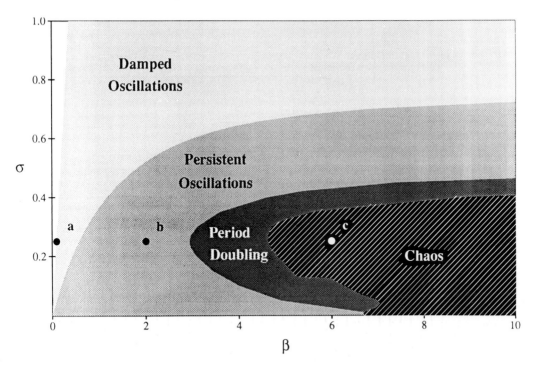

Fig. 2. Behavioral phase diagram as a function of normalized time delay $\beta \equiv \alpha\tau$ and uncertainty parameter σ for system of non-predictive agents. Every agent has the same payoffs: $G_1 = 4 + 7f_1 - 5.333f_1^2$ for resource 1 and $G_2 = 7 - 3f_2$ for resource 2, where f_i is the fraction of agents using resource i. In the unshaded region, the system exhibits overdamped (non-oscillatory) relaxation to the fixed point. The chaotic region is punctuated by numerous tiny windows of periodic behavior (not shown). Results were obtained by integrating eq. (1) numerically.

β increases to values much greater than β_2. Some of these behaviors are illustrated in fig. 3[#2].

The appearance of oscillations for sufficiently long time delays is associated with anti-correlation between the predicted and actual system behavior. As the delay is increased from $\beta = 0$ to $\beta = \beta_1$ (the phase boundary above which damped oscillations appear), one can show that the correlation R, introduced in section 3, decreases monotonically from 1 to $-\beta_1 e^{-2(1+\beta_1)}/\rho'(f_0)$[#3]. As β increases further, R decreases further to negative values, reaching $1/\rho'(f_0)$ at $\beta = \beta_2$ (the phase boundary above which persistent oscillations appear). This is consistent with simulation results reported in ref. [5], which show that the magnitude of fluctuations grows monotonically with the delay. At $\beta = \beta_2$, the denominator of eq. (6) is no longer proportional to A, indicating that, in the limit as $A \to \infty$, the fluctuations become macroscopic, which is indeed consistent with the onset of persistent oscillations.

In fig. 3a, the optimality gap between f_{opt} (represented by the dashed line) and the equilibrium $f_N(\sigma)$ is fairly small because the uncertainty σ is small. If we normalize the performance to 1 if the system is at f_{opt} and 0 if the agents make completely random decisions (i.e. $f = 0.5$), the normalized performance is 0.95. In figs. 3b and 3c, the oscillations cause f to deviate from f_{opt} significantly, and their normalized performances are -0.18 and -0.43, respectively – *worse* than a system of agents making completely random choices!

The behavioral phase diagram of fig. 2 suggests one method for quelling these undesirable oscillations: increasing σ by an amount sufficient to

---

[#2] If there are three or more resources in the system, it is possible to observe chaotic behavior even for linear payoff functions [5].

[#3] This can be obtained using the characteristic equation and related expressions found in ref. [5].

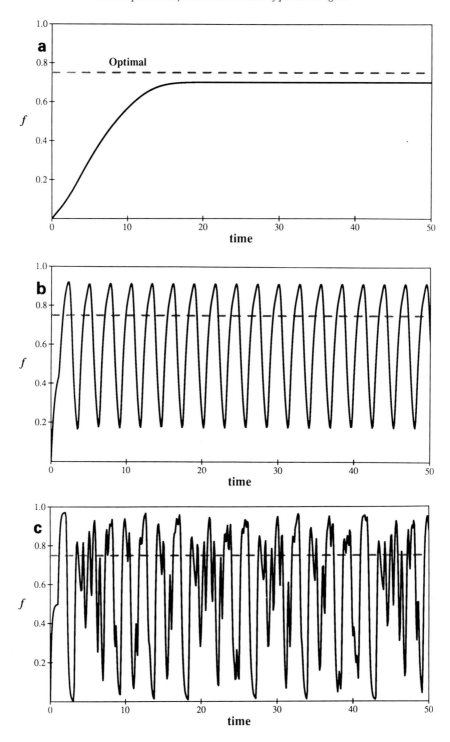

Fig. 3. Typical behaviors of non-predictive agents corresponding to points marked in behavioral phase diagram of fig. 2, obtained by integrating eq. (1) numerically. (a) Overdamped approach to stable fixed point. (b) Simple persistent oscillations. (c) Chaotic oscillations. Time scale is in units of the delay τ. $f_{opt} = 0.75$ is represented by light dashed line.

allow the system to reach an equilibrium. The drawback is that $f_N(\sigma)$ drifts away from f_{opt} as σ is increased. The best compromise is to increase σ by just enough to place the system on the boundary between damped oscillations and persistent oscillations [5]. This corresponds to an intentional degradation of the information available to the agents by the addition of random noise. This general principle has been discovered independently by several other researchers and has found its way into some important applications, most notably in the design of computer network protocols (e.g. exponential backoff algorithms used in the Ethernet [7]) and in load-balancing algorithms [8]. However, it would seem more desirable to find ways of avoiding the shift of the equilibrium away from the optimal value. To that end, we shall devote the remainder of this paper to investigating more sophisticated prediction strategies in the hope of improving the ability of the agents to make use of their information, rather than degrading their information because we lack faith in their ability to make proper use of it.

5. Technical analysts

One basic scheme that an agent might use for predicting the behavior of the system is to extrapolate from previous observations of the aggregate quantity $f(t)$, a method which is analogous to technical analysis of market behavior. In this section, we investigate two methods of extrapolation. The first is simple linear extrapolation based upon f and its derivative at time $t - \tau$, while the second incorporates a somewhat more sophisticated knowledge of the dynamics of the system.

5.1. Linear extrapolation

Linear extrapolation from f and its derivative is perhaps the most obvious and easily analyzed type of prediction. In this case, we simply substitute

$$f^*(t) = f(t - \tau) + \tau \dot{f}(t - \tau) \qquad (9)$$

into eq. (1) and perform linear stability analysis about f_N. After some algebra, one obtains an implicit expression for the persistent oscillation threshold β_2:

$$\beta_2 = \sqrt{\frac{1 - (\beta_2 \rho')^2}{\rho'^2 - 1}} \cos^{-1}\left(\frac{1 + \beta_2 \rho'^2}{\rho'(1 + \beta_2)}\right), \qquad (10)$$

where the principal value of the arccos is to be taken and ρ' is shorthand for $\rho'(f_N)$. Numerical comparison with the corresponding threshold for non-predictive agents [5] shows that the threshold for persistent oscillations is shifted to smaller delays for linear extrapolaters, i.e. a system of linear extrapolaters is less stable. Interestingly, linear extrapolaters never exhibit damped oscillations.

5.2. Cyclical trend analysis

We now explore a more sophisticated extrapolation technique which incorporates some knowledge of the system dynamics. For simplicity, we shall deliberately set the parameters of the system such that its range of possible behaviors is restricted to either fixed-point stability or oscillation with a well-defined period and amplitude [4, 5]. (This is accomplished by using monotonically decreasing payoff functions $G_1(f)$ and $G_2(1 - f)$.) Knowing this, a lone technical analyst could correctly predict $f(t)$ from values of f at times $t' \leq t - \tau$ by using the following simple algorithm: Find the two maxima of $f(t')$ which occurred most recently, set τ' equal to the time difference between the two maxima, then make the prediction $f^*(t) = f(t - \tau')$.

Fig. 4 demonstrates the behavior of a system in which just 10% of the agents use the technique described above, while the remainder make the naive prediction $f^*(t) = f(t - \tau)$. After a few periods, the technical analysts determine the period of the oscillations in $f(t)$, using this information to make better predictions. The oscillations are in turn noticeably affected by the behavior of the technical agents, diminishing in magnitude by 24%

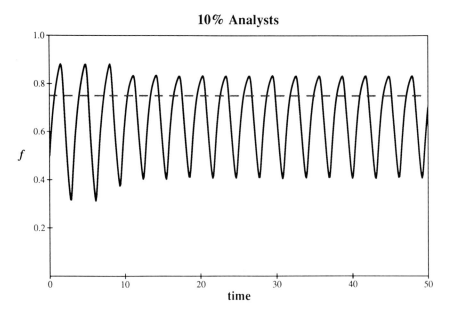

Fig. 4. $f(t)$ obtained by integrating eq. (1) numerically for system containing 10% cyclical trend analysts and 90% naive agents. Each agent uses a payoff $G_1 = 7 - f_1$ for resource 1 and $G_2 = 7 - 3f_2$ for resource 2. The uncertainty $\sigma = 0.125$, the reevaluation rate is $\alpha = 1.0$, and the time delay $\tau = 1.0$. $f_{opt} = 0.75$ is represented by light dashed line. The monotonic decrease of the payoffs with increasing resource usage and the presence of only two resources guarantees that the most complicated behavior which could occur in a system of non-predictive agents would be simple persistent oscillations.

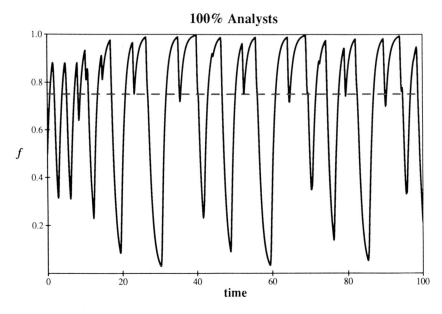

Fig. 5. 100% cyclical trend analysts. All other parameters are same as in fig. 4.

and decreasing in period by about 6%. However, when *all* of the agents use technical analysis (fig. 5), the system exhibits complex oscillations which are much greater in amplitude than those which occurred in the system of non-predictive agents.

The problem with the technical analysts is that their technique for detecting periodicity is based on the simple oscillations that occur when all of the other agents in the system are non-predictive. This technique is adequate when only 10% of the agents use technical analysis. However, the presence of technical agents in the system induces a qualitative change in the dynamics which they do not themselves take into account, so the technical agents perform abysmally when they constitute a substantial fraction of the agent population.

One method which might ameliorate this problem is to provide the technical agents with a richer description of the behavioral modes which the system can display. For example, the agents might try to Fourier-analyze $f(t)$ or use a chaotic time-series predictor [9-12]. However, having isolated a critical weakness of these particular technical analysts which is possibly shared even by more sophisticated ones, it is incumbent upon us to turn now to a radically different approach to prediction, in which the agents are given some notion of the activity that underlies the aggregate quantity $f(t)$ and an understanding of how that activity affects the system dynamics.

6. System analysts

In this section we explore the consequences of giving agents knowledge about both the individual characteristics of other agents in the system and how those characteristics are related to overall system dynamics. We shall refer to agents of this type as "system analysts".

In an ideal, infinitely large system, every agent could make perfect predictions, and the effective information time delay would be reduced to zero. This would require that each agent know the parameters describing all of the other agents' utilities. In addition, this perfect knowledge would have to be *common knowledge* (i.e. every agent knows that every agent knows that... *ad infinitum*...every agent knows the parameters [13]).

However, this ideal scenario could never be realized in practice. It is very unlikely that system analysts could possess sufficiently detailed knowledge about the nature of all of the other agents in the system to make perfect predictions, particularly in large systems which continually change in structure and composition. In practice, each agent must make predictions based upon uncertain information about parameters which characterize the other agents' goals (embodied as payoff parameters) and the other agents' knowledge of other agents' goals. Furthermore, even if the other agents' goals and knowledge are completely known, there are two sources of uncertainty which render exact prediction impossible in systems of finite size. First, the exact times at which decisions are to be made by other agents are not assumed to be known; only their average rate α. Second, the other agents' actions are not completely determined by the state of the system; they are described by the probabilistic function $\rho(f)$.

In order to analyze the behavior of system analysts, we shall present a mathematical framework for describing the assumptions that they make about one another and the means by which they predict the current system behavior from old information. Then, we shall present three examples which illustrate the behavior of systems of such agents under the conditions of both perfect and imperfect assumptions about one another's characteristics.

6.1. Theoretical framework

For mathematical convenience, we shall assume that there are a finite number S of different species of agents, each of which constitute a fraction g_s of the total number of agents. Each member of species s has reevaluation rate α_s, and probability for choosing resource 1, $\rho_s(f_s^*)$. Furthermore, all members of s are identical in their beliefs about the behavior of all of the other species of agents. As a result, they all share the same estimate $f_s^*(t)$.

Eq. (1) generalizes in such a case to

$$\frac{df(t)}{dt} = \sum_{s=1}^{S} \alpha_s g_s [\rho_s(f_s^*(t)) - f(t)]. \quad (11)$$

System analysts use a differential equation of the *form* of eq. (11) to predict $f_s^*(t)$, i.e. they assume the validity of the theory presented in section 3. However, since an agent of species s may make incorrect assumptions about the parameters used by agents of another species $s' \neq s$, it integrates eq. (11) with *assumed* parameters:

$$\frac{df_s^*(t)}{dt} = \sum_{s_1=1}^{S} \alpha_{ss_1} g_{ss_1} [\rho_{ss_1}(f_{ss_1}^*(t)) - f_s^*(t)] \quad (12)$$

starting from the initial condition $f_s^*(t-\tau) = f(t-\tau)$. In eq. (12), the parameters α_{ss_1} and g_{ss_1} represent the beliefs of agents of species s about the reevaluation rate and fraction of agents of species s_1, while ρ_{ss_1} is the belief of agents of species s about the decision-probability function used by species s_1 and $f_{ss_1}^*(t)$ is the belief of agents of species s about $f_{s_1}^*(t)$. In order to calculate $f_{ss_1}^*(t)$, a system analyst of species s must also make assumptions about the assumptions that all other species s_1 make about all other species s_2. Thus, $f_{ss_1}^*(t)$ is governed by a dynamical equation of the same form as eq. (12), but with even more indices.

Of course, these are just the first few steps in an infinite recursion of assumptions and predictions. Since the agents possess finite computational resources, they must truncate this infinite chain of deliberation in some reasonable way. One method is to enforce restrictions on the parameters $\alpha_{ss_1 \ldots s_n}$ and $g_{ss_1 \ldots s_n}$ and the functions $\rho_{ss_1 \ldots s_n}$. Alternatively, one could assume the value of $f_{ss_1 \ldots s_{n-1}}^*(t)$ at some level $n-1$ (rather than trying to calculate it from level n) and then substitute iteratively into the next lowest level, finally bottoming out at eq. (11), from which $f(t)$ could be obtained. Typical assumptions for $f_{ss_1 \ldots s_{n-1}}^*(t)$ might include $f(t-\tau)$, a random function, or $f_{ss_1 \ldots s_m}^*(t)$, where $m < n-1$. In the latter case, the chain of non-linear differential equations is effectively terminated by looping it back onto itself – a procedure which can be thought of as an enforcement of self-consistency. For $n = 2$ and $m = 0$, which is tantamount to each assuming that all of the agents in the system will calculate the same estimate that it does, eq. (12) reduces to

$$\frac{df_s^*(t)}{dt} = \sum_{s_1=1}^{S} \alpha_{ss_1} g_{ss_1} [\rho_{ss_1}(f_s^*(t)) - f_s^*(t)], \quad (13)$$

a simple non-linear differential equation which can be solved independently for each component s and then substituted into eq. (11) to yield $f(t)$.

6.2. Perfectly informed system analysts

Let us suppose that each agent is perfectly informed about the parameters and the knowledge of all of the other agents, i.e. it knows the relevant parameters which describe all of the other agents and this fact is common knowledge. Then each agent will calculate the same estimate $f_s^*(t)$, so eq. (13) applies. Since the beliefs α_{ss_1}, g_{ss_1}, and ρ_{ss_1} are correct (i.e. they equal α_s, g_s, and ρ_s, respectively), $f_s^*(t)$ satisfies the same equation as $f(t)$ and thus is equal to $f(t)$. Thus, for an infinitely large system, the effective time delay is zero, and the system will relax to $f_N(\sigma)$.

However, as has been mentioned, the lack of complete specification of when other agents make their decisions and the inherent non-determinism of those decisions result in statistical fluctuations which render perfect prediction impossible in systems of finite size. In order to understand how fluctuations affect the performance of a system of perfectly informed system analysts, we now consider a case in which there is just one species of agent. Then the predictions are generated by solving

$$\dot{f}^*(t) = \alpha[\rho(f^*(t)) - f^*(t)]. \quad (14)$$

It can be shown that, in the neighborhood of the equilibrium $f_N(\sigma)$, the prediction is given by

$$f^*(t) = f_N + [f(t-\tau) - f_N] e^{-\beta(1-\rho')}. \quad (15)$$

Using the definition given in section 3, the correlation R is

$$R = \frac{\langle \delta f(t) \, \delta f(t-\tau) \rangle}{s_{eq}^2} e^{-\beta(1-\rho')}. \quad (16)$$

In the limit of long time delays, the exponential factor becomes very small, so $R \to 0$, and the agents predict $f^*(t) = f_N$ regardless of $f(t-\tau)$. Intuitively, a long time delay gives the system plenty of time to recover from any deviation from equilibrium which may have been present at time $t - \tau$. Further statistical fluctuations may have occurred since time $t - \tau$, but they are inherently unpredictable, so the best estimate that the agents can make of the current state of the system at time t is to assume equilibrium. For relatively short time delays, the effects of deviations from equilibrium at $t - \tau$ are still present at time t, so the correlation R is positive, reaching 1 in the limit of zero time delay.

Evaluating eq. (6), we find that

$$s_{eq} \approx \sqrt{\frac{f_N(1-f_N)}{A}} \quad (17)$$

for $R = 0$ (long time delays) and

$$s_{eq} \approx \sqrt{\frac{f_N(1-f_N)}{A[1-\rho'(f_N)]}} \quad (18)$$

for $R = 1$ (zero time delay). Thus, due to the poor statistical correlation between predicted and actual system behavior when there are long time delays, the fluctuations about the equilibrium are a factor of $\sqrt{1-\rho'}$ larger than when there is no time delay. Regardless of the time delay, however, the fluctuations decrease in magnitude as $1/\sqrt{A}$, so the predictions are an adequate, albeit imperfect, substitute for up-to-date information in sufficiently large systems.

As a simple example, fig. 6 compares simulations of systems of 100 non-predictive agents and

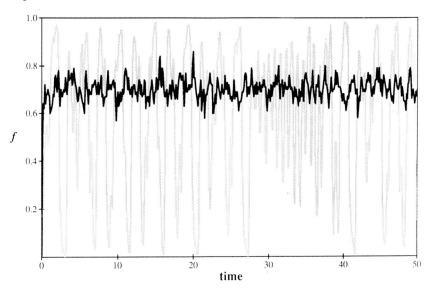

Fig. 6. System of 100 system analysts (dark curve) versus system of 100 non-predictive agents (gray curve). All parameters for these simulations are the same as were used in fig. 3c. The non-predictive agents exhibit chaotic behavior, spending very little time in the vicinity of $f_{opt} = 0.75$. The system analysts quickly reach $f_N(\sigma = 0.125) = 0.700$, which is reasonably close to $f_{opt} = 0.75$ and is identical to the equilibrium obtained by non-predictive agents with a much smaller time delay in fig. 3a.

100 systems analysts with the same parameters as in fig. 3. The time delay is so large that the system of non-predictive agents is in the chaotic regime, thus spending very little time in the vicinity of f_{opt}. However, the perfectly informed system analysts quickly reach the equilibrium $f_N(\sigma)$, which is reasonably close to f_{opt}. The fluctuations about the equilibrium are approximately three times larger than they would be if there were no time delay.

6.3. Mixture of perfectly informed system analysts and non-predictive agents

How well does the system perform when only a fraction g of the agents are perfectly informed system analysts? More explicitly, let us suppose that there are two species: No. 1, the perfectly informed system analysts, and No. 2, the non-predictive agents. Substituting the relevant parameters into the formalism introduced earlier in this section, we obtain

$$\dot{f}(t) = \alpha[gp(f(t)) + (1-g)\rho(f(t-\tau)) - f(t)]. \quad (19)$$

Starting from eq. (14) of ref. [5], one can show that the dependence of the persistent oscillation threshold β_2 upon g is given by

$$\beta_2 = \frac{\cos^{-1}[(1-g\rho')/(1-g)\rho']}{\sqrt{(1-\rho')(2g\rho'-\rho'-1)}}. \quad (20)$$

As illustrated in fig. 7, the time delay at which the system becomes unstable can be increased by increasing the fraction g of perfectly informed system analysts. If g is increased above the critical faction $g_c = (1+\rho')/2\rho'$, the system is guaranteed to be stable at the Nash equilibrium regardless of the time delay. Numerical solution of eq. (19) demonstrates that system analysts reduce the amplitude of the oscillations even when they are not present in large enough numbers to completely quell the oscillations.

6.4. Slightly misinformed system analysts

We have seen two examples in which system analysts perform quite well. In this example, we shall illustrate a weakness of theirs which threatens their practicality. Suppose that all of the agents

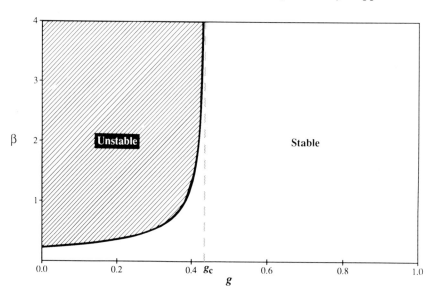

Fig. 7. System behavior as a function of β and fraction g of system analysts using parameters of fig. 3, for which $\rho' = -7.57$. The boundary between unstable and stable behavior is the curve $\beta_2(g)$, given by eq. (20). Above the critical fraction $g_c = 0.434$, the system is stable regardless of the delay.

in the system use reevaluation rate α and decision-probability function ρ, but that each agent believes that all other agents use reevaluation rate α^* and decision-probability function ρ^*. From eq. (13), we find that each agent's prediction $f^*(t)$ is given by solving

$$\dot{f}^*(t') = \alpha^*[\rho^*(f^*(t')) - f^*(t')], \quad (21)$$

with the initial condition $f^*(t-\tau) = f(t-\tau)$. Having determined $f^*(t)$, we can then determine the actual dynamics of the system by solving

$$\dot{f}(t) = \alpha[\rho(f^*(t)) - f(t)]. \quad (22)$$

Note that, for each infinitesimal step in eq. (22), one must integrate eq. (21) over a finite time interval τ.

We now examine eq. (21) in the limits of extremely small and large assumed reevaluation rates, showing that both limits can yield undesirable behavior. If the agents believe that the other agents will not make decisions often ($\alpha^* \to 0$), they will predict that the state of the system has not changed since $t - \tau$: $f^*_{eq} = f(t - \tau)$. Thus, they are identical to naive agents, which, as was demonstrated in section 4, can exhibit unstable behavior. On the other hand, if the agents believe that the other agents will reevaluate their status very frequently ($\alpha^* \to \infty$), the prediction converges to a value defined by

$$f^*_{eq} = \rho^*(f^*_{eq}), \quad (23)$$

which is completely uncorrelated with the actual value of f. Therefore, according to eq. (22), $f(t)$ approaches the fixed value $f_{eq} = \rho(f^*_{eq})$, with fluctuations about that equilibrium given by eq. (17) (i.e. they decrease as $1/\sqrt{A}$). In this case, f_{eq} can be quite far from $f_N(\sigma)$ (the equilibrium which would be obtained with correct assumptions about the other agents) and thus is likely to be far from

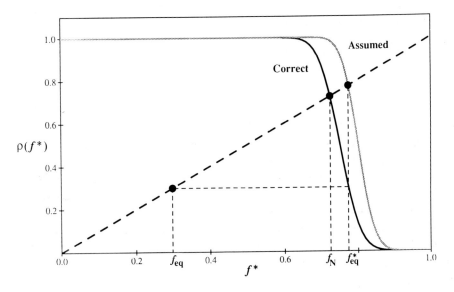

Fig. 8. Correct and assumed decision-probability functions for choosing resource 1: $\rho(f^*)$ (dark curve) and $\rho^*(f^*)$ (gray curve). $\rho(f^*)$ is based upon the payoff parameter of fig. 4, while $\rho^*(f^*)$ is based upon a modification of G_2 from $7 - 3f_1$ to $7 - 4f_1$. $f_N = 0.724$ is close to $f_{opt} = 0.75$ because σ is small. If the assumed reevaluation rate α^* is sufficiently large, the agents will believe that the system is currently at the assumed equilibrium, $f^*_{eq} = \rho^*(f^*_{eq}) = 0.773$, regardless of the most recent information $f(t-\tau)$. Thus, all of the agents will believe that resource 1 is somewhat more crowded than they would like, paradoxically inducing a majority of them to choose resource 2: $f_{eq} = \rho(f^*_{eq}) = 0.298$.

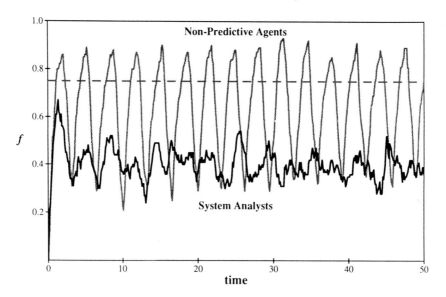

Fig. 9. Simulation of system of 100 system analysts using imperfect information. All agents have the same reevaluation rate $\alpha = 1.0$ and the same actual payoffs: $G_1 = 7 - f_1$ and $G_2 = 7 - 3f_2$. However, all of the agents believe that all of the other agents have the payoff $G_{12} = 7 - 4f_2$. Light curve: assumed reevaluation rate $\alpha^* = 0$, equivalent to using naive prediction. Fluctuations are periodic with magnitude $s = 0.079$ about an average value $\bar{f} = 0.693$. Dark curve: $\alpha^* = 4$. The equilibrium is reduced far below $f_{opt} = 0.75$ (light dashed line) to $f_{eq} = 0.291$. Fluctuations are no longer periodic, and their magnitude is reduced to $s_{eq} = 0.029$.

$f_{opt} = f_N(\sigma = 0)$ as well. As illustrated in fig. 8, a small difference between the assumed equilibrium f_{eq}^* as given by eq. (23) and $f_N(\sigma)$ can be magnified by the steep slope of the decision boundary into a large discrepancy between f_{eq} and $f_N(\sigma)$.

Fig. 9 illustrates the behavior of a system of agents with the correct and assumed decision-probability functions of fig. 8 for small and large values of α^*. When the assumed evaluation rate $\alpha^* = 0$, the system analysts behave exactly as non-predictive agents, and $f(t)$ exhibits large periodic oscillations which spend very little time in the vicinity of the optimal value f_{opt}. When $\alpha^* = 4$, $f(t)$ is stable about a value of f_{eq} which is so far below the optimal value that the system analysts perform even worse than the non-predictive agents.

The problem with the system analysts is that they have no mechanism for adjusting their assumptions about the nature of the other agents in the system. They blissfully continue to make grossly incorrect predictions even when there is strong evidence to the contrary.

7. Adaptive system analysts

Technical analysts are responsive to the behavior of the system, but suffer from an inability to take into account the strategies of other agents. On the other hand, system analysts are designed to take into account the strategies of other agents, but pay no heed to the actual behavior of the system. This suggests combining the strengths of both methods to form a hybrid – the adaptive system analyst – which modifies its assumptions about other agents in response to feedback about the success of its own predictions. We shall show that feedback based upon hindsight can be used to bring f_{eq} and f_{eq}^* into agreement, coalescing at f_N, which is typically close to f_{opt} when the uncertainty is small.

Let us suppose that each agent in the system makes predictions using eq. (21). The goal of the agents is to set the parameter α^* to α and the function ρ^* to ρ, in which case they will make the best possible predictions. For simplicity, ρ^* is

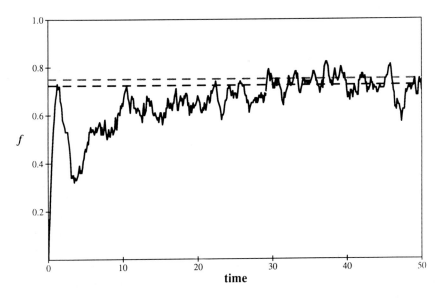

Fig. 10. Simulation run for system of 100 adaptive agents with learning rate $r = 0.1$, demonstrating that they eventually reach $f_N = 0.724$ (dark dashed line), which is close to the optimal value $f_{opt} = 0.75$ (light dashed line). The correct and initially assumed probability functions $\rho(f^*)$ and $\rho^*(f^*)$ are those of fig. 8.

parametrized by two parameters which govern the location and width of the decision boundary, respectively: the coefficient of f_2 in the payoff G_2 and the uncertainty parameter σ. Thus the agents' model of the system can be described by a three-dimensional parameter vector, $\boldsymbol{P}$. At time t, when the agent just becomes aware of what happened at time $t - \tau$, it evaluates the prediction error

$$\delta f \equiv f^*(t-\tau) - f(t-\tau) \qquad (24)$$

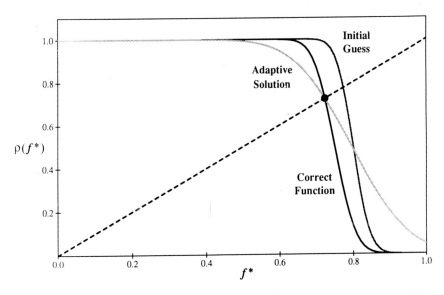

Fig. 11. Adaptive solution for $\rho^*(f^*)$ (light gray) compared to initially assumed (medium gray) and actual (black) decision-probability functions for system described in fig. 10.

and the gradient $\nabla_P \delta f$. Then, gradient descent is employed to minimize the prediction error δf:

$$P \to P - r\Delta t \, \delta f \, \nabla_P \delta f, \tag{25}$$

where r is the learning rate and Δt is the time interval between decisions.

Fig. 10 shows the results of a typical simulation run in which all of the agents initially use the incorrect decision-probability function $\rho^*(f^*)$ as displayed in fig. 8 and the incorrect reevaluation rate $\alpha^* = 4.0 \neq \alpha = 1.0$, but then evolve according to eq. (25). At first, $f(t)$ heads towards the same disastrously non-optimal value as in fig. 9. However, as the agents adapt their parameter, $f(t)$ climbs towards f_N at a rate which is determined by the learning rate r. If r is chosen to be too large, the system can exhibit persistent oscillations in much the same manner as a system of non-predictive agents with a reevaluation rate α which is greater than the persistent oscillation threshold.

Interestingly, the asymptotic adaptive solution for P corresponds to an incorrect model of the system, as illustrated in fig. 11. However, the adaptive solution is good enough to satisfy the two criteria of section 3: the agents locate the equilibrium value f_N, which is close to f_{opt}, and the correlation R between the predictions and the actual behavior is reduced only slightly from the value it would have for perfectly informed system analysts. In general, the agents can make sufficiently good predictions if they estimate f_N accurately; failure to match the slope of $\rho(f^*)$ in the vicinity of f_N just has the effect of increasing the fluctuations slightly.

In this example, it was assumed implicitly that each agent could share its experience with the other agents through the globally available information P. In a completely distributed system, each agent might update its own private P based upon its own experience. In this case, the appropriate learning rate would be decreased by a factor of the order of the number of agents in the system.

8. Conclusion

We have studied the role of predictions in a model of a coevoluationary system which was originally inspired by computational ecosystems. The dynamics has been studied on two different time scales: that on which the agents make their decisions ($1/\alpha$) and (in the last example) the much longer time scale over which they adapt their models of the other agents ($1/r$). For technical analysts, a good predictive strategy for a single agent may be disastrous if applied on a global scale. On the other hand, predictions of system behavior (and therefore decisions and actions) which are based upon system analysis are highly sensitive to perceptions about the utilities and perceptions held by the other agents – so much so that slight inaccuracies can lead to extremely poor collective performance. When the technical analysts' responsiveness to system behavior is combined with the system analysts' knowledge of the dynamical equations which govern the evolution of the system, the resultant hybrid is able to perform well.

This work suggests a number of future directions for the study of predictive agents. As was mentioned at the end of section 5, technical analysts might perform much better if they were to use a chaotic time-series predictor. To date, all such algorithms have operated upon externally generated time series. It would be extremely interesting to determine whether a coevolutionary system of such predictors could successfully predict dynamics which are generated by the system itself. Decision-theorist agents, which would average the payoffs over several possible scenarios weighted by an assumed probability distribution, might prove to be a worthwhile alternative to technical analysts and system analysts. Regardless of the details of their prediction strategy, agents must adapt both to a changing external environment and to other agents, and this undoubtedly will require adaptive methods which are more sophisticated than the gradient descent method studied here.

Several other interesting issues remain to be explored. The observed sensitivity of the global behavior to expectations might provide a mechanism for making transitions between metastable states at a much more rapid rate than can be achieved by non-predictive agents, leading to both greater adaptibility and greater volatility. In many situations (e.g. when there is overhead for changing a decision), the *future* state of the system may influence current decisions. Preliminary studies of the consequences of predicting the future state of the system, $f^*(t + \tau)$, lead to differential-advanced equations describing the dynamics of such systems, the resolutions of which are unstable. The ramifications of this are unclear at present, but promise to be interesting. Finally, other types of models of coevolutionary systems, particularly economic models in which the resources participate in the dynamics by charging prices which maximize their own expected profits [14–16], may prove to be even richer in their dynamics than the model we have studied here.

Acknowledgement

This work was partially supported by the Office of Naval Research under contract No. N00014-82-0699.

References

[1] W.B. Arthur, Self-reinforcing mechanisms in economics, in: The Economy as an Evolving Complex System, SFI Studies in the Sciences of Complexity, eds. P.W. Anderson and K.J. Arrow (Addison–Wesley, Reading, MA, 1988) pp. 9–31.

[2] S. Kauffman, The evolution of economic webs, in: The Economy as an Evolving Complex System, SFI Studies in the Sciences of Complexity, eds. P.W. Anderson and K.J. Arrow, (Addison–Wesley, Reading, MA, 1988) pp. 125–145.

[3] B.A. Huberman and T. Hogg, The behavior of computational ecologies, in: The Ecology of Computation, ed. B.A. Huberman (North-Holland, Amsterdam, 1988) pp. 77–115.

[4] J.O. Kephart, T. Hogg and B.A. Huberman, Dynamics of computational ecosystems: Implications for DAI, in: Distributed Artificial Intelligence, Vol. 2, ed. M.N. Huhns (Kaufmann, Los Altos, CA, 1989).

[5] J.O. Kephart, T. Hogg and B.A. Huberman, Dynamics of computational ecosystems, Phys. Rev. A 40 (1989) 404–421.

[6] G. Hardin, The tragedy of the commons, Science 162 (1968) 1243.

[7] R.M. Metcalfe and D.R. Boggs, Ethernet: Distributed packet switching for local computer networks, Commun. ACM 19 (1976) 395–404.

[8] J.C. Pasquale, Intelligent decentralized control in large distributed computer systems, Ph.D. Thesis, University of California, Berkeley, CA (1988).

[9] A. Lapedes and R. Farber, Nonlinear signal processing using neural networks: Prediction and system modelling, Technical Report LA-UR-87-2662, Los Alamos National Laboratory (1987).

[10] J.D. Farmer and J.J. Sidorowich, Exploiting chaos to predict the future and reduce noise, Technical Report LA-UR-88-901, Los Alamos National Laboratory (1988).

[11] H.D.I. Abarbanel, Prediction in chaotic nonlinear systems: Time series analysis for nonperiodic evolution, Technical Report INLS-1020, Institute for Nonlinear Science, UCSD (1989).

[12] L. Smith, private communication (1989).

[13] J.Y. Halpern, Using reasoning about knowledge to analyze distributed systems, Ann. Rev. Comput. Sci. 2 (1987) 37–68.

[14] Y.C. Ho, L. Servi and R. Suri, A class of center-free resource allocation algorithms, Large Scale Syst. 1 (1980) 51–62.

[15] J.F. Kurose and R. Simha, A microeconomic approach to optimal resource allocation in distributed computer systems, IEEE Trans. Comput. 38 (1989) 705–717.

[16] C.A. Waldspurger, A distributed computational economy for utilizing idle resources, Master's Thesis, Massachusetts Institute of Technology (1989).

TOWARD DIAGNOSIS AS AN EMERGENT BEHAVIOR IN A NETWORK ECOSYSTEM

Roy A. MAXION

Department of Computer Science, Carnegie-Mellon University, Pittsburgh, PA 15213, USA

This paper describes progress toward a system for autonomous diagnosis as an emergent behavior. The work is based on the idea that a computational organism consisting of a collection of primitive, cooperative processes acts to adapt to temporally local external behavior. Patterns of behavior and feedback among primitive processes and external stimuli form an internal representation of normal behavior against which external events can be compared. Functional stimuli are regarded as deviations of physical stimuli from this internal norm. Diagnosis, appearing as an emergent property of collective adaptivity at primitive levels, is the organism's response to environmentally significant anomalies. A system embodying these ideas was implemented and run at the primitive level only, and was applied in diagnosis of a communication network. Preliminary results are given, demonstrating the system's effectiveness in a real-time environment.

1. Introduction

Diagnosis is an important element of a system's or organism's response to failure or anomalous conditions. Diagnosis will play a critical role in our quest to build highly dependable, highly available, safe systems for the future, particularly those to be used in applications whose failure could result in significant economic loss or loss of life. Many of the systems of the future will be extremely complex, widely distributed systems in which control from a global perspective will be difficult or impossible to achieve. An example of such a system is a nationwide or global heterogeneous telecommunication network whose performance and configuration characteristics are constantly changing. Detecting, diagnosing and remediating anomalous conditions in such a system will be an unimaginably difficult task. Even now our technological artifacts are so complex as to preclude their complete testing and diagnosis.

One possibility for overcoming the immense complexity and the intrinsic nonlinearity of these kinds of systems is emergent computation. By this tactic a diagnostic regime could arise as a consequence of the structure of a performance-monitoring subsystem, and could adapt itself to extant conditions as the environment changes. A hierarchical symbol structure, based on a nonsymbolic substrate, could be constructed (and revised) automatically, and henceforth employed in diagnosing unusual conditions in the system. The symbol structure, when used by the organism itself, facilitates closing the loop in which outside intervention is usually required, thereby eliminating the usual requirement for human interpretation of interim results. Such a plan mirrors natural systems like those exemplified by social insect colonies and animal immune and nervous systems.

This paper explores some of these ideas, and demonstrates their application in a telecommunication environment consisting of a very large, highly heterogeneous computer network with several thousand nodes. First we provide a motivation for the work, followed by a brief review of recent fault-tolerance approaches to diagnosis. We then characterize the network domain, particularly with respect to its unruly and difficult behavior, and explain why a network can be regarded as an ecological organism sharing many characteristics of biological organisms. Diagnosis is then reexamined from those perspectives, and it is suggested that diagnosis may be one emergent property of certain complex systems. Architectural elements facilitating this emergence, consisting of adaptation-level-sensitive sensory/monitoring devices and symbol hierarchy, are described, as is an experimental implementation. Results are presented which demonstrate the present efficacy of a real-time, self-adapting diagnostic system based on the proposed architecture.

While this paper should be regarded more as a progress report than a product report, the results are encouraging. The system described herein has been successful in detecting and diagnosing a number of difficult network problems, and has, in its modest sphere of expertise, achieved a level of performance surpassing that of human experts.

We have been encouraged in the writing of this paper to be a bit speculative; hence our description of diagnostic symbol hierarchies as work in progress. In the reporting of results, however, we have presented only nonspeculative empirical evidence to substantiate the claims made here.

2. Why diagnosis?

Our dependence on computing systems has grown so great that it is now becoming impossible to return to less sophisticated mechanisms. As more and more daily tasks fall to automation, the life-threatening consequences and significant economic impact of computer failure become ever more apparent. One of the principal mechanisms for achieving greater system availability and reliability is fault tolerance, the survival attribute of digital systems. A goal of fault tolerance is to handle exceptional or anomalous conditions swiftly enough so that clients of the system will be unaware that an exceptional condition existed. Fault tolerance provides an important triad of responses to failure: detection, diagnosis and recovery.

Diagnosis is an important element of machine and human response to failure or other unusual conditions. An understanding of how diagnostic functions can be replicated on a computer is important for many reasons. First, many of the technological artifacts being developed today are of such complexity that no person can comprehend these devices fully enough to troubleshoot and fix them when they malfunction. Networks, in particular, tend to be so large, complex and decentralized that it is not possible to maintain a careful watch over all system components. Note that networks closely resemble biological systems such as insect colonies, immune systems and nervous systems in that they are large, complex, distributed entities with no global control. Automated diagnosis is a necessity for such systems.

Second, complex computer systems are now being used for mission-critical functions from automated banking and nuclear plant control to hospital patient monitors and space navigation. In these areas computer downtime of only a few seconds can bring serious service disruptions, catastrophe and sometimes death. If a critical system contained its own mechanism for automatic detection, diagnosis and compensation of faults, system downtime and its concomitant side effects could be substantially mitigated, providing a greater measure of availability and safety to system clients.

Finally, diagnosis is a ubiquitous activity, and well it should be for any organism, biological or computational. Many people are surprised to learn that they are performing diagnoses many times per day. Why did the toast burn? Why isn't there any hot water? Why won't the car start? Why is the traffic so heavy today? Given the ubiquity of the diagnostic phenomenon, understanding it and deploying it in artificial organisms – such as computers – would appear to be a worthwhile endeavor.

3. Diagnosis: definition and approaches

Diagnosis is one response to an organism's sensation of ill being, abnormalcy or significant change from the organism's level of accommodation to prevailing conditions in the environment, or adaptation level. The term adaptation is used here in a quasi-biological sense in which it is taken broadly to suggest adjustment to the conditions under which an organism, biological or artificial, must exist in order to survive or otherwise meet its objectives. An adaptation level is that level of activity in the environment to which the organism accommodates at a zero-reaction level, i.e., a condition in which the stimuli impinging on the organism elicit no reaction from the organism; the organism has accommodated to the current conditions in the environment. For example, if one is accustomed to a high level of traffic on a daily commuting route, then one will have adjusted one's activities to account for the delays incurred by the heavy traffic, and one will not react specially to (or be "surprised" by) heavy traffic. People will deal with that high level of activity (or inconvenience) and adjust to it as if it were normal; its habitual occurrence will elicit no particular new reaction (other than wishing it were otherwise, perhaps); the normal state of affairs becomes a zero-reaction state. Whenever the traffic activity changes substantially from the adapted zero-reaction-level, either up or down, people will notice it, and perhaps even be surprised by it. If the change persists, people will accommodate to the new level. A similar scenario ensues in people's getting used to living next to the train tracks or near a noisy bus route, or in users becoming accustomed to traffic levels on a congested computer network. Diagnosis thus regarded, as a reaction to shifts from a zero-reaction level, could be exploited in ways that are similar to the ways that diagnosis arises in nature as a consequence of environmental shifts. This idea will be explored more thoroughly in a subsequent section.

Diagnosis is understood by most people to be the inferential and step-wise process of identifying a disease, situation, condition or problem from its signs, symptoms or distinguishing characteristics. It is a multistage, cyclic process of hypothesis generation and evaluation, and demands for information are continuously changing along with hypotheses. The diagnostic experience is a form of pattern recognition in the sense of recognizing a high-level pattern, symptom or symbol, and knowing the mapping from the symbol to the causal mechanism. When the mapping is not immediately apparent, diagnosis becomes a problem-solving process in which the high-level symbol is broken down, level by level, into its component symbols, until a sufficiently primitive level is found at which an assignable cause is located. The cause can be interpreted as a significant shift from normal conditions. Hence, diagnosis is both a process and a result. The *process*, initiated by detection of change, consists of the recursive decomposition of high-level symbols. The *result* is the isolation of the locus of change deviating from a zero-response adaptation level somewhere in an abstraction hierarchy of symbols empirically constructed from experience.

Diagnosis always requires a standard or a set of expectations against which observations can be compared. For example, you cannot recognize that your car sounds "funny" or abnormal if you have no expectation (of what the car normally sounds like) against which to compare the current sounds. This basis of relative judgement is the zero-reaction level or the level of normal behavior in the environment. The review of diagnostic approaches in subsection 3.1 identifies the source of the expectations for each approach.

3.1. Approaches to diagnosis

Two basic fault-tolerance approaches to diagnosis are specification-based and symptom-based. Each approach is predicated on detecting some sort of change in the normal operation of the device under diagnostic scrutiny. The principal difference between the two approaches lies in the way that *normal* is defined.

Specification-based. In a specification-based approach, system design specifications (e.g., schematics or performance requirements) are used to determine in advance the ways in which systems will fail. Diagnostic test tools are then based on these determinations. Traditional diagnostic programs, of the sort that are employed to help isolate a fault after evidence of failure has surfaced, are exemplary of the specification-based approach. Another example is the use of simulation and fault insertion to develop a fault dictionary [5] which is subsequently used to map symptoms back to faults, often manually. Fault dictionaries can be incomplete and time consuming to generate, and they, as well as other specification-based approaches, are only as accurate as the simulators and fault models they employ.

The change being detected is a deviation from what was specified to be expected or normal behavior. Most often this is synonymous with *correct* behavior, since that is the only thing that can usually be determined using the specification-based approach. Correct, however, does not always mean normal; when the system violates "correct" performance envelopes, even if there is nothing otherwise wrong, the specification-based approach will fail[#1]. It will also fail in the many cases that were left unspecified, and for a reasonably complex system the number of such cases is likely to be extraordinarily large.

Symptom-based. If diagnosis is understood as the identification of a condition based on its signs, symptoms, or distinguishing characteristics, then the phrase "symptom-based diagnosis" seems redundant. Symptom-based diagnosis uses data captured by system-event monitoring. General fault diagnosis, however, can be based on factors other than the symptoms directly exhibited by the machine. In particular, diagnostic tests designed to reproduce failures are often used to provide information about a problem; users or operators may furnish additional symptomatic information by running standard test regimes such as memory tests. This information is both artificial and external, in the sense that factors are introduced which would not ordinarily be noted in a system's internal record of events.

An example of symptom-based diagnosis is trend analysis of system event logs. One event common to such logs is a soft (recoverable) read error on a disk. Under normal conditions it is expected that some soft read errors will occur; the normal rate of these errors per unit time (or per disk access) varies from system to system and from disk to disk. Empirical evidence shows that when soft errors cluster into certain patterns, a hard failure is imminent [18]. Consequently it is possible to predict catastrophic failures, based on event patterns [11, 12, 6].

The sense of normalcy or expectation used in symptom-based diagnosis is rooted in two assumptions. First, there is the assumption that the particular symptom employed (e.g. disk soft error) is a proper or useful symptom that will indict some portion of the system when appropriate; this assumption may not always be justified. Second, there is the assumption that *any* clustering into unusual patterns indicates faults in the system. This assumption is not necessarily true, because while the unusual pattern may be associated with an assignable cause, that cause might not be a fault in the classical sense; it may simply be abnormal behavior attributable to unanticipated interactions in the system. In this case an investigation would determine whether or not the anomaly was assignable to a specific cause.

There are several reasons for pursuing symptom-based diagnosis. Strategies for symptom-based diagnosis closely mirror the strategies employed by human technicians performing troubleshooting, thus making symptom-based diagnosis easier to understand and follow. The symptoms on which the approach is based are frequently more accurate indicators of the actual state of a system than are the indicators in a specification-based approach. Moreover, it is a common observation that standard diagnostic programs cannot stress a system the way an actual workload can, and hence are often incapable of replicating a failure. As opposed to relying on external factors such as verbal reports or test results, the symptom-based approach to diagnosis seeks to get the

---

[#1] Consider, for example, that it is not correct for the fender on a car to rattle, but if it does rattle as a benign matter of course, then rattling is normal behavior for *that car*.

most out of existing symptoms. It is not suggested that testing is never needed, but that in many cases a fault can be isolated without testing, and that in other cases the range of tests used for fault isolation can be narrowed significantly. Thus, the symptom-based approach is centered on the nature and limits of diagnosis based *solely* on existing symptoms as monitored by a diagnostic subsystem and included in a system's internal event log.

While symptom-based diagnosis represents an advance from pure specification-based diagnosis, it still relies on certain specification-based information. For example, the act of selecting disk soft errors is itself a specification; moreover, the distinction between soft and hard disk errors is essentially arbitrary, and is at best specification based, because the soft/hard distinction is based on a rule of thumb according to what system analysts' normal experiences have been. The diagnostic problem requires a method of selecting event definitions as well as the magnitude of events sufficiently different from normal to merit maintenance attention. It also requires a mechanism for determining what is normal at any particular time.

We have discussed some inadequacies of the specification-based approach and the symptom-based approach: in the first, specifications are simply inadequate; in the second, the very choice of symptoms has a specification-based flavor/influence, and there is no appeal to principle in the clustering of symptoms. What is needed is a mechanism for selecting not only the symptom, but also the level of severity at which the symptom is beyond the bounds of normal behavior. In other words, what is needed is a method for generating expectations that is dependent only on the operational characteristics of the device under scrutiny, and not on its specifications.

Even disregarding these concerns, there are systems for which it is not possible to specify symptoms in advance (correctly or not) or for which the stability of the symptoms cannot be guaranteed. For example, in nonlinear, highly dynamic systems like networks, the symptom "load-too-high" may have different definitions from time to time or from network to network, hence the symptom is unstable. Subsection 3.2 takes a deeper look at the diagnostic process with an eye toward eliminating some of the difficulties just discussed.

3.2. Diagnosis: Another look

Atomic changes. We have already said that diagnosis is a process initiated as a reaction to perceived change. The phrase "a diagnosis" signifies the result of the process of diagnostic reasoning: a concise technical description of a taxonomic entity giving its distinguishing characteristics. In computer systems (and possibly most other systems) an atomic (single-point or single-parameter) change at some low level of the system is propagated, possibly with some fanout, to a higher level at which point the manifestation of the change is detected. An example of an atomic change is a gate failure in an integrated circuit. The gate failure causes a circuit-board failure which, in turn, causes a disk error that is manifested by the file-open failure observed by the client. At several levels of the system a behavioral change could be observed, each of which could be blamed for the high-level manifestation of the atomic change. Simply knowing what changed at the atomic level is equivalent to knowing the diagnosis. Hence a diagnosis consists of knowing the locus of atomic change in a system[#2].

Level sensitivity and zero-reaction level. The diagnostic process can be viewed as being embedded in some organism or entity whose reaction to unusual conditions is to inquire after, or search for, the cause of the surprisingly abnormal situation. The inquiry is the invocation of the diagnostic process. Unusual, abnormal or surprise conditions are those conditions differing from prevailing conditions. This suggests that the organism does not react, either positively or negatively, to

---

[#2] We have discussed atomic change as if every fault manifestation could be tracked to a single change. While there certainly are cases of multiple simultaneous changes, they are so rare at the atomic level that their effects are negligible [15].

prevailing (i.e. normal) conditions. Since prevailing conditions do not evoke a reaction from the organism, we say that the organism is *adapted* to prevailing conditions. The level of external stimulus or environment that will not evoke a reaction or response from the organism is the zero-reaction level or the adaptation level. It is a neutral position on a continuum against which all measurements, positive and negative, are compared. This neutral position can float, or adapt, in accordance with drifting conditions in the external environment. Thus the operant perceptual mechanism is attuned to the evaluation of changes or differences rather than to the evaluation of absolute magnitudes. Past experience and present context define the level of adaptation.

A simple biological example of this is adaptation to lighting conditions. Consider the three lighting environments surrounding a movie house during a matinee: the bright outdoors, the dim lobby and the dark theatre. An employee at the lobby's confection counter adjusts to the dim light there, and for that person this is the neutral or zero-reaction level, because his/her eyes do not react positively or negatively (e.g. by pupil dilation or constriction) to the lighting level. If that person goes either outdoors into the bright light or into the theatre's darkness, the visual system will react because the zero-reaction-level threshold envelopes have been violated; given sufficient time the eyes will adapt to the new conditions, afterwhich those new conditions will become the new zero-reaction level[#3]. This suggests that the organism must adapt to changing or drifting conditions in the environment, and must also react to changes from prevailing conditions. In the luminance example, the environmental change induces a physical response in the organism; it might also invoke an inquiry about what caused the change. This invocation or inquiry, in response to a sensitivity to the level of external stimuli, is the beginning of the diagnostic process. Diagnosis, then, could be said to be *level sensitive* because its invocation depends on a change from the normal level of external events.

Symbol hierarchies and primitive levels. Changes in a system are almost never observed directly at the locus of atomic change, but rather as they are manifested at some higher abstract level. For example, a gate going bad in a computer workstation would not be observed directly; it might instead be observed as it manifested itself in terms of the workstation's inability to continue to operate at full capacity. To have low-level changes be diagnostically useful they must be symbolically represented in an abstraction hierarchy.

Almost any system can be regarded as a hierarchy of entities, one entity connected to others in various ways. From a symbolic perspective this hierarchy is a set of abstractions, any one of which may be a satisfactory point from which to observe the system. For example, a computer can be seen at any of these (and other) abstraction levels: circuit, logic-design, program and processor-memory-switch (PMS) [16]. An anomaly observed at the PMS abstraction level may be traceable to the circuit level, in which case the anomaly itself symbolizes a series of changes in all the levels below it. To be diagnosed, the anomaly/symbol must be decomposed, level by level, until the level of atomic change (e.g. the circuit level) has been reached. A symbol hierarchy of multiple levels would be constructed originally as follows: an atomic change occurs in some entity at some primitive level in the physical system; the positive or negative change away from the zero-reaction level for that entity is propagated upwards to the next higher level of the symbol hierarchy at which point the "sensory" element at that level registers the change from the level below; the change at that level would again

---

[#3] There will almost always be some absolute floor or ceiling, as in luminance being either too low/dark or too high/bright for the organism to function successfully. Note that adaptation as described is different from the better known homeostasis in that the zero-reaction level is a neutral or equilibrium point from which all change is measured. There is no implication that a particular zero-reaction state is a *goal* of the organism as the regulation of body temperature is. Homeostasis, as in temperature maintenance, has reference to a single point, while adjustment to zero-reaction levels is completely independent of any particular point.

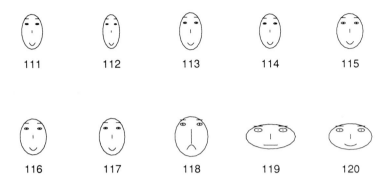

Fig. 1. The transition of a network system from good to poor health is represented by a series of multidimensional graphic symbols called Chernoff faces. Each face represents five dimensions of network performance.

be propagated upwards to the next higher level, and so on, until the top of the symbol hierarchy, consisting of a single element, is reached; at the uppermost level of the symbol hierarchy the symbol residing there would be an abstraction reflecting the entire chain of events that began as an atomic change at the primitive level[#4]. By this mechanism every symbol in the hierarchy would be grounded in the real events being sensed at the primitive level, as opposed to being defined only in terms of other symbols. To determine what event began the chain, the top-level symbol would be stepwise decomposed until the primitive level was reached, at which point the atomic change would indicate the cause of the series of changes. Note that while symbol grounding may also occur by inference, as in consultative expert systems, a critical difference is that environmental conditions are usually interpreted by the user before being provided to the expert system. Hence the system relies on a symbol created and grounded by the user instead of by the system itself.

The general state of health of a system is represented by a single, abstract, top-level, highly multidimensional symbol, much as a car's noises tend to represent the functioning of the entire car. One can imagine such a high-level symbol being represented graphically by Chernoff faces [4] like those shown in fig. 1[#5]. The figure shows a series of expression changes depicting the transition of a network system from good health to poor health (faces 118 and 119 are bad)[#6]. Each face represents five dimensions of network performance.

The primitive level of the symbol hierarchy is lowest level at which external controllability is effective. For example, if one has repair control at the level of circuit boards, but not at the level of chips or gates, then the circuit board level is the primitive level; it makes no sense to push the primitive level any lower, because it is of no use to know about an anomalous circuit when there is no control at the circuit level. From a common-sense perspective, the primitive level is always the level at which control will provide adequate compensation for anomalies, even if sensing could be done at lower levels.

[#4] Two special cases of atomic change should be discerned. (1) Some changes are too subtle to be observed. This is a matter of sensory resolution, which can always be increased, but usually not without some attentional cost to the organism. If a parameter drifts very slowly over a long enough period it can be detected not as an anomaly with respect to the current representation, but as a difference from past conditions as may be stored in memory. Humans, as well as artificial systems, are subject to this problem. (2) Some high-level anomalies will occur when two low-level parameters drift mutually away from each other. While the movement of either parameter alone would be too subtle to detect (as above), their relative difference would be detected. In these cases the resultant diagnosis will be ambiguous, and both nodes will be in the ambiguity set.

[#5] Humans are especially sensitive to small changes in facial expressions, hence their usefulness in representing nuances in multidimensional stimuli.

[#6] This particular transition was due to unusually heavy traffic between a particular host–server pair just before 2 o'clock in the morning.

In section 4 we describe a computational environment in which a level-sensitive approach to diagnosis would seem to work better than a specification-based or symptom-based approach.

4. Characteristics of networks

We have chosen to apply diagnosis to networks because of the technical need to do so and because of networks' resemblance to many real-world situations. This section describes networks in general, and the CMU/Andrew/CSD networks in particular. We examine the network environment and the challenges it presents to diagnostic regimes. Salient characteristics are given which indicate why the problem of network diagnosis is not straightforward. A rationale for characterizing networks as ecologies is provided.

4.1. Description of networks

The network system on which the present work was performed is the Andrew network at Carnegie-Mellon University [14]. It is a large, heterogeneous network typical of the kinds of networks that are expected to emerge at educational and commercial/business campuses around the world in the next decade.

The CMU Andrew network is composed of roughly 129 network spurs and about 5000 host addresses, with more than 2000 of those addresses in the Computer Science Department (CSD) alone. There are 40 Apple Talk subnets, 34 Ethernets, 43 token rings, and several experimental networks. There are 15 public file servers, three private file servers, and numerous similar facilities hidden behind various departmental routers, gateways and other boundaries. The network is distributed over 90 acres and 55 buildings. It is highly heterogeneous, with equipment from many vendors operating in various capacities, both permanent and experimental. The principle workstations are provided by DEC (MicroVAXes and PMAXes), IBM (ATs and RTs), NeXt and Sun (various). In addition, there are dozens of mainframe, parallel-processing, systolic-processing and experimental machines on the network. The network is primarily TCP/IP, though XNS and DECnet are also supported. The CSD network, on which all the experimental monitoring was conducted, is primarily 10 MB Ethernet combined with several token rings. The CSD network is configured as a spine or backbone with ten spurs. Each floor of our eight-story building has at least one dedicated spur. Clients are segmented to some extent by floor, depending on the laboratories and projects being operated on each floor. From floor to floor or spur to spur, there is considerable heterogeneity, just as in Andrew in general.

4.2. Salient characteristics

The salient characteristics of these networks are their dynamic character, their intrinsic noisiness, their inherent lack of global control, their wide geographic distribution, their nonlinearity, and their nonstationarity. Because there is no global control in an Ethernet network, clients (human or machine) may initiate network transactions at any time with little or no regard for extant traffic, thus contributing to a dynamic and noisy network character. The heterogeneity of most networks is a function of various technologies being provided by whatever vendor could supply them at the time their need was recognized, hence the opportunity for global control is limited despite any desire to the contrary. Global control over components is the exception rather than the rule; more likely is a localized cooperation out of which may emerge a global order which itself is unpredictable (and often unattainable). The characteristics of a network tend to drift as workloads and configurations change, producing a nonstationary behavior. Network protocols attempt to arbitrate among contentious clients. Arbitration is not perfect, however, hence over- and under-corrections contribute to the nonlinear character of the network. One network will resemble another only vaguely, suggesting that having an understanding of one network is of only marginal utility in understanding another; hence expertise does not transfer well among networks. The wide geographic distribution of networks makes manual interrogation of individual components difficult. Failure conditions are often manifested as performance degradations. Diagnostic programs often cannot recreate a failure situation, and even the concept of a network fault is fuzzy and ill defined. Networks are poorly understood particularly with respect to scale and troubleshooting. There are few, if any, network troubleshooting and management experts[#7]. Troubleshooting is mostly an ad hoc exercise, typically involving the disconnection of entire spurs just to see if a problem disappears, thereby crudely "localizing" the problem to that spur.

4.3. Network ecosystems

We find it useful to recognize similarities between networks, as computational ecosystems, and biological ecosystems. An ecosystem is an assemblage of organisms that occur together in an area. These assemblages are knitted into organized systems, and the properties of assemblages that manifest this organization are seen as emergent. The ecosystem is a highly organized community whose members are inextricably and complexly linked to one another and to the physical environment such that characteristic patterns recur and properties arise that can be neither understood nor predicted based on knowledge of the component members. These properties, known as emergent properties, represent new levels of organization in the system.

In a biological ecosystem the environment affects the organism, and the organism affects the environment. In biological ecosystems the evolution of relationships, the sense of equilibrium with the prevailing environment and functional interactions between co-occurring organisms are of primary interest. The major concern is with the processes involved in the interactions between organisms and their environment, with the mechanisms responsible for those processes, and with the origin, through evolution, of those mechanisms.

Ecological systems develop in accord with the regional environment, reaching a dynamic steady state within the limiting conditions of the environment. Although these systems have evolved to resist the normal expected perturbations encountered in the environment, unusual disturbances and catastrophic events can upset and even destroy the system. In this case, recovery can occur after the disturbance is compensated for or stopped.

A system can be defined as a group of interacting and interdependent elements (or organisms) that form a collective entity. Network systems (or simply networks) are "living" entities. They form an ecosystem which may comprise hundreds or thousands of independent entities, each of which may interact with the network community either at will or in close cooperation with one or more other entities. The term "ecosystem" is used here because it not only considers the adjustments of individuals (e.g. nodes) and species (e.g. workstations, servers, routers, etc.) to their environments and to one another, but it also considers the interrelationships of computational organisms and their environments, especially as manifested by natural cycles and rhythms, community development and structure, interaction between different kinds of organisms, geographic distributions and population alterations. In a computational ecology, a term first introduced by Hubermann and Hogg, what is important is the totality or pattern of relations between organisms and their environment. Hubermann and Hogg [8] have said that

"incomplete knowledge and delayed information are the novel features of ... computational systems which have no central controls. These large networks, emerging from the increasing connectivity between diverse processors, are becoming self-regulating entities very different in their nature from their individual components. The possibility of spawning remote processes and the consequent inability to explicitly predict resource utilization, give rise to a society of computational

---

[#7] There *are* people who are better than others at troubleshooting networks, and in this sense are experts, but in the more general sense of the word, there are few experts.

agents which, in their interactions, are reminiscent of biological and social organizations."

In these terms networks are ecosystems in every sense of the word. This is an important recognition, because in any ecosystem it is expected that some kind of natural adaptation, in a quasi-Darwinian sense, will be observed among the inhabitants of the system. While present networks are largely controlled by humans, the complexity, diversity and geographic distribution of networks may make it necessary for networks to maintain themselves in a sort of evolutionary sense, just as biological organisms do. In that regard it would be expected that networks would exhibit emergent properties similar to those exhibited by biological organisms.

One such emergent property that would be especially useful would be a self-healing or self-repairing property which might otherwise be called fault tolerance. Fault tolerance is the survival attribute of digital systems [1][#8]. Fault tolerance is the special architectural element of a system that maintains the logical integrity of the system in the face of physical failure. Fault tolerance in a network environment would certainly need to arise as a consequence of the entities and activities of the network itself, since the ill-structured nature of the environment and the general lack of specifications for other emergent properties of the network would seem to prohibit an explicit construction of such a system.

In subsection 4.4 we discuss network diagnosis from several points of view, and suggest how diagnosis could manifest itself as an emergent fault tolerance in a network ecosystem.

4.4. Network diagnosis

Given the salient characteristics of networks as being dynamic, noisy, nonlinear and nonstationary, we now examine network diagnosis from the perspectives of the diagnostic approaches discussed above. We recall that the primary requirements for diagnosis are observations of current behavior and levels of expected behavior. Current observations (levels of stimuli impinging on the organism from the environment) will be compared against current levels of adaptation. A stimulus value outside the current adaptation level, or zero-reaction level, will initiate a search for the atomic change that occurred, and will also cause the organism to move its adaptation level in the direction of the stimulus.

Diagnosis can now be regarded as a process, or a collection of locally cooperative processes performing primitive computations, initiated in reaction to an anomaly or shift in the level of external (or internal) stimuli away from the zero-reaction or adaptation level. When such a shift is detected by the organism, two things occur. First, the organism initiates a search for the locus of atomic change in the ecosystem by successively decomposing internally represented symbols starting at the symbolic level at which the anomaly was detected; the change is the cause of the stimulus anomaly. Second, the organism adjusts its internal representation of the environment to accommodate to the new prevailing conditions; it adapts to the new environment via patterns of interactions produced by the primitive processes. This creates a new zero-reaction level against which subsequent stimuli or observations will be compared. Two requirements for these activities are: an internal model or set of expectations for normal behavior in the environment; and an abstraction hierarchy for storing symbols representing behaviors at the levels below them. This will be discussed presently. Note how these ideas fold back on themselves in that consequential shifts in high-level patterns are interpreted as anomalous events which, in turn, trigger diagnosis. Inconsequential shifts are handled on as as-needed basis by the local, primitive processes that maintain the adapting representation.

Specification-based network diagnosis. Diagnosis of networks as specification-based devices seems at this point a frivolous notion, given that the emergent behavior of networks is almost com-

---

[#8] For a review of fault-tolerant computing, see ref. [13].

pletely unpredictable. Hence no specification exists that will provide either an expectation of behavior or a standard of comparison. This approach should be trivially dismissed, because there is no way that a specification can provide an expectation for normal or zero-reaction-level behavior in an environment that is always changing[#9].

Symptom-based network diagnosis. Diagnosis of networks as symptom-based devices is a slippery notion, because though the symptoms themselves have been predefined, the definitions are somewhat arbitrary. For example, in disk systems the difference between a soft error and a hard error is that a hard error is not recoverable in 16 retries. Where did the "16" come from? Disregarding the constraints of binary arithmetic, it was an arbitrary choice in that some system designer picked 16 as a number that often provided acceptable performance; 16 is not necessarily sensitive to a particular device, nor is it sensitive to changes in the disk environment. While trends in symptom-based diagnosis suggest that the environment is changing, the notion of trend analysis does not go far enough either in altering the zero-reaction level or in isolating the atomic change. Furthermore, the predefined symptoms are in fact symbols that need to be mapped into causal mechanisms, but because the symptoms/symbols are predefined at a high level with no descending hierarchy, they are frequently not grounded in a nonsymbolic substrate. Hence the high-level symptoms detected in symptom-based diagnosis cannot always be mapped to causes through hierarchical symbol decomposition[#10].

Level-sensitive network diagnosis. Diagnosis of networks as level-sensitive ecosystems does appear to answer the objections raised above. The mechanism required for level-sensitive diagnosis would comprise the following functions: sensation, orientation, habituation, discrimination, classification and memory. Sensation is a complex of entities for perceiving changes in the condition of the external environment (or within the organism itself). Sensation in a computational organism provides sensory projections of objects and events to the organism; they are the analog of iconic representations. Orientation – or the orienting reflex – is a phenomenon which causes us (as animal organisms) to turn toward or pay particular attention to an unusual stimulus; it provides a gating mechanism for attentional processes in the organism. Habituation is a learning phenomenon which provides a facility for "getting used to" a stimulus that is benign in terms of the organism's survival; it enables us to ignore repeated stimuli which have no value as new information. Discrimination is a distinguishing mechanism for picking out invariate features from sensory inputs or from among internal representations of sensory inputs. These can take the form of functions or combinations of features, differences, similarities, regularities, ratios and so on. Classification is a set of processes which assigns objects or instances to groups within a system of categories distinguished by structure, origin, similarities, differences and so on. Classification forms event categories, and assigns to these categories the names (symbols) of objects or events. As the category memberships expand, a hierarchical reorganization may be effected from time to time, the result of which is that higher-order symbols can be resolved in terms of their more elementary representations which, at the most primitive level, are grounded in empirical observations. Finally, memory is a repository in which we may store observations described at any level of classification we may choose.

The kinds of emergent properties we seek are those which enhance the survival opportunities of the organism in the face of a novel, changing environment. Survival can be cast as a goal in terms of retaining a certain level of usefulness or as maintaining a prespecified level of resources. In dependable computing the specific goal of the system is twofold: maintain availability resources,

---

[#9] In practice the specification-based approach can be used to provide a floor or ceiling for stimuli, hence setting absolute bounds on acceptable system behavior. This would form part of a hybrid diagnostic system that included situation- or context-dependent mechanisms for diagnosis.

[#10] See Harnad's article on symbol grounding in this volume [7].

and detect and diagnose failures or performance degradations. In either case, emergent properties should reflect behavioral shifts in the computational organism to facilitate its adaptation to changing conditions. In the sections to follow we focus on the adaptive aspects of the single-element mechanisms just described, as implemented for the lowest level only. We have implemented the sensory element of the architecture as well as the orientation and habituation aspects. These components work together in an autonomous, self-adapting network-diagnosis system.

5. Network diagnosis – first steps

In previous sections we described an adaptive sensory element with habituation and orientation characteristics. These elements are linked to the external environment at the lowest (primitive) level of a constantly evolving symbol hierarchy. At the primitive level the sensors respond to real-time, nonsymbolic, real-world data; at higher levels the same kinds of elements are used to sense shifts in the primitive-level sensor outputs, thus forming a second (or higher) level in an abstraction hierarchy in which symbols are created and modified in response to changes in the external environment. In the sections to follow we describe the implementation of a single adaptive sensory element, and we show its diagnostic performance in a single dimension.

The present work takes the perspective of a technician faced with the task of troubleshooting a problematic device. A study of cognitive aspects of diagnosis [11] showed that technicians look for patterns in performance data when isolating equipment troubles. In particular, they look for anomalous, or unusual patterns, and they use those patterns first to search their memories about troubles they have seen before, and second to help localize the current trouble by problem solving. Device failures are almost always brought to the technician's attention by an anomaly or change noticed in the device's normal level of operational behavior; i.e. there has been a significant shift away from the zero-reaction level for the device. Once the shift has been detected, the reaction is no longer zero, or neutral; rather, it is positive or negative, signaling the direction of compensation for the anomaly. The anomaly, which is usually symptomatic of failure or performance degradation, is in the form of a high-level symbol which must be incrementally broken down into its constituent lower-level parts until the atomic, single-element change which initiated the perturbation is located. In this more real-world context we should note that the word "normal" is considered to be synonymous with the term "zero-reaction level" because we do not typically react to conditions that are normal.

The following is a simplified schema for technicians' troubleshooting activities:

(i) make observations of the device or system;
(ii) verify that those observations are normal;
(iii) explain/diagnose anomalous observations;
(iv) remember the scenario for future reference.

A basic model of troubleshooting consists of several simple steps. First one observes, or samples, the environment. The observation is compared to an internal representation of what constitutes normal conditions in that environment. If the sample observation tests normal, then there is no reason to devote extra processing cycles to that stimulus. If the sample observation tests abnormal, then extra processing resources can be devoted to explaining (diagnosing) the nature of the anomaly. The occurrence of the anomalous event can be regarded as a *cue* to invoke diagnostic problem solving. Diagnosis at the current level decomposes the anomaly (high-level symbol) into its constituents at the next lower level. If this level is the primitive level, then a diagnosis is indicated by the change in some atomic element at that level; if this is not the primitive level, then the process continues recursively until the primitive level is reached. Having arrived at a diagnosis, the agent/system may choose to remember the event; a future occurrence of that event might not require the processing overhead to resolve the source of the anomaly, since information about it could be looked up in memory. After explanation of the anomaly, the system/agent invokes appropriate recovery procedures and returns to sampling its environment[#11].

While this model seems straightforward enough, a key question is: Where does the internal representation of normalcy come from? It is learned. Humans, through continued observation of their environment, learn what is normal, and hence are able to classify future observations as being either normal or anomalous, based on a comparison of external observations against internal representations. This is achieved through habituation learning [10] in which the (possibly artifactual) organism "gets used to" normal situations. An unusual event, in contrast, evokes orientation towards the event [17]. It seems fitting that an automated, mechanical system should act similarly.

6. Data monitoring

We have developed our own network-monitoring tools for tracking 10 MB Ethernet network behavior within the CMU Computer Science Department. These consist of noninvasive, passive hardware devices that are attached to network spurs, reporting their results on demand to a diagnostic server on a VAX minicomputer. The network is configured as ten spurs off a single spine. The primary monitoring parameters are packets, collisions, heartbeats and load. Secondary parameters are packet sources, packet destinations, source-destination pairs, packet lengths, error packets, etc. These parameters are at the first and second levels of the ISO network protocol [9]. The monitors run autonomously, in real time, 24 hours a day. Each primary monitor (one per spur) reports network activity every 200 ms. These reports are amalgamated into one-minute reports. Forty parameters are monitored every minute, yielding 57 600 data points, and requiring approximately 0.5 MB of disk storage per spur per day. Due to the fault-tolerance control structure for the monitors, it is rare that any data are missing or corrupted, though any data of this type are immediately detected and handled by the monitoring substrate.

The primitive level for the present system was selected to be the first and second levels of the ISO network protocol, namely the physical layer and the data-link layer. These were chosen for the primitive level because they are the lowest-level places in the network for which external control is effective. We could have selected the bit level or the electron level on the physical network cable, but this would not have been effective because we cannot exert control of the network at those levels. At the physical and data-link layers, we can issue commands that will have a positive effect on network control. We could also have selected a higher level (e.g. the transport layer or the session layer), but higher levels do not offer the resolution in nonsymbolic observations that the physical and network layers do. In other words, at higher levels of the network protocol the data are already so abstract that there is not enough information in them to be useful in determining what might have happened at lower levels where control must be effected. This is another version of the symbol grounding problem which, in common sense parlance, simply says that one should sample the environment at a suitably detailed level.

7. Establishing zero-reaction level

This section describes briefly the process for building a daily[#12] representation of normal be-

---

[#11] The model suggested here is an abstract one for the purpose of simplifying the presentation; of course observations continue in parallel with diagnostic problem solving, both in the system being described as well as in real-world diagnosis.

[#12] The selection of an epoch over which to construct a representation is somewhat arbitrary, and usually depends on environmental trends. In the example presented here, a period of 24 h was selected because it is a good reflection of external influences, such as peoples' work habits, on the network environment. A shorter or longer period could just as easily be represented.

havior in a particularly ill-behaved environment – communication networks. All the figures in the following discussion illustrate measures taken from the CMU campus computing network described earlier.

There are three main operations involved in forming and using a representation of normal behavior, each of which will be discussed below:

(i) extract general shape of epoch;
(ii) evolve weighted shape of normal model;
(iii) establish range of variation for normal.

7.1. Extracting general shape

Extracting the general shape of the local epoch (e.g., a day's observations) has the effect of retaining the general trend of the data while removing noise and significant variation. The shape or trend of the epoch should reflect general rises, falls, cyclic behaviors and sequences in the data without including small-scale fluctuations or rapid oscillations. A primary goal is to be resistant to outliers in the data, because outliers tend to have an overbearing influence which often skews shape in an unsatisfactory manner. General shape extraction is accomplished by use of a series of robust, outlier-resistant smoothing operations adapted from the techniques of exploratory data analysis [19], and resulting in a smooth curve. Once the fast oscillations and noise have been removed from the data, a representation of the general trend or shape for the day is formed as shown superimposed on the scatterplot data in fig. 2. Note that the smoothing techniques employed here are consistent with the idea of blurring or defocusing, often used in image processing [3].

7.2. Evolving normal model

The general shape obtained for each epoch/day will be representative of the observations for that epoch. Provided that each day is like the preceding day, the general shape obtained for one day will describe adequately any other day. Most real-

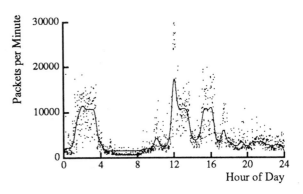

Fig. 2. Raw and smoothed data superimposed: the scatterplot shows minute-by-minute counts of Ethernet packets over the course of 24 h; the curve depicts a smooth, resistant model of the general trend of the scatterplot for this 24 h period.

world environments, however, are nonstationary, and conditions change at least slightly from one day to the next. An example of this is the gradual shift in weather conditions from one season to the next; the general profile of weather patterns for one summer day will roughly match the profile for another summer day, but will match that of a fall day less well. It is necessary to combine the successive daily profiles, one at a time, into a general model of normalcy that reflects the recent past more closely than the distant past, but contains less variation than the daily shapes. This is done by exponential smoothing [2], using a weighting factor that accepts the influence of recent events more heavily than distant events. The weighting factor of the exponential smoother determines how quickly the normal model will reflect a persistent change in behavior. This provides a method by which the model of normalcy can habituate to, or "get used to", repeated stimuli while ignoring transient fluctuations. Using weather conditions as an example again, one's model of a normal summer day should not be unduly influenced by a single day's freak temperature of 30 degrees higher than expected. Alternatively, if many days passed with such elevated temperatures, then one's normal model (or expectation) would gradually adapt and come to expect the elevated temperature as being normal. The swiftness with which the environment is changing provides a metric for deter-

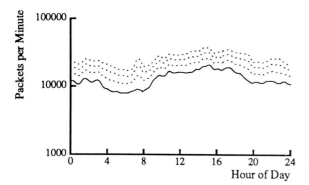

Fig. 3. June normal model (solid curve) and tolerance envelopes (dotted curves).

mining the extent to which recent observations will influence the normal model, and hence the rate of habituation. Figs. 3 and 4 depict normal models as solid curves. Notice the difference in the models over the four-month period. An advantage of using exponential smoothing, as opposed to standard averaging methods, is that it employs a recurrence relation requiring only the current model and the new data, both of which can be represented as vectors.

7.3. Establishing range of variation

If any future stimulus/observation does not fall precisely on the curve representing normal behavior, it would be judged anomalous, triggering a false orienting response or false alarm. Avoiding excess false alarms can be achieved by permitting some tolerance about the model of normalcy. This

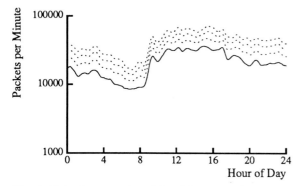

Fig. 4. November normal model (solid curve) and tolerance envelopes (dotted curves).

is consistent with the idea that the curve representing normal behavior is in fact the zero-reaction level. Any stimulus that is not precisely the same as the current zero-reaction-level value would cause the organism to react. There should be a tolerance envelope, or range of acceptable variation, within which a stimulus will cause the organism to adapt in the direction of the stimulus but not evoke an external reaction; this permits the organism to make small adaptive shifts without invoking corrective procedures that might cause the organism to go into oscillation.

Establishing a range of acceptable variation about the model of normalcy is accomplished in the same way that the normal model is determined. The only difference is that while the data for the normal model consist of measurements of the environment throughout the epoch or day, the tolerance model uses the variances of those data. The curve representing acceptable variation is then multiply added to the curve representing normalcy in a manner reminiscent of standard deviations. The result is a series of tolerance envelopes within which observations may be judged as being within some degree of normalcy. Decision rules, which may include space–time averaging, can then be used to determine when to activate a response (e.g. a diagnosis) to an anomalous condition.

Tolerance envelopes are depicted in figs. 3 and 4 as broken curves; the solid curves represent the models of normal behavior. Notice that these curves changed substantially over the period of time between the two figures. It can be seen that if a stimulus observed at noon in June had a magnitude of 100 000 it would lie outside all the envelopes and hence would be judged anomalous. Fig. 4 is markedly different, illustrating the change in network behavior from June to November, and also illustrating that learning occurred in adapting from the June model to the November model. It is clear that many observations judged anomalous in June would be judged normal in November. The system has learned the new behavior through repeated observation and habituation to higher-valued stimuli.

The effect achieved by the procedures outlined above is that of a level-sensitive anomaly detector which will achieve detection/orienting and habituating responses by: (1) providing an updating/learning mechanism so that the model for normal behavior will gradually adapt to repeated stimuli; and (2) establishing tolerance envelopes around the model within which a stimulus/observation can fall and still be termed normal, hence mitigating the false alarms to an acceptable level.

8. Examples from a network domain

This section shows results from a diagnostic server system embodying the above mechanisms in the network diagnostic-monitoring environment described earlier. During the analysis of the minute-by-minute data the models of normal behavior are updated to reflect the most recent activity on the respective network. If there is no existing model of normal behavior, then one is initialized along with a range of variation for the tolerance envelopes. It takes ten days for the system to learn the general trend of normal behavior. That the learning rate is set to ten days is a function of the nonstationarity of the application environment. Experience shows that the CMU network changes often enough that models over two weeks old tend to retain information that is no longer useful; hence the learning rate (and concomitantly, the forgetting rate) was set to two weeks.

The figures below illustrate both the detection (orientation) and habituation responses. Notice in fig. 5 that very few events were termed anomalous, because the model was well adapted to current network behavior. Anomalies are depicted by vertical lines drawn from the anomalous points to the edge of the third tolerance envelope. None of these surpassed alarm threshold. If a model from June is used as an internal representation for normalcy, the match against November data brings some surprises. Fig. 6 shows the anomalies; the dark regions between noon and 4 PM correspond to alarm-level detection responses. There are many anomalies because the June model reflects the lower level of activity expected during the summer at American universities, as opposed to the expected level during the academic months. Needless to say, these detections would have resulted in false alarms. Note that a single anomaly may not be sufficient to raise an alarm. This is due to the space–time averaging nature of the organism.

On June 30th a file server failed on the CMU network. The result was that hundreds of machines attempted to access files using a broadcast protocol, increasing network traffic significantly, as shown in fig. 7. Since the diagnostic server had not habituated to any stimuli remotely resembling this, it amounted to a surprise for the diagnostic server, which detected the anomaly immediately. The failure was detected 1.5 h before the first user

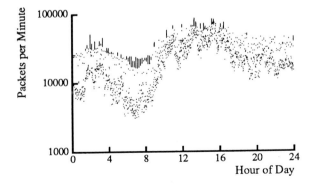

Fig. 5. November data matched against November model. Anomalies are depicted as vertical lines.

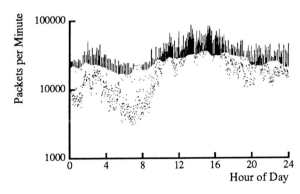

Fig. 6. November data matched against June model. Anomalies are depicted as vertical lines.

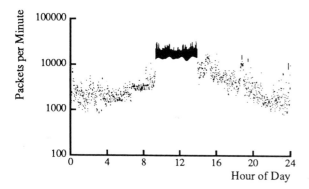

Fig. 7. June 30th data against June 30th model – file-server failure caused anomaly shown as dark region at center.

complaint, and it was repaired at about 2 PM; the repair is reflected in the diminished response shown in the figure.

If the June file-server failure were contrasted against the expectation model for November, would any anomalies be detected? No. The reason is that the elevated traffic levels resulting from the file-server failure would not have risen past November's threshold. This does not mean that a file-server failure would not be detected if it occurred in November; it would have been. The November traffic would have been perturbed to the same extent, relative to extant network activity, as was the June traffic relative to extant activity at that time.

9. Performance and practical experience

The system is in daily real-time operation at CMU. It detects 100% of all human-detected anomalies, based on a study of trained technical personnel attempting to detect network anomalies. It surpasses human expertise in terms of better noise resistance, finer representation of normal behavior or zero-reaction levels and more sensitive and reliable anomaly detection. It operates autonomously in real time, 24 hours a day, which no human could do, and it survives other failures through utilization of a fault-tolerant substrate – the layer of software upon which the system was built. It operates at a setpoint of a 2.5% error rate with 99.87% confidence that all normal events will lie within the predicted bounds. In over a year's operation the system has experienced no known misses (every problem detected by human technicians was also detected and alarmed by the system) and no false alarms (though this may be due to conservatively adjusted parameters).

The system has also been used in discovering design flaws in network routers; these flaws were subsequently corrected. It has been used in discovering errors in network software; also corrected. It has been employed in a network management role to determine a near-optimal splitpoint on a network; a bridge was installed. It has alerted facilities personnel to the fact that a particular workstation was serving off the wrong server, a configuration error; this was corrected. It has detected a major network disturbance on the order of once every two weeks for the last six months. Finally, it has begun to change the way the CMU CS department does network troubleshooting.

10. Conclusions

In this paper we have proposed a mechanism for facilitating diagnosis as an emergent property of a system of adaptive sensory elements, each nonprimitive element of which accommodates to local changes in the symbol hierarchy environment, and each primitive element of which adapts to changing conditions in the external environment. The symbol structure is built based on a nonsymbolic substrate (data from the real world), and it terminates in a single, high-level symbol which represents the condition of the external environment as a whole. A change in this high-level symbol can be decomposed recursively in terms of its constituent symbols until the locus of atomic change is isolated, at which point the cause of the original system perturbation has been identified. Diagnosis, defined as a search through the symbol hierarchy for the locus of atomic change in the environment, can be seen as an emergent property of a community of locally cooperative adaptive elements composing the symbolhierarchy. This concept can be extended to groups of such symbol hierarchies covering dozens of networks and thousands of nodes, all culminating in a single, high-level, multidimensional representation of the state of the entire network system. It is suggested that given the highly nonlinear, nonstationary characteristics of communication network environments, such an architecture for emergent diagnosis may be the only practical way to maintain and manage large, heterogeneous network ecosystems. It should be noted that nowhere in the system is there a place at which "the diagnosis" is performed – diagnostic behavior emerges as a consequence of the cooperative behavior of many autonomous processes.

Results of a real-time network diagnostic system running one level of the proposed architecture were shown. This demonstrated that an adaptive component of an automated diagnostic system can acquire and employ dynamic, internal representations of normal operating behavior to (1) detect anomalies in dynamic, high-noise, nonlinear, nonstationary environments, and (2) accommodate to the changes in the concept of normal behavior. The locus of atomic change in the demonstration system was verified to be the cause of several anomalous events on the network system employed as a research vehicle.

One drawback to the approach propounded here is that if the system is trained initially on abnormal data (without knowing it, during a period of anomalous operating conditions) then it will learn the wrong behavior as being normal. The nature consequence is that an ensuing period of actual normal/correct behavior will be flagged as different from the usual, since the usual behavior has always been, in this example, wrong. As the target system is brought under control, the normal model will gradually adapt to the correct mode. Another way to check for this condition is to compare two or more similar monitoring processes; if a difference is noted, then attention can be called to the situation. Yet another method is to include behavioral limits based on system specifications, irrespective of the possibility that the specifications may themselves be prone to error.

The techniques presented here are not limited to network performance data. They are amenable for treatment of many types of data. For example, process control data are easily accommodated by the techniques shown. Work is in progress to transfer the system to other domains and to extend its capabilities.

Acknowledgements

It is a pleasure to acknowledge the contributions and comments of Mark Detweiler, Frank Feather, Jack Goldberg, Marc Green, Jeff Hansen, Pamela Hayes, Sandy Ramos, Dan Siewiorek and Tom Ziomek. Thanks, too, to three anonymous reviewers for their helpful comments.

References

[1] A. Avizienis, Fault tolerance: The survival attribute of digital systems, Proc. IEEE 66 (1978) 1109–1125.
[2] G.E.P. Box and G.M. Jenkins, Time Series Analysis: Forecasting and Control (Holden-Day, San Francisco, 1976).
[3] J.F. Canny, A computational Approach to edge detection, IEEE Trans. Pattern analysis Machine Intelligence (PAMI) 8 (1986) 679–698.
[4] H. Chernoff, The use of faces to represent points in K-dimensional space graphically, J. Am. Statist. Assoc. 68 (1973) 361–368.
[5] R.W. Downing, J.S. Nowak and L.S. Tuomenksa, No. 1 ESS maintenance plan, Bell System Techn. J. 43 (1964) 1961–2019.
[6] J.P. Hansen, Analysis and modeling of uni/multiprocessor event logs, Master's Thesis, Carnegie Mellon University (1988).
[7] S.R. Harnad, Physica D 42 (1990) 335–346, these Proceedings.
[8] B.A. Huberman and T. Hogg, The behavior of computational ecologies, in: The Ecology of Computation, ed. B.A. Hubermann (North-Holland, Amsterdam, 1988).
[9] S.M. Klerer, The OSI management architecture: An overview, IEEE Network 2 (March 1988).
[10] J.W. Kling and J.G. Stevenson, Habituation and extinction, in: Short-Term Changes in Neural Activity and

Behavior, eds. G. Horn and R. Hinde (Cambridge Univ. Press, Cambridge, 1970) pp. 41–61.

[11] R.A. Maxion, A study of diagnosis, Ph.D. Thesis, University of Colorado (December 1985).

[12] R.A. Maxion and D.P. Siewiorek, Symptom based diagnosis, in: Proceedings of the IEEE International Conference on Computer Design (1985) pp. 294–297.

[13] R.A. Maxion, D.P. Siewiorek and S.A. Elkind, Techniques and architectures for fault-tolerant computing, Annual Review of Computer Science, eds. J.T. Traub, B.J. Gross, B.W. Lampson and N.J. Nillson (Annual Reviews, Inc., Palo Alto, CA, 1987) pp. 469–520.

[14] J. Morris, M. Satyanarayanan, M.H. Conner, J.H. Howard, D.S.H. Rosenthal and F.D. Smith, Andrew: A distributed personal computing environment, Commun. ACM 29 (1986) 184–201.

[15] D.P. Siewiorek and R.S. Swarz, The Theory and Practice of Reliable System Design (Digital Press, Bedford, MA, 1982).

[16] D.P. Siewiorek, C.G. Bell and A. Newell, Computer Structures: Principles and Examples (McGraw-Hill, New York, 1982).

[17] E.N. Sokolov, Higher neuron functions: the orienting reflex, Ann. Rev. Physiology 25 (1963) 545–580.

[18] M.M. Tsao, Trend analysis and fault prediction, Ph.D. Thesis, Carnegie-Mellon University (1983).

[19] J.W. Tukey, Exploratory Data Analysis (Addison–Wesley, Reading, MA, 1977).

THE COMPLEX BEHAVIOR OF SIMPLE MACHINES

Rona MACHLIN
Relational Technology, Park 80 West Plaza 1, Saddle Brook, NJ 07662, USA

and

Quentin F. STOUT[1]
Electrical Engineering and Computer Science, University of Michigan, Ann Arbor, MI 48109-2122, USA

This paper interprets work on understanding the actions of Turing machines operating on an initially blank tape. While this is impossible for arbitrary machines, complete characterizations of behavior are possible if the number of states is sufficiently constrained. The approach combines normalization to drastically reduce the number of machines considered, human-generated classification schemes, and computer-generated proofs of behavior. This approach can be applied to other computational systems, giving complete characterizations in sufficiently small domains. This is of interest in the area of emergent systems since the properties of such systems are often difficult to determine. By using computers to eliminate multitudes of machines with well understood behavior, some unanticipated exotic machines with complex behavior were discovered. These exotic machines show that it is quite difficult to estimate the number of states needed to produce a given behavior, and hence subjective estimates of complexity may be poor approximations of the true complexity.

1. Introduction

This expository paper discusses work on understanding the possible actions of a single simple machine interacting with a simple input. The machines are Turing machines, defined below, which have only a few states, and the input is an all-blank tape. Depending on one's background, this may either seem to be a very easy task, since the machines have very simple descriptions, or an impossible task, since among computer scientists it is well known that one cannot even decide whether or not an arbitrary Turing machine will halt. We show that it is instead a possible, but difficult, task, as long as the number of states is suitably restricted.

We believe that the techniques used to understand small Turing machines may prove to be useful in understanding other "simple" systems,

especially if one wants to produce provably complete classifications of behavior in suitably restricted classes. Since this work is largely unknown outside of computer science, and in fact is not even well known within computer science, we have taken a mainly expository approach in order to reach a wider range of researchers. This work may also revise notions of interesting or desirable behavior in Turing machines. Further, while we are successful in characterizing sufficiently small Turing machines, we show that a single Turing machine can be viewed as an emergent system, and thus any attempt at an unrestricted classification of the behavior of all emergent systems in any sufficiently powerful class is doomed to failure. This limitation needs to be more widely understood.

Finally, we show that the behavior of small Turing machines is far more complicated than most people would guess, and that exhaustive search can locate machines that are exceedingly difficult to create on one's own. Because such

[1] Partially supported by *Incentives for Excellence* grant from Digital Equipment Corp.

machines are quite unexpected, people tend to significantly overestimate the number of states needed to produce their behavior. This creates a false impression of complexity, leading one to believe that a system has many more components that it really does, rather than understanding that the complexity can come from the repeated application and interaction of a few simple, carefully chosen rules.

In section 2, we define Turing machines, and define the busy beaver and halting probability problems. These problems motivated work in classifying the behavior of small Turing machines. In section 3 we introduce the notion of tree normalization, which is used to drastically reduce the number of cases that must be considered. In section 4 we show the techniques used to classify Turing machines that are in infinite loops, which completes the computation of small busy beaver numbers. Section 5 shows how to apply this work to estimate the halting probability, and in section 6 we offer some concluding remarks.

2. Background

Turing machines are an attempt to formalize the notion of effective computation. While it is impossible to prove that one has correctly captured the intuitive notion of effective computation, all other attempts have yielded systems that can compute only functions computable by Turing machines, and hence there is fairly widespread acceptance of the Church–Turing thesis that Turing machines do indeed compute all functions that are effectively computable [14].

For our purposes, a (deterministic) *Turing machine* has an input–output *alphabet* of $\{0,1\}$, which writes and reads from a 2-way infinite tape of *squares*. The 0's and 1's are called *symbols*, and each tape square contains exactly one symbol. The 0 is thought of as being equivalent to blank.

A Turing machine has some finite number, k, of internal states, labeled $1,\ldots,k$, and a read/write head connecting it to the tape. At each time unit the read/write head is positioned under some square of the tape, and based on the symbol read and the current state, the Turing machine will write a (perhaps different) symbol at the square, move the read/write head left or right one square, and switch into a (perhaps different) state. See fig. 1.

For each possible pair of current state and symbol read there is a unique instruction specifying the symbol printed, head movement, and new state. Such instructions will be given as

(state, symbol, new symbol,

    head movement, new state),

where L or R are used to indicate head movement to the left or right, respectively. We assume that the tape is initially all 0 (all blank), and that the Turing machine is initially in state 1.

A Turing machine continues to execute its instructions until it encounters an instruction specifying a new state of 0, in which case it prints the symbol, moves the head, and then halts. Fig. 2 illustrates this action, which the subscripts on the tape indicate the state of the Turing machine and the location of the head.

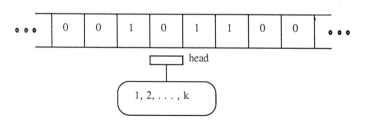

Fig. 1. A Turing machine.

Turing machine instructions		
	Symbol read	
State	0	1
1	1R2	1L3
2	1L1	1R2
3	1L2	1R0

n	Tape status after n steps					
1					1	0_2
2					1_1	1
3				0_3	1	1
4			0_2	1	1	1
5		0_1	1	1	1	1
6		1	1_2	1	1	1
7		1	1	1_2	1	1
8		1	1	1	1_2	1
9		1	1	1	1	1_2
10		1	1	1	1	1	0_2
11		1	1	1	1	1_1	1
12		1	1	1	1_3	1	1
13		1	1	1	1	1_0	1

Subscripts denote machine state and position of read/write head

Fig. 2. Turing machine execution.

As fig. 2 shows, even though the individual instructions have simple, precise definitions, the overall behavior of the Turing machine can be quite complicated and not readily apparent from the individual instructions. Thus a single Turing machine, viewed as a collection of cooperating individual instructions, is an emergent structure according to the definition in ref. [7].

2.1. Halting problems and halting probability

Many problems have been posed involving Turing machines. The best known of these, the *Halting Problem*, asks for an algorithm with input consisting of a Turing machine M and an initial tape T, and which outputs "true" if M will eventually halt when started on input T, and outputs "false" otherwise. This problem is well known to be impossible in the sense that no Turing machine can provide such an algorithm (see any standard text in computability, such as ref. [14]). Assuming that the Church–Turing thesis is correct, this shows that no effective algorithm of any form can solve the halting problem. It is also well known that there is a fairly simple machine U, the universal Turing machine, such that it is impossible to provide an algorithm to decide if U halts on input T for arbitrary T. Similarly, the *restricted halting problem*, in which the machine varies but the initial tape is all blank, is also unsolvable.

A related problem, which we call the *halting probability problem*, can be intuitively phrased as "What is the probability that a random Turing machine will halt when started on an all-blank tape?" To formally define this probability, denoted Ω, one must assign probabilities to Turing machines. Since there are infinitely many Turing machines it seems that no assignment is completely natural, and we postpone such an assignment until section 3. However, one can show that for any nontrivial assignment Ω cannot be computed, where by an algorithm computing Ω we mean that given any $\epsilon > 0$, the algorithm will return a rational number which is within ϵ of Ω.

Apparently Chaitin was the first to formally define Ω [5], though his definition differs from that given in section 3. Ω has many interesting properties [9], and recently Chaitin used it in his significant transformation of Gödel's incompleteness theorem into a statement about the solutions of a specific exponential Diophantine equation [6].

2.2. Busy beaver problems

Tibor Rado felt that the arguments used to prove the uncomputability of the halting problem were not sufficiently intuitive, and posed the Busy Beaver Problem as a more concrete variation. To define this, let $H(k)$ denote the set of all k-state Turing machines which eventually halt when started on a blank tape. Note that $H(k)$ is finite since it is a subset of the set of all k-state Turing machines, and there are exactly $(2 \cdot 2 \cdot (k+1))^{2k}$ different sets of instructions for k-state Turing machines, i.e. there are $2k$ instructions that need to be supplied, one for each (state, symbol) pair, and for each instruction there are 2 choices of new symbol, 2 choices for head direction, and $k+1$ choices for new state.

For a Turing machine M in $H(k)$, let $\sigma(M)$ denote the number of 1's left on the tape when M halts after starting on an all-blank tape, and let $s(M)$ denote the number of steps performed by M before halting. For example, for the machine in fig. 2, $\sigma(M) = 6$ and $s(M) = 13$. Rado [16] defined the kth *busy beaver number*, denoted $\Sigma(k)$, by

$$\Sigma(k) = \max\{\sigma(M): M \in H(k)\},$$

that is, $\Sigma(k)$ is the maximum number of 1's left on the tape by any halting k-state Turing machine. Similarly, he defined $S(k)$ by

$$S(k) = \max\{s(M): M \in H(k)\}.$$

The *busy beaver problem* is to give an algorithm which computes $\Sigma(k)$ for all values of k. Note that for any k, this problem merely asks for the maximum among a finite set of values.

Rado provided a nice proof that Σ could not be effectively computed by showing that if f is any function computable by a Turing machine, then there is a number n, depending on f, such that $\Sigma(n) > f(n)$. This also shows that S cannot be effectively computed since $S(k) \geq \Sigma(k)$ for all k. While Rado emphasized computing Σ, we will instead concentrate on the function S, for reasons which will become clearer in section 2.3. Work on computing S and Σ is discussed in refs. [1–3, 7, 10–13, 15–17].

2.3. Problem relationships

The halting problem, halting probability problem, and busy beaver problems are closely related, in that a solution to any one of them would yield a solution to each of the others (although the transformations may not have any practical usefulness). To illustrate this, suppose we have an algorithm which computes S, and want to solve the restricted halting problem. To decide if machine M halts on blank tape, merely count the numbers of states in M, call this k, and then simulate the running of M for $S(k)$ steps. If M has not halted in $S(k)$ steps then it must be that it will never halt, by the definition of S. Notice that $\Sigma(k)$ would not have been as useful since it may be that M sometimes writes 1's and sometimes erases them, making it difficult to guarantee that it will not suddenly erase all but $\Sigma(k)$ or fewer 1's and then halt.

To see how the halting probability problem can be used to solve the restricted halting problem, let M be some specified Turing machine, and suppose it has probability p. (It suffices to merely know that p is a nonzero lower bound on the probability of M.) Using any effective enumeration of the Turing machines, simulate running the first Turing machine for one step, then simulate the first Turing machine for two steps, then the second Turing machine for two steps, then the first Turing machine for three steps, then the second Turing machine for three steps, then the third Turing machine for three steps, and so on. (This stimulation process is known as *dove-tailing*.) Whenever a machine halts, add its probability to a running total. Eventually, either M halts, or else the running total becomes large enough so that if p were added to it then the total would exceed the halting probability. In this latter case M cannot halt.

All the other possible choices of using a solution of one problem to solve another can be done similarly, with the exception that it may not be obvious that an algorithm which computes $\Sigma(k)$ can be used to compute $S(k)$ (and hence to solve any of the other problems). Rado [16] noted that one could prove that

$$S(k) \leq (k+1)\Sigma(5k)2^{\Sigma(5k)}$$

(much better bounds are possible), and any upper bound T for $S(k)$ can be used to compute $S(k)$. To do this, one merely runs all k-state Turing machines for T steps. Any machine which runs for T steps without halting must be in an infinite loop, so the largest number of steps used by a machine which halts in T or fewer steps must be $S(k)$.

A solution to any of these problems would, at least theoretically, provide an effective means of solving mathematical problems. For example, to solve Fermat's last theorem, which asserts that there are no positive integers x, y, z, i such that $i > 2$ and $x^i + y^i = z^i$, one could write a program which uses dove-tailing to try all possible choices of x, y, z, and i, and which halts if it ever finds equality. Therefore the solution to Fermat's last theorem is reduced to the restricted halting problem for this program.

A more general reduction of mathematics to halting problems can be obtained by noticing that a proof is merely a finite sequence of symbols which can be generated, and verified, by a computer. Given any mathematical statement one is curious about, construct a program which generates all possible proofs and then verifies if it is a proof of the desired statement, halting whenever such a proof is found. Using this procedure, the provability of any mathematical statement is "reduced" to the problem of deciding if a specific Turing machine halts. This fact is central to the work in ref. [6].

3. Tree normalization

Despite the fact that the halting probability and the busy beaver problem cannot be solved, one can ask about partial solutions. For the halting probability problem one could ask for upper and lower bounds on the probability, and for the busy beaver problem one might determine exact values for some values of k, or bounds on the values. This approach was taken by Rado in his classes, and has been pursued by many others since [1, 2, 7, 10–13, 17]. We will emphasize the approaches to the busy beaver problem since that is where most of the work has been performed, though we were first attracted to working on the halting probability problem.

To evaluate $\Sigma(k)$ or $S(k)$ for small values of k, one immediately encounters the problem of having a large number of possible machines. As was noted above, there are $[4(k+1)]^{2k}$ k-state Turing machines, which, for example, is 25 600 000 000 when $k = 4$. However, many of these are equivalent or their behavior is readily apparent. Lin and Rado [12] noted that, since one starts in state 1 reading a zero, if the instruction is to go to state 0 then the machine will halt after only one step, while if the instruction is to go to state 1 then the program will be an infinite loop. This observation alone classifies the behavior of 10 240 000 000 4-state machines. Therefore the only unknown behavior is to go to a new state, and since the labels of the states are arbitrary we may as well call it state 2. Further, if the machine makes its first head movement to the left it will just be a mirror image of an equivalent machine with left and right head movement reversed for all instructions. Therefore one can assume that the first head movement is to the right.

Since Lin and Rado emphasized calculating $\Sigma(k)$, they could make a final reduction, namely that the first step prints a 1. This is because if it does not print a 1, imagine following the machine's operation until it first prints a 1, and starting it instead at the instruction that printed that 1. This new start will eventually halt if and only if the original one did, and both will produce the same number of 1's. Thus Lin and Rado could assume that the original instruction was (1, 0, 1, R, 2). However, the new start will use fewer steps than the original, and hence this normalization may underestimate $S(k)$ by as much as $k - 1$. This can be corrected by first using the Lin and Rado normalization to obtain a lower bound S' of $S(k)$. Then, noting that $S' + k - 1$ is an upper bound on $S(k)$, one can utilize the approach described in section 2.3 to use this upper bound to determine the exact value of $S(k)$.

The Lin and Rado approach can be extended (though they did not do so) to the viewpoint that

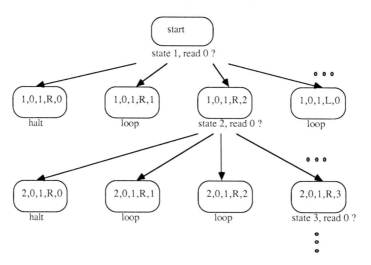

Fig. 3. Tree-normal programs.

one deals with incompletely specified Turing machines, only filling in instructions as they are needed. For example, if the initial instruction was (1, 0, 1, R, 2), then the machine will go to state 2 and again encounter a 0, so now a new instruction is needed. For this, either a 0 or 1 can be printed, the head can be moved either L or R, and either the machine halts (state 0), goes to a state already used (1 or 2), or goes to a new state, which we can relabel as 3. Thus there are $2 \cdot 2 \cdot 4 = 16$ effectively different choices for being in state 2 and seeing a 0, given the prior choice of instruction (1, 0, 1, R, 2). Four of these choices halt, four go into infinite loops (printing 0 or 1, moving R, and going to state 1 or 2), while the remaining eight each will then encounter a situation requiring that yet another instruction be generated. This approach was utilized in refs. [1, 2, 11].

The generation of instructions as they are needed yields a tree representation of the machines generated, as illustrated in fig. 3. The machines generated are said to be in *tree-normal* form. Notice that a single machine in tree-normal form may represent many Turing machines. For example, the Turing machine with instructions (1, 0, 1, R, 2) and (2, 0, 1, R, 2) represents $[4(k+1)]^{2k-2}$ k-state Turing machines, each of which will go into an infinite loop.

3.1. Probabilities

Tree-normal form can be used to assign probabilities to Turing machines. We say that the root of the tree has probability 1, and whenever a node has children (i.e. when it eventually encounters a situation where a new instruction is needed) then its probability is evenly divided among all of its children. In general, if a node with probability p represents a partial Turing machine where states $1, \ldots, i$ have been explicitly referenced so far (with the convention that the root explicitly references state 1), if a new instruction is needed then there are exactly $2 \cdot 2 \cdot (i + 2)$ children, each of which has a probability $p/[4 \cdot (i + 2)]$.

This is the notion of Turing machine probability that we will use for the halting probability problem, so to determine Ω we need to find the sum of the probabilities of all leaf nodes corresponding to a new instruction sending the machine to state 0. Similarly one could define the *infinite loop probability* as the sum of the probabilities of all leaf nodes corresponding to machines in infinite loops. There is a slight technical question of whether the sum of these two probabilities is 1, since the tree has infinite height and one can show that there exist such trees, with probabilities assigned in the same manner, for which the sum of the probabilities of the leaves is less than 1. However, it is easy to show that in our case these two probabilities do indeed sum to 1.

It is important to note that our definition of Ω differs somewhat from that in refs. [5, 6, 9], and that this definitional change alters the value. While the definition in refs. [5, 6, 9] has some properties that make it simpler to use for theoretical purposes, we believe our definition is somewhat more natural and easier to understand.

3.2. Variations

It was the problem of defining, and then estimating, the halting probability that lead us to tree normalization. However, we discovered that it had been used earlier by Brady [1] in his work on the busy beaver problem, and so his terminology was incorporated into the work reported in ref. [11] (written by the first author under her maiden name). The work in ref. [11] is an independent confirmation of Brady's results, which is important since the sheer volume of human and computer work involved raises the possibility of error. The work in ref. [1] was eventually published in ref. [2], while the work in ref. [11] has not been previously published.

There are four differences between tree normalization used to find $\Sigma(k)$ or $S(k)$ and the version used to define Ω. In each of these, the trees used for busy beaver problems are smaller than the tree for the halting probability. First, as was noted earlier, the initial instruction for finding $\Sigma(k)$ can be taken to be (1, 0, 1, R, 2), ignoring the other 11 possibilities. As was stated above, this may slightly underestimate $S(k)$, but a post-mortem check can be used to correct it. Second, no node need be generated which sends the machine into a state labeled $k + 1$ or higher. Third, if a node represents a partial machine in which only states $1, \ldots, i$ have been explicitly referred to so far, and if $2i - 1$ instructions have already been defined, then the new instruction is the last instruction involving states $1, \ldots, i$. If the new instruction sends the machine to states $1, \ldots, i$ then the machine will be

Table 1
Number of Turing machines to be analyzed.

k	Tree-normal	$[4(k + 1)]^{2k}$
2	41	20 736
3	3 936	16 777 216
4	603 712	25 600 000 000

an infinite loop, so to compute busy beaver numbers one need only consider instructions sending the machine to states 0 or $i + 1$. Finally, fourth, the instructions sending the machine into the halt state need only print a 1 and move R, since any other options would produce the same value for $S(k)$ and either the same or smaller values for $\Sigma(k)$.

All counts of numbers of nodes will be in terms of the tree normalization used to find $S(k)$, though estimates for Ω will use the correct tree normalization for it. All counts are taken from ref. [11], and when infinite loops are classified these counts differ slightly from those in ref. [2]. These small differences are due to slight variations in the definitions used, number of steps stimulated, and the order in which the tests were applied.

Table 1 shows the number of tree-normal machines generated for the busy beaver problems, as opposed to the number of Turing machines which are formally different. This clearly shows the significant reductions accomplished through the use of tree normalization. This table was made by a back-tracking program which simulated each node until it halted, was in an infinite loop, or reached a situation where another definition was needed, in which case the appropriate children were generated. The task of determining when a program is in an infinite loop is discussed in section 4.

4. Infinite loops

The major effort in calculating busy beaver numbers and estimates for the halting probability lies in proving that large numbers of machines are in infinite loops. The approach taken in refs. [1, 11, 12] is to examine some of these machines by hand, elicit a common behavior which insures that a machine is in an infinite loop, and then write a program which examines candidate machines and proves that some of them do indeed have that behavior. This process tends to iterate, with the researcher constantly trying to reduce the number of unclassified machines by either generalizing types of behavior earlier searched for, or by discovering new types of behavior.

In the end, a small enough number of machines remain so that they can be manually examined and verified to be in infinite loops. In ref. [12], only one type of behavior was needed in determining $\Sigma(3)$, with only 40 machines which needed to be verified by hand. In refs. [1, 11], four types of behavior were used in determining $\Sigma(4)$ and $S(4)$, along with a couple of hundred machines which were verified by hand.

One byproduct of this approach is that, while the busy beaver and halting probability problems are defined in terms of machines that halt, the interesting work involves machines that do not halt. In a certain sense, this approach treats all halting programs as equivalent, only needing to record probabilities, number of ones produced, or number of steps used, while the programs that do not halt must be more carefully examined and characterized.

The remainder of this section is based on the work in ref. [11], which was in turn based on ref. [1]. At each stage, machines not yet classified are called *holdouts*. Using tree normalization and allowing machines to run a couple of hundred steps produces 1364 3-state machines which halt, and 2572 3-state holdouts, and 182 604 4-state machines which halt, along with 421 108 4-state holdouts. The holdouts were run through a program to see if they could be proven to have a behavior known as a simple loop. The holdouts from the simple loop test were then put through a back-tracking analysis to see if it could be shown they were in infinite loops without determining which type of loop they were. (Actually, a fast test for

Table 2
Classification of tree-normal machines.

	3-state	4-state
total	3936	603 712
halted	1364	182 604
infinite loop	2572	421 108
simple loop	2495	404 733
back-track	50	10 363
Christmas tree	25	5 144
shadow Christmas		241
counter	2	417
holdout		210

simple loops was combined with the simulation program, and a more thorough simple loop test was run after back-tracking analysis, but logically there was no need to do so. The numbers reported are the sum of those found by the different simple loop tests, with the vast majority found by the fast test.) The remaining holdouts were run for a while, and based on the rate at which new tape squares were visited, they were tentatively classified as being a Christmas tree or a counter. For each of these classes, a program was developed which in most cases could prove a candidate was of the indicated type and was indeed in an infinite loop. Finally, the remaining holdouts were examined by hand.

The following subsections explain this process in more detail, and table 2 shows the number of machines classified at different states. Given a Turing machine M, we use M^c to denote the machine formed from M by changing all R moves to L, and vice versa. By a *word* we mean a (perhaps empty) finite string of 0's and 1's. For a word W and state r we use W_r to mean that the machine is in state r examining the rightmost symbol of W; $_rW$ means that the machine is in state r examining the leftmost symbol of W; W_r means that the machine is in state r examining the first symbol to the right of W, and $_rW$ means that the machine is in state r examining the first symbol to the left of W. We use 0^* to mean infinite occurrences of 0, and W^i to mean i concatenated copies of W.

State	Symbol read	
	0	1
1	1R2	1L3
2	1R3	
3	0L1	0R1

1$\underline{0}$	११००$\underline{1}$	१११११०१$\underline{0}$
11$\underline{0}$	11०0$\underline{1}$	1111110११$\underline{0}$
1$\underline{1}$	111$\underline{0}$1	11111101$\underline{1}$
1$\underline{1}$	1111$\underline{1}$	1111110$\underline{1}$1
$\underline{1}$	11110$\underline{0}$	11111100$\underline{1}$
$\underline{0}$1	111101$\underline{0}$	1111100$\underline{1}$
$\underline{0}$01	1111011$\underline{0}$	111111$\underline{0}$01
1$\underline{0}$1	111101$\underline{1}$	1111110$\underline{1}$
11$\underline{1}$	11110$\underline{1}$1	1111111$\underline{1}$
110$\underline{0}$	1111$\underline{0}$01	1111111$\underline{0}$0
1101$\underline{0}$	1111$\underline{0}$01	11111111$\underline{0}$0
11011$\underline{0}$	1111$\underline{0}$01	11111111$\underline{0}$10
1101$\underline{1}$	1111$\underline{1}$01	1111111$\underline{1}$011
110$\underline{1}$1	1111$\underline{1}$11	11111111$\underline{1}$011
1100$\underline{1}$	1111100	11111111$\underline{0}$01

Underlining indicates position of read/write head

Fig. 4. A simple loop.

State	Symbol read	
	0	1
1	1R2	1L1
2	0L1	1L3
3	1R4	1L3
4		0R2

Fig. 5. A machine analyzable via back-tracking.

recent occurrence of each state–symbol pair. After each step the table was consulted to see if a simple loop condition could be detected. Holdouts from the initial check were run through a second version which maintained a table of the conditions each time the Turing machine scanned a square at the edge of the critical portion of the tape (i.e. the nonzero portion). Of the initial 2572 3-state holdouts, all but 77 were proven to be simple loops, while of the initial 421 108 4-state holdouts, all but 16 375 were simple loops.

4.1. Simple loops

Fig. 4 shows a simple loop in operation. A Turing machine M is called a *simple loop* if either M or M^c satisfy one of the following:

(1) Some tape configuration is repeated infinitely often. That is, there is a nonzero state s and words X and Y such that at some time step the tape configuration is $0^* X_s Y 0^*$, and the same tape configuration is reached at some later time step.

(2) M periodically moves to the right, in that there is some nonzero state s, words X and Y and a nonempty word V, such that at one time the tape configuration is $0^* X_s Y 0^*$, at some later time the tape configuration is $0^* V X_s Y 0^*$, and between these times M never moved left of the left edge of the initial X.

It is clear from the definition that a machine classified as a simple loop is indeed in an infinite loop.

Initial checks for simple loops were done by maintaining a table containing the tape configuration, position, time step, and state of the most

4.2. Back-tracking

One straightforward way to prove that a program is in an infinite loop is to directly prove that it cannot reach the halt state. For example, consider the tree-normalized machine in fig. 5, which has only one unspecified instruction, namely being in state 4 while scanning a zero. This machine can halt only if it reaches this instruction, i.e. it must reach a local tape configuration of 0_4. To get there, it must have been in state 3 scanning a zero to the right of this zero, i.e. it must have been in configuration $0_3 0$. The only instructions which move to state 3 are for state 2 input 0, or for state 3 input 1, so the tape must have been in the configuration $0 1_2$ or $0 1_3$. However, both of these would produce $0_3 1$, which is not what was needed. Therefore the configuration 0_4 can never be reached, and the machine must be in an infinite loop.

While back-tracking can be useful, it cannot be guaranteed to always stop since otherwise it would supply a solution to the halting problem. As with all of the heuristics we discuss, one must make some decision as to how long to run this technique

	Symbol read	
State	0	1
1	1R2	1L1
2	1L1	1R3
3		1R1
1<u>0</u>	11111<u>0</u>	1111111<u>1</u>
1<u>1</u>	11111<u>1</u>	1111<u>1</u>111
<u>0</u>11	1111<u>1</u>1	111<u>1</u>1111
1<u>1</u>1	111<u>1</u>11	11<u>1</u>11111
11<u>1</u>	11<u>1</u>111	1<u>1</u>111111
111<u>0</u>	1<u>1</u>1111	<u>1</u>1111111
111<u>1</u>	<u>0</u>111111	<u>0</u>11111111
111<u>1</u>	1111111	1<u>1</u>1111111
1<u>1</u>11	1<u>1</u>11111	11<u>1</u>1111111
<u>0</u>1111	11<u>1</u>1111	111<u>1</u>111111
1<u>1</u>111	111<u>1</u>111	1111<u>1</u>11111
11<u>1</u>11	1111<u>1</u>11	11111<u>1</u>1111
111<u>1</u>1	11111<u>1</u>1	111111<u>1</u>111
1111<u>1</u>	111111<u>1</u>	1111111<u>1</u>11
1111<u>1</u>	1111111<u>0</u>	11111111<u>1</u>1
	11111111	111111111<u>1</u>

Underlining indicates position of read/write head

Fig. 6. A Christmas tree.

before abandoning it. When applied with a 15-step limit to the 3-state holdouts only 27 holdouts were left. When applied with a 10-step limit on the 4-state holdouts only 6012 remained as holdouts.

4.3. Christmas trees

When Lin and Rado analyzed 3-state Turing machines, they applied some of the initial stages of tree normalization, and wrote programs to detect simple loops. They ended up with 40 machines that they analyzed by hand. Most of these exhibited the back-and-forth sweeping motion shown by the machine in fig. 6. Brady called this behavior a *Christmas tree*. While the behavior is more complex than that of a simple loop, it is still clearly repetitive and in an infinite loop.

Formally, a Turing machine M is a *Christmas tree* if either M or M^c satisfy the following conditions for some nonzero state s:

(1) There are nonempty words U, V, and X such that the tape configuration at some time is $0^*UV_s0^*$, and at some later time is $0^*UXV_s0^*$.

(2) The following conversions hold, where X, X', Y, Y', Z, V, V', V'', U, and U' are nonempty words and q and r are nonzero states (the symbol $\Rightarrow$ means that M transforms the left-hand side into the right-hand side after some numbers of steps):

(a) $XV_s0^* \Rightarrow_q X'V'0^*$;
(b) $X_qX' \Rightarrow_q X'Y$;
(c) $0^*U_qX' \Rightarrow 0^*U'Y'_r$;
(d) $Y'_rY \Rightarrow ZY'_r$;
(e) $Y'_rV' \Rightarrow ZV''_s$;

(3) $U'Z^iV'' = UX^{i+1}V$ for all $i \geq 1$.

While this definition is somewhat complicated, it just guarantees that the machine sweeps back and forth, growing a periodic middle part of the tape configuration.

Again it can be proved that any Christmas tree must be in an infinite loop. To detect these, a program was written which ran a holdout for a couple of hundred steps to overcome startup effects, and which then cut the nonblank part of the tape in half to obtain candidates for V and U. Then it ran the machine until a back-and-forth sweep was observed. If after the sweep the new tape had a right-hand portion that matched V and a left-hand that matched U, then the remainder in the middle was taken to be X. This process was continued to find values for X', Y, etc., and to verify the conditions. If the program ever ran too many steps without finding the desired behavior, or could not successfully determine appropriate words, then the machine remained a holdout.

There are many variations of Christmas trees, so the initial program was modified to detect more of the variants. Brady called one variation an *alternating Christmas tree*, for it takes two back-and-forth sweeps to complete its cycle. Another variation, a *shadow Christmas tree*, illustrated in fig. 7, creates an increasing "shadow" at one edge, which it never scans past.

After running the 3-state holdouts through the various Christmas tree programs, only 2 holdouts remained, while for the 4-state machines only 627 holdouts remained.

	Symbol read	
State	0	1
1	1R2	1L1
2	0L1	0R3
3	1R4	1R3
4	0R2	

1̲0	11010	111011100	11110111100
1̲	110110	11101110	1111011110
01̲	1101100	111011110	11110111110
11̲	110110	11101111	1111011111
100̲	1101110	11101111	1111011111
101̲0	110111	11101111	1111011111
10100̲	110111	11101111	1111011111
101̲0	110111	11101111	1111011111
10110̲	110111	11111111	1111011111
101̲1	111111	11110111	1111111111
101̲1	111011	11110111	1111101111
101̲1	111011	11110111	1111101111
1111̲	1110110	111101110	1111101111
1101̲	11101110	111011110	1111101111

Underlining indicates position of read/write head

Fig. 7. A shadow Christmas Tree.

	Symbol read	
State	0	1
1	1R2	
2	1L3	1R1
3	0R1	0L3

1̲0	0101	0010001	110101
1̲1	1101	010001	111101
001̲	1101	110001	111101
01̲	1111	110001	111111
11̲	11110	111001	1111110
110̲	111110	111101	11111110
1110̲	111111	110101	11111111
1111̲	111101	100101	11111101
1101̲	111001	0000101	11111001
1001̲	110001	000101	11110001
00001̲	100001	100101	11100001
0001̲	0000001	110101	11000001
1001̲	000001	0010101	10000001
1101̲	100001	010101	000000001
00101̲	110001	110101	00000001

Underlining indicates position of read/write head

Fig. 8. A counter.

4.4. Counters

The final class of loops for which programs were written were called counters by Brady. Fig. 8 illustrates a counter, and it is obvious that it is indeed acting as a type of binary counter.

Formally, a Turing machine M is a *counter* if either M or M^c satisfies the following conditions:

(1) There are nonempty words E, X, Y, Z, and Z', a nonzero state s, and a positive integer n such that at some time the tape configuration is $0^* E Y_s Z' Z^n X 0^*$.

(2) The following conversions hold, for some nonempty word X':

(a) $Y_s Z' \Rightarrow_s Z' Z$;
(b) $0^* E_s Z' \Rightarrow 0^* E X'_q$;
(c) $X'_q X \Rightarrow Y X'_q$;
(d) $X'_q Z \Rightarrow_s Z' X$;
(e) $X'_q 0^{|X|} \Rightarrow_s Z' X$

where $|X|$ denotes the length of X.

In this definition, the X acts as a "one", and the Z acts as a "zero", in a binary counter.

Using an approach similar to that used for Christmas trees, a counter detector program was developed. This program successfully classified the final two 3-state holdouts as counters, and when run on the 4-state machines left only 210 holdouts.

4.5. Final holdouts

The final 210 holdouts were examined by hand to verify that they were in infinite loops (ref. [2] reported 218 final holdouts). More than half were variations of counters, including base-3 and base-4 counters. Also discovered were further Christmas tree variations, such as alternating shadow trees and triple and quadruple sweep trees.

Brady noted an additional class of machines, which he called *tail-eating dragons*. They have a back-and-forth sweep, limited by the end of the "tail" they create. After each sweep they "bite off" a piece of the tail, and when it is completely consumed they create a new, larger tail. As with other classes, there are also variations on this behavior. Fig. 9 gives the instructions of a tail-eating dragon.

	Symbol read	
State	0	1
1	1R2	1L1
2	1R3	0R4
3	1L1	
4	0L1	1R4

Fig. 9. Instructions for a tail-eating dragon.

Table 3
Turing machine probabilities.

Probability of halting	0.465
Probability of infinite loop	0.529
Uncertainty	0.007

5. Halting probability

The known values of S can be used to estimate Ω. Tree normalization appropriate for Ω is used (see section 3.2) to generate a portion of the tree. A node corresponding to a machine of k states is simulated for at most $S(k)$ steps. If it does not halt in this many steps then it is in an infinite loop and its probability is added to a running total of infinite loop probabilities. If it reaches a point were a new instruction is needed, then the probabilities of all children which halt immediately (i.e. those in which the new instruction has a new state of 0) are added to a running total for the halting probability, and the remaining children are simulated.

For a node corresponding to a machine with a number of states for which the S value is unknown the machine is simulated for some predetermined number of steps. If the machine does not halt or need a new instruction, then it is abandoned, and its probability is part of the uncertainty in the knowledge of the halting probability. Because no node is simulated for more than the predetermined upper limit (counting all the steps leading to the node), only a finite portion of the tree is explored. However, to reduce the stack requirements, one may also abandon nodes which have more than some (significantly smaller) predetermined number of defined instructions. While such nodes add to the uncertainty, they have relatively small probability since the probability of a single node decreases rapidly with a number of defined instructions.

As in the use of the tree for finding busy beaver numbers, some additional simplifications can be incorporated. For example, suppose a node corresponds to a machine with k states, and all instructions but one have been defined. If the node reaches a point where this instruction is needed then any definitions which do not go to a new state (or 0) must yield an infinite loop, and hence their probabilities can be immediately added to the infinite loop probability. One can also use the classification routines discussed in section 4 to prove that machines are infinite loops, rather than just adding their probability to the uncertainty total.

Using the above techniques, including the knowledge of $S(k)$ for $k \leq 4$, but not using the classification routines on machines of more than four states, yielded the results in table 3.

6. Final comments

As the *Emergent Computations* conference demonstrated, there is a significant interest in the general problem of understanding the behavior of simple systems. Further, researchers working on such problems have a wide range of backgrounds. Because of this, we felt it useful to describe work that led to the complete characterization of Turing machines of four or fewer states, and which has also produced results such as provable bounds on the halting probability. We note that a single Turing machine is an emergent system, in that it satisfies all of the conditions set forth in ref. [8] for an emergent system, and it can indeed have a very complex observed behavior as it moves through time and space (along the tape). Thus this work shows that complete, provable characterization of a suitably restricted nontrivial class of emergent systems has been achieved.

This work is completely rigourous, as opposed to, say, mere statistical sampling. Further, by making extensive use of computers to prove that certain machines have well-understood behavior, the researchers were able to focus their attention on the final holdouts, discovering unexpected behavior such as base-4 counters and tail-eating dragons. Most people would be hard pressed to develop a 4-state machine with such behavior, just as they would be unlikely to develop a 4-state machine which moves for 107 steps before halting. The 210 final holdouts exhibit significant diversity, and such machines would probably not be found other than through careful use of computers to sift out machines with known behavior. The holdouts represent only about 0.3% of the tree-normal 4-state machines, and only about 0.0002% of the unnormalized 4-state machines. This approach may prove to be generally useful for researchers seeking simple emergent systems with unusual properties. Further, the success in "decompiling" all 4-state machines and provably deciding their behavior may make it an attractive approach for other researchers trying to "decompile" simple systems to obtain an understanding of their behavior.

One crucial step in reducing the computational workload was the introduction of tree normalization. Tree normalization is a form of "lazy evaluation" which adds just those instructions which are needed, simulating the effects of a collection of instructions until a situation is encountered where a new instruction is needed. Classes of emergent systems other than Turing machines may also be numerically reduced through normalizations.

One way in which emergent systems research can impact upon work on Turing machines is by intensifying interest in the behavior of infinite loops. Computer scientists usually try to produce programs that rapidly complete their task and finish, rather than continue forever. (Operating systems are an important exception.) However, emergent systems research is most concerned with systems that have infinite, or very long, lifespans. Systems with short lifespans are usually easier to understand, just as it is trivial to see that a Turing machine which just moves right 10 steps and then halts (when given all-blank input) must have at least 10 states. It is much more complicated to understand or design infinite behavior, such as finding a minimal state Turing machine that has visited $\Theta(t^{1/5})$ tape squares after t steps. Even the work in refs. [1, 2, 11] did not completely classify all 4-state infinite loops, but rather resulted in hand analysis of a few holdouts. This analysis satisfied the authors that the machines were indeed in infinite loops, and they noted some interesting behavior, but they did not carefully describe all behaviors encountered, nor the number of machines with each behavior.

One caution for emergent systems researchers is that, while we believe formal approaches can be applied to other emergent systems models, we must emphasize that there are limits as to when exhaustive, provable characterization can be performed. Such characterizations must be tried with care and within appropriate parameter constraints. For example, as was noted previously, it is impossible to write a program which determines $S(k)$ for arbitrary k, and it is even impossible for any program to provide an upper bound for infinitely many k. Therefore the classification of all k-state Turing machines cannot be completed for arbitrary k, and similar statements can be made for almost all sufficiently general models.

This leaves, however, the interesting question of determining how far formal approaches can be pushed within emergent systems research. While one can easily prove that many problems involving infinite domain are unsolvable, it is not easy to delimit subdomains of solvable or feasible subproblems. For example, it is interesting to predict how far S will be determined. Such predictions are perilous, since, for example, in 1962 Rado felt that no known approach would yield $S(3)$, and that $S(4)$ was "entirely hopeless at present" [15]. Only two years later he and Lin published the solution for $S(3)$ [12], and by 1974 Brady had determined $S(4)$ [1].

In 1983 the largest known lower bound for $S(5)$ (i.e. the largest number of steps taken by a halting 5-state machine yet discovered) was 7707, and the

(a)		Symbol read	
	State	0	1
	1	1R2	1R1
	2	1L3	1L2
	3	1R1	1L4
	4	1R1	1L5
	5	1L0	0L3

(b)		Symbol read	
	State	0	1
	1	1R2	0L4
	2	1L3	1R4
	3	1L1	1L3
	4	1R0	1R5
	5	1R1	0R2

Fig. 10. Busy 5-state machines. (a) A machine that leaves 4098 1's. Discovered by Marxen and Buntrock [4]. (b) A machine that performs 23 554 768 steps before halting. Discovered by Marxen and Buntrock [4].

largest known lower bound for $S(6)$ was 13 488 [2]. In 1985, Uhling showed $S(5) \geq 2\,358\,063$, and $\Sigma(5) \geq 1915$ [7]. In 1989 Buntrock and Marxen [4] discovered that $S(5) \geq 23\,554\,768$ and $\Sigma(5) \geq 4098$ (see fig. 10). Based on Uhling's results, Brady [3] predicted that there will never be a proof of the values of $\Sigma(5)$ and $S(5)$. We are just slightly more optimistic, and are lead to recast a parable due to Erdös (who spoke in the context of determining Ramsey numbers): suppose a vastly superior alien force lands and announces that they will destroy the planet unless we provide a value of the S function, along with a proof of its correctness. If they ask for $S(5)$ we should put all of our mathematicians, computer scientists, and computers to the task, but if they ask for $S(6)$ we should immediately attack because the task is hopeless.

References

[1] A.H. Brady, UNSCC Technical Report 11-74-1 (November 1974).
[2] A.H. Brady, Math. Comp. 40 (1983) 647.
[3] A.H. Brady, in: The Universal Turing Machine: A Half-Century Survey, ed. R. Herken (Oxford Univ. Press. Oxford, 1988) p. 259.
[4] J. Buntrock and H. Marxen, personal communication (1989).
[5] G.J. Chaitin, J. Ass. Comput. Mach. 22 (1975) 329.
[6] G.J. Chaitin, Algorithmic Information Theory (Cambridge Univ. Press, Cambridge, 1987).
[7] A.K. Dewdney, Sci. Am. 251 (2) (August 1984) 19; 251 (5) (November 1984) 28; Sci. Am. 252 (4) (April 1985) 30.
[8] S. Forrest, Physica D 42 (1990) 1–11, these Proceedings.
[9] M. Gardner, Sci. Am. 241 (1979) 20.
[10] M.W. Green, Fifth IEEE Symposium on Switching Theory (1964) p. 91.
[11] R.J. Kopp, The busy beaver problem, M.A. Thesis, Mathematical Sciences, State University of New York at Binghamton (1981).
[12] S. Lin and T. Rado, J. Ass. Comput. Mach. 12 (1972) 196.
[13] D.S. Lynn, IEEE Trans. Computers 21 (1972) 894.
[14] M. Machtey and P. Young, An Introduction to the General Theory of Algorithms (North-Holland, Amsterdam, 1978).
[15] T. Rado, Symposium on Mathematical Theory of Automata (1962) p. 75.
[16] T. Rado, Bell Systems Tech. J. 91 (1962) 877.
[17] T.R.S. Walsh, Ass. Comput. Mach. SIGACT News 14 (1982) 38.

COMPUTER ARITHMETIC, CHAOS AND FRACTALS

Julian PALMORE
Department of Mathematics, University of Illinois, 1409 W. Green Street, Urbana, IL 61801, USA

and

Charles HERRING
US Army Construction Engineering Research Laboratory, P.O. Box 4005, Champaign, IL 61824-4005, USA

In this paper we explore aspects of computer arithmetic from the viewpoint of dynamical systems. We demonstrate the effects of finite precision arithmetic in three uniformly hyperbolic chaotic dynamical systems: Bernoulli shifts, cat maps, and pseudorandom number generators. We show that elementary floating-point operations in binary computer arithmetic possess an inherently fractal structure. Each of these dynamical systems allows us to compare the exact results in integer arithmetic with those obtained by using floating-point arithmetic.

1. Introduction

In this paper we view the computer as a dynamical system. We investigate the effects of computer arithmetic on computations in deterministic chaotic dynamical systems.

We show that the computation of orbits in highly unstable systems – deterministic chaotic dynamical systems – can be affected greatly by the use of the floating-point machine representations to mimic the real number field. We have found that the use of finite precision arithmetic in binary floating-point representations introduces another sensitivity into calculations in these highly unstable systems. Finite machine representations preclude the exact specifications of numbers that are not dyadic rationals. For this reason, even the act of entering data into a program may induce errors from the very beginning of the calculation. Differences in the least significant bit between the data entered and the machine representation begin to propagate immediately into the mainstream calculation of the chaotic orbit. This error propagation destroys validity of conclusions based on an inferred relation between the data entered and the output achieved.

A fundamental difference between the usual form of calculation in a chaotic dynamical system and what we describe here is that we couple the orbit calculation with conditional branching whenever a prescribed condition is met. The use of conditional branching in orbit computations is usually not considered in dynamical systems theory for those systems having uniqueness.

We can generate a pseudo-orbit using branching by the following mappings and conditions:

$$x \to 3x \mod 1$$

when $n \otimes 1.0/n = 1.0$ and

$$x \to 3x + \varepsilon \mod 1$$

when $n \otimes 1.0/n \neq 1.0$.

Here ⊗ represents floating-point multiplication (a combination of arithmetic multiplication and rounding) and $1.0/n$ represents a floating-point divide. Pseudo-orbits produced in this way can be nonperiodic because $x_n = x_k$ for $n > k$ does not imply that n and k trigger branching in the same way.

We have found that an underlying fractal structure exists when finite precision is imposed on real arithmetic. By finite precision we mean that the computer word length (in bits) is fixed and roundings are imposed on all arithmetic operations. By a fractal structure we mean a self-similar structure that remains when scaling to all levels of description within the finite precision.

In this paper we show that floating-point operations in binary computer arithmetic have an inherently fractal structure. We demonstrate the effects of finite precision arithmetic by simple experiments on orbits in three uniformly hyperbolic chaotic dynamical systems: Bernoulli shifts, Cat maps and pseudorandom number generators. Each of these dynamical systems, described in section 2, allows us to compare the exact results in integer arithmetic with those obtained by using floating-point arithmetic. In section 3 we discuss in detail the representation format for floating-point numbers. The numeric processor extension used in the numerical experiments is described in section 4. Conditional branching along orbits of chaotic systems and the results of our experiments are found in section 5. Finally, we summarize the implications of our investigations into the dynamics of computer arithmetic in section 6.

2. Three chaotic systems

In this section we detail the exact orbit structure for Bernoulli shifts and Cat maps on binary and decimal lattices. Random number generators as multiplicative linear congruential generators are defined on integer lattices. The exact calculations are made possible by the use of integer arithmetic.

2.1. Bernoulli shift

The simplest chaotic dynamical system is the Bernoulli shift on the unit interval, described in ref. [1]:

$$x_{n+1} = Dx_n \mod 1,$$

where D is an integer larger than 1. In base D arithmetic, x is written as

$$x = .a_1 a_2 \ldots,$$

where a_i takes values $[0, 1, \ldots, D-1]$.

Thus, the mapping in base D arithmetic is a shift of the base point to the right by one digit with each application and mod 1 recovers the fractional part. This dynamical system on the interval exhibits sensitive dependence on initial conditions (x_0) and is characterized by disorder and geometric growth with each iteration.

In this section we consider the Bernoulli shift given by

$$f(x) = 3x \mod 1$$

for x in $[0, 1)$, and its restriction to various lattices.

For a uniform binary lattice with 2^N sites, the period of f is 2^{N-2} and there are two orbits with this period, $N \geq 4$. For $N = 0$, there is a fixed point; for $N = 1$, there are two fixed points; for $N = 2$, there are two fixed points and a 2-cycle. For $N = 3$, there are two fixed points and three 2-cycles.

For a uniform quinary lattice with 5^N sites the period of f is $4 \cdot 5^{N-1}$, $N \geq 1$. For $N = 0$, there is a fixed point; for $N = 1$, there is a fixed point and a 4-cycle.

For a uniform decimal lattice with 10^N sites the period of f is $4 \cdot 5^{N-1}$. For $N = 0$, there is a fixed point; for $N = 1$, there are two fixed points and two 4-cycles. For $N = 2$, there are two fixed points, a 2-cycle, four 4-cycles and four 20-cycles.

The results above give the orbit decomposition of the various lattices by the Bernoulli shift. In section 5 we will compare these results with those obtained by floating-point arithmetic.

2.2. Cat map

The Cat map [2] is a linear symplectic automorphism of the torus T^2 that is given by the formula

$$F(x, y) = (x + y \mod 1, x + 2y \mod 1)$$

for all $0 \leq x, y < 1$.

Theorem 1. The orbit decomposition of a uniform binary lattice with 2^{2N} sites on the torus T^2 by the Cat map is:
  one fixed point,
  five 3-cycles,
  2^k cycles of length $3 \cdot 2^{k-2}$,
for $3 \leq k \leq N$.

Theorem 2. The orbit decomposition of a uniform quintary lattice with 5^{2N} sites on the torus T^2 by the Cat map is:
  one fixed point,
  two 2-cycles,
  three 10-cycles,
  $12 \cdot 5^{k-2}$ cycles of length $2 \cdot 5^k$,
for $2 \leq k \leq N$.

The proofs of theorems 1 and 2 use elementary algebra and will be published separately.

On a uniform decimal lattice, with mesh 0.1 in both x and y, the Cat map decomposes the lattice into cycles of period: 1, 2, 3, 6, 10, and 30.

Example 3. The decomposition of the uniform decimal lattice with 100 sites by the Cat map is:

One fixed point:
  (0, 0);

Two 2-cycles:
  (2, 6) → (8, 4) → (2, 6),
  (4, 2) → (6, 8) → (4, 2);

One 3-cycle:
  (5, 0) → (5, 5) → (0, 5) → (5, 0).

Two 6-cycles:
  (2, 1) → (3, 4) → (7, 1) → (8, 9)
    → (7, 6) → (3, 9) → (2, 1),
  (1, 3) → (4, 7) → (1, 8) → (9, 7)
    → (6, 3) → (9, 2) → (1, 3);

Two 10-cycles:
  (2, 0) → (2, 2) → (4, 6) → (0, 6) → (6, 2) → (8, 0)
    → (8, 8) → (6, 4) → (0, 4) → (4, 8) → (2, 0),
  (4, 0) → (4, 4) → (8, 2) → (0, 2) → (2, 4) → (6, 0)
    → (6, 6) → (2, 8) → (0, 8) → (8, 6) → (4, 0);

Two 30-cycles:
  (1, 0) → (1, 1) → (2, 3) → (5, 8) → (3, 1) → (4, 5)
    → (9, 4) → (3, 7) → (0, 7) → (7, 4) → (1, 5)
    → (6, 1) → (7, 8) → (5, 3) → (8, 1) → (9, 0)
    → (9, 9) → (8, 7) → (5, 2) → (7, 9) → (6, 5)
    → (1, 6) → (7, 3) → (0, 3) → (3, 6) → (9, 5)
    → (4, 9) → (3, 2) → (5, 7) → (2, 9) → (1, 0),
  (3, 0) → (3, 3) → (6, 9) → (5, 4) → (9, 3) → (2, 5)
    → (7, 2) → (9, 1) → (0, 1) → (1, 2) → (3, 5)
    → (8, 3) → (1, 4) → (5, 9) → (4, 3) → (7, 0)
    → (7, 7) → (4, 1) → (5, 6) → (1, 7) → (8, 5)
    → (3, 8) → (1, 9) → (0, 9) → (9, 8) → (7, 5)
    → (2, 7) → (9, 6) → (5, 1) → (6, 7) → (3, 0).

Fig. 1 shows the orbit decomposition of the uniform decimal lattice of 100 points that is described in example 3.

2.3. Pseudorandom number generator

In 1951, Lehmer [3] proposed an algorithm that has become a de facto standard pseudorandom number generator. As it is usually implemented, the algorithm is known as a "prime modulus multiplicative linear congruential generator" or Lehmer generator. The form of the Lehmer generator is

$$x_{n+1} = ax_n \mod m.$$

Park and Miller [4] propose a portable minimal standard Lehmer generator. Their choice of m, the prime modulus, is $2^{31} - 1$, and their choice of a, the multiplier, is 7^5. Their minimal standard Lehmer generator is given by

$$x_{n+1} = 16807 x_n \mod 2147483647.$$

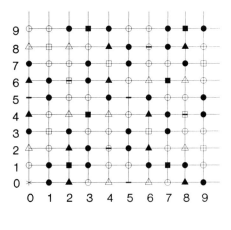

Fig. 1. Orbit decomposition of a uniform decimal lattice by the Cat map.

A comparison of this minimal standard pseudorandom number generator to a Bernoulli shift is given in ref. [5]. In section 5 we use this minimal standard pseudorandom number generator to produce fractals and compare its "orbits" with those of the Cat map.

3. Floating-point arithmetic

To understand the description of the numerical experiments described later in this paper it is necessary to describe in some detail the actual representation of numbers in floating-point format. The Institute of Electrical and Electronic Engineers (IEEE) adopted a standard for Binary Floating-Point Arithmetic (ANSI/IEEE Std 754-1985) [6]. This standard defines a binary arithmetic system which includes representation formats, a limited number of arithmetic operations, specifications for rounding and format conversions between representations.

The standard defines two basic sizes for representing floating-point numbers: single precision (32 bits) and double precision (64 bits). Optional extended-precision size formats (42 and 80 bits) are also described. The extended-precision formats were added to accommodate intermediate temporary results with the goal of reducing rounding error and preventing overflow.

A floating-point number is defined by the relation: $(-1)^S \times 1.F \times 2^{(E-B)}$, where S is the sign bit, F is the fractional part of the significand (mantissa), E is the biased exponent, and B is the exponent bias [7]. The bit patterns for these formats are shown below in fig. 2. The width in bits of each part of the format is numbered from 0 starting with the sign bit, S. The most significant bit of the exponent, E, is numbered starting with bit 1 to the least significant bit to the right. The bit positions of the fractional part of the significand, F, is numbered from left to right beginning with the most significant bit.

Floating-point numbers are stored in normalized form. A biased exponent is used to ensure that the exponent is nonnegative. Thus the true

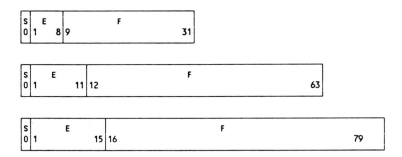

Fig. 2. 32, 64 and 80-bit floating-point formats.

exponent, e, is $E - B$, where, for the single precision case, $1 \leq E \leq 2^8 - 1$ and $B = 2^7 - 1$, and e has the range $-126 \leq e \leq 127$. The bit patterns shown in fig. 2 are decoded according to the relation given above as follows. The sign bit, S, is either 0 or 1 and determines the sign of the number when multiplied by -1. The true significand is $1.F$, i.e. the fractional part F plus 1. The value of the floating-point number is formed by multiplying the true significand by 2 raised to the true exponent and setting the sign.

The standard specifies that four rounding modes will be implemented. The default rounding mode is to the nearest representable number with rounding to even in the case of a tie. The standard implies that when rounding to nearest occurs the result shall be accurate to one half in the least-significant digit position. The other three rounding modes are called "directed" rounding in that the user may select to round toward positive infinity, negative infinity, and toward zero (truncate, also called "chop"). Rounding is usually required for conversion from an extended format (80-bit) to the destination format (64-bit).

There are considerable differences in the implementations offered by the various manufacturers [8]. Furthermore, many of the technical details, e.g. exactly how floating-point numbers are multiplied, may not be easily obtained and in that case can be inferred by inspection of the results.

4. Numeric processor extension

In this section we describe the particular hardware and software configuration used in our investigation. The computer hardware used in the numerical experiments described in section 5 consists of a Compaq Computer Corporation DeskPro 386. This machine is based on the 32-bit Intel 80386 microprocessor and includes an Intel 80387 Numeric Processor Extension [9]. The 80387 numeric processor implements the IEEE Standard. The software used is Microsoft's MS-DOS 3.31 and Borland International's Turbo-C version 2.0.

The 80387 numeric processor has an 80-bit format (extended precision) that is capable of using four rounding modes for floating-point operations. The 80-bit format has a sign bit, a 15-bit exponent and a 64-bit significand. The first bit in the significand is a "1" followed by an implicit binary point. The various modes available in the 80387 can be set by the user through the Control Word, a 16-bit field. The four rounding controls in the 80387 control word are: 00 – round to nearest or even, 01 – round down (toward $-\infty$), 10 – round up (toward $+\infty$), and 11 – chop (truncate toward zero). These last three roundings are called directed roundings. These rounding controls affect the instructions that perform rounding at the end of arithmetic operations. The precision controls on the 80387 control word are: 00 – 24 bits (single precision), 01 – reserved, 10 – 53 bits (double precision), and 11 – 64 bits (extended precision). By default, the 80387 is set to extended precision and round to nearest.

With these possibilities of rounding and precision control there are 12 combinations that must be investigated. Table 1 gives the rounding results of floating-point division in the default rounding mode. These results were obtained using the extended precision format.

5. Chaos and fractals in computer arithmetic

In this section we describe several numerical experiments in floating-point arithmetic on orbits

Table 1
Examples of rounding to nearest on floating-point divide.

Floating point divide	Hexadecimal	Rounding to nearest 80387 processor
1.0/3.0	.AAAAAAAAAAAAAAAb	increases
1.0/5.0	.CCCCCCCCCCCCCCCd	increases
1.0/11.0	.bA2E8bA2E8bA2E8C	increases
1.0/19.0	.d79435E50d79435E	decreases
1.0/23.0	.b21642C8590b2164	decreases

in the three chaotic systems that are written down in section 2.

5.1. Branching in floating-point arithmetic

Several self-similar scaled structures result when floating-point arithmetic is used to compute orbits. The following branching condition is used: if N satisfies $N \otimes 1.0/N = 1.0$, then continue; else, distinguish (print) N. For example, when computing an orbit of the Bernoulli shift, $x \to 3x \bmod 1$, $\{x_0, x_1, \ldots, x_n, \ldots\}$, the branching condition is: if x_n satisfies $x_n \otimes 1.0/x_n = 1.0$, then continue; else, print x_n and continue. This process yields a subsequence of the orbit. In table 2 we provide evidence of the extreme sensitivity to branching in the chaotic orbits of the Bernoulli shift $f(x) = 3x \bmod 1$ as computed using floating-point arithmetic. Although the orbit itself is sensitive to the initial data, the table demonstrates that the branching sequence produced by changing the initial data by as little as 10^{-15} is far different than the original branching sequence.

5.2. Scaling in floating-point arithmetic

By examining the integer binades, i.e. $2^p \leq N < 2^{p+1}$, and distinguishing the set $S_p = \{N, N \otimes 1.0/N \neq 1.0\}$ we have the following results: if N is in S_p, then it follows that $2 \otimes N$ is in S_{p+1}. This is a consequence of the floating-point representation that multiplication by 2 results in a change of exponent only, the significand remains the same.

The rounding modes yield the following results. In modes 01 (round up), 10 (round down) and 11 (chop) only $N = 2^k$ satisfies the equality $N \otimes 1.0/N = 1.0$, for all N, $1 \leq N \leq 8192$. All other N fail in all three precisions – 24, 53 and 64. The rounding mode 00 (round to nearest or even) gives three distinct sequences for the three precisions. Fig. 3 shows the relative positions of members of the sequence, normalized to $(0,1)$ for three precisions. Specifically, we call this display a "linear" fractal. The sequence for precision control 53 appears more sporadic in the low values of N. For

Table 2
Sensitivity of branching to changes in chaotic orbits.

0.1	$0.1 + 10^{-15}$	$0.1 + 10^{-14}$	$0.1 + 10^{-13}$
4	4	4	4
8	8	1134	182
12	286	2888	2151
1610	287	4084	8316
2197	403	9183	8526
3570	9229	13567	14085
4711	9521	17144	14534
9458	9741	19226	15519
15450	9877	21728	20765
15451	10080	25452	25759
15657	10530	34415	25760
16255	14651	37416	27261
18838	15277	37922	28058
21597	16977	41302	32270
21650	18420	44961	34520
22099	18800	48875	35166
22100	19537	50491	38570
22101	21837	53113	42104
24693	26853	53114	42682
24867	29362	53841	42711
29126	31469	54713	47939
32867	35226	54714	50068
33394	35705	58148	50069
34468	35738	59560	50070
35174	35885	59656	51145
35175	36361	65557	55517
37381	39905	67039	56253
38417	45131	67551	59543
41011	46063	69696	63289
43620	49562	81210	68496
44719	51610	85921	70898
46604	51611	87355	73891
48528	53045	89855	75436
49916	53104	94342	76429
53438	58533	97444	79317
60440	65066	99771	84513
60890	66907	99772	86057
64911	69518	1 = 37	91084
65153	73324	0 = 99962	93022
67117	76458		97270
70160	76459		97513
71229	76926		1 = 41
72186	77831		0 = 99958
72187	77832		
74528	77833		
78474	82634		
79397	86521		
79598	86986		
81068	90296		
83526	92986		
88725	1 = 50		
93193	0 = 99949		
1 = 52			
0 = 99947			

Fig. 3. Graphical display of integer fractal structure in single, double and extended precisions.

positive numbers the rounding modes 10 (round down) and 11 (chop) yield identical displays in all three precisions. For $1 \le N \le 2^{21}$, in precision 64, we find that there are two values for the product $N \otimes 1.0/N$. These are 1.0 and $1.0 - 1$ unit in the last place (ulp).

Fig. 4 shows the two-dimensional display of these "linear" fractals when 250 fractals are stacked vertically. The fractals are adjusted line by line for conditional branching on rational numbers. The apparent hyperbolas are a global feature of the change in branching algorithm with the computation of each linear array for precision 64.

Table 3 displays the binary bit patterns for the floating-point divide $1.0/N$, for all N for which the product $N \otimes 1.0/N$ takes the value $1.0 - 1$ ulp. The precision is 64. The table displays the normalized bit patterns from 1 to 66 bits. The significand in extended precision has 64 bits. The 65th and 66th bits for all of these N are the pair 01. Similar patterns are found for precisions 24 and 53. That is, for precision 24 the 25th and 26th bits in the

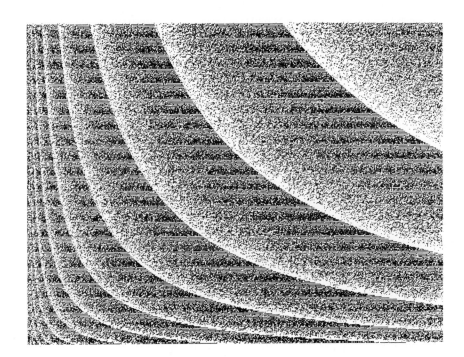

Fig. 4. Two-dimensional graphic display of rational fractal structure in extended precision.

Table 3
Normalized binary bit patterns in floating-point divide in extended precision.

```
         1         2         3         4         5         6
1234567890123456789012345678901234567890123456789012345678901234567
```

Bit pattern	N
11000111110011100000110001111100111000001100011111001110000011 0001	41
10110110000010110110000010110110000010110110000010110110000010 1101	45
10010100111100100000100101001111001000001001010011110010000010 0101	55
10000110010010111000101001111101111001101101000111010110000010 0001	61
11000111110011100000110001111100111000001100011111001110000011 0001	82
10110110000010110110000010110110000010110110000010110110000010 1101	90
10100101011111101011010100000010100101011111101011010100000010 1001	99
10010100111100100000100101001111001000001001010011110010000010 0101	110
10001110011110000011010101011011010001010000001000111001111000 1101	115
10000110010010111000101001111101111001101101000111010110000010 0001	122
10000101001101000000100001010011010100000010100110100000001010 0011	123
10000011000100100110111010010111100011010100111110101111001110 1101	125
11010000101101101001111111100101111010010010110000001101000010 1101	157
11001110000101101000101001101100100101000010000001100111000010 1	159
11000111110011100000110001111100111000001100011111001110000011 0001	164
11000110100110000000110010100110000001100011010010010000000110 0001	165
10110110000010110110000010110110000010110110000010110110000010 1101	180
10101011000111101101001111000101000001101011001110011010001000 0101	191
10100101011111101011010100000010100101011111101011010100000010 1001	198
10011000011010001100100000000100110000110100011001000000010011 0001	215
10010111000000100101110000000100101110000000100101110000000100 0101	217
10010100111100100000100101001111001000001001010011110010000010 0101	220
10001110011110000011010101011011010001010000001000111001111000 1101	230
10001011011100000011010001010000100110011011110001101011010 1001	235
10001001000110101100011100111010110100110000011001101101010 0001	239
10000110010010111000101001111101111001101101000111010110000010 0001	244
10000100110100000100001010011010000010000101001101000000010 0011	246
10000011000100100110111010010111100011010100111110101111001101 1101	250
10000001100001001000101101010001111010111000010100100111011 1101	253
11011100101010001110001010110001100011111111001000110101011100 0001	297
11010110110111101000011111110010100010000010111100000000011 0101	305
11010000101101101001111111101010010010110000001101000010 1101	314
11001110000101101000101001101100100101000010000001100111000010 1	318
11000111110011100000110001111100111000001100011111001110000011 0001	328
11000110100110000000110001101001100000001100011010010000000110 0001	330
11000011010000010011101111001100000010010101011100011101110 1101	335
10111000010011011110000110110011100011111000001000010011011 1101	355
10110110000010110110000010110110000010110110000010110110000010 1101	360
10110000101001011001101101000001100011010111010010011010101001 1101	371
10101111011001100100001101000010100100101101111101111110000111 1001	373
10101101110101011100110001100100011111110101001000101010000110 0001	377
10101011000111101101001111000101000001101011001110011010001000 0101	382
10100101011111101011010100000010100101011111101011010100000010 1001	396

binary bit patterns for the floating-point divide $1.0/N$ for all N for which the product $N \otimes 1.0/N$ takes the value $1.0 - 1$ ulp are the pair 01. For precision 53 the 54th and 55th bits in the binary bit patterns for integers N satisfying the branching condition are the pair 01.

These facts suggest the following interpretation based on the standard for rounding to nearest and format comparisons. In the floating-point divide operation in the 80387 numeric processor extra bits are kept prior to implementing rounding: guard, round and sticky bits. The pairs 00, 01, 10, 11 can appear in the guard and round bit positions in the 68 bit significand of the normalized result prior to rounding to the extended precision format. The standard for binary floating-point arithmetic implies that an error of at most $\frac{1}{2}$ ulp is made by each floating-point operation in the rounding to nearest or even mode. Rounding on bit pairs 00 and 11 cause an error of $\frac{1}{4}$ ulp (at most). Subsequent multiplying by N does not cause the error for the pair of operations to exceed $\frac{1}{2}$ ulp. Thus, all numbers N having bit pairs 00 or 11 in the guard and round bit positions satisfy the equality $N \otimes 1.0/N = 1.0$. Rounding to nearest or even on bit pair 10 causes an increase in the significand or causes a small relative error; i.e. a 0 in the least significant bit position triggering round to even leads to a relative error of at most $\frac{1}{4}$ ulp and this satisfies the equality $N \otimes 1.0/N = 1.0$. This leaves only the bit pair 01 as the candidate pair for triggering inequality. This bit pair causes truncation to occur. Evidence that this does trigger inequality is seen in table 3 for precision 64. That this bit pair 01 may not trigger an inequality is found, for example, for $N = 19, 31, 37, 38, \ldots$, for precision 64.

5.3. Fractal structures of the Cat map

From example 3 in section 2 we know that an orbit of the Cat map on the decimal lattice with 100 elements and spacing 0.1 in x and y that starts at $(0.1, 0.0)$ has period 30. In a floating-point calculation, the number $1/10$, calculated with a floating-point divide, is given a value that is the nearest binary machine number. By entering the point $(0.1, 0.0)$ a change is made from the data entered to the machine representation. The orbits of all points of the Cat map are unstable. Therefore, the machine produces an orbit of very long period. Plate I shows the orbit that results after many iterations of the Cat map with initial point $(0.1, 0.0)$. The number of iterations is 307 200. Plate II shows the "orbit" produced by the pseudorandom number generator described in section 2.3. The list of pseudorandom numbers is normalized to $(0, 1)$ and plotted successively in pairs: $(x_n, y_n) = (N_n, N_{n+1})$. The number of points used in this display is the same as for the Cat map in plate I. A count of the number of pixels lighted in plate I differs from those lighted in plate II by 0.2%. The total number of pixels in each figure is 480 by 480.

The visual comparison of the outcome of two extremely different processes – (1) the pseudorandom number generator uses integer arithmetic to produce its sequence and (2) the Cat map operates by integer multiplication on floating-point numbers – shows how the results of floating-point arithmetic can mimic random number generation.

In this case we know the comparison in advance. In other cases, the spurious results of floating-point calculations may not be detected.

Using conditional branching on an orbit of the minimal standard random number generator in section 2 we are able to generate a fractal image on the torus $[0, 1) \times [0, 1)$. This fractal image is produced by first generating a sequence of integers N which satisfy the branching condition $N \otimes 1.0/N \neq 1.0$. We then plot the sequence by taking successive pairs in the sequence as above and placing each pair as a point on the torus. The binade structure is apparent. The result is given in plate III.

We have produced a "fuzzy fractal" using branching on an orbit of the Cat map in floating-point arithmetic. For the orbit of the Cat map starting at $(0.1, 0.0)$ and by branching on either $x \otimes 1.0/x \neq 1.0$ or $y \otimes 1.0/y \neq 1.0$, we can produce a fractal image in the binade scale. This sequence produced as a result of branching on this orbit is shown graphically in plate IV. Observe that as many as eight binades are captured by each image in plates III and IV.

6. Conclusions

Computer arithmetic rounding decisions are made at the conclusion of floating-point operations. In a stable system, differences introduced into the computation by rounding decisions may be of no consequence. In a highly unstable system, each difference between the orbit and the computation contributes to a change that can propagate into the bulk computation. It is evident from the examples given above that floating-point errors can propagate into a mainstream calculation in a chaotic system and, within 30–50 steps, destroy all accuracy of a result.

In the experiments reported on in section 5 we have shown that a "fractal" (self-similar structure) exists in computer arithmetic when conditional branching is used to investigate the set of values taken by $x \otimes 1.0/x$, $x \neq 0$. That a set of values

Plate I. Cat map floating-point orbit with initial data (0.1, 0.0).

Plate II. Graphical plot of random vector (X, Y).

Plate III. Fractal structure of random vector (X, Y) with branching.

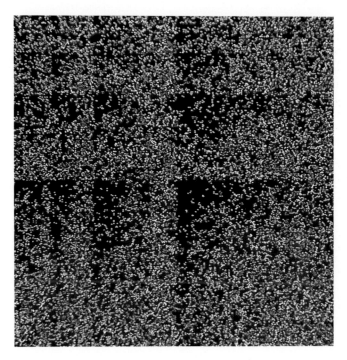

Plate IV. Fractal structure of Cat map orbit with branching.

exists for any arithmetic operation that is evaluated over the domain of machine representable numbers is a consequence of the failure of computer arithmetic to represent a field [10]. The scaling in the fractal structure is limited by the exponent range in floating-point representation and cannot be extended indefinitely. The fractal structure is dependent on the base of arithmetic (and therefore, the period of the rational numbers), the language employed, the machine and the precision used. Furthermore, different software implementations of a mathematical operation, format conversions, and assignments can change the structure. We are currently researching this area.

Acknowledgement

The authors gratefully acknowledge discussions on floating-point arithmetic with Dr. Aaron Averbuch and Mr. Mark Ginsberg of USACERL.

Note added in proof

The fractal structure described in this article is an artifact of the IEEE Standard 754 round to nearest/even rounding mode.

References

[1] J.L. McCauley and J. Palmore, Computable chaotic orbits, Phys. Lett. A 115 (1986) 433–436.
[2] V.I. Arnol'd, Ordinary differential equations (MIT Press, Cambridge, MA, 1973).
[3] D.H. Lehmer, Mathematical methods in large-scale computing units, Ann. Computing Lab. Harvard Univ. 26 (1951) 141–146.
[4] S.K. Park and K.W. Miller, Random number generators: good ones are hard to find, Comm. ACM 31 (1988) 1192–1201.
[5] C. Herring and J. Palmore, Random number generators are chaotic, ACM SIGPLAN, No. 11 (1989).
[6] IEEE Standard for Binary Floating-Point Arithmetic (ANSI/IEEE Std 754-1985), The Institute of Electrical and Electronics Engineers.
[7] S. Mor, Introduction to floating-point systems, Progress 10 (1982) 23.
[8] W.J. Cody, Floating-Point Standards – Theory and Practice. Conf. – 8709382/Dec. 89 003598, US Department of Energy (1987).
[9] Intel – Microprocessor and Peripheral Handbook, Vol. 1. Microprocessor 80387, 80386 and 80387 Sections, Intel Literature (1987).
[10] R.T. Gregory and E.V. Krishnamurthy, Methods and Applications of Error-Free Computation (Springer, Berlin, 1984), pp. 1–3.

THE COREWORLD:
EMERGENCE AND EVOLUTION OF COOPERATIVE STRUCTURES IN A COMPUTATIONAL CHEMISTRY

Steen RASMUSSEN[a,b,1], Carsten KNUDSEN[a,b],
Rasmus FELDBERG[a] and Morten HINDSHOLM[a]

[a]*Physics Laboratory III, The Technical University of Denmark, DK-2800 Lyngby, Denmark*
[b]*Center for Nonlinear Studies and Theoretical Division (T-13), MS-B258, Los Alamos National Laboratory, Los Alamos, NM 87545, USA*

We have developed an artificial chemistry in the computer core, where one is able to evolve assembler-automaton code without any predefined evolutionary path. The core simulator in the present version has one dimension, is updated in parallel, the instructions are only able to communicate locally, and the system is continuously subjected to noise. The system also has a notion of local computational resources. We see different evolutionary paths depending on the specified parameters and the level of complexity, measured as distance from initially randomized core. For several initial conditions the system is able to develop extremely viable cooperative structures (organisms?) which totally dominate the core. We have been able to identify seven successive evolutionary epochs each characterized by different functional properties. Our study demonstrates a successive emergence of complex functional properties in a computational environment.

1. Introduction

Emergent computation can be defined as a subset of self-organizing processes where the new macroscopic dynamics has a clear computational interpretation. In this study the interacting microscopic primitives are modified von Neumann machine code instructions. This means that in fact all levels of interaction in this system have a clear computational interpretation.

The motivation for this work is associated with the problem of the origin and the evolution of life. In the theoretical tradition quantitative descriptions of evolution and the origin of life have generally a priori outlined a specific evolutionary route. It has been assumed that the chosen path is relevant, so modeling and calculations have only analyzed the different evolutionary consequences of this particular predefined route [4, 5, 7, 14, 19, 20]. Such an approach neglects an infinity of alternative possibilities, and it also misses the single most characteristic property of real evolutionary processes: the ability to utilize historical events to choose a particular mechanism among a number of qualitatively different possibilities.

In these theoretical studies of the origin of life, selection of a single possible path seems necessary both due to the number and the complexity of all possible chemical reactions here on Earth. All possibilities cannot be feasibly taken into account.

The history of science also indicates that much progress has been achieved using exactly this selective approach: an observed phenomenon is explained by taking a possible mechanism which is consistent with the laws of Nature and is able to model the phenomenon. Experiments then help to eliminate irrelevant models. Unfortunately, experimental work in the area of evolution is very cumbersome, because we can only guess under

---

[1]Address for correspondence is Los Alamos National Laboratory.

which conditions the products of evolution initially developed, and also because evolutionary experiments by definition demand large "populations" observed over many generations. In wet carbon chain chemistry, the creation of even the simplest living entity has not yet been possible in the test tube due to the incredible number of different initial conditions such an experiment can have. Each experiment can only test one of them.

Another quantitative approach *is* possible which does not require any predefined evolutionary route using a much simpler "artificial" chemistry. By defining a universal set of low-level rules, an open evolutionary process can be started within a computer. Depending on the initial structure, the parameters, and the historical events along the way, the system can choose its own evolutionary direction. The goal for such an approach is not to model the properties of real chemistry, but merely to capture the basic principles of self-organization, which apparently enable the real chemistry to develop functional properties in an open-ended way. As we believe that the essence of "life" basically is associated with the functional organization of the different parts in an organism, and hence does not depend on the properties of the hardware in which it is implemented, such an approach should relate the *processes* we create to the real physical phenomena.

The different theoretical approaches are further discussed in fig. 1.

Inspired by the computer game "Core Wars" [2], we have chosen to build our chemistry as a core simulator system in which only a few basic instructions are operating. The object of the Core Wars game is to cause the opposing player to terminate abnormally. The object for our system is to create new computational properties. In this first version of our simulator, we have followed some of the specifications for a core simulator given in ref. [12], quite closely, although there are a number of significant differences, which allows us to evolve the core. Since the name of the core simulator in the Core War game is MARS, after the Roman god of war, we named our simulator

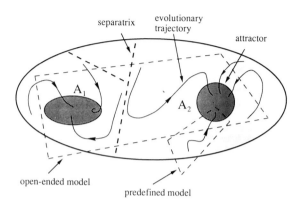

Fig. 1. The predefined model can only tell something about one out of a possibly infinite number of alternative evolutionary pathways. In a more open-ended approach the model is able to find its own evolutionary path based on a set of low level "chemical rules" and some initial conditions. Note that such a model in general has many evolutionary paths and more than one attracting area in the rule area. We define an artificial possibility space W_i, which compared to the "real" chemical possibility space W_r, is very simple. We believe that the *processes* of emergence and evolution of properties in W_i relate to the corresponding physical *processes* in W_r.

VENUS after the Roman goddess of natural productivity, love and beauty, with the hope that it would create interesting life-like properties.

The basic instructions together with their interaction rules define our artificial chemistry. With these low-level local rules we hope to evolve *functional properties* in our system, similar to the evolution of Dawkins' [1] "bimorph" *forms*, but without any external human selection mechanism.

Related work in progress in this area includes refs. [8, 15, 16]. For a discussion of open-ended evolution, see for instance refs. [6, 18]. For a general introduction to the field of artificial life, see ref. [13].

2. The core simulator VENUS

The core in our simulator has a one-dimensional address array with periodic boundary con-

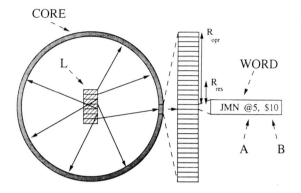

Fig. 2. The central part of VENUS is the core. The core has 3584 addresses and each address contains one word. The core is cyclic modulus the core size. Each word consists of an instruction and up to two operands, A and B. At the magnified core area also the resource radius R_{res} and the operation radius R_{opr} are shown. The queue of execution pointers is symbolized by L. Note that one of the execution pointers is pointing at the address where the JMN(@5, $10) instruction is located.

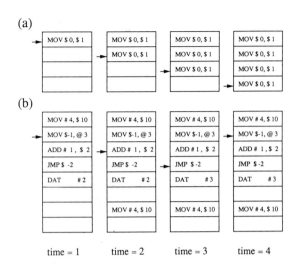

Fig. 3. Examples showing (a) a simple self-propagating instruction, and (b) a simple loop structure. All the addressing is relative to the executing instruction. Note that a loop structure like this is very powerful in multiplying any instruction. Similar loops are responsible for the phase transitions we sometimes meet in the distribution of instructions in the core.

ditions. Each address in the core is occupied by one of the ten basic instructions. Each address is associated with a certain amount of computational resources. An instruction is executed if its address has a pointer, and if its resource neighborhood has sufficient computational strength. Unless the executed instruction tells the pointer to move to a specified location, the virtual machine moves the pointer to the next address in the memory, as the core is updated. The core is further described in fig. 2, and some simple examples of programs are described in fig. 3. The basic instruction set, called red-code, is arbitrary chosen to be the same as that used in the Core Wars game. In table 1, the ten instructions and their addressing modes are listed.

In many respects the core operates as an ordinary multitasking von Neumann machine.

At time t each address x in the core is associated with a certain amount of computational resources $r(x, t)$, which is measured in fractions of one execution, one *exec*. An *exec* is equivalent to what is used for the execution of a single interaction. Thus, an instruction with a pointer can only be executed if its neighborhood has computational resources equivalent to at least one *exec*. The resource neighborhood is defined by a radius R_{res}. The number of addresses each instruction can obtain resources from is therefore $2R_{res} + 1$. The simulator executes instructions and hereby removes resources in a sequence determined by the order in which the pointers occur in the execution queue. Resources are renewed by an amount Δr at each core update. However, the resources are never allowed to exceed a maximum resource level, r_{max}. r_{max} is a global parameter always smaller than one *exec*. Δr ($\leq r_{max}$) is also a global parameter.

Besides the limited computational resources associated with each core address, the system also has a limited number L of active pointers. This means that each core update executes at most L instructions. The current system has 220 places in the execution queue ($L = 220$).

The VENUS core also has another important locality parameter, the operation radius R_{opr}. This radius defines how far away each instruction is allowed to access and alter data relative to its own address. The operation radius R_{opr} and the re-

Table 1
The red-code instructions and their opcode and their addressing mode (#, $, @, <): Each statement operand consists of an addressing mode and a value. The actual contents of A and B depends upon the addressing mode used.

Opcode	Word		Description
0	DAT	B	non-executable statement reserves space for data
1	MOV	A, B	move the content of A to B
2	ADD	A, B	add the content of A to B and put result in B
3	SUB	A, B	subtract the content of A from B and put result in B
4	JMP	A	transfer program pointer to A
5	JMZ	A, B	transfer program pointer to A if B equals 0
6	JMN	A, B	transfer program pointer to A if B differs from 0
7	DJN	A, B	decrement B and execute JMN A, B
8	CMP	A, B	compare A and B and skip next statement if unequal
9	SPL	B	split execution between B and next statement

Addressing mode	Name	Description
#	immediate	the actual operand is the value
$	direct	the actual operand is the statement pointed to by the value
@	indirect	the actual operand is the statement pointed to by the value pointed to by the value
<	auto-decrement indirect	the actual operand is the statement pointed to by the decremented value pointed to by the value

Table 2
Parameters defining a desert and a jungle.

	Desert	Jungle
R_{res}	5	3
Δr	0.10	0.50
r_{max}	0.50	0.50
P_{point}	0.05	0.05

source radius R_{res} allow us to achieve locality in our core.

The resource level at any address x at any time t is determined by

$$r(x, t) = \min[r_{max}, r_L(x, t-1) + \Delta r], \qquad (1)$$

where $r_L(x, t)$ is recursively defined as an iteration over the entire pointer execution queue of length L. $r_L(x, t)$ needs to be calculated in this way because of different pointers with overlapping resource areas take resources from the same address at each core update. Remember that the pointer execution queue is not organized after addresses but after history. Also note that the execution queue of pointers for each core update is only passed through once giving $r(x, t)$ for every x. The actual resource updating is a little complicated due to the sequential emulation of an inherently parallel process. The details of the updating scheme are given in appendix A.

Large Δr (and r_{max}) defines "jungle" conditions opposite to small Δr (and r_{max}) which defines the "desert". A small r_{max} can of course be compensated by a large R_{res} but this is not computationally effective for our simulations. The jungle and the desert we shall use as a reference are defined in table 2.

The core is updated in parallel. When the system is running it has many pointers active at the same time. All the instructions associated with these pointers are updated before changes in the core are made. In this way each instruction "sees" the same core, when we are simulating on a sequential machine. In case of conflict between two instructions the instruction with the lowest number in the execution queue will have its changes effected. However, specific details on how conflicts are resolved do not affect the system's ability to evolve.

As described so far our system still misses a fundamental property: the creative aspect of evolution or the notion of *noise*. We have introduced random fluctuations in two different ways. Whenever a MOV instruction is executed there is a certain probability P_{mut} that the copy of the source operand of the MOV instruction mutates. Each word has in this situation an equal probability to change into any of the ten legal instructions. The operands for the new instruction are also chosen at random. Changes are thereby restricted always to yield legal syntax.

A pointer dies whenever it meets a DAT instruction. Although a pointer is duplicated whenever it meets a SPL instruction, there is a finite probability that every pointer in the core dies, because every instruction initially is equally distributed. In order to ensure that the system is always active, we disturb the core by introducing new pointers at random with a low pointer appearance frequency P_{point}, each time the core is updated. This is an additional way of driving the system besides the influx of computational resources, Δr.

The details of how the noise is introduced seem to be of major importance for what can happen. With our present experience with the system we do not believe that the current implementation of the noise is optimal for the open-ended evolutionary potential in our system. We fixed this implementation of noise at an early point in order to have consistency in the different simulations.

Looking at the functional properties of a particular instruction and the instructions with which it communicates, almost any mutation in an instruction will cause a significant change in the functional properties of the involved communicating instructions. Any change will normally cause the instruction and its interaction neighborhood to "jump" to another location in the space of functional properties (not in the instruction space). Thus, the system primitives miss a fundamental notion of continuity of functional properties with respect to perturbations. In this respect it is similar to real chemistry. For instance the reaction $O_2 + 2H_2 \rightarrow 2H_2O$ has the same property. Here two gases, hydrogen and oxygen, interact and form water, which has properties very different from the reactants. As will become clear from the experiments, the continuity of functional properties emerges as one of the *products* of the evolution process.

Given the above properties, the low-level rules for our system define a system somewhere between a cellular automaton (CA) and a von Neumann machine. We shall refer to it as an assembler automaton. It is a compact and powerful form for writing a CA with a very high number of possible states where the dynamics interacts with the rules.

In the present version of our system we are at any time able to take the core out and analyze it in detail after it has been grown for some time. For this purpose we have developed a number of statistical tools useful for inspecting the cores. Interesting cores can at a later point be taken into the system and developed further. When we start a simulation we are also able to "seed" an "engineered" program sequence at a random location in the randomized core. Thereby we can bias the initial conditions in different ways. We can also investigate the dynamics of the evolution interactively: (1) A graphical representation of the whole core can be shown where each instruction has a different color, and where the active pointers are shown as highlighted underscores. (2) A selected part of the core can be followed graphically in more detail, where successive generations of the selected core area can be seen simultaneously (as for a one-dimensional cellular automaton). (3) The sequence of current executing instructions can be written to the screen. The simulations to be described in the following are all performed on a

IBM PS2 model 80, where a typical simulation follows the execution of ~ 22 000 000 instructions over 100 000 core updates, and takes about twelve hours.

3. Evolutionary field studies

The initial work with this new Coreworld was both exciting and very frustrating. On the one hand it was obvious that something was happening as the simulation proceeded; the core was changing. On the other hand the basic physics of this universe is so different from our "real" physics, that we had no clue about what would develop in the system. We did not know what to look for in the core. At the same time a core contains a very large amount of data, which we had to analyze at different times in order to see what was changing as the evolution proceeded. Since we did not have any automated methods for interpreting these data, we had no other choice than to dump many, many cores, where each hardcopy occupies several meters of paper, and hope that the human eye was able to catch anything of interest. It took a long time before we were ready to test evolution of different environments in a more systematic way.

In this period, we learned to distinguish between a core developed under desert conditions and a core grown as a jungle. We also learned to date a core, i.e. to guess for how many generations it had been active. From these "field studies" a pattern of different interesting structures slowly emerged. We named some of the structures and were later able to recognize them when they appeared in other cores. In the desert with no initial structure (a randomized core), we normally found a number of "simple loops" after a few thousand core updates. In the jungle we also found loops, but some of them were more complicated than the ones we found in the desert. The jungle seems to be characterized by the development of "SPL-falls": pointer-dense structures kept alive by one or more SPL instructions. These structures were never found in the desert, due to their high density of pointers. Looking very carefully, we also became aware of the existence of "fossils" in cores which had been simulated for some time. These fossils could for instance be traces of the simple self-propagating instruction (see fig. 3a) and various loops, where for instance a DJN instruction, earlier closing the loop, eventually was counted down to zero, and thereby allowed the pointer to pass. We could locate these loops when some instruction inside the loop earlier had a significant effect on the surroundings. We also learned that any "human engineered" organism – both programs we designed and the programs designed to participate in the Core Wars – were too brittle to survive in the noisy and chaotic VENUS universe. They all die after a number of generations, depending on the noise level. Later, however, we learned that the system was able to develop its own stable organisms.

In the following, we shall describe some of the evolutionary processes in detail. The processes are later summarized in table 3. In section 4 we shall discuss the systematic properties of these processes.

VENUS has a high number of independent parameters and the initial conditions in the core can be defined in many different ways. From the very beginning we therefore had to choose, quite arbitrarily, a certain number of parameters and initial conditions to vary as the rest were kept constant.

We have used two different initial conditions in the core for the simulations to be discussed here.

(A) The randomized core, which is both randomized with respect to instructions and addresses, is probed with a single active JMP($0) at a random location. This instruction does not affect any of the other addresses in the core (recall table 1), and is only placed there in order to prevent an empty execution queue which will cause the simulation to terminate. The randomly introduced pointers, which appear with frequency P_{point}, here initiate the core development.

(B) The same kind of randomized core is probed with a simple self-replicating program at a random location. The major effect such a program has is the production of a highly inhomogeneous core. The full self-replicating program is shown in appendix B. It consists of 8 instructions and has a cycle of 18 core updates. It has, however, an instruction-copying loop similar to the loop shown in fig. 3a. The presence of this loop is important. After the program has replicated a number of times it falls apart partly due to the noise and partly because the different programs start to copy on top of each other. It turns out that the self-replicating program has a number of copying loops as nearest "Hamming" neighbors in the space of functional properties (not in the address space). These loops copying a single instruction stay active long after the original replication has failed. It is the activity of these loops which is responsible for the inhomogeneity. It is important to note that the self-replicating program is only used to create a certain kind of inhomogeneity in the randomized core.

The resource parameters in the core are either set as a jungle or as a desert as defined in table 2. Besides these two resource settings and the initial conditions we have also varied the mutation frequency P_{mut} and the operation radius R_{opr}.

A desert with a small operation radius and no initial structure typically develops into a relatively stable core with (a) many fixed points for the pointers, (b) some simple loops, (c) a few more complicated loops. The fixed points are jump-instructions like; JMP(#X), JMZ(#X, 0), JMN($0, Y), and DJN(#X, Y), where X can be anything and Y is different from zero. These instructions are all characterized by pointing to themselves. The JMZ(#X, 0) is rare because it is unlikely that the second argument will be exactly zero. They can only interact with other instructions, by changing their B-fields. A simple loop structure typically consists of one of the jump instructions, which sends the pointer upstream to a certain address in the core; where the virtual machine again moves it downstream until it meets the jump instruction. After some time a loop may disappear, due to an instruction either inside or outside the loop altering data inside the loop. More complicated loops such as overlapping loops, nested loops, and loops in series are also found.

The evolution in the desert, initiated with instruction-copying loops, and parameters as shown in table 2, does not seem to be very different from the above situation. Slightly more complicated loops are seen here, caused by the remains of the altered and mutated self-replicating programs. The evolution has only changed the global chemical composition a little. The global distribution of the different instructions is almost identical with the initial instruction distribution.

The jungle simulated with a small operation radius R_{opr} and no initial structure is able to develop more complicated structures. The significant new feature appearing in the jungle is clusters of dense structures, driven by pointers from SPL instructions. Earlier we named these structures "SPL falls".

For a large R_{opr} (larger than 500) and no initial structure in a jungle, the typical evolutionary path is to develop a number of loops with a jump instruction at the end to return the pointers. These loops will dominate the core for some thousand core updates. More and more pointers, however, will be trapped at simple fixed points. In one simulation ($R_{opr} = 2000$, $\Delta r = 0.25$, and otherwise a jungle defined as in table 2) we found 7 JMP(#), 3 JMN(#), and 202 DJN(#), which means that 212 out of 220 possible pointers after 455 000 generations (~96 million instructions) were trapped at the different jump instructions. This state of the core is very stable.

When no initial structure is present, the jungle seems to develop more interesting structures for smaller R_{opr}. However, the story is different if some initial structure is used.

The dynamics of the core under the different evolutionary conditions is summarized in table 3.

Starting with a self-replicating program in a jungle with $R_{opr} = 100$, we are able to develop

Table 3
Summary of the characteristics for the core dynamics under different evolutionary conditions: jungle/desert, R_{opr} small/large, initially with (s)/without self-replicating program (⌐s). Obviously, the combination: jungle, large R_{opr}, and self-replicating program, provides the most interesting dynamics.

	Initially	Small R_{opr}	Large R_{opr}
desert	⌐s	fixed points for pointers, many simple loops, and few more complicated loops	fixed points for pointers and few simple loops
	s	fixed points for pointers, simple loops, and several complicated loops	fixed points for pointers, simple loops, and several complicated loops
jungle	⌐s	fixed points for pointers, complicated loops, and SPL falls	fixed points for pointers
	s	multiple evolutionary paths with different pointer dense structures	multiple evolutionary paths, very complicated structures

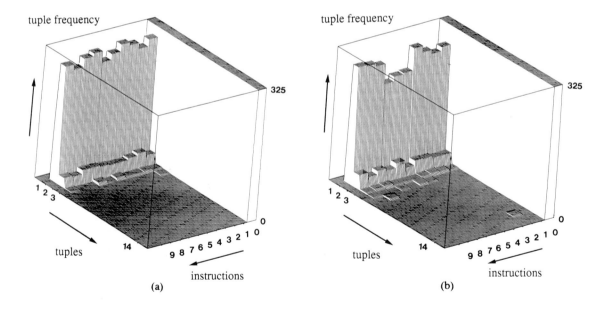

Fig. 4. The history of a specific core, a jungle with initial structure $R_{opr} = 20$ and $P_{mut} = 0.05$, told by its distribution of instructions at two different times, (a) at generation 0 and (b) at generation 72 700. The histogram shows the frequency of each instruction as it is found alone, as it is found in pairs, triples, etc., up to 14-tuples. Higher-order tuples may occur but are not shown. The total number of instructions in the core is 3584. The scale of the box is 325 instructions. The macroscopic composition of such a system only changes very little over many generations. However, several large programs kept alive by SPL instructions together with a number of complicated loops are found in this garden. This shows that significant changes have occurred at the micro level.

larger areas in the core where the instructions interact in an interesting way. After some ten thousand core updates we often find stationary "organisms", which primarily consist of SPL instructions. These programs are typically 20 to 100 instructions long. They are very robust because of the many pointers continuously being produced in the area. Since almost every pointer is occupied by these SPL structures very few external interactions are able to alter these structures.

In another simulation it takes the same jungle about 100 000 generations ($\sim$ 18 000 000 instructions) to develop two huge structures mainly consisting of a mixture of SPL and MOV instructions. The core has in this case more than 800 copies of both the SPL and the MOV instruction. This means that these two instructions occupy more than one third of the universe.

For R_{opr} smaller (< 40) the interesting structures also become smaller and occur less frequently. In fig. 4 the evolutionary history for such a core is shown. The choice of graphical representation contains the most important information in a compact form which we need in order to understand what the state of the core is. The x-axis for the graph represents the ten different instructions given by their opcode (see table 2), starting with DAT and ending with SPL. The y-axis represents the number of different tuples of the same instruction, starting by singles, pairs, triples, etc., up to 14-tuples. Higher-order tuples may occur but are not shown. The z-axis shows the frequency of the different tuples. The scale of the box is 325 instructions. By summing an instruction up over all its tuples (the y-axis) we get the total number of this particular instruction in the core. However, a state vector of the total instruction frequencies does not tell very much about the functional properties in the core. Note that the macroscopic conditions as described in fig. 4 barely change over the time of evolution whereas the microscopic dynamics has changed significantly compared to the initial core dynamics.

It turns out that we need to have R_{opr} of at least order one hundred, to allow the system to explore evolutionary paths leading to really interesting structures. The small SPL structures and the larger MOV–SPL structures mentioned above give a glimpse of what will come. Expanding R_{opr} to the core size, lowering P_{mut} to 10^{-3}, and still using a self-replicating program as an initiator, allows the system to create novelty in an even more radical way. Eleven such evolutionary histories are shown in table 4.

Every second core grown under these conditions develops a core universe dominated by a mixture of SPL and MOV instructions. In table 4 cores 2, 3, 4, 5, 8, and 9 are of this type. However, only

Table 4
Eleven evolutionary processes. The dominating instructions are listed at each sampling time (number of instructions). Only the noise sequences used to randomize the initial cores and to induce mutations and spontaneous pointers were different in these processes. All cores are jungles (recall table 2) with $R_{opr} = 3584$ and $P_{mut} = 10^{-3}$ initiated with a self-replicating program.

		1 000 000	11 000 000	21 000 000
1	MOV (DJN)		dead	
2	MOV, SPL (ADD)		SPL (MOV)	SPL (ADD, SUB, one MOV)
3	MOV, SPL		SPL (solely)	SPL (solely)
4	MOV, SPL		MOV, SPL (few SUB)	MOV, SPL (SUB)
5	MOV, SPL		DJN (+mix)	DJN (+mix)
6	JMN (MOV, DJN)		JMZ (+mix)	JMZ (+mix)
7	MOV, SUB (ADD)		dead	
8	MOV, SPL (SUB)		dead	
9	MOV, SPL (ADD, JMZ)		dead	
10	JMP, DJN (+mix)		MOV (+mix)	dead

			80 200 000
11		continuation of core #3 from time 1 000 000	SPL (JMZ, MOV)

half of these cores have the necessary organization for perturbation stability. In one kind of mature MOV–SPL structure a belt of active pointers normally sweeps through the core, altering the details of the core, but conserving the macroscopic mixture of the two instructions. This organizational structure is able to move in the core, as it copies SPL and MOV instructions outside itself. On the evolutionary path which creates these MOV–SPL organisms, the change of chemical composition is obvious. The system here goes through something we may characterize as a phase transition in the distribution of different instructions. From a fairly randomized core the system later solely consists of the MOV and the SPL instruction indeed indicating a form of self-organization.

The mature MOV–SPL structure together with part of the process leading to it is shown in plate I.

This self-organizing process also demonstrates an evolutionary path where the system finds a stable area in the rule space. The mature MOV–SPL combination is extremely stable, and almost any perturbation caused by the noisy copying part of MOV and every perturbation caused by the introduction of new pointers in this system are damped. The MOV instruction usually copies either a MOV instruction or a SPL instruction, guaranteeing the reproduction. The SPL instruction hands out pointers either to another SPL instruction or to a MOV instruction, thereby guaranteeing that the organism is kept alive.

The pathway leading to the MOV–SPL organism is only one among many possible directions the evolution can take. Another common phenomenon found in cores with a large operation radius R_{opr} is the development of a complicated transient, which eventually dies out and stabilizes after a large number of instructions. The relaxed system has relatively few active pointers, either stationary or moving in rather trivial patterns. This is probably a phenomenon similar to that found on a small lattice with Conway's CA rule Life [9]. In Life, the complicated behavior always seems to be transient.

The situations where we see an activity collapse in the core associated with a virtual extinction of the SPL instruction, combined with a very high number of MOV instructions. Such a core is very sensitive to the presence of DAT instructions since new pointers only appear as part of the noise given by P_{point}. A multiplication of DAT instructions in such a core kills all pointers. Cores 1, 7, 8, 9, and 10 in table 4 eventually all suffer this death. Another common activity collapse is seen when a core is heavily populated with jump instructions having the immediate addressing mode at the A field. These jump instructions are fixed points for the pointers. In table 4 cores 5 and 6 are such fixed point cores.

However, table 4 does not contain the full spectrum of organizational forms we meet in the VENUS jungle. It only gives a typical picture. We shall close these evolutionary field studies by inspecting some old cores grown as jungles with R_{opr} = core size and P_{mut} = 0.05. The statistical properties of these cores are found in fig. 5, where the frequency of a single instruction surrounded by other instructions is shown together with pairs of the same instruction surrounded by other instructions, of triples, etc., up to 14-tuples of the same instruction.

One interesting evolutionary product is found after 110 000 generations (~ 24 000 000 instructions), where 2859 CMP instructions (out of 3584 possible) are present. In this core, a funny discontinuous motion of small moving and "jumping" pointer structures is seen everywhere (see the core statistics in fig. 5a). Yet another evolutionary path (182 000 generations, ~ 40 000 000 instructions) led the system to develop a core with 1276 SPL instructions and only 42 MOV instructions. In this core large areas were boosted with pointers for some time, thereafter the activity died out locally, only to re-emerge at another location. These organisms indeed have an irregular pointer dynamics! Core statistics are shown in fig. 5b.

A jungle which is silent to watch after 100 000 core updates, but still has a potential to develop interesting behavior, is shown in fig. 5c. With the

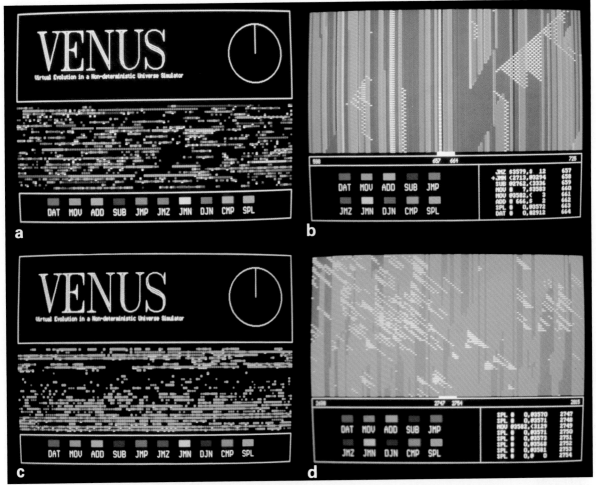

Plate I. (a) Screen dump of a core evolved under jungle conditions (recall table 4) after 1000 generations. The word at address 0 is shown in the upper left corner and the addresses increase horizontally towards the right. The last word in row 1 has the address 127, the first word in row 2 has the address 128, etc., and the word in the lower right corner is associated with the address 3583. Each instruction is shown with a particular color, following the color code given in the bottom of the picture. The black squares are not empty, but nothing has been written to these addresses since the screen was cleared approximately 100 core updates ago. This means that only the core addresses which have been active at least once over the last 100 generations are colored and hereby show which of the ten instructions these addresses are occupied by. Some of the squares have a white underscore, which indicates the presence of a pointer awaiting execution at that address for the next core update. An important characteristic of the core at this evolutionary stage is the builtup of many successive core addresses with identical instructions. This indicates that the system is somewhere on the transition between the second and third epoch as discussed in connection with fig. 6. (b) Time trace (CA view) of a small part of 128 addresses (here addresses: 598–725) of the same core at 70 consecutive generations later. Again, each instruction has its own color and the underscores indicate pointers (color code for the instructions in the lower left). In this mode all addresses are shown independent of them being active or not. Time progresses downward from the top. The color changes indicate the spatio-temporal re-organization of the instructions in the core. In the lower right we can see a zoom of the detailed instruction activity at 8 addresses (here addresses: 657–664) at the current generation. Note the pointer at address 658. (c) The same core at generation 5400. Here we have one of the fully developed MOV–SPL structures. The whole core consists almost solely of MOV and SPL instructions. As in (a) only addresses which have been active during the last 100 core updates are colored. Note the existence of something we may call a belt of activity where most of the pointers are located. The viability of this structure stems from a cooperation between the MOV and the SPL instructions. The MOV instruction with very high probability either copies another MOV instruction or a SPL instruction, guarantying the reproduction. The SPL instruction hands out pointers either to another SPL instruction or to a MOV instruction, hereby guarantying that the structure is kept alive. (d) Time trace (CA view) of addresses 2688–2815. We see how the microscopic spatial organization of the instructions change radically as time passes. Even though the system is changed continuously on the micro level it keeps its macroscopic properties; the MOV–SPL cooperation. The core is now in the fourth evolutionary stage (see also fig. 6). The erratic pointer patterns are caused by a finite execution queue, $L = 220$, combined with a very high pointer production rate due to many SPL instructions. Many pointers updated by the virtual machine cannot get into the floated execution queue and must therefore die.

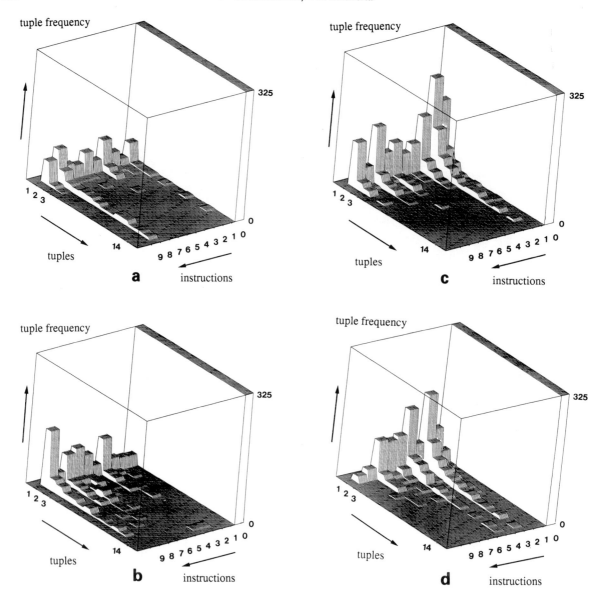

Fig. 5. Histograms telling how the frequency of the different instructions in the core has changed after approximately one hundred thousand generations. All cores shown originated from jungles with R_{opr} = core size and $P_{mut} = 0.05$. They are all on the same scale; box corresponding to 325 instructions. Compare with the randomized core in fig. 4. In (a) 2859 CMP instructions are developed after 110 000 generations. In this core, a funny discontinuous motion of small moving and "jumping" pointer structures is seen everywhere. In (b) the jungle developed 1276 SPL instructions after 182 000 core updates. In this core large areas are boosted with pointers for some time, thereafter the activity died out locally, only to re-emerge at another location. (c) Shows a core with 1471 MOV instructions. With that many MOV instructions one must expect that the core is still developing after 100 000 generations. Almost all pointers are active in this core (204 out of 220 possible). In (d) the core is floated with 1323 ADD instructions after 145 000 generations. The current activity in this core is very low, since only 78 of the 220 possible pointers are present. However, this core earlier had a very active period with all 220 pointers in use. This kind of transient behavior is further discussed in the text.

high number of MOV instructions we must expect a change at some point. If the DAT instructions are amplified further the system has a high probability of dying.

In another of the runs where we experience an activity collapse in our system, the transient core, as in table 4 (5), at one point had a high number of active MOV and SPL instructions. The core activity eventually collapsed to a silent state where only 78 of the 220 possible pointers are active. The core has at this time, after 145 000 generations (~32 000 000 instructions), only 15 SPL instructions left in the whole core, which is totally floated with 1323 ADD instructions (core statistics in fig. 5d). A similar transient behavior results in 2654 JMZ instructions after 30 000 000 instructions. In this core only 60 pointers are active. In these silent cores almost all the remaining pointers are trapped at fixed points, i.e. at one of the jump instructions.

4. Process properties

From the systematic simulations we have performed with VENUS, we can make some statements relating our processes to what we know about biological evolution. These conclusions are stated as *postulates* which we elaborate as we proceed:

(i) A certain flow of energy and resources is an imperative.

Unless we pump the system, the second law of thermodynamics prevents any build up of negentropic structures. Without energy no life.

(ii) Local fluctuations in the chemical composition seems to be of major importance to the emergence and evolution of interesting structures.

Not every chemical environment is able to develop life, although the fundamental chemical laws are the same everywhere. Apparently, even with optimal energetic and interactional conditions the system is not able to take off unless there exists chemical potentials built up by some other processes. Our core needs to have a heterogeneous distribution of the different instructions.

(iii) Different evolutionary conditions in general cause different evolutionary paths, although some continuity is seen for most parameter combinations. In these parameter ranges a slight change of the parameters does not seem to affect the path.

For some parameter settings, the system always seems to develop the same features independently of the detailed structure of the initial randomized core and the noise induced during the dynamics. This is the case for most of the evolutionary parameter combinations we have seen in the core. In table 3 it is only the jungle with initial structure which is heavily dependent on the details of the noise. This ties in to the next postulate.

(iv) Some evolutionary conditions make the system very sensitive to initial conditions or historical perturbations at certain phases of development.

Under certain conditions the evolutionary path is extremely sensitive to details of the initial random core and the noise.

It is not surprising that we meet this high degree of sensitivity to historical events in rich environments where the system has the energy and the resources to build up structures. This ability is needed in order to amplify microscopic importance. Instabilities in poor environments with only little energy flow do not have enough power for this amplification process. As we know from dynamical systems theory, we only meet sensitivity to initial conditions in flows with at least one direction of expansion.

The high sensitivity to microscopic fluctuations is particularly seen when a jungle is evolved with some initial structure and a large operational radius. In such a situation, we have a high sensitivity to initial conditions and noise, and the system has many attracting paths where most of them exclude one another. The major reason for this sensitivity is to be found in the mechanism for the amplification of single instructions. This instruction multiplication is a consequence of the presence of copying loops. Depending on which instruction the copying is directed towards and how many loops of this particular kind are present in the core, we see a build up of particular instructions. Since the details of these loops both depend on the details of the initial core composition and on the details of the noise, the later evolutionary stages also become very dependent on these details. This leads us directly to the next postulate.

(v) Optimal environmental conditions are not enough to ensure the evolution of cooperative structures. Chance or historical events plays a central role in the outcome of a particular process.

Every time the system re-organizes itself through an instability, the detailed properties of the new macroscopic states become very sensitive to historical events. Once the new stable state is reached the system enters another phase where it normally is more robust to perturbations. At some later point this state may become unstable due to another instability again altering the system. This is evolution through successive instabilities, and this is exactly what we see in our core under these conditions: First an exponential growth of self-replicating programs, followed by an exponential growth of instruction-copying loops, which is succeeded by an exponential growth of single identical instructions. This is then succeeded by the MOV–SPL structures (at least every second time), which themselves can survive for a long time. But these structures can still be poisoned by some other instructions, which then can be multiplied by the structure altering the structure of the core through yet another instability. Maybe this forms a MOV–SUB dominated core, which later turns into a MOV dominated core. Such a core can die if it coincidently multiplies to many DAT instructions. It can also, however, multiply one of the jump (immediate) instructions and then turn itself into a fixed point score.

Self-organizing systems of this kind indeed have altering phases of high sensitivity to detailed microscopic events.

(vi) Life did probably not emerge in the desert. The deserts were presumably populated by organisms originated elsewhere.

Obviously, jungles develop more interesting structures than deserts in our system. We have not been able to develop any really viable cooperative structures in the desert environment. We therefore believe that an adaptation to desert conditions for a jungle organism is more likely than an independent emergence of life in the desert.

(vii) An organism and the environment for the organism is an integrated system. The more primitive the life form is, the more difficult it is to distinguish between the two.

This is an important point since we are operating in the most primitive end of the scale, where the organism/environment distinction is fuzzy. As our evolutionary process proceeds, the local environment changes. The very chemical activity of the developed structures in the Coreworld changes the universe, exactly like the chemical activities do in our "real world". For instance, the oxygen in our modern atmosphere is believed to be created by photosynthetic activity of organisms previously evolved under anaerobic conditions.

The reason why the Coreworld structures do not develop into more closed organizational forms, in a geometrical sense, is probably partly due to the special geometrical properties of VENUS, which we shall return to in section 6, and probably partly due to the modest size of the system. This brings us to the next point.

(viii) A system needs to be large enough to guarantee the presence of different environments preventing the system from getting "stuck" as it develops into non-evolvable structures.

Many of the evolutionary processes in our simulator are caught in stable configurations which have a non-interesting organization and dynamics (see also next postulate). Recall for instance the cores with fixed points for pointers. If the system were large enough for compartmentalization it could afford to make different environments where some of them were not interesting by themselves. However, combinations of different organizational forms evolved under different conditions could give rise to new cooperative structures on higher levels.

(ix) Evolution of efficient organizational structures presumably sets some restrictions on the universal set of functional primitives, and on how noise is realized in the system, e.g. how different functional properties can be changed due to some external processes.

We arbitrarily picked the red-code as our instruction primitives. This is a redundant computational universal set, but we do not believe that it is in any way optimal for our purpose.

The major problem we have for the cooperative structures developed in the core is that these structures apparently are stuck and are not able to develop further into more advanced organizational forms. We may be able to develop more advanced structures even within this modest system if the noise was implemented in a different way. This is an important issue we shall return to in section 6.

(x) Cooperative structures seem to be important for the evolution of anything of viable nature.

The cooperative structures we are able to evolve in the Coreworld are different in several aspects from the living things we know in the "real world". An interesting difference is the way they reproduce and develop. For instance the mature MOV–SPL structure does not make a true copy of itself. It expands as it moves through the core, and it interacts with whatever it meets on its way. It does not have a well defined geno/pheno-type distinction, like the more advanced, modern life forms have. We may interpret the specific SPL instruction(s) and the specific MOV instruction(s) to be part of a gene pool, because they only change a little as the structure grows, whereas the actual mixture of these instructions, i.e. the sequence in which they appear in the organism and the size may be interpreted as the phenotype. We may also interpret this structure as a MOV–SPL instruction colony.

If we want to relate this proto-organism to a corresponding organizational form built from our "real world" chemistry, it could for instance be a cooperative, probably autocatalytic, chemical network which reproduces its elements through cooperative chemical reactions. Such a system also interacts with everything it meets in its environment and it also does not have a clear geno/pheno-type distinction. At a higher level the cooperative MOV–SPL structure may also relate to some of the primitive multicellular organisms: the colonies. These organisms constitute a symbiosis of different cells, where the overall bauplan for the structure apparently is a collective phenomenon.

(xi) It seems easier to evolve a "reproductive network" than a clean "genetic" system.

As described in section 3 none of the interesting self-sustaining structures has a clean genome, but they are still able to reproduce. In a reproductive network, the "template information" is part of the interaction structure.

(xii) Even a very brittle computational chemistry is able to find a stable area in its rule space, and from there, if it is rich enough, evolve highly stable and viable organizations.

The assembler-automaton simulator inherited some of the unfortunate properties of its ancestors: Both the von Neumann code and the simple cellular automaton suffer from computational brittleness. However, the brittleness of more sophisticated structures is not a problem peculiar to our assembler-automaton system. Functional continuity as we see it in modern life is a product of evolution and not a property of the underlying chemistry. If we imbed any modern biochemical pathway in a random chemical environment it will surely collapse. The individual processes and subsystems in the modern biochemical organisms are only robust because they are part of a beautifully optimized parallel system with a large error margin for most of its subsystems. We shall return to attractors for evolutionary systems in section 6.

(xiii) It may take a chemistry a long time to create an environment in which, from an evolutionary point of view, the right subset of chemical rules are active. To find the "right" area in rule space may be the crucial problem for any evolutionary process.

Obviously the wet carbon chain chemistry is the stable area in the chemical rule space in which life has emerged. In VENUS the mixture of MOV and SPL instructions form a basis on which chemical cooperation can take form. However, the current Coreworld is apparently not rich enough to allow for higher-level processes.

Finally we shall stress some features of the core in the significant parameter regime – the jungle with large R_{opr}, initiated with a self-replicating program. This system is able to evolve autonomously through several successive macroscopic states characterized by very different functional properties. (1) The first state is the randomized core with the initial seed of the self-replicating program. This seed causes the core to be populated with copying loops. Eventually the copying process is slowed down, because more and more copies of the program lose their ability to replicate in a proper manner. (2) Some of them turn into instruction-copying loops similar to the one shown in fig. 3b. As a result of that process more and more of the core is overwritten by the single instructions these loops copy. (3) Eventually this process also saturates as more and more of the loops disappear. (4) The new instruction mixture resulting from these processes may then start actively to develop, eventually actively taking over the core, as we have seen in several examples. (5) The new cooperative structure may then later evolve into something different, maybe a pure SPL structure as seen in table 4, or maybe first into a MOV–SUB–SPL structure, (6) which then later can turn into a SPL–SUB core. Other successions have even more epochs before the system eventually dies (recall the evolution process described in postulate (v)).

We cannot claim that the Coreworld exhibits a truly open-ended evolution, since each evolutionary process seems to get trapped after a number of successions. The system apparently always reaches some kind of attractor. However, one has to conclude that the system is autonomously able to evolve through a number of macroscopic states, characterized by very different functional properties. This is a step toward open-ended evolution.

The evolutionary succession process is summarized in fig. 6.

5. Evaluation of the developed functional properties

To compare the different structures developed in the VENUS cores, we have to measure their functional properties in some way. An ultimate goal is to be able to order the structures on some adaptability scale, or maybe more appropriately, on a scale of distance relative to non-interesting functional properties.

In order to develop such a measure we can adopt some basic properties required of a complexity measure. Grassberger [10, 11] argues that a complexity measure must favor: (1) a large number of interacting entities, and (2) diversity among the interacting entities. We add (3) a notion of stability to perturbations among the interacting entities.

As a first approximation we can, for a given interacting part χ of the core over a certain period of time τ, take the product of the number of interacting instructions $\mathscr{I}_{all}$, the number of different instructions $\mathscr{I}_{diff}$ minus one, and a non-brittleness index β, to obtain an evaluation number. This procedure assumes that we beforehand know which physical core addresses are sharing pointers or exchanging data, and over how

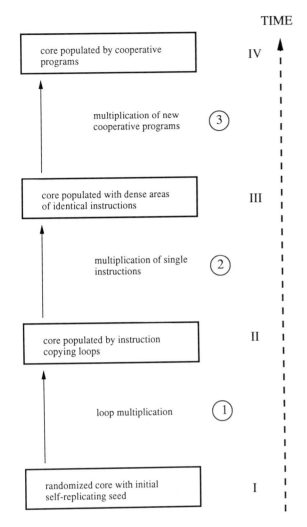

Fig. 6. A step toward open-ended evolution. Our core is able to evolve through several functionally very different macroscopic states, or epochs.

long a time it is suitable to observe the structure under investigation.

The robustness index β is here defined as the probability $P_{\neg c}$ that a given structure S conserves its functional properties during the run time τ. The index is therefore context sensitive, which means that a part of β also depends on the properties of the environment for S. It is therefore not easy to calculate β in the general case. To estimate β we have to evaluate the probability $P_{\neg h}$ that S is not hit, together with the probability $P_{\mathrm{hi}\neg c}$ that S is hit by internal noise but conserves its functional properties, together with the probability $P_{\mathrm{hx}\neg c}$ that S is hit by external interactions and conserve its functional properties: $P_{\neg h} \cup P_{\mathrm{hi}\neg c} \cup P_{\mathrm{hx}\neg c} \setminus (P_{\mathrm{hi}\neg c} \cap P_{\mathrm{hx}\neg c})$, which is equivalent to

$$\beta = P_{\neg c} = P_{\neg h} + P_{\mathrm{hi}\neg c} + P_{\mathrm{hx}\neg c} - (P_{\mathrm{hi}\neg c} P_{\mathrm{hx}\neg c}). \tag{2}$$

However, the intrinsic part of β, denoted β_{in}, also gives important information about S, and is easier to calculate. The intrinsic β_{in} is defined as

$$\beta_{\mathrm{in}} = P_{\mathrm{hi}\neg c} = P_{\mathrm{hi}} \beta_0, \tag{3}$$

where P_{hi} is the probability that S is hit by internal noise during τ, and where β_0 is defined as the number of non-fatal perturbations $N_{\mathrm{non\text{-}fatal}}$, over the number of possible perturbations N_{possible} in S. β_0 is therefore the probability that S does not change functional properties given it is hit. Since S is not able to survive unless $P_{\mathrm{hi}} \ll 1$, we can use β_0 as an upper bound for β_{in}:

$$\beta_{\mathrm{in}} \leq \beta_0 = \frac{N_{\mathrm{non\text{-}fatal}}}{N_{\mathrm{possible}}}. \tag{4}$$

Our complexity measure Γ can now be expressed by

$$\Gamma = \mathscr{I}_{\mathrm{all}}(\mathscr{I}_{\mathrm{diff}} - 1)\beta, \tag{5}$$

where $\mathscr{I}_{\mathrm{all}}$ is the number of interacting instructions, $\mathscr{I}_{\mathrm{diff}}$ is the number of different instructions among the interacting instructions and β is the non-brittleness index.

Let us use (2) through (5) to evaluate three of the structures we have discussed earlier: (i) the self-propagating structure (given in fig. 3a), (ii) the self-replicating program (given in appendix C), and (iii) the MOV–SPL structure discussed in sections 3 and 4. The results are summarized in table 5.

Table 5
The complexity measure Γ together with β_0 for three different structures.

Self-propagating instruction	0.000 (2.438 × 10^{-9})
Self-replicating program	48.160 (0.249)
Cooperative MOV–SPL structure	2400.0 (1.000)

(i) To evaluate the self-propagating instruction MOV($0,$1) as shown in fig. 3a, we need to observe the structure for a single core update ($\tau = 1$) at a single address ($\chi = 1$). Since the structure consists of only a single instruction, $\mathscr{I}_{\text{diff}} = 1$, $\Gamma_{(i)}$ immediately becomes zero. Therefore we do not need to calculate β. This holds for any single instruction being able to survive in the core like JMP($\#Y$), JMZ($\#Y,0$), etc., where Y can be any number. The intrinsic part of β for S can easily be estimated, since $P_{\text{hi}} = P_{\text{mut}}$ and $\beta_0 = [10 \times 4 \times (2R_{\text{opr}} + 1) \times 4 \times (2R_{\text{opr}} + 1)]^{-1}$. With $P_{\text{mut}} = 10^{-3}$, $R_{\text{opr}} = 800$, we get $\beta_0 = 2.438 \times 10^{-9} > \beta_{\text{in}} = 2.438 \times 10^{-12}$.

(ii) The simple self-replicating program from appendix C needs $\tau = 18$ and $\chi = 8$. Here $\mathscr{I}_{\text{diff}} = 8$ and $\mathscr{I}_{\text{all}} = 8$.

S can either change properties if it fails to make a correct copy of itself due to internal mutations or if the structure is hit by external interactions or external noise.

Let us first calculate β_0. The total number of possible instruction changes in S, given that $R_{\text{opr}} = 800$, is: $8 \times 10 \times 4 \times 1601 \times 4 \times 1601$. Since the external noise also implies the introduction of a new pointer at one of the 8 locations, we get: $N_{\text{possible}} = 8 \times 10 \times 4 \times 1601 \times 4 \times 1601 + 8$. The sum of non-fatal changes for the 8 instructions is: $N_{\text{non-fatal}} = 10 \times 4 \times 1601 \times 4 \times 1601 + 1$ (the same instruction) + 1 (the same instruction) + 2 (the same instruction and JMP($\$-1, < -3$)) + 1 (the same instruction) + 3166 (an ADD($\#;\$2$) and a SUB($\#;\$2$) where the A-field address is different from $[-8,9]$) + 1 (the same instruction) + 10 × 4 × 1601 × 4 × (1601 − 18) (the A-field address is different from $[-8,9]$) + 3 (three pointers; one at each of the DAT instructions and one at the first MOV instruction). This gives us $\beta_0 = 0.249$.

The detailed calculations leading to β as expressed in (2) are shown in appendix C. The result of these calculations yield $\beta = 0.860$ and $\beta_{\text{in}} = 0.249$.

Since β and β_0 have the same order of magnitude and since it is much simpler to calculate β_0, this measure might be good enough as an approximation for β.

The resulting estimate for the complexity measure for the simple self-replicating structure becomes $\Gamma_{(ii)} = 8 \times (8 - 1) \times 0.860 = 48.160$. Using β_0 instead yields $\Gamma_{(ii)} = 13.922$. A low score will be true for almost any human "engineered" organism, with nice, well defined properties.

(iii) We know from the simulations that the emerging MOV–SPL structures have a very complicated microscopic dynamics. We shall therefore make some simplifications in order to ease the computations for its stability to perturbations. If we extract some of the major characteristics from the developed MOV–SPL structures and condense these into a simpler version of the MOV–SPL structure we are able to obtain some estimates on β.

Assume that the MOV instructions on average copy from negative addresses towards positive addresses. To be a little more precise, assume that the A-field in the MOV instruction is taken from a uniform distribution over the relative addresses $[-R_{\text{opr}}, 0]$ and that the B-field in the MOV instruction is taken from a uniform distribution over the relative addresses $[0, R_{\text{opr}}]$. Further assume that the SPL instructions on average distribute pointers locally ($\ll R_{\text{opr}}$) with an equal probability of positive and negative addresses. Since the system has a finite pointer execution queue, the local generation of pointers results in an overall pointer movement towards positive addresses facilitated by the virtual machine. Another result of these properties is a system tendency to create connected core areas, in a pointer interaction sense, where all the system's L pointers are located. Let us assume that we only have one such area and that this area has an address extension of R_{opr}. This defines $\chi = R_{opr}$. Finally, let us assume that we have a random mixture of MOV and SPL instructions and that the number of MOV instructions equals the number of SPL instructions.

Since we have assumed that S has all the L pointers in the system, we only need to look at the internal dynamics.

We shall now compare the probability P_{by} that S creates at least one "strange" instruction Y, in a core update, and the probability P_{dy} that S erases a strange instruction in a core update (this defines $\tau = 1$). We hereby approximate the dynamics by a linear Markov chain, a birth and death process for Y. P_{by} is determined by

$$P_{by} = 1 - P_{\neg by}$$
$$= 1 - (1 - P_{mut})^{L/2}, \qquad (6)$$

where we have assumed that the MOV and SPL instructions share the L pointers. Likewise for P_{dy}, we get the expected probability for removing a single strange instruction Y,

$$P_{dy} = 1 - \left(1 - \frac{1}{R_{opr}}\right)^{L/2}. \qquad (7)$$

Inserting $R_{opr} = 800$ and $P_{mut} = 10^{-3}$, we get $P_{by} = 0.104 < P_{dy} = 0.129$. This guarantees that S is stable to all perturbations in the core: $\beta = 1$. We therefore get $\Gamma_{(iii)} = 3 \times 800 \times (2-1) \times 1 = 2400$.

The linear Markov chain is, however, only an approximation since both the birth and the death processes are state dependent. There is a positive feedback between the number of strange instructions Y and P_{by} as well as between the number of Y's and P_{dy}. This means that fluctuations above a certain size have a tendency to explode, which we also observe in the rare situations where the MOV-SPL structure develops into something else.

Note that the measure in the considered examples reflects an evolutionary arrow of time, since structures emerging later (iii) also have a higher measure than the earlier ones (ii).

McCaskill [16] has suggested another way of evaluating the functional properties of structures developed in such a system. Instead of calculating β, we could place S in a sequence of predefined environments controlled by us. An evaluation of how the structure performs in these environments could then tell us something about its adaptability or evolutionary sophistication.

6. Discussion

Introducing the simple artificial chemistry that constitutes this Coreworld created a system where *processes* are able to self-organize interesting co-operative structures like we see real world *processes* do. The key point here is apparently not the actual details of a chemistry and the physical and geometrical properties associated with such a chemistry, but that the system has a rich range of possible functional interactions between the basic elements.

The choice of the fundamental set of instructions, the red-code instructions, was arbitrary. These instructions constitute a redundant computationally universal set, which means that any computational task can be formulated within it in a number of different ways. We believe, however, that a more suitable universal instruction set, from an evolutionary point of view, could be constructed using specifications given by the kind of functional interaction primitives such a system ideally should have.

The noise is currently implemented only in the execution of a MOV instruction where the addressing mode for the A field is not immediate (#). Since the MOV instructions sometimes are eliminated by the dynamics, the noise is sometimes eliminated. In order to prevent this and also induce more variability, noise could be introduced each time an instruction writes to memory. By the construction of another instruction set one could also build in a functional ordering among the different instructions and thereby only allow noise to alter an instruction into one of its Hamming neighbors. Such an approach may actually reduce the computational brittleness in the core and thereby create a smoother fitness landscape. However, using the assembler-automata approach, as we do here, the computational brittleness is difficult to escape. But since the system is able to develop a stable composition autonomously, this may not be a major issue from a practical or constructive point of view.

A fundamental theoretical question to ask a system like this is: given a core updating map $T(\cdot)$,

$$\text{core}(t+1) = T(\text{core}(t)) \qquad (8)$$

what are the dynamical properties of the iterated map? Here the computationally universal system maps the core at time t, $\text{core}(t)$, onto itself to give the core at time $t+1$, $\text{core}(t+1)$. What kind of attractors do we find for $T(\cdot)$? We are currently trying to understand the structure of the dynamics for (8) and related iterated maps.

Our system does not obey any conservation law known from physics. Neither energy nor matter is conserved in VENUS. In order to implement such conservation laws in the Coreworld we probably need to consider empty core addresses and associate address changes with energy transformations in a more specific way than we do now.

The geometry an observer experiences located at one of the core addresses is not easy to define. Firstly the memory addresses constitute a cyclic one-dimensional array along which the virtual machine advances pointers by one address at each core update. Therefore this part of the dynamics is one dimensional. However, each instruction potentially "sees" all addresses within R_{opr} as having the distance one relative to itself, although at most two addresses within R_{opr} are actually the immediate interaction neighbors for this instruction. The connection geometry for the instruction interactions can thus be expressed by a directed graph. The structure of this graph implies a multidimensionality for the instruction interactions. Further, this directed graph is a function of the core dynamics, which implies a feedback between the dynamics and the interaction geometry in the system. Finally, each instruction also "sees" computational resources within R_{res} as having distance one to itself. With R_{res} small the directed connectivity graph for the resources implies that the computational resources can be viewed as located inside a two-dimensional torus. To summarize, the different means of communication in the core defines different geometries.

Since the pointer dynamics dominates the core activity and since it is by default controlled by the virtual machine, the essential dynamics in the current Coreworld is one-dimensional. This has a serious disadvantage. For instance it prevents two different propagating structures from "passing" each other. When such structures meet they immediately interact changing both of them. This property can partly be eliminated by going to a two- or higher-dimensional addressing array.

In order to obtain a natural compartmentalization in the core we may be able to define an instruction or instruction combination which protects a certain set of addresses from being written to by instructions outside this set. By only allowing interactions within such a set or allowing a reduced environmental interaction, and at the same time providing a feedback between the core dynamics and the formation and deterioration of such sets, the formation of "membrane-like" structures can develop.

A variety of the above discussed possibilities are currently being tested in a new and more general core simulator, we have named EARTH.

This work also relates to *computer viruses* in the following sense: what kind of structures can live in a computer environment? Computer viruses as they are known today, are human-engineered "smart" programs, which either via the network or via transported disks are able to invade or infect other computers and via a replication process occupy available memory and available computational resources. Until now, no known computer viruses have truly adapted to new computer environments by variation combined with selection or some other form of learning. All the properties these known viruses have are given to them by their creators.

The environment in which current computer viruses "live" is noiseless, in order to facilitate the accurate execution of any machine code program the programmer wants executed. In this respect, a "real" computer core is different from our VENUS core. However, the competition for computational resources between any illegal computer virus program in the core and any of the legal user's program is very similar to the competition we see between different programs in our core.

From any computer user's point of view it is of course desirable to be able to kill any unauthorized pointer executing in the computer. However, it is very difficult a priori to distinguish between legal and illegal processes. Put another way: it is very difficult to build an immune system for a computer, because such a system needs to be able to discriminate itself from nonself, and afterwards it needs to be able to kill the intruders. How this actually is going on in modern biological systems is not yet known in detail [17].

Besides the idea of internally encrypting all the legal processes in a computer, and in this way being able to discriminate illegal processes by the lack of the correct encryption, it may actually be possible to distinguish between wanted and nonwanted processes in the computer memory by looking at the qualitatively different trajectories a virus program and a "normal" program have in the core. It should be possible to learn the computational patterns of "good" and "bad" programs using similar approaches to those described in the previous sections.

In a few years it may well be the case (as in biology) that there is an "arms race" between the computer's immune system and parasites wanting access to the computational resources "inside".

7. Conclusion

We shall not try to judge whether the cooperative structures we have evolved in the core are alive or not. However, it seems clear that the simple Coreworld is a good vehicle for studying fundamental properties of emergent computation and open-ended evolution.

Despite the brittleness of the individual instructions and the modest core size, our system is indeed able to evolve stable cooperative structures. The system develops into a computational robust system through a number of successive macroscopic core epochs, each characterized by different functional properties.

The interplay between chance and necessity changes significantly for different parameters in our chemistry. In some parameter regimes the evolutionary path seems quite deterministic, whereas other regimes support multiple exclusive attractors with very delicate borders between the basins of attraction. Historical events are of major importance in the selection of evolutionary path at certain phases for these processes.

We have also developed a crude adaptability or complexity measure with which we are able to evaluate the different functional properties emerging and evolving in VENUS. This measure also reflects an evolutionary arrow of time.

Finally the approach we have used to understand the computational properties of our core is discussed in relation to computer viruses. This approach may also be appropriate in the future construction of immune systems for computers.

Acknowledgements

We would like to thank Doyne Farmer and Chris Langton, with whom we have discussed some of the later parts of the development of VENUS and the interpretation of some of our results, and who also have criticized earlier versions of this paper. Peter Grassberger is acknowledged for a number of discussions on how to measure complexity, and John McCaskill and Walter Fontana are acknowledged for discussions on the evaluation of functional properties in artificial chemistries. Finally Y.C. Lee is acknowledged for discussions on computer viruses.

Appendix A

Recall that each address x in the core is associated with a certain amount of computational resources, $r(x, t)$, which is measured in fractions of one execution, one *exec*, which is equivalent to what is used in the execution for a single instruction. An instruction with a pointer can therefore only be executed if its neighborhood has computational resources equivalent to at least one *exec*. The resource neighborhood is defined by a radius R_{res}. The number of addresses each instruction can obtain resources from is therefore $2R_{res} + 1$. The simulator executes instructions and hereby removes resources in a sequence determined by the order in which the pointers occur in the execution queue. Resources are renewed by an amount Δr at each core update. However, the resources are never allowed to exceed a maximum resource level, r_{max}. r_{max} is a global parameter always smaller than one *exec*. Δr, $\leq r_{max}$, is also a global parameter.

Besides the limited computational resources associated with each core address, the system also has a limited number of active pointers. At each core update at most L instructions are executed.

The operation radius R_{opr} defines how far away each instruction is allowed to access and alter data relative to its own address. The operation radius R_{opr} and the resource radius R_{res} allow us to achieve locality in our core.

Let x_j refer to the core location of the jth pointer in the execution queue, and let $N(x_j)$ be the resource neighborhood for the jth pointer. $\alpha(x_j)$ is then defined as

$$\alpha(x_j) = \left(\sum_{k=x_j-R_{res}}^{x_j+R_{res}} r_{j-1}(k, t) - 1 \right) \Big/ \\ \times \sum_{k=x_j-R_{res}}^{x_j+R_{res}} r_{j-1}(k, t) \quad (A.1)$$

$$\text{for } \sum_{k=x_j-R_{res}}^{x_j+R_{res}} r_{j-1}(k, t) \geq 1$$

and gives the fraction by which the computational resources at each address in $N(x_j)$ is decremented. $r_j(x, t)$ can now be defined as the jth iteration over the pointer execution queue

$$r_j(x, t) = \alpha(x_j) r_{j-1}(x, t),$$

if $x \in N(x_j)$

and if $\sum_{k=x_j-R_{res}}^{x_j+R_{res}} r_{j-1}(k, t) \geq 1$.

Else $r_j(x, t) = r_{j-1}(x, t)$. \quad (A.2)

$r_0(x, t)$ is defined as $r(x, t-1)$. In the case where

$$\sum_{k=x_j-R_{res}}^{x_j+R_{res}} r_{j-1}(k, t) < 1$$

the j's pointer disappears and the instruction will not be executed.

Note that the execution queue for each t is only passed through once giving the $r(x, t)$'s for all x.

The resource level at any address x at any time t is therefore determined by

$$r(x, t) = \min[r_{max}, r_L(x, t-1) + \Delta r]. \quad (A.3)$$

Appendix B

Listing of the self-replicating program MICE, originally written by Chip Wendell [3]. The program executes the second instruction first.

DAT #7
MOV #7, $ − 1
MOV @ − 2, < 5
DJN $ − 1, $3
SPL @3
ADD #417, $2
JMZ $ − 5, $ − 6
DAT $714

Appendix C

(1) $P_{hx\neg c}$: First we note that $P_{hx\neg c} = P_{\neg c|hx} P_{hx} = \beta_0 P_{hx}$. The probability P_{hx} that S is hit by external perturbations, of course, depends on the rest of the core environment. We assume that the instructions are homogeneously distributed in the environment S can see. We also assume that all external perturbations are possible. This means that the external MOV instruction potentially can interact with any of the fields; opcode, A-field, and B-field. S can be perturbed with pointers from the following instructions: SPL, JMP, JMZ, JMN, and DJN. The ADD and the SUB instructions are only able to perturb the B field of the instructions in S. The DAT and the CMP instruction does not have any effect on S. Since the relative number of non-fatal perturbations is almost identical for field interactions and pointer interactions, we can treat all interactions in the same way.

If we assume that $\chi \ll R_{opr}$ and there are $L/2$ active pointers in the core, the probability P_{hx} that S is hit by external interactions during τ core updates is

$$P_{hx} = 1 - P_{\neg hx}^{\#pointers(R_{opr})}, \qquad (C.1)$$

where $P_{\neg hx}$ is the probability that S is not hit by external interactions during τ and #pointers- (R_{opr}) is the expected number of external pointers within two operation radians of S.

Denoting the core size by C, we get

$$P_{\neg hx} = \frac{2R_{opr} + 1 - \chi}{2R_{opr} + 1}, \qquad (C.2)$$

and

$$\#pointers(R_{opr}) = \left(\frac{L}{2} - 1\right)\frac{2R_{opr}}{C}\frac{8}{10}. \qquad (C.3)$$

Inserting $C = 3584$, $L = 220$, $R_{opr} = 800$, $\tau = 18$, and $\chi = 8$, we get $P_{hx} = 0.1773$.

Under the given assumptions the probability that S does not change during τ if it is hit by external perturbations can now be estimated: $P_{hx\neg c} = 0.0441$.

(2) $P_{hi\neg c}$ is easier to calculate. Only the second MOV instruction in the program can mutate. Since this MOV instruction is executed 7 times during τ, we get a probability of $(1 - P_{mut})^7$ that no mutation occurs during the run time. One minus this probability has to be multiplied by the probability $P_{hx\neg c}$ that a mutation is non-lethal for the new structure, which we know from the previous calculation. With $P_{mut} = 10^{-3}$ we get the following estimate: $P_{hi\neg c} = 0.007 \times 0.0441 = 0.309 \times 10^{-4}$.

(3) $P_{\neg h}$: The probability that S is not hit by any interactions or noise during τ is given by

$$P_{\neg h} = 1 - (P_{hx} + P_{hi}) + P_{hx}P_{hi} = 0.816. \qquad (C.4)$$

Now inserting $P_{hx\neg c}$, $P_{hi\neg c}$, and $P_{\neg h}$ into (2) we obtain $P_{\neg c} = \beta = 0.860$.

References

[1] R. Dawkins, The Evolution of Evolvability, in: Artificial Life, SFI Studies in the Sciences of Complexity, Vol. III, ed. C. Langton (Addison–Wesley, Reading, MA, 1989) pp. 201–220.
[2] A. Dewdney, In the game called Core War hostile programs engage in the battle of bits, Sci. Am. (May 1984) 15–19.

[3] A. Dewdney, A program called MICE nibbles its way to victory at the first Core-War tournament, Sci. Am. (January 1987) 8–11.
[4] M. Eigen, Self-organization of matter and evolution of biological macromolecules, Naturwissenschaften 58 (1971) 465.
[5] M. Eigen and P. Schuster, The hypercycle – A principle of natural self-organization (Springer, Heidelberg, 1979).
[6] J.D. Farmer and N.H. Packard, Evolution, games, and learning: models for adaptation in machines and nature, Physica D 22 (1986) vii–xii.
[7] J.D. Farmer, S.A. Kauffman and N.H. Packard, Autocatalytic replication polymers, Physica D 22 (1986) 50.
[8] W. Fontana, Digital chemistry, preprint (1989).
[9] M. Gardner, The fantastic combinations of John Conway's new solitaire game "Life", Sci. Am. 223 (1970) 120–123.
[10] P. Grassberger, Toward a quantitative theory of self-generated complexity, Int. J. Theor. Phys. 25 (1986) 907.
[11] P. Grassberger, Complexity and forecasting in dynamical systems, preprint (1988).
[12] Note on the Core Wars simulator, International Core Wars Society, 8619 Wassall, Wichita, KS 67210-1934, USA (1986).
[13] C. Langton, Artificial life, in: Artificial Life, SFI Studies in the Sciences of Complexity, Vol. VI, ed. C. Langton (Addison–Wesley, Reading, MA, 1989) pp. 1–47.
[14] S. Kauffman, Autocatalytic sets of proteins, J. Theor. Biol. 119 (1986) 1–24.
[15] J. McCaskill, A minimal integrated recognition-processing model for macromolecular evolution, preprint (1988).
[16] N.H. Packard, Intrinsic adaptation in a simple model for evolution, in: Artificial Life, SFI Studies in the Sciences of Complexity, Vol. VI, ed. C. Langton (Addison–Wesley, Reading, MA, 1989) pp. 141–155, and some later developments of the code on Evolving bugs in an artificial ecology.
[17] A. Perelson, Theoretical Immuniology, I, II, SFI Studies in the Sciences of Complexity, Vols. II, III, ed. A. Perelson (Addison–Wesley, Reading, MA 1988).
[18] S. Rasmussen, Aspects of instabilities and self-organizing processes, Ph.D. Thesis, Physics Laboratory III, The Technical University of Denmark (1985) (in Danish).
[19] S. Rasmussen, Toward a quantitative theory of the origin of life, in: Artificial Life, SFI Studies in the Sciences of Complexity, Vol. VI, ed. C. Langton (Addison–Wesley, Reading, MA, 1989) pp. 79–104.
[20] S. Rasmussen, B. Bollobás and E. Mosekilde, Elements of a quantitative theory of prebiotic evolution, LA-UR-89-1881, J. Theor. Biol., to appear.

REQUIREMENTS FOR EVOLVABILITY IN COMPLEX SYSTEMS: ORDERLY DYNAMICS AND FROZEN COMPONENTS

Stuart A. KAUFFMAN

Department of Biochemistry and Biophysics, School of Medicine, University of Pennsylvania, Philadelphia, PA 19104-6059, USA and Santa Fe Institute, 1120 Canyon Road, Santa Fe, NM 87501, USA

This article discusses the requirements for evolvability in complex systems, using random Boolean networks as a canonical example. The conditions for crystallization of orderly behavior in such networks are specified. Most critical is the emergence of a "frozen component" of the binary variables, in which some variables are frozen in the active or inactive state. Such frozen components across a Boolean network leave behind *functionally isolated islands* which are not frozen.

Adaptive evolution or learning in such networks via near mutant variants depends upon the structure of the corresponding "fitness landscape". Such landscapes may be smooth and single peaked, or highly rugged. Networks with frozen components tend to adapt on smoother landscapes than those with no frozen component. In coevolving systems, fitness landscapes themselves deform due to coupling between coevolving partners. Conditions for optimal coevolution may include tuning of landscape structure for the emergence of frozen components among the coadapting entities in the system.

1. Introduction

The dynamical behavior of complex information processing systems, and how those behaviors may be improved by natural selection, or other learning or optimizing processes, are issues of fundamental importance in biology, psychology, economics, and, not implausibly, in international relations and cultural history. Biological evolution is perhaps the foremost example. No serious scientist doubts that life arose from non-life as some process of increasingly complex organization of matter and energy. A billion years later we confront organisms that have evolved from simple precursors, that unfold in their own intricate ontogenies, that sense their worlds, categorize the states of those worlds with respect to appropriate responses, and in their interactions form complex ecologies whose members coadapt more or less successfully over ecological and evolutionary time scales. We suppose, probably rightly, that Darwin's mechanism, natural selection, has been fundamental to this astonishing story. We are aware that, for evolution to "work", there must be entities which in some general sense reproduce, but do so with some chance of variation. That is, there must be both heritability and variation. Thereafter, Darwin argues, the differences will lead to differential success, culling out the fitter and leaving behind the less fit.

But, for at least two reasons, Darwin's insight is only part of the story. First, in emphasizing the role of natural selection as the Blind Watchmaker, Darwin and his intellectual heritors have almost come to imply that without selection there would be no order whatsoever. It is this view which sees evolution as profoundly historically contingent; a story of the accidental occurrence of useful variations accumulated by selection's sifting: evolution as the Tinkerer. But secondly, in telling us that natural selection would cull the fitter variants, Darwin has implicitly assumed that successive cullings would be able to *successively accumulate* useful variations. This assumption amounts to presuming what I shall call *evolvability*. Its assumption is essential to a view of evolution as a tinkerer which cobbles together ad hoc but remarkable solutions to design problems. Yet "evolvability" is not itself a self-evident property in complex systems. Therefore, we must wonder what the construction requirements are which permit evolvability, and whether selection itself can achieve such a system.

Consider the familiar example of a standard computer program on a sequential von Neumann universal Turing machine. If one were to randomly exchange the order of the instructions in a program, the typical consequence would be catastrophic change in the computation performed.

Try to formulate the problem of evolving a *minimal program* to carry out some specified computation on a universal Turing machine. The idea of a minimal program is to encode the program in the shortest possible set of instructions, and perhaps initial conditions, in order to carry out the desired computation. The length of such a minimal program would define the *algorithmic complexity* of the computation. Ascertainment that a given putative minimal program is actually minimal, however, cannot in general be carried out. Ignore for the moment the problem of ascertainment, and consider the following: Is the minimal program itself likely to be evolvable? That is, does one imagine that a sequence of minimal alterations in highly compact computer codes could lead from a code which did not carry out the desired computation to one which did?

I do not know the answer; nevertheless, it is instructive to characterize the obstacles. Doing so helps define what one might mean by "evolvability". In order to evolve across the space of programs to achieve a given compact code to carry out a specified computation, we must first be able to ascertain that any given program actually carries out the desired computation. Think of the computation as the "phenotype", and the program as the "genotype". For many programs, it is well known that there is no short cut to "seeing the computation" carried out beyond running the program and observing what it "does", that is, observing its "phenotype". Thus, to evolve our desired program, we must have a process which allows candidate programs to exhibit their phenotypes, and then a process which chooses variant programs and "evolves" towards the target minimal compact program across some defined program space. Since programs, and if need be their input data, can be represented as binary strings, we can represent the space of programs in some high-dimensional Boolean hyperspace. Each vertex is then a binary string, and evolution occurs across this space to or toward the desired minimal target program.

Immediately we find two problems. First, can we define a "figure of merit" which characterizes the computation carried out by an arbitrary program which can be used to compare how "close" the phenotype of the current program is to that of the desired target program? This requirement is important since, if we wish to evolve from an arbitrary program to one which computes our desired function, we need to know if alterations in the initial program bring the program closer or further from the desired target program. The distribution of this figure of merit, or to a biologist, "fitness" across the space of programs defines the "fitness landscape" governing the evolutionary search process. Such a fitness landscape may be smooth and single peaked, with the peak corresponding to the desired minimal target program, or may be very rugged and multipeaked. In the latter case, typical of complex combinatorial optimization problems, any local evolutionary search process is likely to become trapped on local peaks. In general, in such tasks, attainment of the global optimum is an *NP*-complete problem, and an evolutionary search will not attain the global optimum in reasonable time. Thus, the second problem with respect to evolvability of programs relates to how rugged and multipeaked the fitness landscape is. The answers are not known, but the intuition is clear. The more compact the code becomes, the more *violently* the computation carried out by the code changes at each minimal alteration of the code. That is, long codes may have a variety of internal sources of redundancy which allows small changes in the code to lead to small changes in the computation. By definition, a minimal program is devoid of such redundancy. Thus, inefficient redundant codes may occupy a landscape which is relatively smooth and highly correlated in the sense that nearby programs have nearly the same fitness by carrying out similar computations. But as the programs become shorter, small changes in the programs induce ever more pronounced changes in the phenotypes. That is, the landscapes become ever more *rugged and uncorrelated*. In the limit where fitness landscapes are entirely uncorrelated, such that the fitness of "1-mutant" neighbors in the space are random with respect to one another, it is obvious that the fitness of a neighbor carries no information about which directions are good directions to move across the space in an evolutionary search for global, or at least good, optima. Evolution across fully uncorrelated landscapes amounts to an entirely random search process where the landscape itself provides no information about where to search [29]. In short, since minimal programs almost surely "live on" fully uncorrelated landscapes in program space, one comes strongly to suspect that *minimal programs are not themselves evolvable*.

Analysis of the conditions of evolvability, therefore, requires understanding: (1) What kinds of systems "live on" what kinds of "fitness landscapes"; (2) what kinds of fitness landscapes are "optimal" for adaptive evolution; and (3) whether there may be selective or other adaptive processes in complex systems which might "tune" (1) and (2) to achieve systems which are able to evolve well.

Organisms are the paradigm examples of complex systems which patently have evolved, hence now do fulfill the requirements of evolvability. Despite our fascination with sequential algorithms, organisms are more adequately characterized as complex *parallel-processing* dynamical systems. A single example suffices to make this point. From among about 100 000 distinct "structural" genes which code for protein products in higher metazoans such as humans, some genes' products *regulate* the activity of other genes in a complex regulatory web which I shall call the genomic regulatory network. Different cell types in an organism differ from one another because different subsets of genes are active in the different cell types. During ontogeny from the zygote, genes act in parallel, synthesizing their products, and mutually regulating one another's synthetic activities. Cell differentiation, the production of diverse cell types from the initial zygote, is an expression of parallel processing among the 100 000 genes in each cell lineage. Thus the metaphor of a "developmental program" encoded by the DNA and controlling ontogeny is more adequately understood as pointing to a parallel-processing genomic dynamical system whose dynamical behavior unfolds in ontogeny. Understanding development from the zygote, and the evolution of development, hence the *evolvability* of ontogeny, requires understanding how such parallel-processing dynamical systems might give rise to an organism, and be molded by mutation and selection.

Other adaptive features of organisms, ranging from neural networks to the anti-idiotype network in the immune system, are quite clearly examples of parallel-processing networks whose dynamical behavior and changes with learning, or with antigen exposure, constitute the "system" and exhibit its evolvability.

The hint that organisms are to be pictured as parallel-processing systems leads me to focus the remaining discussion on the behavior of such networks, and the conditions for their evolvability. Central to this is the question of whether even random, disordered, parallel-processing networks can exhibit sufficiently ordered behavior to provide the raw material upon which natural selection might successfully act. This discussion therefore serves as an introduction to the following topics:

(1) What kinds of random, disordered, parallel-processing networks exhibit strongly self-organized behavior which might play a role in biology and elsewhere?

(2) What kinds of "fitness landscapes" do such systems inhabit?

(3) What features of landscapes abet adaptive evolution?

(4) Might there be selective forces which "tune" the structures of fitness landscapes by tuning the structure of organisms, and tune the couplings

among fitness landscapes, such that coevolutionary systems of coupled adapting organisms coevolve "well"?

In section 2 I discuss random Boolean networks as models of disordered dynamical systems. We will find that such networks can exhibit powerfully ordered dynamics. In the third section I discuss why such systems exhibit order. It is due to the percolation of a "frozen" component across the network. In such a component, the binary elements fall to fixed active or inactive states. The frozen component breaks the system into a percolating frozen region and isolated islands which continue to change, but cannot communicate with one another. In section 4 I discuss the basic features of evolution on rugged fitness landscapes and the bearing of landscape structure on the evolvability of Boolean networks. Networks with frozen components evolve on more correlated landscapes than those without frozen components. In section 4 I also briefly discuss *coevolution* where the adaptive moves by one partner *deforms* the fitness landscapes of its coevolutionary partners. The analogue of frozen components remerges in this coevolutionary context. In addition, selective forces acting on individual partners can lead them to tune the structure of their own fitness landscapes and couplings to other landscapes to increase their own sustained fitness, and that these same adaptive moves "tune" the entire coevolutionary system towards an optimal structure where all partners coevolve "well". Thus, we have a hint that selection itself may in principle achieve systems which have optimized evolvability.

2. Introducing Boolean dynamical networks

I have now asked what kinds of complex disordered dynamical systems might exhibit sufficient order for selection to have at least a plausible starting place, and whether such systems adapt on well-correlated landscapes. This is an extremely large problem which goes to the core of the ways in which complex systems must be constructed such that improvements by accumulation of improved variants by mutation and selection, or any analogue of mutation and selection, can take place. In this section I shall confine this question to one coherent approach by asking what kinds of "discrete" dynamical systems whose variables are limited to two alternative states, "on" and "off", adapt on well-correlated landscapes.

2.1. The state space dynamics of autonomous Boolean networks

Boolean networks are comprised of binary, "on/off" variables. A network has N such variables. Each variable is *regulated* by some of the variables in the network which serve as its "inputs". The *dynamical behavior* of each variable, whether it will be active (1) or inactive (0) at the next moment, is governed by a *logical switching rule*, or *Boolean function*. The Boolean function specifies the activity of the regulated variable at the next moment for each of the possible combinations of current activities of the input variables.

Let K stand for the number of input variables regulating a given binary element. Since each element can be active or inactive, the number of combinations of states of the K inputs is just 2^K. For each of these combinations, a specific Boolean function must specify whether the regulated element is active or inactive. Since there are two choices for each combination of states of the K inputs, the total number of Boolean functions, F, of K inputs is

$$F = 2^{2^K}. \tag{1}$$

An autonomous Boolean network is specified by choosing for each binary element which K elements will serve as its regulatory inputs, and assigning to each binary element one of the possible Boolean functions of K inputs. If the network has no inputs from "outside" the system, it is considered to be "autonomous". Its behavior depends upon itself alone.

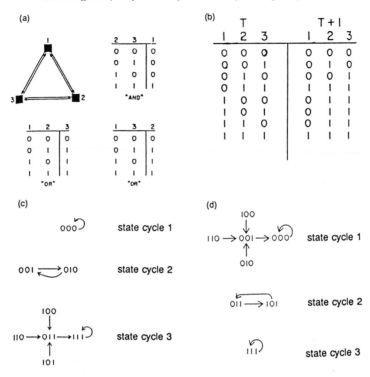

Fig. 1. (a) The wiring diagram in a Boolean network with three binary elements, 1, 2, 3, each an input to the other two. One element is governed by the Boolean AND function, the other two by the OR function. (b) The Boolean rules of (a) rewritten showing for all $2^3 = 8$ states of the Boolean network at time T, the activity assumed by each element at the next time moment, $T + 1$. Read from left to right this figure shows, for each state, its successor state. (c) The state transition graph, or behavior field, of the autonomous Boolean network in (a) and (b), obtained by showing state transitions to successor states, (b), as connected by arrows, (c). This system has three state cycles. Two are steady states (000) and (111), the third is a cycle with two states. Note that (111) is stable to all single Hamming unit perturbations, e.g. to (110), (101), or (011), while (000) is unstable to all such perturbations. (d) Effects of mutating rule of element 2 from OR to AND. From *Origins of Order: Self Organization and Selection in Evolution* by S.A. Kauffman [32]. Copyright © 1990 by Oxford University Press, Inc. Reprinted by permission.

Fig. 1 shows a Boolean network with three elements, 1, 2, and 3. Each receives inputs from the other two. 1 is governed by the AND function, 2 is governed by the OR function, and 3 is governed by the OR function. The simplest class of Boolean networks are *synchronous*. All elements update their activities at the same moment. To do so each element examines the activities of its K inputs, consults its Boolean function, and assumes the prescribed next state of activity. This is summarized in fig. 1b. Here I have rewritten the Boolean rules. Each of the 2^3 possible combinations of activities of the three elements corresponds to one *state* of the entire network. Each state at one moment causes all the elements to assess the values of their regulatory inputs, and, at a clocked moment, assume the proper next activity. Thus, at each moment, the system passes from a state to a unique successor state. Over time the system passes through a succession of states, called a *trajectory*, as shown in fig. 1c.

Since there is a finite number of states, any autonomous Boolean network must eventually reenter a state previously encountered; thereafter, since the system is deterministic and must always pass from a state to the same successor state, the system will *cycle repeatedly* around this *state cycle*. These state cycles are the *dynamical attractors* of the Boolean network. The set of states flowing into one state cycle or lying on it constitute the *basin of attraction* of that state cycle. The *length* of a state cycle is the number of states on the cycle, and can range from 1 for a steady state to 2^N. The simple Boolean network in fig. 1a has three state cycle attractors. Left to its own, the system eventually settles down to one of its state cycle attractors and remains there.

The following properties of autonomous Boolean networks are of immediate interest:

(1) The number of states around a state cycle, called its length. The length can range from one state for a steady state to 2^N states.

(2) The number of alternative state cycles. At least one must exist, but a maximum of 2^N might occur. These are the permanent asymptotic alternative behaviors of the entire system. The sizes of the basins of attraction drained by the state cycle attractors.

(3) The stability of attractors to minimal perturbation, flipping any single element to the opposite activity value.

(4) The changes in dynamical attractors and basins of attraction due to mutations in the connections or Boolean rules. These changes will underlie the character of the *adaptive landscape* upon which such Boolean networks evolve by mutation to the structure and rules of the system.

Boolean networks are discrete dynamical systems. The elements are either active or inactive. The major difference between a continuous and a discrete deterministic dynamical system is that two trajectories in a discrete system can merge. To be concrete, fig. 1c shows several instances where more than one state converge upon the same successor state.

2.2. The NK Boolean network ensemble: conditions for orderly dynamics

The behaviors of Boolean networks as a function of N, the number of elements in the net, K, the average number of inputs to each element in the net, and "P", which will measure particular biases on the set of all $(2^2)^K$ Boolean functions used in the net are summarized here.

In order to assess the expected influence of these parameters, I have analyzed the *typical behavior* of members of the entire *ensemble* of Boolean networks specified by any values of the parameters N, K, and P. The first results I describe allow no bias in the choice of Boolean functions; hence N and K are the only parameters. I further simplify and require that each binary element have exactly K inputs.

In order to analyze the typical behavior of Boolean networks with N elements each receiving K inputs, it is necessary to sample at random from the ensemble of all such systems, examine their behaviors, and accumulate statistics. Numerical simulations to accomplish this, therefore, construct exemplars of the ensemble entirely at random. Thus, the K inputs to each element are chosen at random, then fixed, and the Boolean function assigned to each element is chosen at random, then fixed. The resulting network is a specific member of the ensemble of NK networks.

NK Boolean networks are examples of *strongly disordered systems* [4, 6, 7, 12, 14, 15, 20–22, 24, 25, 27, 42, 43]. Both the connections and Boolean functions are assigned at random. Were any such network examined, its structure would be a complex tangle of interactions, or "input wires", between the N components, and the rule characterizing the behavior of one element will typically differ from its neighbors in the network. Such Boolean networks are spiritually similar to spin glasses, and the NK family of landscapes, described elsewhere [30, 31, 33].

2.3. Major features of the behavior of random Boolean networks

Table 1 summarizes the salient features for the following cases: $K = N$, $K > 5$, $K = 2$, $K = 1$.

(1) $K = N$. In these networks, each element receives inputs from all elements. Hence there is only one "wiring diagram" among the elements. Each element is assigned at random one of the $(2^2)^N$ Boolean functions. In these *maximally disordered systems*, the successor to each state is a

Table 1
Properties of random Boolean nets for different values of K [a]

	State cycle length	Number of state cycle attractors	Homeostatic stability	Reachability among cycles after perturbation
$K = N$	$0.5 \times 2^{N/2}$	N/e	low	high
$K > 5$	0.5×2^{BN} ($B > 1$)	$\sim \frac{1}{2} N \log[(1/2 \pm \alpha)^{-1}]$	low	high
$K = 1$	$\exp(\frac{1}{8} \log^2 N)$	$(2/\sqrt{e})^{N[1 + \mathcal{O}(1)]}$	low	high
$K = 2$	$\sqrt{N}$	$\sqrt{N}$	high	low

[a] Column 1: state cycle length is the median number of states on a state cycle. Column 2: number of state cycle attractors in behavior of one net. ($\alpha = P_K - 1/2$, where P_K is mean internal homogeneity of all Boolean functions on K inputs; see text). Column 3: homeostatic stability refers to tendency to return to the same state cycle after transient reversal of activity of any one element. Column 4: reachability is the number of other state cycles to which a net flows from each state cycle after all possible minimal perturbations, due to reversing activity of one element.

completely *random choice* among the 2^N possible states.

Table 1 shows that the lengths of state cycles average $0.5 \times 2^{(N/2)}$, the number of state cycle attractors averages N/e, state cycles are unstable to almost all minimal perturbations, and state cycles are all totally disrupted by random replacement of the Boolean function of a single variable by another Boolean function.

State cycle lengths of $0.5 \times 2^{(N/2)}$ are vast as N increases. For $N = 200$, the state cycles average $2^{100} = 10^{30}$. At a microsecond per state transition, it would require billions of times the history of the universe to traverse the attractor. Here is surely a "big" attractor wandering through state space before finally returning. I will call such attractors, whose length increases *exponentially* as N increases, "*chaotic*". This does *not mean* that flow "on" the attractor is divergent, as in the low-dimensional chaos discussed in an earlier section. A state cycle is the analogue of a one-dimensional limit cycle.

Because the successor to each state is randomly chosen, each element is equally likely to assume either activity 1 or 0 at the next moment; hence virtually all elements "twinkle" on and off around the long attractor.

The number of cycles, hence basins of attraction, however is small, N/e. Thus a system with 200 elements would have only about 74 alternative asymptotic patterns of behavior. This is already an interesting intimation of order, even in extremely complex disordered systems. A number of workers have investigated this class of systems [3, 6, 7, 20–22, 25–27, 45].

(2) $K > 5$. Networks in this class have an enormous number of alternative connection patterns among the N elements. As shown in table 1, the most essential feature of these systems is that their attractors remain "chaotic"; they increase in length exponentially as N increases. The exponential rate at which attractors grow is low for small values of K, and increases to $N/2$ as K approaches N. This implies that even for $K = 5$, state cycle lengths eventually become huge as N increases. Similarly, along any such attractor, any element "twinkles" on and off around such an attractor [26, 27]. The expected number of alternative attractors is N/e or less [26, 27]. Hence these extremely complex, arbitrarily constructed networks have only a few alternative modes of asymptotic behavior. The stability of such attractors to minimal perturbations remains low.

(3) $K = 2$. Nets crystallize spontaneous order. Random Boolean networks with $K = 2$ inputs exhibit unexpected and powerful collective spontaneous order. As shown in table 1, the expected length of state cycles is only $N^{1/2}$; similarly the

number of alternative state cycle attractors is also $N^{1/2}$; each state cycle is stable to almost all minimal perturbations; and mutations deleting elements or altering the logic of single elements only alter dynamical behavior slightly [4, 6, 7, 20–22, 23, 25–27, 42, 43].

Each property warrants wonder. State cycles are only $\sqrt{N}$ in length. Therefore a system of 10 000 binary elements with $2^{10000} = 10^{3000}$ alternative states, settles down and cycles among a mere 100 states. The attractor "boxes" behavior into a tiny volume of 10^{-2998} of the entire state space. Here is spontaneous order indeed! At a microsecond per state transition, the system traverses its attractor in 100 μs, rather less than billions of times the history of the universe.

The number of alternative attractors is only $\sqrt{N}$. A system with 10 000 elements and 10^{3000} combinations of activities of its elements has only 100 alternative asymptotic attractor patterns of integrated behavior. Ultimately, the system settles into one of these.

Along these state cycle attractors, many elements are "*frozen*" into either the active or inactive value. I return to this fundamental property below. It governs the correlated features of the adaptive landscapes in these systems. More critically, this property points to a new principle of collective order.

Another critical feature of random $K = 2$ networks is that each attractor is stable to most minimal perturbations. Small state cycles are therefore correlated wtih *homeostatic return to an attractor after perturbation*.

And in addition we will find shortly that most "mutations" only alter attractors slightly. $K = 2$ networks adapt on highly correlated landscapes.

The previous properties mean that this class of systems simultaneously exhibit *small attractors, homeostasis, and correlated landscapes* abetting adaptation. Further, these results demonstrate that random parallel-processing networks exhibit order without yet requiring any selection.

(4) $K = 1$. In these networks, each element has only a single input. The structure of the network falls apart into separate loops with descendant tails. If the network connections are assigned at random, then most elements lie on the "tails" and do not control the dynamical behavior, since their influence "propagates" off the ends of the tails. On the order of $\ln N(N^{1/2})$ of the number of elements lie on loops. Each separate loop has its own dynamical behavior and cannot influence the other structurally isolated loops. Thus such a system is *structurally modular*. It is comprised by separate isolated subsystems. The overall behavior of such systems is the product of the behaviors of the isolated systems. As table 1 shows, the median lengths of state cycles increase rather slowly as N increases, the number of attractors increases exponentially as N increases, and their stability is moderate. There are four Boolean functions of $K = 1$ input, "yes", "not", "true", and "false". The last two functions are constantly active, or inactive. The values in table 1 assume that only the Boolean functions "yes" and "not" are utilized in $K = 1$ networks. When all four functions are allowed, most isolated loops fall to fixed states, and the dynamical behavior is dominated by those loops with no "true" or "false" functions assigned to elements of the loop. Flyvbjerg and Kjaer [12] and Jaffe [19] have derived detailed results for this analytically tractable case.

The results summarized here are discussed elsewhere [20, 23–25, 27], where I interpret the binary elements as *genes switching one another on and off*, while the Boolean network models the cybernetic genetic regulatory network underlying ontogeny and cell differentiation.

3. A new principle of order in disordered Boolean model systems

3.1. Percolation of frozen clusters

We investigated in section 2 the behavior of randomly constructed disordered Boolean networks with N binary variables, each regulated by

K other variables. We found that fully random networks with $K=2$ spontaneously exhibit extremely small, stable attractors which adapt on highly correlated landscapes.

What principles of order allow $K=2$ networks to exhibit such profound order? The answer appears to be that such networks develop a connected mesh or *frozen core* of elements, each frozen either in the 1 or the 0 state. The frozen core creates percolating walls of constancy which partition the system into a frozen core and isolated islands of elements which continue to change activities from 1 to 0 to 1. But these islands are *functionally isolated* from one another. Alterations of activities in one island cannot propagate to other islands through the frozen core. The emergence of such a frozen core is a sufficient condition for the emergence of orderly dynamics in random Boolean networks.

Two related means of forming such percolating walls are now established. The first are called *forcing structures* [13–15, 16, 21–23, 27]. The second have, as yet, no specific name. I propose to call them *internal homogeneity clusters*. These two kinds of structure constitute the only known means by which orderly dynamics arises in disordered Boolean networks. Because Boolean networks are the logical skeletons for a wide range of continuous non-linear systems, there are good grounds to suppose that the same principles will account for order in an extremely wide class of systems.

Forcing structures are described next. Consider the Boolean "OR" function. This function asserts that if *either one or the other* of the two regulating inputs is active at a given moment, then the regulated element will be active at the next moment. Notice that this Boolean function has the following property: if the *first input is currently active*, that alone guarantees that the regulated element will be active at the next moment, regardless of the current activity of the second input. That is, this Boolean function has the following property: the regulated element can be *fully insensitive* to variation in the activity of the second input, if the first input is active.

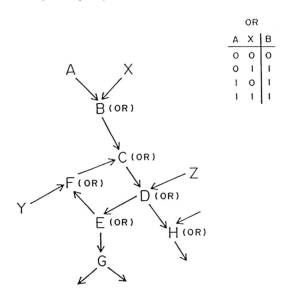

Fig. 2. Forcing structure among binary elements governed by the Boolean OR function. The forcing "1" value propagates down the structure and around the forcing loop, which eventually is "frozen" into the forced state with "1" values at all elements around the loop. The loop then radiates fixed forced values downstream. From *Origins of Order: Self Organization and Selection in Evolution* by S.A. Kauffman [32]. Copyright © 1990 by Oxford University Press, Inc. Reprinted by permission.

I will call a Boolean function *canalyzing* if the Boolean function has at least one input having at least one value, 1 or 0, which suffices to guarantee that the regulated element assumes a specific value, 1 or 0. OR is an example of such a function. So is AND, fig. 1, since if either the first or the second input is 0, the regulated locus is guaranteed to be 0 at the next moment. By contrast, the EXCLUSIVE OR function, in which the regulated locus is active at the next moment if one or the other but not both inputs are active at the present moment, is not a canalyzing Boolean function. No single state of either input guarantees the behavior of the regulated element.

- Next consider a system of several binary variables, each receiving inputs from two or three of the other variables, and each active at the next moment if any one of its inputs is active at the current moment (see fig. 2). That is, each element is governed by the OR function on its inputs. As shown in fig. 2, this small network has feedback loops. Now the consequence of the fact that all elements are governed by the OR function on their inputs is that if a specific element is currently in the "1" state, at the next moment all of those elements that it regulates are *guaranteed or forced* to be in the "1" state. Thus the "1" value is guaranteed to propagate from any initially active element in the net, iteratively to all "descendents" in the net. But the net has loops; thus the guaranteed "1" value cycles around such a loop. Once the loop has "filled up" with "1" values at each element, the loop remains in a fixed state with "1" at each element in the loop, and cannot be perturbed by outside influences. Further the "fixed" "1" values propagate to all descendents of the feedback loop, fixing them in the "1" value as well. Such circuits are called *forcing loops* and *descendent forcing structures* [13–15, 21–23, 27]. Note that the fixed behavior of such a part of the network provides walls of constancy. No signal can pass through elements once they are frozen in their forced values.

The limitation to the OR function is here made only to make the picture clear. In fig. 3 I show a network with a forcing structure in which a "1" state at some specific elements forces a descendant element to be in the "0" state, which in turn forces its descendent element to be in the "1" state. The key, and defining feature, of a forcing structure in a Boolean network is that, at each point, a single element has a single state which can force a descendent element to a specific state regardless of the activities of other inputs. Propagation of such guaranteed states occurs via the forcing connections in the network. For a connection between two regulated elements to be classed as "forcing", the second element must be governed by a canalyzing Boolean function, and the first element, which is an input to the second element, must itself directly or indirectly (i.e. via $K = 1$ input connections) be governed by a canalyzing Boolean function, and the value of the first element which can be "guaranteed" must be the value of the first element, which itself guarantees the activity of the second element. Clearly a net-

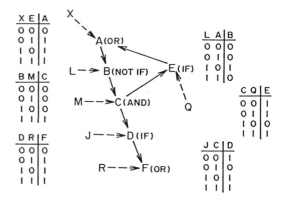

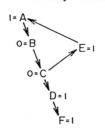

Fig. 3. Forcing structure among binary elements governed by a variety of Boolean functions. Forced values propagate downstream through the forcing structure and around forcing loops, which eventually fall to a "frozen" forced state. The loop then radiates fixed forced values downstream into a forcing structure. From *Origins of Order: Self Organization and Selection in Evolution* by S.A. Kauffman [32]. Copyright © 1990 by Oxford University Press, Inc. Reprinted by permission.

work of elements governed by the OR function meets these requirements. More general, they create a transitive relation such that if A forces B and B forces C, then A indirectly forces C via B. Guaranteed values must propagate down a connected forcing structure.

Large networks of N switching elements, each with $K = 2$ *inputs* drawn at random from among the N, and each assigned at random one of the $(2^2)^K$ Boolean switching functions on K inputs, are random disordered systems. Nevertheless, they can exhibit markedly ordered behavior with small attractors, with homeostasis and, as we see below, with highly correlated fitness landscapes. The reason for this is that large forcing structures exist in such networks. The forcing structures form a large connected interconnected web of components which stretches or *percolates* across the entire network [14, 15, 18, 21–23, 25, 27]. This web falls to a fixed state, each element frozen in its forced value and leaves behind *functionally isolated islands of elements* which are not part of the forcing structure. These isolated islands are each an interconnected cluster of elements which communicates internally. But the island clusters are functionally isolated from one another because signals cannot pass through the walls of constancy formed by the percolating forcing structure.

The occurrence of such walls of constancy due to the percolation of extended forcing structures depends upon the character of the switching network, and in particular on the number of variables which are inputs to each variable, that is, upon the connectivity of the dynamical system. Large connected forcing structures form, or "percolate", spontaneously in $K = 2$ networks because a high proportion of the 16 possible Boolean functions of $K = 2$ inputs belong to the special class of "canalyzing Boolean functions". If two elements regulated by canalyzing Boolean functions are coupled, one as the input to the second, then the probability that the connection is a "forcing connection" is 0.5. This means that in a large network all of whose elements are regulated by canalyzing Boolean functions, on average half of the connections are forcing connections.

The expected size and structure of the resulting forcing structures is a mathematical problem in random graph theory [10, 11, 17, 23, 24, 26, 27]. Percolation "thresholds" occur in random graphs and determine when large connected webs of elements will form. Below the threshold such structures do not form, above the threshold they do. The percolation threshold for the existence of extended forcing structures in a random Boolean network requires that the ratio of forcing connections to elements be 1.0 or greater [22, 24, 27]. Thus in large networks using elements regulated by canalyzing functions on two inputs, half the $2N$ connections are forcing. Therefore the ratio of forcing connections to elements, $N/N = 1$, is high enough that extended large forcing structures form. More generally, for $K = 2$ random networks and networks with $K > 2$, but restricted to canalyzing functions, such forcing structures form and literally crystallize a frozen state which induces orderly dynamics in the entire network.

Because the percolation of a frozen component also accounts for the emergence of order due to *homogeneity clusters* discussed just below, I defer for a moment describing how the frozen component due to either forcing structures or homogeneity clusters induces orderly dynamics.

3.2. Percolation of "homogeneity clusters"

Low connectivity is a sufficient, but not a necessary, condition for orderly behavior in disordered switching systems. In networks of high connectivity, order emerges with proper constraints on the class of Boolean switching rules utilized. The class of canalyzing Boolean functions and the percolation of forcing structures across the network provides one possible constraint. Another is provided by the concept of internal homogeneity.

Consider a Boolean function of four input variables. Each input can be on or off; hence the Boolean function must specify the response of the regulated switching element for each of the 2^4 combinations of values of the four inputs. Among the 16 "responses", the "1" or the "0" response might occur equiprobably, or one may occur far more often than the other. Let P be the fraction of the 2^K positions in the function with a "1" response. If P is well above 0.5, and approaches 1.0, then most combinations of activities of the four variables lead to a "1" response. The deviation of P above 0.5 measures the "*internal homogeneity*" of the Boolean function.

Derrida and Stauffe [5], Weisbuch and Stauffer [46, 47] and de Arcangelis [2] summarized in refs. [42, 43] and in ref. [48] have studied two-dimensional and three-dimensional lattices with nearest-neighbor coupling, and found that if P is larger than a critical value, P_c, then the dynamical behavior of the network breaks up into a connected "frozen" web of points fixed in the "1" value, and isolated islands of connected points which are free to oscillate from "0" to "1" to "0", but are functionally cut off from other such islands by the frozen web, see fig. 4.

In contrast, if P is closer to 0.5 than P_c, then such a percolating web of points fixed in "1" values does not form. Instead small isolated islands of frozen elements form, and the remaining lattice is a single connected percolating web of elements which oscillate between "1" and "0" in complex temporal cycles. In this case, transiently altering the value, "1" or "0", of one point can propagate via neighboring points and influence the behavior of most of the oscillating elements in the lattice.

These facts lead us to a new idea: *The critical value of P, P_c, demarks a kind of "phase transition"* in the behavior of such a dynamical system.

For P closer to 0.5 than P_c, the lattice of on/off variables, or two state "spins", has no percolating frozen component. For P closer to 1.0 than P_c, the lattice of on/off variables, or two state "spins", does have a large frozen component which percolates across the space.

The arguments for the percolation of a frozen component for $P > P_c$ do not require that the favored value of each on/off "spin" variable in the lattice be "1". The arguments carry over perfectly if half the on/off variables responds with high probability, $P > P_c$, by assuming the "1" value and the other half responds with $P > P_c$ with the "0" value. In this generalized case, in the frozen web of "spins" in the lattice, each frozen spin is frozen in its more probable value, "1" or "0". Thus, for arbitrary Boolean lattices, $P > P_c$ provides a criterion which separates two drastically different behaviors, chaotic versus ordered.

The value of P for which this percolation and freezing out occurs, depends upon the kind of lattice, and increases as the number of neighbors

```
           8    8    1   1228228228228228228228   1   1   1   1   1   1   1   1   1   1   1
           8    8    1    1    1    1228228228228   1   1   1   1   1   1   1   1   1   1   1
           8    8   8456456456228228228228228228   1   1   1'  1   1  10  10  10   1   1   1
           1    8    1   1228228228228   1   1   1   1   1   1   1   1  10  10  10   1   1   1
           1    1   1228228228228228228228   1   1   1   1   1   1   1   1   1   1   1   1
           1    1    1   1228228228228228228228   1   1   1   1   1   1   1   1   1   1   4
           1    1    1    1    1    1   1228228228228   1   1   1   1   1   1   1   1   1   1
           1    1    1    1    6    1   1228228228228   1   1   4   1   1   1   1   1   1   1
           1    4    1    6    6    6    1   1228228228228228228   4   1   4   1   1   1   1   1
           1    4    1    1    6    6   6228228228228   1   1   4   1   4   1   1   1   1   1
           4    4    1    6    6    6    6   6228228228   1   1   1   1   1   1   1   1   1   4
           1    4   12    6    6    6   1228228228228   1   1   1   1   1   8   8   8   1   1   1
         220    1    1    1    1    1    1   1228228228   1   1   1   1   8   8   8   1   1  1220
         220220    1    1    1    1    1   1228228228228   1   1   1   8   8   4   8   1   1
         220220    1    1    1    1    1   1228228   1   1   1   1   1   1   1   1   1  1220110
           1220110110   1   1   1   1   1228228   1   1   1   1   1   1   1   1  20  20110110
           1110110110   1    1    4    1   1228   1   1   2   4   1   1   1   1   1  20  20110110
         110110110110110   1    4    1   1   1   1   2   4   1   1   1  20  20  20  20   1110
         110110110  22    1    1    1    1    1    4  4228   1   1   1  20  20  20  20  20  20110
         110110    1    1    1    1    1    1    1   1   1228   1   4   1  20  20  20  20  20  20110
         110   22   22   22   22    1   1228228   1   1228228   1   4   4   1   1   1   1   4  20   2  22
          22   88   22   22    1    1    1   1228   1228228228   1   1   1   1   1   1   1  20   2
           1   88    1    1   1228228228228228228228228228   1   1   1   1   1   1   4   4   4
           1    8    1   1228228228228228228228228228228   1   1   1   1   1   1   1   1   1
```

Fig. 4. Two-dimensional lattice of sites, each a binary state "spin" which may point up or down. Each binary variable is coupled to its four neighbors and is governed by a Boolean function on those four inputs. The number at each site shows the periodicity of the site on the state cycle attractor. Thus "1" means a site frozen in the active or inactive state. Note that frozen sites form a frozen cluster that percolates across the lattice. Increasing the measure of internal homogeneity, P, the bias in favor of a "1" or a "0" response by any single spin, above a critical value, P_c, leads to percolation of a "frozen" component of spins, fixed in the 1 or 0 state, which spans the lattice leaving isolated islands of spins free to vary between 0 and 1.

to each point in the lattice increases. On a square lattice for $K = 4$, P_c is 0.28 [42–44]. On a cubic lattice, each point has six neighbors, and P_c is greater than for square lattices. This reflects the fact that the fraction of "bonds" in a lattice which must be in a fixed state for that fixed value to percolate across the lattice, depends upon the number of neighbors to each point in the lattice.

Let me call such percolating frozen components for $P > P_c$ *homogeneity clusters* to distinguish them from extended forcing structures. I choose this name because freezing in this case depends upon the internal homogeneity of the Boolean functions used in the network. That the two classes of objects are different in general is clear: In forcing structures the characteristic feature is that at each point a single value of an element alone suffices to force one or more descendent elements to their own forced values. In contrast, homogeneity clusters are more general. Thus, consider two pairs of elements, A1, A2, B1 and B2. A1 and A2 might receive inputs from both B1 and B2 as well as other elements, while B1 and B2 receive inputs from A1 and A2 as well as other elements. But due to the high internal homogeneity $P > P_c$ of the Boolean functions assigned to each, *simultaneous "1" values by both A1 and A2* might jointly guarantee that B1 and B2 each be active regardless of the activities of other inputs to B1 and B2. At the same time, simultaneous "1" values by both B1 and B2 might jointly guarantee that A1 and A2 be active regardless of the activities of other inputs to A1 and A2. Once the four elements are jointly active, they mutually guarantee their continued activity regardless of the behavior of other inputs to the four. They form a frozen component. Yet it is not a forcing component since the activity of two elements, A1 and A2, or B1 and B2, must be jointly assured to guarantee the activity of any single element.

While there appear to be certain differences between forcing structures and homogeneity clusters, those differences are far less important than the fact that, at present, *the two are the only established* means to obtain orderly dynamics in large, disordered Boolean networks.

4. Why frozen structures lead to correlated fitness landscapes

Whether percolation of a frozen phase is due to an extended forcing structure or to a homogeneity cluster due to $P > P_c$, the implications include these:

(1) If a frozen phase does *not form*:

(a) The attractors in such a system are very large, and grow *exponentially* as the number of points in the lattice, N, increases. Indeed, the attractors are so large that the system can be said to behave *chaotically*.

(b) As indicated, a minor alteration in the state of the lattice, say, "flipping" one element from the "1" to the "0" value at a given instant, propagates alterations in behavior throughout the system. More precisely, consider two identical lattices which differ only in the value of one "spin" at a moment T. Let each two lattices behave dynamically according to their identical Boolean rules. Define the "damage" caused by the initial "spin flip" to be the total number of sites in the lattices which at the succession of time moments are now induced to be in *different states*, "1" or "0". Then for P closer to 0.5 than P_c, such damage propagates across the lattice with a finite speed, and a large fraction of the sites are damaged [5, 36, 41–43, 46]. Propagation of "damage" from a single site difference implies that dynamical behavior is highly sensitive to small changes in initial conditions.

(c) Consequently, many perturbations by single flips drive the system to an entirely different attractor.

(d) Damage by "mutations" deleting an element or altering its Boolean function tends strongly to alter many attractors. Thus such systems adapt on very rugged fitness landscapes.

(2) In contrast, if the ratio of forcing connections to elements is greater than 1.0, or if internal homogeneity P is closer to 1.0 than P_c:

(a) Then a large frozen component and percolating walls of constancy do form, leaving behind functionally isolated islands which cannot communicate with one another.

(b) The result is that *attractors are small*, typically increasing as the number of nodes to some, *fractional power* [20, 23, 27, 34, 35, 42, 43, 46]. This means that the sizes of attractors increase less than linearly as the number of points in the lattice, N, increases. Such attractors are small indeed, for the entire state space is the 2^N possible combinations of the on/off "spins" in the lattice. An attractor comprised by less than N states is tiny compared to 2^N. Thus, either the existence of a frozen component due to forcing structures or due to "homogeneity clusters" for P greater than P_c implies that such systems spontaneously "box" themselves into very small volumes of their state space and exhibit high order.

(c) Further, damage does not spread. Transient reversal of the state of a "spin" propagates alterations in behavior only locally if at all [20, 23, 27, 42, 43]. This means that attractors tend strongly to exhibit *homeostatic return after perturbation*.

(d) For both frozen components due to forcing structures and homogeneity clusters, the system is typically not much altered by "mutations" deleting single elements or altering their Boolean rules. Any element which is itself "buried" in the frozen component cannot propagate alterations to the remainder of the network. A mutated element within one functionally isolated island communicates only within that island. Damage does not spread. Thus such systems adapt on *correlated fitness landscapes*.

To summarize: the percolation of a frozen component yields disordered Boolean systems which nevertheless exhibit order. They have small attractors precisely because large fractions of the variables remain in fixed states. Furthermore, due to this frozen component of the lattice, minor modifications of connections, or "bits" in a Boolean function, or substitution of one for another Boolean function at one point, or alterations in other parameters will lead to only minor modifications of dynamical behavior. Thus, such networks have attractors which *adapt on correlated fitness landscapes*. This is not surprising; the properties of the system which give it small attractors, and hence homeostasis, tend to make it insensitive to small alterations in the parameters affecting any one part of the system. Selection for one of these connected set of properties is selection for all. Self-organization for one, bootstraps all.

5. Landscape structure governs adaptive evolution

The correlation structure of rugged multipeaked fitness landscapes is critical to the way adaptive search by mutation and selection, or other learning algorithms, occurs. By "correlation structure" I mean the similarity of "nearby" "mutant variants" of the initial system with respect to a "phenotype" of interest. Fitness landscapes range from smooth and single peaked to fully random. In the latter case, the fitness of one entity carries no implications about the fitness of its nearest neighbors.

A central question therefore is whether an adaptive process which is constrained to pass via fitter 1-mutant or "few mutant" variants of a network, can "hill climb" to networks with desired attractors. Adaptive evolution leads us to examine the statistical features of fitness landscapes, including correlation structure, the numbers of local optima, the lengths of walks to optima via fitter 1-mutant variants, the number of optima accessible from any point, and so forth [29]. In particular, in considering adaptation in Boolean network space, any specific measurable property of such networks yields a fitness landscape over the space of systems. Again we can ask what the structure of such landscapes looks like.

Previous work [28] confirms the intuition drawn from the existence of frozen components. Fitness landscapes were defined with respect to selection for a prespecified attractor. We found that the fitness landscapes are multipeaked but are far more correlated in $K = 2$ networks with frozen components than are those of $K = 5$ or $K = 10$ landscapes, which do not have frozen components. In all these cases, due to the rugged multipeaked character of the landscapes, adaptive evolution fails to reach global optima, and remains trapped on or near suboptimal solutions.

Adaptive evolution on such rugged landscapes exhibits a number of general properties. The most important are these:

(1) In "smooth" landscapes with few peaks, as the mutation rate increases, or the complexity of the system under selection increases, an adapting population at first can search locally and "climb" to fitness peaks and cluster as a small cloud near the peak. But as the mutation rate increases further, or the complexity of the system under selection increases further, the adapting population *melts* from the adaptive peak and "diffuses" along ridges of roughly equal fitness values [8, 9, 32, 40]. Thus, selection cannot achieve and maintain the highest adaptive peaks if the mutation rate is too high, or the complexity of the entities under selection is too great.

(2) As landscapes become more rugged and multipeaked, an adapting population tends to be "localized" into a small subregion of the entire space over very long time periods, then suddenly escape to a new region. Local search and learning mechanisms can be expected to be quite ineffective in *finding global optima*. The way the adapting population melts and flows across such a landscape is strongly reminiscent of the slow relaxation behavior of an electron in a complex potential surface at a finite temperature. Over short time scales the electron rapidly explores a local region. Over longer scales it escapes from the local region and explores wider reaches of the space.

(3) The "melting" from peaks in smooth landscapes, and becoming trapped in small regions of landscapes in rugged multipeaked landscapes, are twin pincers which tend to limit adaptive evolution. One limitation arises inevitably in smooth landscapes, the other in rugged ones. The evolution of complex entities has therefore been bound by and guided by these limitations. Smooth adaptive landscapes are generated, roughly, by *modular entities*, whose parts can be changed without propagating changes to the entire system. Boolean networks with frozen components and isolated unfrozen islands fall into this category. Very rugged landscapes are generated by entities with no isolable parts, in which alteration of any part impinges on the functional consequences of many other parts. Boolean networks without frozen components exemplify such systems.

(4) The evolution of evolvability almost certainly has occurred by changing the kinds of entities which are adapting, hence their landscape structure. Thus we must consider a "metadynamics" in evolution, in which the structure of fitness landscapes themselves might be optimized. Among the central issues are what features of fitness landscapes might optimize adaptive evolution.

(5) The "optimal" structure of a fitness landscape for adaptive evolution depends upon the mutation rate itself. Higher sustained fitness may occur in more rugged fitness landscapes because the "sides" of fitness peaks are steeper, hence allow more rapid increase in fitness by single "mutations", than do single mutations on smoother landscapes. Thus, assessment of "evolvability" is subtle, and must take account not only of the structure of fitness landscapes, but the sustained fitness of entities perturbed by continuing mutations or their analogues.

(6) Fitness landscapes are not fixed. In the simplest case, the non-biological environment changes, deforming landscapes. Sustained fitness can be higher on a more rugged landscape subject to external deformations, than on a smooth one. Again, this is because on more rugged landscapes single-mutant variants can increase fitness more dramatically than on smooth landscapes.

(7) For a similar reason, entities evolving on more rugged landscapes can sustain higher fitness than entities on very smooth landscapes when the entities in question are *coevolving*. In coevolving systems, the "fitness landscape" of one entity depends upon the positions of the other entities on their own fitness landscapes. As the coevolving partners move, their landscapes persistently deform. For example, both the fitness and the fitness landscape of the fly are altered by the evolution of a frog with a sticky tongue. In coevolving systems

there is no fixed fitness landscape, hence no analogue of a potential function with peaks as point attractors of the evolutionary dynamics. Assessment of evolvability now requires analysis of the *coupling among landscapes*, and how badly the landscape of one partner is deformed when its partners move. The generalization of adaptive peaks becomes closely related to game theory, and Nash equilibria, in which each partner is at an adaptive peak consistent with peaks attained by its coevolving partners. Of course, these peaks are peaks only with reference to the mutant search range of each partner. Such Nash equilibria may involve *some or all of the coevolving partners* in the "ecosystem". At such equilibria, some or all of the partners therefore stop changing. Therefore, these Nash equilibria correspond to a natural generalization of *frozen components* in Boolean networks.

Attainment of Nash equilibria in coevolving systems depends upon the ruggedness of fitness landscapes and how the coupling between landscapes leads to deformation as each partner moves on its own landscape. Strikingly, simulations based on the "NK" spin-glass-like model of rugged fitness landscapes mentioned above suggest that sustained fitness of coevolving partners may be maximized when Nash equilibria for the coevolving "ecosystem" just begin to percolate across the ecosystem [32]. Percolation of such frozen components in coevolving systems, in turn, depends upon achieving a just sufficient ruggedness of landscapes. Among the interesting, and potentially testable, features of such systems is the fact that alterations of the external environment affecting a single partner can unleash an *avalanche* of coevolutionary change which propagates through the ecosystem. Based on the spin-glass-like model, it appears that when sustained fitness is optimized, hence frozen Nash equilibria are tenuously percolating across the ecosystem, then the distribution of avalanche frequencies versus sizes falls to a power-law distribution with many small avalanches and few large ones. Two features are of interest. First, this distribution is reminiscent of the distribution of sizes of extinction events in the evolutionary record [39]. Avalanches of coevolutionary change might be expected to be associated with drops in fitness leading to extinction events. Second, Bak et al. [1] pointed out that such a distribution is characteristic of a poised self-critical state in which avalanches propagate on all length scales.

These preliminary results begin to suggest that the capacity to coevolve may be optimized when landscape ruggedness, and couplings among the landscapes of coevolving partners, is tuned to achieve a poised state in which Nash equilibria just percolate and tenuously "freeze" the coevolving system.

6. Concluding remarks

What kinds of dynamical systems have the capacity to accumulate useful variations, hence evolve and coevolve? How do such systems interact with their "worlds" in the sense of categorizing their worlds, acting upon those categorizations, and evolving as their worlds with other players themselves evolve? No one knows. The following is clear. Adaptive evolution, whether by mutation and selection, or learning, or otherwise, occurs on some kind of "fitness landscape". This follows because adaptation is some kind of local search in a large space of possibilities. Further, in any coevolutionary context, fitness landscapes deform because they are coupled. The structure and couplings among landscapes reflect the kinds of entities which are evolving and their couplings. Natural selection or learning may tune both such structures and couplings to achieve systems which are evolvable.

A further point is clear. Complex parallel-processing Boolean networks which are disordered can exhibit ordered behavior. Such networks are reasonable models of a large class of non-linear dynamical systems. The attractors of such networks are natural objects of interest. In a Boolean network receiving inputs from an external world, the attractors of a network are the natural classifications the network makes of the external world. Thus, if the world can be in a single state, yet the network can fall to different attractors, then the network can categorize that state of the world in alternative ways and respond in alternative ways to a single fixed state of the external world. Alternatively, if the world can be in alternative states, yet the network falls to the same attractor, then the network categorizes the alternative states of the world as identical, and can respond in the same way. In brief, non-linear dynamical systems which interact with external worlds classify and "know" their worlds.

If we could find natural ways to model coevolution among Boolean networks which received inputs from one another and external worlds, we would find that such systems tuned their internal structures and couplings to one another so as to optimize something like their evolvability. An intuitive bet is that such systems would achieve internal structures in which the frozen components were nearly melted. Such structures live on the edge of chaos, in the "liquid" interface suggested by Langton [37], where complex computation can be achieved. In addition, I would bet that couplings among entities would be tuned such that the frozen Nash equilibria are tenuously held to optimize fitness of all coevolving partners in the face of exogenous perturbations to the coevolving system. But a tenuous frozen component in a coevolutionary context would be a repeat of "the edge of chaos" on this higher level. Perhaps such a state corresponds to something like Bak's self-organized critical state [1]. It would be exciting indeed if coadaptation in mutually categorizing dynamical systems tended to such a definable state, for the same principles might recur on a variety of levels in biology and beyond.

References

[1] P. Bak, C. Tank and K. Wiesenfeld, Self-organized criticality, Phys. Rev. A 38 (1988) 364–374.

[2] L. De Arcangelis, Fractal dimensions in three-dimensional Kauffman cellular automata, J. Phys. A 20 (1987) L369–L373.

[3] B. Derrida and H. Flyvbjerg, Multivalley structure in Kauffman's model: analogy with spin glasses, J. Phys. A 19 (1986) L1003–L1008.

[4] B. Derrida and Y. Pomeau, Random networks of automata: a simple annealed approximation, Biophys. Lett. 1 (1986) 45–49.

[5] B. Derrida and D. Stauffer, Phase-transitions in two-dimensional Kauffman cellular automata, Europhys. Lett. 2 (1986) 739–745.

[6] B. Derrida and H. Flyvbjerg, The random map model: a disordered model with deterministic dynamics, J. Phys. (Paris) 48 (1987) 971–978.

[7] B. Derrida and H. Flyvbjerg, Distribution of local magnetizations in random networks of automata, J. Phys. A 20 (1987) L1107–L1112.

[8] M. Eigen, New concepts for dealing with the evolution of nucleic acids, in: Cold Spring Harbor Symposia on Quantitative Biology, Vol. LII, Cold Spring Harbor Laboratory (1987) pp. 307–320.

[9] M. Eigen and P. Schuster, The Hypercycle, A Principle of Natural Self-Organization (Springer, Berlin, 1979).

[10] P. Erdos and A. Renyi, On the Random Graphs 1, Vol. 6 (Inst. Math. Univ. DeBreceniens, Debrecar, Hungary, 1959).

[11] P. Erdos and A. Renyi, On the evolution of random graphs, Math. Inst. Hung. Acad. Sci., Publ. No. 5 (1960).

[12] H. Flyvbjerg and N.J. Kjaer, Exact solution of Kauffman's model with connectivity one, J. Phys. A 21 (1988) 1695–1718.

[13] F. Fogelman-Soulié, Frustration and stability in random Boolean networks, Discrete Appl. Math. 9 (1984) 139–156.

[14] F. Fogelman-Soulié, Ph.D. Thesis, Université Scientifique et Medical de Grenoble (1985).

[15] F. Fogelman-Soulié, Parallel and sequential computation in Boolean networks, Theor. Comp. Sci. 40 (1985).

[16] A.E. Gelfand and C.C. Walker, Ensemble Modeling (Dekker, New York, 1984).

[17] F. Harary, Graph Theory (Addison–Wesley, Reading, MA, 1969).

[18] H. Hartman and G.Y. Vichniac, in: Disordered Systems and Biological Organization, eds. E. Bienenstock, F. Fogelman-Soulié and G. Weisbuch (Springer, Berlin, 1986).

[19] S. Jaffe, Kauffman networks: cycle structure of random clocked Boolean networks, Ph.D. Thesis, New York University (1988).

[20] S.A. Kauffman, Metabolic stability and epigenesis in randomly connected nets, J. Theor. Biol. 22 (1969) 437–467.

[21] S.A. Kauffman, Cellular homeostasis, epigenesis and replication in randomly aggregated macro-molecular systems, J. Cybern. 1 (1971) 71–96.

[22] S.A. Kauffman, Gene Regulation Networks: A Theory for Their Global Structure and Behavior, Current Topics in Developmental Biology, Vol. 6, eds. A. Moscana and A. Monroy (Academic Press, New York, 1971) pp. 145–182.

[23] S.A. Kauffman, The large-scale structure and dynamics of gene control circuits: an ensemble approach, J. Theor. Biol. 44 (1974) 167–190.

[24] S.A. Kauffman, Development constraints: internal factors in evolution, in: Developmental Evolution, eds. B.C. Goodwin, N. Holder and C.G. Wylie (Cambridge Univ. Press, Cambridge, 1983) pp. 195–225.

[25] S.A. Kauffman, Emergent properties in random complex automata, Physica D 10 (1984) 145–156.

[26] S.A. Kauffman, A framework to think about regulatory systems, in: Integrating Scientific Disciplines, ed. W. Bechtel (Nijhoff, The Hague, 1986) pp. 165–184.

[27] S.A. Kauffman, Boolean systems, adaptive automata, evolution, in: Disordered Systems and Biological Organization, eds. E. Bienenstock, F. Fogelman-Soulié and G. Weisbuch (Springer, Berlin, 1986) pp. 338–360.

[28] S.A. Kauffman and R.G. Smith, Adaptive automata based on Darwinian selection, Physica D 22 (1986) 68–82.

[29] S.A. Kauffman and S. Levin, Towards a general theory of adaptive walks on rugged landscapes, J. Theor. Biol. 128 (1987) 11–45.

[30] S.A. Kauffman, E.D. Weinberger and A.S. Perelson, Maturation of the Immune Response Via Adaptive Walks on Affinity Landscapes, Theoretical Immunology, Part 1, SFI Studies in the Sciences of Complexity, Vol. II, ed. A.S. Perelson (Addison–Wesley, Reading, MA, 1988) pp. 349–382.

[31] S.A. Kauffman and E.D. Weinberger, Application of NK model to maturation of immune response, J. Theor. Biol., in press.

[32] S.A. Kauffman, Origins of Order: Self-Organization and Selection in Evolution (Oxford Univ. Press, Oxford, 1990).

[33] S.A. Kauffman and D. Stein, Application of the NK model of rugged landscapes to protein evolution and protein folding, Abstract AAAS Meeting on Protein Folding, June, 1989.

[34] K.E. Kürten, Correspondence between neural threshold networks and Kauffman Boolean cellular automata, J. Phys. A 21 (1988) 615–619.

[35] K.E. Kürten, Critical phenomena in model neural networks, Phys. Lett. A 129 (1988) 157–160.

[36] P.M. Lam, A percolation approach to the Kauffman model, J. Stat. Phys. 50 (1988) 1263–1269.

[37] C. Langton, Artificial life, in: Artificial Life, Santa Fe Institute Studies in the Sciences of Complexity, Vol. VI, ed. C. Langton (Addison–Wesley, Reading, MA, 1989) pp. 1–47.

[38] J. Maynard-Smith, Natural selection and the concept of a protein space, Nature 225 (1970) 563.

[39] D.M. Raup, On the early origins of major biologic groups, Paleobiology 9 (1983) 107–115.

[40] P. Schuster, Structure and dynamics of replication–mutation systems, Physica Scripta B 26 (1987) 27–41.

[41] H.E. Stanley, D. Stauffer, J. Kertesz and H.J. Herrmann, Phys. Rev. Lett. 59 (1987) 2326.

[42] D. Stauffer, Random Boolean networks: analogy with percolation, Phil. Mag. B 56 (1987) 901–916.

[43] D. Stauffer, On forcing functions in Kauffman's random Boolean networks, J. Stat. Phys. 40 (1987) 789.

[44] D. Stauffer, Percolation thresholds in square-lattice Kauffman model, J. Theor. Biol., in press.

[45] S. Wolfram, Statistical mechanics of cellular automata, Rev. Mod. Phys. 55 (1983) 601.

[46] G. Weisbuch and D. Stauffer, J. Phys. (Paris) 48 (1987) 11–18.

[47] G. Weisbuch and D. Stauffer, Phase transition in cellular random Boolean nets, J. Phys. (Paris) 48 (1987) 11–18.

[48] G. Weisbuch, Dynamics of Complex Systems. An Introduction of Networks of Automata (Interedition, Paris, 1989).

A ROSETTA STONE FOR CONNECTIONISM

J. Doyne FARMER

*Complex Systems Group, Theoretical Division, Los Alamos National Laboratory, Los Alamos, NM 87545, USA
and Santa Fe Institute, 1120 Canyon Road, Santa Fe NM 87501, USA*

The term connectionism is usually applied to neural networks. There are, however, many other models that are mathematically similar, including classifier systems, immune networks, autocatalytic chemical reaction networks, and others. In view of this similarity, it is appropriate to broaden the term connectionism. I define a connectionist model as a dynamical system with two properties: (1) The interactions between the variables at any given time are explicitly constrained to a finite list of connections. (2) The connections are fluid, in that their strength and/or pattern of connectivity can change with time.

This paper reviews the four examples listed above and maps them into a common mathematical framework, discussing their similarities and differences. It also suggests new applications of connectionist models, and poses some problems to be addressed in an eventual theory of connectionist systems.

1. Introduction

This paper has several purposes. The first is to identify a common language across several fields in order to make their similarities and differences clearer. A central goal is that practitioners in neural nets, classifier systems, immune nets, and autocatalytic nets will be able to make correspondences between work in their own field as compared to the others, more easily importing mathematical results across disciplinary boundaries. This paper attempts to provide a coherent statement of what connectionist models are and how they differ in mathematical structure and philosophy from conventional "fixed" dynamical system models. I hope that it provides a first step toward clarifying some of the mathematical issues needed for a generally applicable theory of connectionist models. Hopefully this will also provide a natural framework for connectionist models in other areas, such as ecology, economics, and game theory.

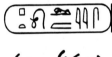

Fig. 1. "Ptolemy", in hieroglyphics, Demotic, and Greek. This cartouche played a seminal role in deciphering hieroglyphics, by providing a hint that the alphabet was partially phonetic [12]. (The small box is a "p", and the half circle is a "t" – literally it reads "ptolmis".)

1.1. Breaking the jargon barrier

Language is the medium of cultural evolution. To a large extent differences in language define culture groupings. Someone who speaks Romany, for example, is very likely a Gypsy; the existence of a common and unique language is one of the most important bonds preserving Gypsy culture. At times, however, communication between subcultures becomes essential, so that we must map one language to another.

The language of science is particularly specialized. It is also particularly fluid; words are tools onto which we map ideas, and which we invent or redefine as necessary. Our jargon evolves as science changes. Although jargon is a necessary feature of communication in science, it can also pose a barrier impeding scientific progress.

When models are based on a given class of phenomena, such as neurobiology or ecology, the terminology used in the models tends to reflect the phenomenon being modeled rather than the underlying mathematical structure. This easily obscures similarities in the mathematical structure. "Neural activation" may appear quite different from "species population", even though relative to given mathematical models the two may be identical. Differences in jargon place barriers to communication that prevent results in one field from being transparent to workers in another field. Proper nomenclature should identify similar things but distinguish those that are genuinely different.

At present this problem is particularly acute for adaptive systems. The class of mathematical models that are employed to understand adaptive systems contain subtle but nonetheless significant new features that are not easily categorized by conventional mathematical terminology. This adds to the problem of communication between disciplines, since there are no standard mathematical terms to identify the features of the models.

1.2. What is connectionism?

Connectionism is a term that is currently applied to neural network models such as those described in refs. [59, 15]. The models consist of elementary units, which can be "connected" together to form a network. The form of the resulting connection diagram is often called the *architecture* of the network. The computations performed by the network are highly dependent on the architecture. Each connection carries information in its weight, which specifies how strongly the two variables it connects interact with each other. Since the modeler has control over how the connections are made, the architecture is plastic.

This contrasts with the usual approach in dynamics and bifurcation theory, where the dynamical system is a fixed object whose variability is concentrated into a few parameters. The plasticity of the connections and connection strengths means that we must think about the entire family of dynamical systems described by all possible architectures and all possible combinations of weights. Dynamics occurs on as many as three levels, that of the states of the network, the values of connection strengths, and the architecture of the connections themselves.

Mathematical models with this basic structure are by no means unique to neural networks. They occur in several other areas, including classifier systems, immune networks, and autocatalytic networks. They also have potential applications in other areas, such as economics, game-theoretic models and ecological models. I propose that the term connectionism be extended to this wider class of models.

By comparing connectionist models for different phenomena using a common nomenclature, we get a clear view of the extent to which these models are similar or different. We also get a *glimpse* of the extent to which the underlying phenomena are similar or different. I emphasize the word glimpse to make it clear that we are simplifying a complicated phenomenon when we model it in connectionist terms. Comparing two connectionist models of, for example, the nervous systems and the immune system, provides a means of extracting certain aspects of their similarities, but we must be very careful in doing this; much richness and complexity is lost at this level of description.

Connectionism represents a particular level of abstraction. By reducing the state of a neuron to a single number, we are collapsing its properties relative to a real neuron, or relative to those of another potentially more comprehensive mathematical formalism. For example, consider fluid dynamics. At one level of description the state of a fluid is a function whose evolution is governed by a partial differential equation. At another level we can model the fluid as a finite collection of spatial modes whose interactions are described by a set of ordinary differential equations. The partial differential equation is not a connectionist model; there are no identifiable elements to connect together; a function simply evolves in time. The ordinary differential equations are *more* connectionist; the nature of the solution depends critically on the particular set of modes, their connections, and their coupling parameters. In fluid dynamics we can sometimes calculate the correct couplings from first principles, in which case the model is just a fixed set of ordinary differential equations. In contrast, for a connectionist model there are dynamics for the couplings and/or connections. In a fully connectionist model, the connections and couplings would be allowed to change, to find the best possible model with a given degree of complexity.

Another alternative is to model the fluid on a grid with a finite difference scheme or a cellular automaton. In this case each element is "connected" to its neighbors, so there might be some justification for calling these connectionist models. However, the connections are fixed, completely regular, and have no dynamics. I will not consider them as "connectionist".

Just as there are limits to what can be described by a finite number of distinct modes, there are also limits to what can be achieved by connectionist models. For more detailed descriptions of many adaptive phenomena we may need models with explicit spatial structure, such as partial differential equations or cellular automata. Nonetheless, connectionism is a useful level of abstraction, which solves some problems efficiently.

The Rosetta Stone is a fragment of rock in which the same text is inscribed in several different languages and alphabets (fig. 1). It provides a key that greatly facilitated the decoding of these languages, but it is by no means a complete description of them. My goal is similar; by presenting several connectionist models side by side, I hope to make it clear how some aspects of the underlying phenomena compare with one another, but I offer the warning that quite a bit has been omitted in the process.

1.3. Organization of this paper

In section 2, I describe the basic mathematical framework that is common to connectionist models. I then discuss four different connectionist models: neural networks, classifier systems, immune networks, and autocatalytic networks. In each case I begin with a background discussion, make a correspondence to the generic framework described in section 2, and then discuss general issues. Finally, the conclusion contains the "Rosetta Stone" in table 3, which maps the jargon of each area into a common nomenclature. I also make a few suggestions for applications of connectionist models and comment on what I learned in writing this paper.

Connectionist models are ultimately dynamical systems. Readers who are not familiar with terms such as automaton, map, or lattice model may wish to refer to the appendix.

2. The general mathematical framework of connectionist models

In this section I present the mathematical framework of a "generic" connectionist model. I make some arbitrary choices about nomenclature, in order to provide a standard language, noting common synonyms whenever appropriate.

To first approximation a connectionist model is a pair of coupled dynamical systems living on a graph. In some cases the graph itself may also have dynamics. The remainder of this section explains this in more detail.

2.1. The graph

The foundation of any connectionist model is a graph consisting of *nodes* (or vertices) and *connec-*

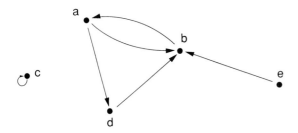

Fig. 2. A directed graph.

tions (also called links or edges) between them as shown in fig. 2. The graph describes the architecture of the system and provides the channels in which the dynamics takes place. There are different types of graphs; for example, the links can be either directed (with arrows), or undirected (without arrows). For some purposes, such as modeling catalysis, it is necessary to allow complicated graphs with more than one type of node or more than one type of link.

For many purposes it is important to specify the pattern of connections, with a *graph representation*. The simplest way to represent a graph is to draw a picture of it, but for many purposes a more formal description is necessary. One common graph representation is a *connection matrix*. The nodes are assigned an arbitrary order, corresponding to the rows or columns of a matrix. The row corresponding to each node contains a nonzero entry, such as "1", in the columns corresponding to the nodes to which it makes connections. For example, if we order the nodes of fig. 2 lexicographically, the connection matrix is

$$C = \begin{bmatrix} 0 & 1 & 0 & 1 & 0 \\ 1 & 0 & 0 & 0 & 0 \\ 0 & 0 & 1 & 0 & 0 \\ 0 & 1 & 0 & 0 & 0 \\ 0 & 1 & 0 & 0 & 0 \end{bmatrix}. \quad (1)$$

If the graph is undirected then the connection matrix is symmetric. It is sometimes economical to combine the representation of the graph and the connection parameters associated with it into a matrix of connection parameters.

A *connection list* is an alternative graph representation. For example, the graph of fig. 2 can also be represented as

$$a \to b, \quad a \to d, \quad b \to a, \quad c \to c, \quad d \to b, \quad e \to b. \quad (2)$$

Note that the nodes are implicitly contained in the connection list. In some cases, if there are isolated nodes, it may be necessary to provide an additional list of nodes that do not appear on the connection list. For the connectionist models discussed here isolated nodes, if any, can be ignored.

A graph can also be represented by an algorithm. A simple example is a program that creates connections "at random" using a deterministic random number generator. The program, together with the initial speed, forms a representation of a graph.

For a *dense* graph almost every node is connected to almost every other node. For a *sparse* graph most nodes are connected to only a small fraction of the other nodes. A connection matrix is a more efficient representation for a dense graph, but a connection list is a more efficient representation for a sparse graph.

2.2. Dynamics

In conventional dynamical models the form of the dynamical system is fixed. The only part of the dynamical system that changes is the state, which contains all the information we need to know about the system to determine its future behavior. The possible ways the "fixed" dynamical form "might change" are encapsulated as parameters. These are usually thought of as fixed in any given experiment, but varying from experiment to experiment. Alternatively we can think of the parameters as knobs that can be slowly changed in the background. In reality the quantities that we incorporate as parameters are usually aspects of the system that change on a time scale slower than those we are modeling with the dynamical system.

Connectionist models extend this view by giving the parameters an explicit dynamics of their own, and in some cases, by giving the list of variables and their connections a dynamics of its own. Typically this also involves a separation of time scales. Although a separation of time scales is not necessary, it provides a good starting point for the discussion. The fast scale dynamics, which changes the *states* of the system, is usually associated with short-term information processing. This is the *transition rule*. The intermediate scale dynamics changes the *parameters*, and is usually associated with learning. I will call this the *parameter dynamics* or the *learning rule*. On the longest time scale, the graph itself may change. I will call this the *graph dynamics*. The graph dynamics may also be used for learning; hopefully this will not lead to confusion.

Of course, strictly speaking the states, parameters, and graph representation described above are just the states of a larger dynamical system with multiple time scales. Reserving the word state for the shortest time scale is just a convenience. The association of time scales given above is the natural generalization of "conventional" dynamical systems, in which the states change quickly, the parameters change slowly, and the graph is fixed. For some purposes, however, it might prove to be useful to relax this separation, for example, letting the graph change at a rate comparable to that of the states. Although all the models discussed here have at most three time scales, in principle this framework could be iterated to higher levels to incorporate an arbitrary number of time scales.

The information that resides on the graph typically consists of integers, real numbers, or vectors, but could in principle be any mathematical objects. The state transition and learning rules can potentially be any type of dynamical system. For systems with continuous states and continuous parameters the natural dynamics are ordinary differential equations or discrete time maps. In principle, the states or parameters could also be functions whose dynamics are partial differential equations or functional maps. This might be natural, for example, in a more realistic model of neurons where the spatio-temporal form of pulse propagation in the axon is important [60]. When the activities or parameters are integers, their dynamics are naturally automata, although it is also common to use continuous dynamics even when the underlying states are discrete.

Since the representation of the graph is intrinsically discrete, the graph dynamics usually has a different character. Often, as in classifier systems, immune networks, or autocatalytic networks, the graph dynamics contains random elements. In other cases, it may be a deterministic response to statistical properties of the node states or the connection strengths, for example, as in pruning algorithms. Dynamical systems with graph dynamics are sometimes called *metadynamical systems* [20, 8].

In all of the models discussed here the states of the system reside on the nodes of the graph[#1]. The states are denoted x_i, where i is an integer labeling the node. The parameters reside at either nodes or connections; θ_i refers to a *node parameter* residing at node i, and w_{ij} refers to a *connection parameter* residing at the connection between node i and node j.

The degree to which the activity at one node influences the activity at another node, or the *connection strength*, is an important property of connectionist models. Although this is often controlled largely by the connection parameters w_{ij}, the node parameters θ_i may also have an influence, and in some cases, such as B-cell immune networks, provide the *only* means of changing the average connection strength. Thus, it is misleading to assume that the connection parameters are equivalent to the connection strengths. Since the connection strength of any given instant may vary depending on the states of the system, and since the form of the dynamics may differ considerably in different models, we need to discuss connection

---

[#1] It is also possible that states could be attached to connections, but this is not the case in any of the models discussed here.

strength in terms of a quantity that is representation-independent, which is well defined for any dynamical model.

For a continuous transition rule the natural way to discuss the connection strength is in terms of the Jacobian. When the transition rule is an ordinary differential equation, of the form

$$\frac{dx_i}{dt} = f_i(x_1, x_2, \ldots, x_N),$$

the *instantaneous connection strength* of the connection from node i to node j (where i is an input to j) is the corresponding term in the Jacobian matrix

$$J_{ji} = \frac{\partial x_j}{\partial x_i} = \frac{\partial f_j}{\partial x_i}.$$

A connection is excitatory if $J_{ji} > 0$ and inhibitory if $J_{ji} < 0$. Similarly, for discrete time dynamical systems (continuous maps), of the form

$$x_j(t+1) = f_j(x_1, x_2, \ldots, x_N),$$

a connection is excitatory if $|J_{ji}| > 1$ and inhibitory if $|J_{ji}| < 1$. In a continuous system, the average connection strength is $\langle J_{ji} \rangle$, where $\langle \ \rangle$ denotes an appropriate average; in a discrete system it is $\langle |J_{ji}| \rangle$. To make this more precise it is necessary to specify the ensemble over which the average is taken.

For automaton transition rules, since the states x_i are discrete the notion of instantaneous connection strength no longer makes sense. The average connection strength may be defined in one of many ways; for example, as the fraction of times node j changes state when node i changes state. In situations where x_i is an integer but nonetheless approximately preserves continuity, if $|\Delta x_i(t)|$ is the magnitude of the change in x_i at time t, the average connection strength can be defined as

$$\left\langle \frac{|\Delta x_j(t+1)|}{|\Delta x_i(t)|} \right\rangle_{|\Delta x_i(t)| > 0}.$$

3. Neural nets

3.1. Background

Neural networks originated with early work of McCulloch and Pitts [42], Rosenblatt [58], and others. Although the form of neural networks was originally motivated by neurophysiology, their properties and behavior are not constrained by those of real neural systems, and indeed are often quite different. There are two basic applications for neural networks: one is to understand the properties of real neural systems, and the other is for machine learning. In either case, a central question for developing a theory of learning is: Which behaviors of real neurons are essential to their information processing capabilities, and which are simply irrelevant side effects?

For machine learning problems neural networks have many uses that go considerably beyond the problem of modeling real neural systems. There are several reasons for dropping the constraints of modeling real neurons:

(i) We do not understand the behavior of real neurons.

(ii) Even if we understood them, it would be computationally inefficient to implement the full behavior of real neurons.

(iii) It is unlikely that we need the full complexity of real neurons in order to solve problems in machine learning.

(iv) By experimenting with different approaches to simplified models of neurons, we can hope to extract the basic principles under which they operate, and discover which of their properties are truly essential for learning.

Because of the factors listed above, for machine learning problems there has been a movement towards simpler artificial neural networks that are less motivated by real neural networks. Such networks are often called "artificial neural networks", to distinguish them from the real thing, or from more realistic models. Similar arguments apply to all the models discussed here; it might also be appropriate to say "artificial immune networks" and "artificial autocatalytic networks". However, this is cumbersome and I will assume that the distinction between the natural and artificial worlds is taken for granted.

Neural networks are constructed with simple units, often called "neurons". Until about five years ago, there were almost as many different types of neural networks as there were active researchers in the field. In the simplest and probably currently most popular form, each neuron is a simple element that sums its inputs with respect to weights, subtracts a threshold, and applies an *activation function* to the result. If we assume that time is discrete so that we can write the dynamics as a map, then we have

$t = 1, 2, \ldots$ = time;
$x_i(t)$ = state of neuron i;
w_{ij} = weight of connection from i to j;
θ_j = threshold;
S = the activation function, often a sigmoidal function such as tanh.

The response of a single neuron can be characterized as

$$x_j(t+1) = S\left(\sum_i w_{ij} x_i(t) - \theta_j\right). \qquad (3)$$

We could also write the dynamics in terms of automata, differential equations, or, if we assume that the neurons have a refractory period during which they do not change their state, as delay differential equations.

The instantaneous connection strength is

$$\frac{\partial x_j(t+1)}{\partial x_i(t)} = w_{ij} S'\left(\sum_i w_{ij} x_i(t) - \theta_j\right), \qquad (4)$$

where S' is the derivative of S. If S is a sigmoid, then S' is always positive and a connection with $w_{ij} > 0$ is always excitatory and a connection with $w_{ij} < 0$ is always inhibitory.

A currently popular procedure for constructing neural networks is to line the neurons up in rows, or "layers". A standard architecture has one layer of input units, one or two layers of "hidden" units, and a layer of output units, with full connections between adjacent layers. For a *feed-forward* architecture the graph has no loops so that the fixed parameters information flows only in one direction, from the inputs to the outputs. If the graph has loops so that the activity of a neuron feeds back on itself then the network is *recurrent*.

For layered networks it is sometimes convenient to assign the neurons an extra label that indicates which layer they are in. For feed-forward networks the dynamics across layers is particularly simple, since first the input layer is active, then the first hidden layer, then the next, etc., until the output layer is reached. If, for definiteness, we choose tanh as the activation function, and let 1 refer to the input layer, 2 to the first hidden layer, etc., the dynamics can be described by eq. (5). Note that because the activity of each layer is synchronized and depends only on that of the previous layer at the previous time step, the role of time is trivial. Since each variable only changes its value once during a given feed-forward step, we can drop time labels without ambiguity:

$$x_{2j} = \tanh\left(\sum_i w_{1ji} x_{1i} - \theta_{1j}\right),$$

$$x_{3k} = \tanh\left(\sum_j w_{2kj} x_{2j} - \theta_{2k}\right),$$

$$x_{4l} = \tanh\left(\sum_k w_{3lk} x_{3k} - \theta_{3l}\right). \qquad (5)$$

From this point of view the neural network simply implements a particular family of nonlinear functions, parameterized by the weights w and the thresholds θ [22]. For feed-forward networks the transition rule dynamics is equivalent to a single (instantaneous) mapping. For a recurrent network, in contrast, the dynamics is no longer trivial; any given neuron can change state more than once during a computation. This more interesting dynamics effectively gives the network a memory, so that the set of functions that can be implemented with a given number of neurons is much larger. However, it becomes necessary to make a decision as to when the computation is completed, which complicates the learning problem.

To solve a given problem we must select values of the parameters w and θ, i.e. we must select a particular member of the family of functions specified by the network. This is done by a learning rule.

The Hebbian learning rules are perhaps the simplest and most time honored. They do not require detailed knowledge of the desired outputs, and are easy to implement locally. The idea is simply to strengthen neurons with coincident activity. One simple implementation changes the weights according to the product of the activities on each connection,

$$\Delta w_{ij} = c x_i x_j. \tag{6}$$

Hebbian rules are appealing because of their simplicity and particularly because they are local. They can be implemented under very general circumstances. However, learning with Hebbian rules can be ineffective, particularly when there is more detailed knowledge available for training. For example, in some situations we have a training set of patterns for which we know both the correct input and the correct output. Hebbian rules fail to exploit this information, and are correspondingly inefficient when compared with algorithms that do.

Given a learning set of desired input/output vectors, the parameters of the network can be determined to match these input/output vectors by minimizing an error function based on them. The back-propagation algorithm, for example, minimizes the least mean square error and is effectively a nonlinear least-squares fitting algorithm. For more on this, see ref. [59].

Since there is an extensive and accessible literature on neural networks, I will not review it further [59, 15].

3.2. Comparison to a generic network

Neural networks are the canonical example of connectionism and their mapping into generic connectionist terms is straightforward.

Nodes correspond to neurons.

Connections correspond to the axons, synapses, and dendrites of real neurons. The average connection strength is proportional to the weight of each connection.

Node dynamics. There are many possibilities. For feed-forward networks the dynamics is reduced to function evaluation. For recurrent networks the node dynamics may be an automaton, a system of coupled mappings, or a system of ordinary differential equations. The attractors of such systems can be fixed points, limit cycles, or chaotic attractors. More realistic models of the refractory periods of the neurons yield systems of delay-differential equations.

Learning rules. Again, there are many possibilities. For feed-forward networks with carefully chosen neural activation functions such as radial basis functions [11, 13, 54] where the weights can be solved through a linear algorithm, the dynamics reduces to a function evaluation. Nonlinear search algorithms such as back-propagation are nonlinear mappings which usually have fixed point attractors. Nondeterministic algorithms such as simulated annealing have stochastic dynamics.

Graph dynamics. For real neural systems this corresponds to plasticity of the synapses. There is increasing evidence that plasticity plays an important role, even in adults [2]. As currently practiced, most neural networks do not have explicit graph dynamics; the user simply tinkers with the architecture attempting to get good results. This approach is clearly limited, particularly for large problems where the graph must be sparse and the most efficient way to restrict the architecture is not obvious from the symmetries of the problem. There is currently a great deal of interest in implementing graph dynamics for neural networks, and there are already some results in this direction [26, 43, 46, 65, 67]. This is likely to become a major field of interest in the future.

4. Classifier systems

4.1. Background

The classifier system is an approach to machine learning introduced by Holland [30]. It was inspired by many influences, including production systems in artificial intelligence [48], population genetics, and economics. The central motivation was to avoid the problem of brittleness encountered in expert systems and conventional approaches to artificial intelligence. The classifier system learns and adapts using a low-level abstract representation that it constructs itself, rather than a high-level explicit representation constructed by a human being.

On the surface the classifier system appears quite different from a neural network, and at first glance it is not obvious that it is a connectionist system at all. On closer examination, however, classifier systems and neural networks are quite similar. In fact, by taking a sufficiently broad definition of "classifier systems" and "neural networks", any particular implementation of either one may be viewed as a special case of the other. Classifier systems and neural networks are part of the same class of models, and represent two different design philosophies for the connectionist approach to learning. The analogy between neural networks and classifier systems has been explored by Compiani et al. [14], Belew and Gherrity [9], and Davis [16]. There are many different versions of classifier systems; I will generally follow the version originally introduced by Holland [30], but with a few more recent modifications such as intensity and support [31].

At its core, the classifier system has a rule-based language with content addressable memories. The addressing of instructions occurs by matching of patterns or rules rather than by the position of the instructions, as it does in traditional von Neumann languages. Each rule or *classifier* consists of a condition and an action, both of which are fixed length strings. One rule invokes another when the action part of one matches the condition part of the other. This makes it possible to set up a chain of associations; when a given rule is active it may invoke a series of other rules, effecting a computation. The activity of the rules is mediated by a *message list*, which serves as a blackboard or short-term memory on which the rules post messages for each other. While many of the messages on the list are posted by other classifiers, some of them are also external messages, inputs to the program posted by activity from the outside world. In the most common implementations the message list is of fixed length, although there are applications where its length may vary. See the schematic diagram shown in fig. 3. You may also want to refer to the example in section 5.

The conditions, actions, and messages are all strings of the same fixed length. The messages are strings over the binary alphabet $\{0,1\}$, while the conditions and actions are over the alphabet $\{0, 1, \#\}$, where $\#$ is a "wildcard" or "don't care" symbol. The length of the message list controls how many messages can be active at a given time, and is typically much smaller than the total number of rules.

The way in which a classifier system "executes programs" is apparent by examining what happens during a cycle of its operation. At a given time, suppose there is a set of messages on the message list, some of which were posted by other classifiers, and some of which are inputs from the external world. The condition parts of all the rules are matched against all the messages on the message list. A match occurs if each symbol matches with the symbol in the corresponding position. The symbol $\#$ matches everything. The rules that make matches on a given time step post their

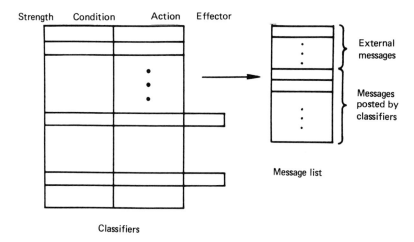

Fig. 3. A schematic diagram of the classifier system.

actions as messages on the next time step. By going through a series of steps like this, the classifier system can perform a computation. Note that in most implementations of the classifier system each rule can have more than one condition part; a match occurs only when both conditions are satisfied.

In general, because of the # symbol, more than one rule may match a given message. The parameters of the classifier system (frequency of #, length of messages, length of message list, etc.) are usually chosen so that the number of matches typically exceeds the size of the message list. The rules then *bid* against each other to decide which of them will be allowed to post messages. The bids are used to compute a threshold, which is adjusted to keep the number of messages on the message list (that will be posted on the next step) less than or equal to the size of the message list. Only those rules whose bids exceed the threshold are allowed to post their messages on the next time step[#2].

An important factor determining the size of the bid is the *strength* of a classifier, which is a real number attached to each classifier rule. The strength is a central part of the learning mechanism. If a classifier wins the bidding competition and successfully posts a message, an amount equal to the size of its bid is subtracted from its strength and divided among the classifiers that (on the previous time step) posted the messages that match the bidding classifier's condition parts on the current time step[#3].

Another factor in determining the size of bids is the *specificity* of a classifier, which is defined as the percentage of characters in its condition part that are either zero or one, i.e. that are not #. The motivation is that when there are "specialists" to solve a problem, their input is more valuable than that of "generalists".

The final factor that determines the bid size is the *intensity* $x_i(t)$ associated with a given message. In older implementations of the classifier system, the intensity is a Boolean variable, whose value is one if the message is on the message list, and zero otherwise. In newer implementations the intensity is allowed to take on real values $0 \le x_i \le 1$. Thus, some messages on the list are "more intense" than others, which means they have more influence on subsequent activity. Under the *support rule*, the intensity of a message is computed by taking the sum over all the matching messages on the previous time step, weighted by the strength of the classifier making the match.

[#2]Some implementations allow stochastic bidding.

[#3]Other variants are also used. Many authors think that this step is unnecessary, or even harmful; this is a topic of active controversy.

The size of a bid is

$$\text{bid} = \text{const} \times w \times \text{specificity} \times F(\text{intensity}). \quad (7)$$

$F(\text{intensity})$ is a function of the intensities of the matching messages. There are many options; for example, it can be the intensity of the message generating the highest bid, or the sum of the intensities of all the matching messages [57].

To produce outputs the classifier system must have a means of deciding when a computation halts. The most common method is to designate certain classifiers as outputs. When these classifiers become active the classifier system makes the output associated with that classifier's message. If more than one output classifier becomes active it is necessary to resolve the conflict. There are various means of doing this; a simple method is to simply pick the output with the largest bid.

Neglecting the learning process, the state of a classifier system is determined by the intensities of its messages (most of which may be zero). In many cases it is important to be able to pass along a particular set of information from one time step to another. This is done by a construction called *pass-through*. The # symbol in the action part of the rule has a different meaning than it has in the condition part of the rule. In the action part of the rule it is used to "pass through" information from the message list on one time step to the message list on the next time step; anywhere there is a # symbol in the action part, the message that is subsequently posted contains either a zero or a one according to whether the *message matched by the condition part* on the previous time step contained a zero or a one.

The procedure described above allows the classifier system to implement any finite function, as long as the necessary rules are present in the system with the proper strengths (so that the correct rules will be evoked). The transfer of strengths according to bid size defines a learning algorithm called the *bucket brigade*. The problem of making sure the necessary rules are present is addressed by the use of *genetic algorithms* that operate on the bit strings of the rules as though they were haploid chromosomes. For example, *point mutations* randomly changes a bit in one of the rules. *Crossover* or *recombination* mimics sexual reproduction. It is performed by selecting two rules, picking an arbitrary position, and interchanging substrings so that the left part of the first rule is concatenated to the right part of the second rule and vice versa. When the task to be performed has the appropriate structure, crossover can speed up the time required to generate a good set of rules, as compared to pure point mutation[#4].

4.2. Comparison to generic network

The classifier system is rich with structure, nomenclature, and lore, and has a literature of its own that has evolved more or less independently of the neural network literature. Nonetheless, the two are quite similar, as can be seen by mapping the classifier system to standard connectionist terms.

For the purpose of this discussion we will assume that the classifiers only have one condition part. The extension to classifiers with multiple condition parts has been made by Compiani et al. [14].

Nodes. The messages are labels for the nodes of the connectionist network. For a classifier system with word length N the 2^N possible messages range from $i = 0, 1, \ldots, 2^N - 1$. (In practice, for a given set of classifiers, only a small subset of these may actually occur.) The state of the ith node is the intensity x_i. The node activity also depends on a globally defined threshold $\theta(t)$, which varies in time.

Connections. The condition and action parts of the classifier rules are a connection list representation of a graph, in the form of eq. (2). Each

---

[#4] Several specialized graph manipulation operators, for example triggered cover operators, have also been developed for classifier systems [57].

classifier rule connects a set of nodes $\{i\}$ to a node j and can be written $\{i\} \to j$. A rule consisting entirely of ones and zeros corresponds to a single connection; a rule with n don't care symbols represents 2^n different connections. Note that if two rules share their output node j and some of their input nodes i then there are multiple connections between two nodes. The connection parameters w_{ij} are computed as the product of the classifier rule strength and the classifier rule specificity, i.e.

w_{ij} = specificity × strength.

When the graph is sparse there are many nodes that have no rule connecting them so that implicitly $w_{ij} = 0$.

Note that only the connections are represented explicitly; the nodes are implicitly represented by the right-hand parts of the connection representations, which give all the nodes that could ever conceivably become active. Thus nodes with no inputs are not represented. This can be very efficient when the graph is sparse.

Although on the surface pass-through appears to be a means of keeping recurrent information, as first pointed out by Miller and Forrest [44], in connectionist terms it is a mechanism for efficient graph representation. Pass-through occurs when a classifier has # symbols at the same location in both its condition and action parts. (If the # is only in the action part, then the pass-through value is always the same, and so it is irrelevant.) The net effect is that the node that is activated on the output depends on the node that was active on the input. This amounts to representing more than one connection with a single classifier. For example, consider the classifier $0\# \to 1\#$. If node 00 becomes active, then the second 0 is "passed through", so the output is 10. Similarly, if 01 becomes active, the output is 11. The net result is that two connections are represented by the same classifier. From the point of view of the network, the classifier $0\# \to 1\#$ is equivalent to the two classifiers $00 \to 10$ and $01 \to 11$. The net effect is thus a more efficient graph representation, and pass-through is just a representational convenience.

Transition rule. In traditional classifier systems a node j becomes active on time step $t+1$ if it has an input connection i on time step t such that $x_i(t)w_{ij} > \theta$. Using the support rule,

$$x_j(t+1) = \sum_i x_i(t) w_{ij}, \qquad (8)$$

where the sum is taken over all i that satisfy $x_i(t)w_{ij} > \theta$. With the support rule the dynamics is thus piecewise linear, with nonlinearity due to the effect of the threshold θ. Without the support rule the intensity is $x_j(t+1) = \max_i\{x_i(t)\}$.

There are two approaches to computing the threshold θ. The simplest approach is to simply set it to a constant value θ. A more commonly used approach in traditional classifier systems is to adjust $\theta(t)$ on each time step so that the number of messages that are active on the message list is less than or equal to a constant, which is equivalent to requiring that the number of nodes active on a given time step is less than or equal to a constant. In connectionist terms this may be visualized as adding a special thresholding unit that has input and output connections to every node.

Learning rule. The traditional learning algorithm for classifier systems is the bucket brigade, which is a particular modified Hebbian learning rule. (See eq. (6).) When a node becomes active, strength is transferred from its active output connections to its active input connections. This transfer occurs on the time step after it was active. To be more precise, consider a wave of activity $x_j(t) > 0$ propagating through node j, as shown in fig. 4.

Suppose this activity is stimulated by m activities $x_i(t-1) > 0$ through input connection parameters w_{ij}, and in turn stimulates activities $x_k(t+1) > 0$ through output connection parameters w_{jk}. Letting H be the Heaviside function

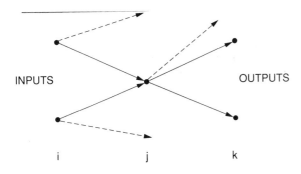

Fig. 4. *The bucket brigade learning algorithm.* A wave of activity propagates from nodes $\{i\}$ at time $t-1$ through node j at time t to nodes $\{k\}$ at time $t+1$. The solid lines represent active connections, and the dashed lines represent inactive connections. Strength is transferred from the input connections of j to output connections of j according to eq. (11). The motivation is that connections "pay" the connections that activate them.

$H(x) = 1$ for $x > 0$, $H(x) = 0$ for $x \leq 0$, the input connections gain strength according to

$$\Delta w_{ij} = \frac{x_j}{m} \sum_k w_{jk} H(x_j w_{jk} - \theta), \quad (9)$$

$$\Delta w_{jk} = -x_j w_{jk} H(x_j w_{jk} - \theta), \quad (10)$$

where

$$\Delta w_{ij} = w_{ij}(t+1) - w_{ij}(t). \quad (11)$$

All the quantities on the right-hand side are evaluated at time t.

This is only one of several variants of the bucket brigade learning algorithm; for discussion of other possibilities see ref. [10].

In order to learn, the system must receive feedback about the quality of its performance[#5]. To provide feedback about the overall performance

---

[#5] It is clearly important to maintain an appropriate distribution of strength within a classifier system, which does not overly favor input or output classifiers and which can set up chains of appropriate associations. Strength is added to classifiers that participate in good outputs, and then the bucket brigade causes a local transfer of feedback, in the form of connection strength, from outputs to inputs. This is further complicated by the recursive structure of classifier systems, which corresponds to loops in the graph. Maintaining an appropriate gradient of strength from outputs to inputs has proved to be a difficult issue in classifier systems.

of the system, the output connections of the system, or the *effectors*, are given strength according to the quality of their outputs. Judgements as to the quality must be made according to a predefined evaluation function. To prevent the system from accumulating useless classifiers, causing isolated connections, there is an activity tax which amounts to a dissipation term. Putting all of these effects together and following ref. [21] we can write the bucket brigade dynamics (the learning rule) as

$$\Delta w_{ij} = \frac{1}{m} \sum_k x_j w_{jk} H(x_j w_{jk} - \theta)$$
$$- x_i w_{ij} H(x_i w_{ij} \theta)$$
$$+ x_i P(t) + k w_{ij}, \quad (12)$$

where k is the dissipation rate for the activity tax, and $P(t)$ is the evaluation function for outputs at time t.

Graph dynamics. The graph dynamics occurs through manipulations of the graph representation (the classifier rules) through genetic algorithms such as point mutation and crossover. These operations are stochastic and are highly nonlocal; they preserve either the input or output of each connection, but the other part can move to a very different part of the graph. The application of these operators generates new connections, which is usually accompanied by the removal of other connections.

4.3. An example

An example makes the graph-theoretic view of classifier systems clearer. For example, consider the classic problem of exclusive-or. (See also ref. [9].) The exclusive-or function is 0 if both inputs are the same and 1 if both inputs are different. The standard neural net solution of this problem is easily implemented with three classifiers:

(i) $0\# \to 10: +1$;
(ii) $0\# \to 11: +1$;
(iii) $10 \to 11: -2$.

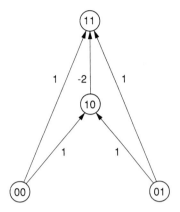

Fig. 5. A classifier network implementing the exclusive-or in standard neural net fashion. The binary numbers, which in classifier terms would be messages on the message list, label the nodes of the network.

Table 1
A wave of activity caused by the inputs (1,1) is shown. The numbers from left to right are the intensities on successive iterations. Initially the two input messages have intensity 1, and the others are 0. The input messages activate messages 10 and 11, and then 10 switches 11 off. For the input (0,0), in contrast, the network immediately settles to a fixed point with the intensities of all the nodes at zero.

node	intensity			
00	1	1	1	1
01	1	1	1	1
10	0	1	1	1
11	0	1	0	0

(The number after the colon is w = strength × specificity.) Although there are only three classifiers, because of the # symbols they make five connections, as shown in fig. 5.

With this representation the node 00 represents one of the inputs, and 01 represents the other input; the state of each input is its intensity. If both inputs are 1, for example, then nodes 00 and 01 become active, in other words, they have intensity > 0, which is equivalent to saying that the messages 00 and 01 are placed on the message list. Assume that we use the support rule, eq. (8), that outputs occur when the activity on the message list settles to a fixed point, and that the message list is large enough to accommodate at least four messages. An example illustrating how the computation is accomplished is shown in table 1.

This example is unusual from the point of view of common classifier system practice in several respects. (1) The protocol of requiring that the system settle to a fixed point in order to make an output. A more typical practice would be to make an output whenever one of the output classifiers becomes active. (2) The message list is rather large for the number of classifiers, so the threshold is never used. (3) There are no recursive connections (loops in the graph).

There are simpler ways to implement exclusive-or with a classifier system. For example, if we change the input protocol and let the input message be simply the two inputs, then the classifier system can solve this with four classifiers whose action parts are the four possible outputs. This always solves the problem in one step with a message list of length one. Note that in network terms this corresponds to unary inputs, with the four possible input nodes representing each possible input configuration. While this is a cumbersome way to solve the problem with a network, it is actually quite natural with a classifier system.

4.4. Comparison of classifiers and neural networks

There are many varieties of classifier systems and neural networks. Once the classifier system is described in connectionist terms, it becomes difficult to distinguish between them. In practice, however, there are significant distinctions between neural nets *as they are commonly used* and classifier systems *as they are commonly used*. The appropriate distinction is not between classifiers and neural networks, but rather between the two design philosophies represented by the typical implementations of connectionist networks within the classifier system and neural net communities. A comparison of classifier systems and neural networks in a common language illustrates their differences more clearly and suggests a natural synthesis of the two approaches.

Graph topology and representation. The connection list graph representation of the classifier system is efficient for sparse graphs, in contrast to the connection matrix representation usually favored by neural net researchers. This issue is not critical on small problems that can be solved by small networks which allow the luxury of a densely connected graph. On larger problems, use of a sparsely connected graph is essential. If a large problem cannot be solved with a sparsely connected network, then it cannot feasibly be implemented in hardware or on parallel machines where there are inevitable constraints on the number of connections to a given node.

To use a sparse network it is necessary to discover a network topology suited to a given problem. Since the number of possible network topologies is exponentially large, this can be difficult. For a classifier system the sparseness of the network is controlled by the length of each message, and by the number of classifiers and their specificity. Genetic algorithms provide a means of discovering a good network, while maintaining the sparseness of the network throughout the learning process. (Of course, there may be problems with convergence time.) For neural nets, in contrast, the most commonly used approach is to begin with a network that is fully wired across adjacent layers, train the network, and then prune connections if their weights decay to zero. This is useless for a large problem because of the dense network that must be present at the beginning.

The connection list representation of the classifier system, which can be identified with that of production systems, potentially makes it easier to incorporate prior knowledge. For example, Forrest has shown that the semantic networks of KL-One can be mapped into a classifier system [23]. On the other hand, another common form of prior knowledge occurs in problems such as vision, when there are group invariances such as translation and rotation symmetry. In the context of neural nets, Giles et al. [25] have shown that such invariances can be hard-wired into the network by restricting the network weights and connectivity in the proper manner. This could also be done with a classifier system by imposing appropriate restrictions on the rules produced by the genetic algorithm.

Transition rule. Typical implementations of the classifier system apply a threshold to each input separately, before it is processed by the node, whereas in neural networks it is more common to combine the inputs and then apply thresholds and activation functions. It is not clear which of these approaches is ultimately more powerful, and more work is needed.

Most implementations of the classifier system are restricted to either linear threshold activation functions or maximum input activation functions. Neural nets, in contrast, utilize a much broader class of activation functions. The most common example is probably the sigmoid, but in recent work there has been a move to more flexible functions, such as radial basis functions [11, 13, 47, 54] and local linear functions [22, 35, 68]. Some of these functions also have the significant speed advantage of linear learning rules[#6]. In smooth environments, smooth activation functions allow more compact representations. Even in environments where a priori it is not obvious the smoothness plays a role, such as learning Boolean functions, smooth functions often yield better generalization results and accelerate the learning process [68]. Implementation of smoother activation functions may improve performance of classifier systems in some problems.

Traditionally, classifier systems use a threshold computed on each time step in order to keep the number of active nodes below a maximum value. Computation of the threshold in this way requires

---

[#6]Linear learning rules are sometimes criticized as "not local". Linear algorithms are, however, easily implemented in parallel by systolic arrays, and converge in logarithmic time.

a global computation that is expensive from a connectionist point of view. Future work should concentrate on constant or locally defined thresholds.

From a connectionist point of view, classifiers with the # symbol correspond to multiple connections constrained to have the same strength. There is no obvious reason why their lack of specificity should give them less connection strength. This intuition seems to be borne out in numerical experiments using simplified classifier systems [66].

Learning rule. The classifier system traditionally employs the bucket brigade learning algorithm, whose feedback is condensed into an overall performance score. In problems where there is more detailed feedback, for example a set of known input–output pairs, the bucket-brigade algorithm fails to use this information. This, combined with the lack of smoothness in the activation function, causes it to perform poorly in problems such as learning and forecasting smooth dynamical systems [55]. Since there are now recurrent implementations of back-propogation [53], it makes sense to incorporate this into a classifier system with smooth activation functions, to see whether this gives better performance on such problems [9].

For problems where there is only a performance score, the bucket brigade is more appropriate. Unfortunately, there have been no detailed comparisons of the bucket brigade algorithm against other algorithms that use "learning with a critic". The form of the bucket brigade algorithm is intimately related to the activation dynamics, in that the size of the connection strength transfers are proportional to the size of the input activation signal (the bid). Although coupling of the connection strength dynamics to the activation dynamics is certainly necessary for learning, it is not clear that the threshold activation level is the correct or only quantity to which the learning algorithm should be coupled. Further work is needed in this area.

5. Immune networks

5.1. Background

The basic task of the immune system is to distinguish between self and non-self, and to eliminate non-self. This is a problem of pattern learning and pattern recognition in the space of chemical patterns. This is a difficult task, and the immune system performs it with high fidelity, with an extraordinary capacity to make subtle distinctions between molecules that are quite similar.

The basic building blocks of the immune system are *antibodies*, "y" shaped molecules that serve as identification tags for foreign material; *lymphocytes*, cells that produce antibodies and perform discrimination tasks; and *macrophages*, large cells that remove material tagged by antibodies. Lymphocytes have antibodies attached to their surface which serve as antigen detectors. (See fig. 6.) For-

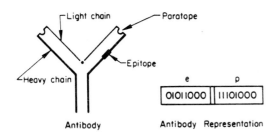

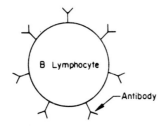

Fig. 6. A schematic representation of the structure of an antibody, an antibody as we represent it in our model, and a B-lymphocyte with antibodies on its surface that function as antigen detectors.

eign material is called *antigen*. A human contains roughly 10^{20} antibodies and 10^{12} lymphocytes, organized into roughly 10^8 distinct *types*, based on the chemical structure of the antibody. Each lymphocyte has only one type of antibody attached to it. Its type is equivalent to the type of its attached antibodies. The majority of antibodies are *free antibodies*, i.e. not attached to lymphocytes. The members of a given type form a clone, i.e. they are chemically identical.

The difficulty of the problem solved by the immune system can be estimated from the fact that mammals have roughly 10^5 genes, coding for the order of 10^5 proteins. An *antigenic determinant* is a region on the antigen that is recognizable by an antibody. The number of antigenic determinants on a protein such as myoglobin is the order of 50, with 6–8 amino acids per region. We can compare the difficulty of telling proteins apart to a more familar task by assuming that each antigenic determinant is roughly as difficult to recognize as a face. In this case the pattern recognition task performed by the immune system is comparable to recognizing a million different faces. A central question is the means by which this is accomplished. Does the immune system function as a gigantic look up table, like a neural network with billions of "grandmother cells"? Or, does it have an associative memory with computational capabilities?

The argument given above neglects the important fact that there are 10^5 distinct proteins only *if we neglect the immune system*. Each antibody is itself a protein, and there are 10^8 distinct antibody, which appears to be a contradiction: How do we generate 10^8 antibody types with only 10^5 genes? The answer lies in combinatorics. Each antibody is chosen from seven gene segments, and each gene segment is chosen from a "family" or set of possible variants. The total number of possible antibody types is then the product of the sizes of each gene family. This is not known exactly, but is believed to be on the order of 10^7–10^8. Additional diversity is created by *somatic mutation*. When the lymphocytes replicate, they do so with an unusually large error rate in their antibody genes. Although it is difficult to estimate the number of *possible* types precisely, it is probably much larger than the number of types that are actually present in a given organism.

The ability to recognize and distinguish self is *learned*. How the immune system accomplishes this task is unknown. However, it is clear that one of the main tools the immune system uses is *clonal selection*. The idea is quite simple: A particular lymphocyte can be stimulated by a particular antigen if it has a chemical reaction with it. Once stimulated it replicates, producing more lymphocytes of the same type, and also secreting free antibodies. These antibodies bind to the antigen, acting as a "tag" instructing macrophages to remove the antigen. Lymphocytes that do not recognize antigen do not replicate and are eventually removed from the system.

While clonal selection explains how the immune system recognizes and removes antigen, it does not explain how it distinguishes it from self. From both experiments and theoretical arguments, it is quite clear that this distinction is learned rather than hard-wired. Clonal selection must be suppressed for the molecules of self. How this actually happens is unknown.

A central question for self–nonself discrimination is: Where is the seat of computation? It is clear that a significant amount of computation takes place in the lymphocytes, which have a sophisticated repertoire of different behaviors. It is also clear that there are complex interactions between lymphocytes of the same type, for example, between the different varieties of T-lymphocytes and B-lymphocytes. These interactions are particularly strong during the early stages of development.

Jerne proposed that a significant component of the computational power of the immune system may come from the interactions of different types of antibodies and lymphocytes *with each other* [33, 34]. The argument for this is quite simple: Since antibodies are after all just molecules, then from the point of view of a given molecule other molecules are effectively indistinguishable from antigens. He proposed that much of the power of the immune system to regulate its own behavior may come from interacting antibodies and lymphocytes of many different types[#7].

There is good experimental evidence that network interactions take place, particularly in young animals. Using the nomenclature that an antibody that reacts directly with antigen AB1, an antibody that reacts directly with AB1 is AB2, etc., antibodies in categories as deep as AB4 have been observed experimentally[#8]. Furthermore, rats raised in sterile environments have active immune systems, with activity between types. Nonetheless, the relevance of networks in immunology is highly controversial.

5.2. Connectionist models of the immune system

While Jerne proposed that the immune system could form a network similar to that of the nervous system, his proposal was not specific. Early work on immune networks put this proposal into more quantitative terms, assuming that a given AB1 type interacted only with one antigen and one other AB2 type. These interactions were modeled in terms of simple differential equations whose three variables represented antigen, AB1, and AB2 [56, 28]. A model that treats immune interactions in a connectionist network[#9], allowing interactions between arbitrary types, was proposed in ref. [21]. The complicated network of chemical interactions between different antibody types, which are impossible to model in detail from first principles, was taken into account by constructing an artificial antibody chemistry. Each antigen and antibody type is assigned a random binary string, describing its "chemical properties". Chemical interactions are assigned based on complementary matching between strings. The strength of a chemical reaction is proportional to the length of the matching substrings, with a threshold below which no reaction occurs. Even though this artificial chemistry is unrealistic in detail, hopefully it correctly captures some essential qualitative features of real chemistry.

A model of gene shuffling provides metadynamics for the network. This is most realistically accomplished with a gene library of patterns, mimicking the gene families of real organisms. These families are randomly shuffled to produce an initial population of antibody types. This gives an initial assignment of chemical reactions, through the matching procedure described above, including rate constants and other parameters[#10]. Kinetic equations implement clonal selection; some types are stimulated by their chemical reactions, while others are suppressed. Types with no reactions are slowly flushed from the system so that they perish. Through reshuffling of the gene library new types are introduced to the system. It is also possible to stimulate somatic mutation through point mutations of existing types, proportional to their rate of replication.

It is difficult to model the kinetics of the immune system realistically. There are five different classes of antibodies, with distinct interactions and properties. There are different types of lymphocytes, including helper, killer and supressor T-cells, which perform regulatory functions, as well as B-cells, which can produce free antibodies. All of these have developmental stages, with different responses in each stage. Chemical reactions include cell–cell, antibody–antibody, and cell–antibody interactions. Furthermore, the responses of cells are complicated and often state dependent. Thus, any kinetic equations are necessarily highly approximate, and applicable to only a subset of the phenomena.

In our original model we omitted T-cells, treating only B-cells. (This can also be thought of as

---

[#7] Such networks are often called *idiotypic networks*.

[#8] This classification of antibodies should not be confused with their type; a given type can simultaneously be AB1 and AB2 relative to different antigens, and many different types may be AB1.

[#9] Another connectionist model with a somewhat different philosophy was also proposed by Hoffmann et al. [29].

[#10] The genetic operations described here are more sophisticated than those actually used in ref. [21]; more realistic mechanisms have been employed in subsequent work [50, 17, 18].

modeling the response to certain polymeric antigens, for which T-cells seem to be irrelevant.) We assumed that the concentration of free antibodies is in equilibrium with the concentration of lymphocytes, so that their populations can be lumped together into a single concentration variable. Since the characteristic time scale for the production of free antibodies is minutes or hours, while that of the population of lymphocytes is days, this is a good approximation for some purposes. It turns out, however, that separating the concentration of lymphocytes and free antibodies and considering the cell–cell, antibody–antibody, and cell–antibody reactions separately give rise to new phenomena that are important for the connectionist view. In particular, this generates a more interesting repertoire of steady states, including "mildly excited" self-stimulated states suggestive of those observed in real immune systems [50, 17, 18].

5.3. Comparison to a generic network

As with classifier systems and neural networks, there are several varieties of immune networks [21, 17, 29, 64], and it is necessary to choose one in order to make a comparison. The model described here is based on that of Farmer, Packard and Perelson [21], with some modifications due to later work by Perelson [50] and De Boer and Hogeweg [17]. Also, since this model only describes B-cells, whenever necessary I will refer to it as a B-cell network, to distinguish it from models that also incorporate the activity of T-cells.

To discuss immune networks in connectionist terms it is first necessary to make the appropriate map to nodes and connections. The most obvious mapping is to assign antibodies and antigens to nodes. However, since antibodies and antigens typically have more than one antigenic determinant, and each region has a distinct chemical shape[#11], we could also make the regions (or chemical shapes) the fundamental variable. Since all the models discussed above treat the concentration of antibodies and lymphocytes as the fundamental variables, I shall make the identification at this level. This leads to the following connectionist description:

Nodes correspond to antibodies, or more accurately, to distinct antibody types. Antigens are another type of node with different dynamics; from a certain point of view the antigen concentrations may be regarded as the input and output nodes of the network[#12]. The free antibody concentrations, which can change on a rapid time scale, are the states of the nodes. They are the immediate indicators of information processing in the network. The lymphocyte concentrations, which change on an intermediate time scale, are node parameters. (Recall that there is a one-to-one correspondence between free antibody types and lymphocyte types.) Changes in lymphocyte concentration are the mechanism for learning in the network.

Connections. The physical mechanisms which cause connections between nodes are chemical reactions between antibodies, lymphocytes, and antigens. The strength of the connections depends on the strength of the chemical reactions. This is in part determined by chemical properties, which are fixed in time, and in part by the concentrations of the antibodies, lymphocytes, and antigens, which change with time. Thus the instantaneous connection strength changes in time as conditions change in the network. The precise way of representing and modeling the connections is explained in more detail in the following.

Graph representation. To model the notion of "chemical properties" we assign each antibody type a binary string. To determine the rate of the chemical reaction between type i and type j, the binary string corresponding to type i is compared

---

[#11]"Chemical shape" here means all the factors that influence chemical properties, including geometry, charge, polarization, etc.

[#12]Future models should include chemical types identified with self as yet another type of node.

to binary string corresponding to type j. A match strength matrix m_{ij} is assigned to this connection, which depends on the degrees of complementary matching between the two strings. Types whose strings have a high degree of complementary matching are assigned large reaction rates. Since the matching algorithm is symmetric[#13] $m_{ij} = m_{ji}$.

There is a threshold for the length of the complementary matching region below which we assume that no reaction occurs and set $m_{ij} = 0$. Since m_{ij} is the connection matrix of the graph, setting $m_{ij} = 0$ amounts to deleting the corresponding connection from the graph. We thus neglect reactions that are so weak that they have an insignificant effect on the behavior of the network. The match threshold together with the length of the binary strings determines the sparseness of the graph. When the system is sparse the matrix m_{ij} can be represented in the form of a connection list. The match strength for a given pair of immune types does not change with time. However, as new types are added or deleted from the system, the m_{ij} that are relevant to the types *in the network* change.

The graph dynamics provides a mechanism of learning in the immune system; as new types are tested by clonal selection, the graph changes, and the system "evolves". Another mechanism for dynamical learning depends on the lymphocyte concentrations, as discussed below.

Dynamics. The m_{ij} are naturally identified as connection parameters for the network. For any given i and j, however, the m_{ij} are fixed. Thus, in B-cell immune networks the parameter dynamics, analogous to the learning rule in neural networks, occurs not by changing connection parameters, but rather by changing the lymphocyte concentration, which is a parameter node. The net reaction flux (or strength of the reaction) is a nonlinear function of the lymphocyte concentrations. Thus changing the lymphocyte concentration changes the effective connection strength. This is a fundamental difference between neural networks and B-cell immune networks; while the connection strength is changeable in both cases, in B-cell immune networks all the connection strengths to a given node change in tandem as the lymphocyte concentration varies. However, since the reaction rates are nonlinear functions, a change in lymphocyte concentration may affect each connection differently, depending on the concentration of the other nodes.

The dynamics of the real immune system are not well understood. The situation is similar to that of neural networks; we construct simplified heuristic immune dynamics based on a combination of chemical kinetics and experimental observations, attempting to recover some of the phenomena of real immune systems. The real complication arises because lymphocytes are cells, and understanding their kinetics requires understanding how they respond to stimulation and suppression by antigens, antibodies, and other cells. At this point our understanding of this is highly approximate and comes only from experimental data. The kinetic equations used in our original paper were highly idealized [21]. The more realistic equations quoted here are due to De Boer and Hogeweg[#14] [17].

Let i label the nodes of the system, x_i the concentration of antibodies, and θ_i the concentration of lymphocytes[#15]. The amount of stimulation received by lymphocytes of type i is approximated as

$$s_i = \sum_j m_{ij} x_j. \tag{13}$$

The rate of change of antibody concentration is

[#13] In our original paper [21] we also considered the case of asymmetric interactions. However, this is difficult to justify chemically, and it is probably safe to assume that the connections are symmetric [28].

[#14] More realistic equations have also been proposed by Segel and Perelson [61], Perelson [51, 50], and Varela et al. [64].

[#15] Note that I use θ to represent lymphocytes because they play the role of node parameters. However, they are not thresholds, but rather quantities whose primary function is to modify connection strength.

due to production by lymphocytes, removal from the system, and binding with other antibodies. The equations are

$$\frac{dx_i}{dt} = \theta_i f(s_i) - kx_i - cx_i s_i. \quad (14)$$

k is a dissipation constant and c the binding constant. f is a function describing the degree of stimulation of a lymphocyte. Experimental observations show that f is bell-shaped. A function with this rough qualitative behavior can be constructed by taking the product of a sigmoid with an inverted sigmoid, for example

$$f(z) = \frac{zk_2}{(k_1 + z)(k_2 + z)}. \quad (15)$$

The production of lymphocytes is due to replenishment by the bone marrow, cell replication, and removal from the system. The equations are

$$\frac{d\theta_i}{dt} = r + p\theta_i f(s_i) - k\theta_i. \quad (16)$$

r is the rate of replenishment and p is a rate constant for replication.

5.4. Comparison to neural networks and classifier systems

There are significant differences between the dynamics of immune networks and neural networks. The most obvious is in the form of the transition and learning rules. The nodes of the immune network are activated by a bell-shaped function rather than a sigmoid function. Since the bell-shaped function undergoes an inflection and its derivative changes sign, the dynamics are potentially more complicated.

B-cell immune networks differ from neural networks in that there is no variable which acts as a connection parameter. Instead, the connection strength is indirectly determined by the node parameters (concentrations and kinetic equations). The instantaneous connection strength is

$$\frac{\partial \dot{x}_i}{\partial x_j} = [\theta_i f'(s_i) - cx_i] m_{ij} - cs_i - k\delta_{ij}, \quad (17)$$

where $\delta_{ij} = 0$ for $i \neq j$, $\delta_{ii} = 1$. All of the terms in this equation except for f' are greater than or equal to zero. For low values of s_i, $f'(s_i) > 0$, but for large values of s_i, $f'(s_i) < 0$. Given the structure of these equations, as s_i increases, at some point before f reaches a maximum, all the connections to a given node change from excitatory to inhibitory. The point at which this happens depends on the lymphocyte concentration of i, the antibody concentration, the concentration of the other antibodies, and on the exact form of the stimulation function. Thus, in contrast to neural networks or the classifier system, a given connection can be either excitatory or inhibitory depending on the state of the system.

The connections in the immune system are chemical reactions. Insofar as the immune system is well stirred, this allows a potentially very large connectivity, as high as the number of different chemical types a given type can react with. In practice, the number of types that a given type reacts with can be as high as about 1000. Thus, the connectivity of real immune networks is apparently of the same order of magnitude as that of real neural networks.

One of the central differences between the B-cell immune networks and neural or classifier networks is that for the immune system there are no independent parameters on the connections. If the average strength of a connection to a given node cannot be adjusted independently of that of other nodes, the learning capabilities of the network may be much weaker or more inefficient than those of networks where the connection parameters are independent. As discussed in section 5.5, this may be altered by the inclusion of T-cells in the models.

5.5. Directions for future research

Whether immune networks are a major component of the computational machinery of the immune system is a subject of great debate. The analogy between neural networks and immune networks suggests that immune networks potentially possess powerful capabilities, such as associative memory, that could be central to the functioning of the immune system. However, before this idea can reach fruition we need more demonstrations of what immune networks can do. At this point the theory of immune networks is still in its infancy and their utility remains an open question.

The immune network may be able to perform tasks that would be impossible for individual cells. Consider, for example, a large antigen such as a bacterium with many distinct antigenic determinants. If each region is chemically distinct, a single type can interact with at most a few of them (and thus a single cell can interact with at most a few of them). Network interactions, in contrast, potentially allow different cells and cell types to communicate with each other and make a collective computation to reinforce or suppress each other's immune responses. For example, suppose A, B, C and D are active sites. It might be useful for a network to implement an associative memory rule such as: If any three of A, B, C, and D are present, then generate an immune response; otherwise do not. Such an associative memory requires the capability to implement a repertoire of Boolean functions. A useful rule might be: "Generate an immune response if active site A is present, or active site B ispresent, but not if both are present simultaneously". Such a rule, which is equivalent to taking the exclusive-or function of A and B, might be useful for implementing self tolerance. Such logical rules are easily implemented by networks. It is difficult to see how they could be implemented by individual cells acting on their own.

Immune memory is another task in which networks may play an essential role. Currently the prevailing belief is that immune memory comes about because of special memory cells. It is certainly true that some cells go into developmental states that are indicative of memory. Although the typical lifetime of a lymphocyte is about five days, there are some lymphocytes that have been demonstrated to persist for as long as a month. This is a far cry, however, from the eighty or more years that a human may display an immune memory. Since cells are normally flushed from the system at a steady rate, it is difficult to believe that any individual cell could last this long. It is only the type, then, that persists, but in order to achieve this individual cells must periodically replicate themselves. However, in order to hold the population stable the replication rate must be perfectly balanced against the removal rate. This is an unstable process unless there is feedback holding the population stable. It is difficult to see how feedback on the population size can be given unless there are network interactions.

In an immune network a memory can potentially be modeled by a fixed point of the network. The concentrations at the fixed point are held constant through the feedback of one type to another type. Models of the form of eqs. (14) and (16) contain fixed points that might be appropriate for immune memory. However, it is clear from experiments that T-cells are necessary for memory, and so must be added to immune networks to recover this effect.

T-cells are a key element missing from most current immune network models. T-cells play an important role in stimulating or suppressing reactions between antibodies and antigens, and are essential to immune memory. From the point of view of learning in the network, they may also indirectly act as specific connection parameters.

One of the most interesting activities of the immune system is "antigen presentation". When a B-cell or macrophage reacts with an antigen it may process it, discarding all but the antigenic determinants. It then presents the antigenic determinant on its surface (as a peptide bound to an MHC molecule). The T-cell reacts with the antigenic determinant and the B-cell, and based on this information may either stimulate or suppress the B-cell. Note that antigen presentation provides information about *both* the B-cell *and* an antigen, and thus potentially about a specific connection in the network.

In a connectionist model, this may amount to a connection strength parameter; a B-cell presenting a given active site contains information that is specific to two nodes, one for the B-cell of the same type as the T-cell, and one for the antigen whose active site is being presented (which may also be another antibody). Due to their interactions with T-cells, the B-cell populations of type i presenting antigenic determinants from type j may play the roles of the connection parameters w_{ij}.

At this point, it is not clear how strongly the absence of explicit connection parameters limits the computational and learning power of immune networks. However, it seems likely that before they can realize their full potential, connection parameters must be included, taking into account the operation of T-cells. T-cells act like catalysts, either suppressing or enhancing reactions. Since catalytic activity is one of the primary tools used to implement the internal functions of living organisms, it is not surprising that it should play a central role in the immune system as well. Autocatalytic activity is discussed in more detail in section 6.

6. Autocatalytic networks

6.1. Background

All the models discussed so far are designed to perform learning tasks. The autocatalytic network model of this section differs in that it is designed to solve a problem in evolutionary chemistry. Of course, evolution may also be regarded as a form of learning. Still, the form that learning takes in autocatalytic networks is significantly different from the other models discussed here.

The central goal of the autocatalytic network is to solve a classic problem in the origin of life, namely, to demonstrate an evolutionary pathway from a soup of monomers to a polymer metabolism with selected autocatalytic properties, which in turn could provide a substrate for the emergence of contemporary (or other) life forms. When Miller and Urey discovered that amino acids could be synthesized de novo from the hypothetical primordial constituents "earth, fire and water" [45], it seemed but a small step to the synthesis of polymers built out of amino acids (polypeptides and proteins). It was hoped that RNA and DNA could be created similarly. However, under normal circumstances longer polymers are not favored at equilibrium. Living systems, in contrast, contain DNA, RNA, and proteins, specific long polymers which exist in high concentration. They are maintained in abundance by their symbiotic relationship with each other: Proteins help replicate RNA and DNA, and DNA and RNA help synthesize proteins. Without the other, neither would exist. How did such a complex system ever get started, unless there were proteins and RNA to begin with? The question addressed in refs. [36, 20, 8] is: Under what circumstances can the synthesis of specific long polymers be achieved beginning with simple constituents such as monomers and dimers?

The model here applies to any situation in which unbranched polymers are built out of monomers through a network of catalytic activity. The monomers come from a fixed alphabet, a, b, c, ... They form one-dimensional chains which are represented as a string of monomers, acabbacbc... The monomer alphabet could be the twenty amino acids, or it could equally well be the four nucleotides. This changes the parameters but not the basic properties of the model. The model assumes that the polymers have catalytic properties, i.e. that they can undergo reactions in which one polymer catalyzes the formation of another. If A, B, C, and E are polymers, and H is water, then the basic reaction is:

$$A + B \overset{E}{\rightleftharpoons} C + H, \qquad (18)$$

where E is written over the arrows to indicate that it catalyzes the reaction.

Our purpose is to model a chemostat, a reaction vessel into which monomers are added at a steady rate. The chemical species that are added to the chemostat are called the *food set*. We assume that the mass in the vessel is conserved, for example, by simply letting the excess soup overflow. For convenience we assume that the soup is well stirred, so that we can model it by a system of ordinary differential equations.

In any real system it is extremely difficult to determine from first principles which reactions will be catalyzed, and with what affinity. Very few if any of the relevant properties have been measured experimentally in any detail, and the number of measurements or computations that would have to be made in order to predict all the chemical properties is hopelessly complex. Our approach is to invent an artificial chemistry and attempt to make its properties at least qualitatively similar to those of a real chemical system. Actually we use one of two different artificial chemistries, based on two different principles:

(i) Random assignment of catalytic properties.

(ii) Assignment of catalytic properties based on string matching.

These two simple artificial chemistries lie on the borders of extreme behavior in real chemistry. In some cases, we know that changing one monomer can have a dramatic effect on the chemical properties of a polymer, either because it causes a drastic change in the configuration of the polymer or because it alters a critical site. If this were always the case, then random chemistry would be a reasonable model.

In other cases, changing a monomer has only a small effect on the chemical properties. Our string matching model is closer to this case; altering a single monomer will only change the quality of matching between two strings by an incremental amount, and should never cause a dramatic alteration in the chemical properties of the polymer.

Another difficulty of modeling real chemistry is that there is an extraordinarily large number of possible reactions. In a vessel with all polymers of length l or less, for example, the total number of polymer species is $\sum_{i=1}^{i=l} m^i$, where m is the number of distinct monomers. For example, with $m = 20$ and $l = 100$, the number of polymer species is in excess of 20^{100}, an extremely large number, and the number of possible reactions is still larger than this. To get around this problem, to first approximation, we neglect spontaneous reactions, and assume that the catalytic properties are sufficiently strong that all catalyzed reactions are much faster than spontaneous reactions[#16].

Once we have assigned chemical properties, we can represent the network of catalyzed chemical reactions as a graph, or more precisely, as a polygraph with two types of nodes and two types of connections [20]. Because of catalysis the graph must be more complicated than for any of the other networks discussed so far. An example is shown in fig. 7. One type of node is labeled by ovals containing the string representation of the polymer species. The other type of node corresponds to catalyzed reactions, and is labeled by black dots. The dark black connections are undirected (because the reactions are reversible), and connect each reaction to the three polymer species that participate in it; the dotted connections are directed, and connect the reaction to its catalysts. All the edges connect polymers to reactions, and each reaction has at least four connections, three connections for the reaction products and one or more for the catalyst(s). In this illustration we have labeled the members of the food set by double ovals.

If we use the random method of assigning chemical properties, then the graph is a random graph and can be studied using standard techniques. The probability p that a reaction selected at random will be catalyzed controls the ratio of connections to nodes. As p increases so does this

---

[#16] In more recent work [7] we make a tractable model for approximate treatment of spontaneous reactions by lumping together all the polymer species of a given length that are not in the autocatalytic network, assuming that they all have the same concentration. These can be viewed as a new type of node in the network. This allows us to include the effect of spontaneous reactions when necessary.

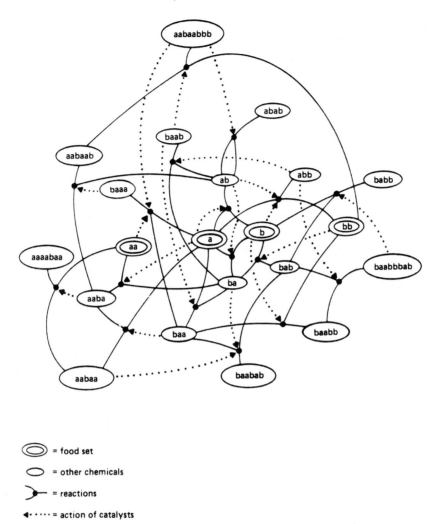

Fig. 7. The graph for an autocatalytic network. The ovals represent polymer species, labeled by strings. The black dots represent reactions. The solid lines are connections from polymer nodes to the reactions in which they participate. The dotted lines go from polymer species to the reactions they catalyze. The double ovals are special polymer nodes corresponding to the elements of the food set, whose concentrations are supplied externally.

ratio. As p grows the graph becomes more and more connected, i.e. more dense.

The graph-theoretic analysis only addresses the question of who reacts with whom, and begs the central (and much more difficult) question of concentrations. Numerical modeling of the kinetics *for any given catalyzed reaction* is straightforward but cumbersome. We introduced a simplified technique for treating catalyzed reactions of this type in ref. [20] that approximates the true catalyzed kinetics fairly well.

Modeling of the complete kinetics for an entire reaction graph is impossible, since the graph is infinite and under the laws of continuous mass action, even if we initialize all but a finite number of the species to zero concentration, an instant later they will all have non-zero concentrations. From a practical point of view, however, it is possible to circumvent this problem by realizing that any chemical reaction vessel is finite, and species whose continuous concentrations are significantly below the concentration corresponding to the presence of a single molecule are unlikely to participate in any reactions. Thus, to cope with this problem we introduce a concentration threshold, and only consider reactions where all the members on either side of the reaction equation (either A, B, and E, or C and E) are above the concentration threshold. This then becomes a metadynamical system: At any given time, only a finite number of species are above the threshold, and we only consider a finite graph. As the kinetics act, species may rise above the concentration threshold, so that the graph grows, or they may drop below the threshold, so that the graph shrinks.

One of the main goals of this model is to obtain closure in the form of an *autocatalytic set*, which is a set of polymer species such that each member of the set is produced by at least one catalyzed reaction involving only other members of the set (including the catalysts). Since the reactions are reversible, a species can be "produced" either by cleavage or condensation, depending on which side of equilibrium it finds itself. Thus an autocatalytic set can be quite simple; for example,

$$A + B \overset{A}{\rightleftharpoons} C + H \qquad (19)$$

is an autocatalytic set, and so is

$$A + B \overset{C}{\rightleftharpoons} C + H. \qquad (20)$$

A, B, and C will be regenerated by supplying either A and B, or by supplying C. Note, however, that such simple autocatalytic sets are only likely to occur when the probability of catalysis is very high. Even for small values of p it is always possible to find autocatalytic sets as long as the food set is big enough. However, the typical autocatalytic set is more complicated than the examples given in eqs. (19) and (20). There is a critical transition from the case where graphs with autocatalytic sets are very rare to that in which they are very common, as described in refs. [20, 36]. The results given there show that it is possible to create autocatalytic sets (in this graph theoretic sense) under reasonably plausible prebiotic conditions.

There are three notions of the formation of autocatalytic sets, depending on what we mean by "produced by" in the definition given above:

(i) *Graph theoretic*. The subgraph defined by the autocatalytic set is closed, so that each member is connected (by a solid connection) to at least one reaction catalyzed by another member.

(ii) *Kinetics*. Each member is produced at a level exceeding a given concentration threshold.

(iii) *Robust*. The autocatalytic set is robust under at least some changes in its food set, i.e. its members are at concentrations sufficiently large and there are enough pathways so that for some alterations of the food set it remains a kinetic autocatalytic set, capable of regenerating removed elements at concentrations above the threshold.

These notions are arranged in order of their strength, i.e. an autocatalytic set in the sense of kinetics is automatically an autocatalytic set in the graph-theoretic sense, and a robust autocatalytic set is automatically a kinetic autocatalytic set.

Describing the details of the conditions under which autocatalytic sets can be created is outside of the scope of this paper. Suffice it to say that, within our artificial chemistry we can create robust autocatalytic sets. Consider, for example, an autocatalytic set based on the monomers a and b, originally formed by a food set consisting of the species a, b, ab, and bb, as shown in fig. 8 and table 2.

We plot the concentrations of the 21 polymer species in the reactor against an index that is arbitrary except that it orders the species according to their length. We compare four different alterations of the original food set, all of which have the same rate of mass input. For two of the altered food sets the concentration of the members of the autocatalytic set remains almost the same; they are all maintained at high concentration. For the other two, the autocatalytic set "dies" in that some of the members of the set fall below the concentration threshold, and most of the concentrations decrease dramatically [7].

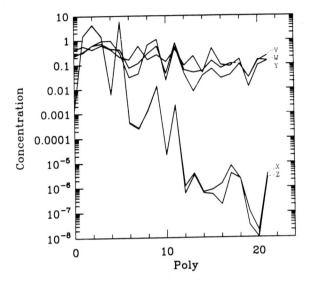

Fig. 8. An experiment demonstrating the robust properties of an autocatalytic set. The food set is originally a, b, ab, and bb. The food set is altered in four different ways, as shown in table 2. For each alteration of the food set the concentrations of all 21 polymers in the autocatalytic set are plotted against the "polymer index". (The polymer index assigns a unique label to each polymer. It is ordered according to length, but is otherwise arbitrary.) Two of the alterations of the food set cause the autocatalytic set to die, while the other two hardly change it. Like a robust metabolism, the autocatalytic set can digest a variety of different foods.

Table 2
An experiment in varying the food set of an autocatalytic set. The table shows the four species of the food set, and the concentration of each that is supplied externally per unit time. Case v is used to "grow" the autocatalytic set, and cases w–z are four changes made once the autocatalytic set is established. x and z kill the autocatalytic set, while w and y sustain it with only minimal alteration, as shown in fig. 8.

	a	b	ab	bb
v	5	5	5	5
w	5	0	5	7.5
x	0	0	10	5
y	10	20	0	0
z	0	10	10	0

Our numerical evidence suggests that any fixed reaction network always approaches a fixed point where the concentrations are constant. However, since spontaneous reactions always take place, there is the possibility that a new species will be created that is on the graph of the autocatalytic set, but which the kinetics did not yet reach. If the catalyzed pathway is sufficiently strong, then the new species may be regenerated and added to the (kinetic) autocatalytic set. This is the way the autocatalytic sets evolve; spontaneous reactions provide natural variation, and kinetics provides selection.

Autocatalytic networks create a rich, focused set of enzymes at high concentration. They form simple metabolisms, which might have provided a substrate for contemporary life.

The results discussed here, as well as many others, will be described in more detail in a future paper [7]. We intend to study the evolution of autocatalytic sets, and to make a closer correspondence to experimental parameter values.

6.2. Comparison to generic network

(i) *Nodes* correspond to both polymer species and to reactions. The states are determined by the concentrations of the polymers.

(ii) *Connections*. The graph connections are quite different in this system, in that there are no direct reaction connections to the same types of nodes. Each reaction node is connected by undirected links to exactly three polymer nodes, and contains one (directed) catalytic link to one or more polymer nodes. A polymer node can be connected by a solid link to any number of reaction nodes, and can have any number of catalytic links to reaction nodes.

(iii) *Dynamics*. The dynamics is based on the laws of mass action. The equations are physically realistic, and are considerably more complicated than those of the other networks we have discussed. Arbitrarily label all the polymer species by an index i, and let x_i represent the concentration of the ith species. Assume that all the forward reactions in eq. (18) have the same rate constant k_f, all the backward reactions have the same rate constant k_r, and that all catalyzed reactions have the same velocity v. Let the quantity m_{ijke} represent the connections in the two graphs, where i and j refer to the two species that join together to form k under enzyme e. $m_{ijke} = 1$ when there is a catalyzed reaction, and $m_{ijke} = 0$ otherwise. $m_{ijke} = m_{jike}$. Let the dissipation constant be k, let the rate at which elements are added to the foodset be d, and let h be the concentration of water. Neglecting the effects of enzyme saturation, the equations can be written

$$\frac{dx_k}{dt} = \sum_{e,i,j} m_{ijke}(1 + \nu x_e)(k_f x_i x_j - k_r h x_k)$$
$$+ 2 \sum_{l,m,e} m_{klme}(1 + \nu x_e)(k_r h x_m - k_f x_k x_l)$$
$$- k x_k + d f(x_k). \tag{21}$$

f is a function whose value is one if x_k is in the food set, and is zero otherwise. More accurate equations incorporating the effect of enzyme saturation are given in ref. [20].

An effective instantaneous connection strength can be computed by evaluating $\partial \dot{x}_k / \partial x_p$. The resulting expression is too complicated to write here. Like the immune network, the instantaneous connection strength can be either excitatory or inhibitory depending on where the network is relative to its steady state value. In contrast to the other networks we have studied, there are no special variables in eq. (21) that explicitly play the role of either node or connection parameters. The concentration of the enzymes x_e that catalyze a given reaction is suggestive of the connection parameters in other connectionist networks. However, since any species can be a reactant in one equation and an enzyme in another, there is no explicit separation of time scales between x_e and the other variables.

(iv) *Graph dynamics.* The separation of time scales usually associated with learning occurs entirely through modification of the graph. The deterministic behavior for any given graph apparently goes to a fixed point. However, in a real autocatalytic system there are always spontaneous reactions creating new species not contained in the catalytic reaction graph. It occasionally happens that one of the new species catalyzes a pathway that feeds back to create that species. Such a fluctuation can be amplified enormously, altering the part of the catalyzed graph that is above the concentration threshold. This provides a mechanism for the evolution of autocatalytic networks.

Autocatalytic networks are interesting from a connectionist point of view because of their rich graph structure and because of the possibilities opened up by catalytic activity. Catalytic activity is analogous to amplification in electronic circuits; it results in multiplicative terms that either amplify or suppress the activity of a given node. The fixed points of the network may be thought of as self-sustaining memories, caused by the feedback of catalytic activity. The dynamical equations that we use here are based on reversible chemical reactions, and lead to unique fixed points. However, other chemical reaction networks can have multiple fixed points, and it seems likely that when we alter the model to study irreversible reactions such as those observed in contemporary metabolisms, we will see multiple fixed points. In this case the computational possibilities of such networks become much more complex.

7. Other potential examples and applications

The four examples discussed here are by no means the only ones where connectionist models have been used, or could be used. Limitations of space and time prevent a detailed examination of all the possibilities, but a few deserve at least cursory mention.

Bayesian inference networks, Markov networks, and constraint networks are procedures used in artificial intelligence and decision theory for organizing and codifying casual relationships in complex systems [49]. Each variable corresponds to a node of the network. Each node is connected to the other variables on which it depends. Bayesian networks are based on conditional probability distributions, and use directed graphs; Markov networks are based on joint probability and have undirected graphs; constraint networks assume deterministic constraints between variables. These networks are most commonly used to incorporate prior knowledge, make predictions and test hypotheses. Learning good graph representations is an interesting problem where further work is needed.

Boolean networks. A neural network whose transition rule is a binary automaton is an example of a Boolean network. In general there is no need to restrict the dynamics to the sum and threshold rules usually used in neural nets (other than the fact that this may make the learning problem simpler). Instead, the nodes can implement arbitrary logical (Boolean) functions. Kauffman studied the emergent properties of networks in which each node implements a random Boolean function [38, 37]. (The functions are fixed in time, but each node implements a different function.) More recently, Miller and Forrest [44] have shown that the dynamics of classifier systems can be mapped into Boolean networks. This allows them to describe the emergent properties of classifier systems. Their work implicitly maps Boolean networks to the generic connectionist framework. The formulation of learning rules for general Boolean networks is an interesting problem that deserves further study. Kauffman has done some work using point mutation to modify the graph [39].

Ecological models and population genetics are a natural area for the application of connectionism. There is a large body of work modeling plant and animal populations and their interactions with their environment in terms of differential equations. In these models it is necessary to explicitly state how the populations interact, and translate this into mathematical form. An alternative is to let these interactions *evolve*. A natural framework for such models is provided by the work of Maynard Smith in the application of game-theoretic models to population genetics and ethology [63]. The interactions of the populations with each other are modeled as game-theoretic strategies. In these models, however, it is necessary to state in advance what these strategies are. A natural alternative is to let the strategies evolve. Some aspects of this have been addressed in the fledgling theory of evolutionary games [24]. A connectionist approach is a natural extension of this work. The immune networks discussed here are very similar to predator–prey models. The strings encoding chemical properties are analogous to genotypes of a given population, and the matrix of interactions are analogous to phenotypes.

Economics is another natural area of application. Again, existing game-theoretic work suggests a natural avenue for a connectionist approach, which could be implemented along the lines of the immune model. The binary strings can be viewed as encoding simple strategies, specifying the interactions of economic agents. Indeed, there are already investigations of models of this type based on classifier systems [4, 5, 41].

Game theory is a natural area of application. For example, Axelrod [6] has studied the game of iterated prisoner's dilemma. His approach was to encode recent past moves as binary variables, and encode the strategy of the player as a Boolean function. He demonstrated that genetic algorithms can be used to evolve Boolean functions that correspond to good strategies. An alternative approach would be to distribute the strategy over many nodes, and use a connectionist model instead of a look-up table. Such models may have applications in many different problems where evolutionary games are relevant, such as economics and ethology.

Molecular evolution models. The autocatalytic model discussed in detail here is by no means the only connectionist model for molecular evolution. Perhaps one of the earliest example is the hypercycle model of Eigen and Schuster [19], which has recently been compared to the Hopfield neural network models [32, 52]. For a review see ref. [27].

8. Conclusions

I hope that presenting four different connectionist systems in a common framework and notation will make it easier to transfer results from one field to another. This should be particularly useful in areas such as immune networks, where connectionist models are not as well developed as they are in other areas, such as neural networks. By showing how similar mathematical structure manifests itself in quite different contexts, I hope that I have conveyed the broad applicability of connectionism. Finally, I hope that these mathematical analogies make the underlying phenomena clearer. For example, comparing the role of the lymphocyte in these models to the role of neurons may give more insight into the construction of immune networks with more computational power.

8.1. Open questions

Hopefully the framework for connectionist models presented here will aid the development of a broader mathematical theory of connectionist systems. From an engineering point of view, the central question is: What is the most effective way to construct good connectionist networks? Questions that remain unclear include:

(i) In some systems, such as neural networks and classifier systems, a connection is always either inhibitory or excitatory. In others, such as immune networks and autocatalytic networks, a connection can be either inhibitory or excitatory, depending on the state of the system. Does the latter more flexible approach complicate learning? Does it give the network any useful additional computational power?

(ii) Is it essential to have independent parameters for each connection? In neural nets, each connection has its own parameter. In classifier systems, the use of the "don't care" symbol means that many connections are represented by one classifier, and thus share a common connection parameter. This decreases the flexibility of the network, but at the same time gives an efficient graph representation, and aids the genetic algorithms in finding good graphs. In B-cell immune networks the parameters reside entirely in the nodes, and thus as a single parameter changes many different connections are effected. Does this make it impossible to implement certain functions? How does this effect learning and evolution? (It is conceivable that the reduction of parameters may actually cause some improvements.)

(iii) What is the optimal level of complexity for the transition rule? Some neural nets and classifier systems employ simple activation functions, such as linear threshold rules. Somewhat more complicated nonlinear functions, such as sigmoids, have the advantage of being smooth; immune networks have even more complicated activation functions. An alternative is to make each node a flexible function approximation box, for example, with its own set of local linear functions, so that the node can approximate functions with more general shapes [22, 68]. However, complexity also increases the number of free parameters and potentially increases the amount of data needed for learning.

(iv) A related question concerns the role of catalysis. In autocatalytic networks, a node can be switched on or off by another node through *multiplicative* coupling terms. In contrast to networks in which inputs can only be summed, this allows a single unit to exert over-riding control over another. A similar approach has been suggested in Σ–Π neural networks [59]; T-cells and neurotransmitters may play a similar role in real biological systems. How valuable is specific catalysis to a network? How difficult is the learning problem when it is employed?

(v) What are the optimal approaches to evolving good graph representations? Most of the work in this area has been done for classifier systems, although even here many important issues remain to be clarified. All known algorithms that can *create* connections and nodes, such as the genetic algorithms, are stochastic; there are deterministic pruning algorithms that can only destroy connections, such as orthogonal projection. Are there efficient deterministic algorithms for *creating* new graph connections?

(vi) What are the best learning algorithms? A great deal of effort has been devoted to answering this question, but the answer is still obscure. A perusal of the literature suggests certain general conclusions. For example, in problems with detailed feedback, e.g. a list of known input–output pairs, deterministic function fitting algorithms such as least-squares minimization (of which back-propagation is an example) can be quite effective. However, if the search space is not smooth, for example because the samples are too small to be statistically stable, stochastic algorithms such as crossover are often more effective [1]. In more general situations where there is no detailed feedback, there seems to be no general consensus as to which learning algorithms are superior.

Thus far, very few connectionist networks make use of nontrivial computational capabilities. In typical applications most connectionist networks end up functioning as stimulus–response systems, simply mapping inputs to outputs without making use of conditional looping, subroutines, or any of the power we take for granted in computer programs. Even in systems that clearly have a great deal of computational power *in principle*, such as classifier systems, the solutions actually learned are usually close to look-up tables. It seems to be much easier to implement effective learning rules in simpler architectures that sacrifice computational complexity, such as feed-forward networks.

It may be that there is an inherent trade-off between the complexity of learning and the complexity of computation, so that the difficulty of learning increases with computational power. At one end of the spectrum is a look-up table. Learning is trivial; examples are simply inserted as they occur. Unfortunately, all too often neural network applications have not been compared to this simple approach. In the infamous NET-talk problem [62], for example, a simple look-up table gives better performance than a sum/sigmoid back-propagation network [3]. Simple function approximation is one level above a look-up table in computational complexity; functions can at least attempt to interpolate between examples, and generalize to examples that are not in the learning data set. Learning is still *fairly* simple, although already the subtleties of probability and statistics begin to complicate the matter. However, simple function approximation has less computational capability than a finite state machine. At present, there are no good learning algorithms for finite state machines. Without counting, conditional looping, etc., many problems will simply remain insoluble.

It is probably more likely that learning *is possible* with more sophisticated computational power, and that we simply do not yet know how to accomplish it. I suspect that the connectionist networks of the future will be full of loops.

Connectionist models are a useful tool for solving problems in learning and adaptation. They make it possible to deal with situations in which there are an infinite number of possible variables, but in which only a finite number are active at any given time. The connections are explicit but changeable. We have only recently begun to acquire the computational capabilities to realize their potential. I suspect that the next decade will witness an enormous explosion in the application of the connectionist methodology.

However, connectionism represents a level of abstraction that is ultimately limited by such factors as the need to specify connections explicitly, and the lack of builtin spatial structure. Many problems in adaptive systems ultimately require models such as partial differential equations or cellular automata with spatial structure [40]. The molecular evolution models of Fontana et al., for example, explicitly model the spatial structure of individual polymers in an artificial chemistry. As a

Table 3
A Rosetta Stone for connectionism.

Generic	Neural net	Classifier system	Immune net	Autocatalytic net
node	neuron	message	antibody type	polymer species
state	activation level	intensity	free antibody/ antigen concentration	polymer concentration
connection	axon/synapse/ dendrite	classifier	chemical reaction of antibodies	catalyzed chemical reaction
parameters	connection weight	strength and specificity	reaction affinity lymphocyte concentration	catalytic velocity
interaction rule	sum/sigmoid	linear threshold and maximum	bell-shaped	mass action
learning algorithm	Hebb, back-propagation	bucket brigade (gen. Hebb)	clonal selection (gen. Hebb)	approach to attractor
graph dynamics	synaptic plasticity	genetic algorithms	genetic algorithms	artificial chemistry rules, spontaneous reactions

result the phenotypes emerge more naturally than in the artificial chemistry in the autocatalytic network model discussed here. On the other hand, the approach of Fontana et al. requires more computational resources. For many problems connectionism may provide a good compromise between accurate modeling and tractability, appropriate to the study of adaptive phenomena during the last decade of this millenium.

8.2. Rosetta Stone

This paper is a modest start toward creating a common vocabulary for connectionist systems, and unifying work on adaptive systems. Like the Rosetta Stone, it contains only a small fragment of knowledge. I hope it will nonetheless lead to a deeper understanding in the future. Table 3 summarizes the analogies developed in this paper.

Acknowledgements

I would like to thank Rob De Boer, Walter Fontana, Stephanie Forrest, André Longtin, Steve Omohundro, Norman Packard, Alan Perelson, and Paul Stolorz for valuable discussions, and Ann and Bill Beyer for lending valuable references on the Rosetta Stone.

I urge the reader to use these results for peaceful purposes.

Appendix. A superficial taxonomy of dynamical systems

Dynamical systems can be trivially classified according to the continuity or locality of the underlying variables. A variable either can be discrete, i.e. describable by a finite integer, or continuous. There are three essential properties:

(i) *Time*. All dynamical systems contain time as either a discrete or continuous variable.

(ii) *State*. The state can either be a vector of real numbers, as in an ordinary differential equation, or integers, as for an automaton.

(iii) *Space* plays a special role in dynamical systems. Some dynamical models, such as automata or ordinary differential equations, do not contain the notion of space. Other models, such as

Table 4
Types of dynamical systems, characterized by the nature of time, space, and state. "Local" means that while this property is discrete, there is typically some degree of continuity and a clear notion of neighborhood.

Type of dynamical system	Space	Time	Representation
partial differential equations	continuous	continuous	continuous
computer representation of a PDE	local	local	local
functional maps	continuous	discrete	continuous
ordinary differential equations	none	continuous	continuous
lattice models	local	discrete or continuous	continuous
maps (difference equations)	none	discrete	continuous
cellular automata	local	discrete	discrete
automata	none	discrete	discrete

lattice maps or cellular automata, contain a notion of locality and therefore space even though they are not fully continuous. Partial differential equations or functional maps have continuous spatial variables.

This is summarized in table 4.

References

[1] D.H. Ackley, An empirical study of bit vector function optimization, in: Genetic Algorithms and Simulated Annealing, ed. L. Davis (Kaufmann, Los Altos, CA, 1987).
[2] D.L. Alkon, Memory storage and neural systems, Sci. Am. 261 (1989) 26–34.
[3] Z.G. An, S.M. Mniszewski, Y.C. Lee, G. Papcun and G.D. Doolen, HI-ERtalker: a default hierarchy of high order neural networks that learns to read English aloud, Technical Report, Center for Nonlinear Studies Newsletter, Los Alamos National Laboratory (1987).
[4] W.B. Arthur, Nash-discovery automata for finite-action games, working paper, Santa Fe Institute (1989).
[5] W.B. Arthur, On the use of classifier systems on economics problems, working paper, Santa Fe Institute (1989).
[6] R. Axelrod, An evolutionary approach to norms, Am. Political Sci. Rev. 80 (December 1986) 1095–1111.
[7] R.J. Bagley and J.D. Farmer, Robust autocatalytic sets (1990), in progress.
[8] R.J. Bagley, J.D. Farmer, S.A. Kaufmann, N.H. Packard, A.S. Perelson and I.M. Stadnyk, Modeling adaptive biological systems, Biocybernetics (1990), to appear.
[9] R.K. Belew and M. Gherrity, Back propagation for the classifier system, Technical Report, University of California, San Diego (1989).
[10] L.B. Booker, D.E. Goldberg and J.H. Holland, Classifier systems and genetic algorithms, Artificial Intelligence 40 (1989) 235–282.
[11] D. Broomhead and D. Lowe, Radial basis functions, multivalued functional interpolation and adaptive networks, Technical Report Memorandum 4148, Royal Signals and Radar Establishment (1988).
[12] E.A. Wallis Budge, The Rosetta Stone (The Religious Tract Society, London, 1929) (reprinted by Dover, 1989).
[13] M. Casdagli, Nonlinear prediction of chaotic time series, Physica D 35 (1989) 335–356.
[14] M. Compiani, D. Montanari, R. Serra and G. Valastro, Classifier systems and neural networks, in: Proceedings of the Second Workshop on Parallel Architectures and Neural Networks, ed. E. Caianiello (World Scientific, Singapore, 1988).
[15] J.D. Cowan and D.H. Sharp, Neural nets, Quart. Rev. Biophys. 21 (1988) 365–427.
[16] L. Davis, Mapping classifier systems into neural networks, in: Neural Information Processing Systems 1, ed. D.S. Touretzsky (Kaufmann, Los Altos, CA, 1989).
[17] R.J. De Boer and P. Hogeweg, Unreasonable implications of reasonable idiotypic network assumptions, Bull. Math. Biol. 51 (1989) 381–408.
[18] R.J. De Boer, Dynamical and topological patterns in developing idiotypic networks, Technical Report, Los Alamos National Laboratory (1990).
[19] M. Eigen and P. Schuster, The Hypercycle (Springer, Berlin, 1979).
[20] J.D. Farmer, S.A. Kauffman and N.H. Packard, Autocatalytic replication of polymers, Physica D 22 (1986) 50–67.
[21] J.D. Farmer, N.H. Packard and A.S. Perelson, The immune system, adaptation and machine learning, Physica D 22 (1986) 187–204.
[22] J.D. Farmer and J.J. Sidorowich, Exploiting chaos to predict the future and reduce noise, in: Evolution, Learning and Cognition, ed. Y.C. Lee (World Scientific, Singapore, 1988).
[23] S. Forrest, Implementing semantic network structures using the classifier system, in: Proceedings of the First International Conference on Genetic Algorithms and their Applications, 1985.
[24] D. Friedman, Evolutionary games in economics, Technical Report, Stanford University (1989).
[25] C.L. Giles, R.D. Griffin and T. Maxwell, Encoding geometric invariances in higher order neural networks, in: Neural Networks for Computing, ed. J.S. Denker (AIP, New York, 1986).

[26] S.A. Harp, T. Samad and A Guha, Towards the genetic synthesis of neural networks, in: Proceedings of the Third International Conference on Genetic Algorithms, ed. J.D. Schaffer (Kaufmann, Los Altos, CA, 1989).
[27] J. Hofbauer and K. Sigmund, The Theory of Evolution and Dynamical Systems (Cambridge Univ. Press, Cambridge, 1988).
[28] G.W. Hoffmann, A theory of regulation and self-nonself discrimination in an immune network, European J. Immunol. 5 (1975) 638–647.
[29] G.W. Hoffmann, T.A. Kion, R.B. Forsyth, K.G. Soga and A. Cooper-Willis, The n-dimensional network, in: Theoretical Immunology, Part II, ed. A.S. Perelson (Santa Fe Institute/Addison-Wesley, Redwood City, CA, 1988).
[30] J. Holland, Escaping brittleness: the possibilities of general purpose machine learning algorithms applied to parallel rule-based systems, in: Machine Learning II, eds. Michalski, Carbonell and Mitchell (Kaufmann, Los Altos, 1986).
[31] J. Holland, K.J. Holyoak, R.F. Nisbett and P.R. Thagard, Induction: Process of Inference, Learning and Discovery (MIT Press, Cambridge, MA, 1986).
[32] J. Hopfield and D.W. Tank, "Neural" computation of decisions in optimization problems, Biol. Cybern. 52 (1985) 141–152.
[33] N.K. Jerne, The immune system, Sci. Am. 229 (1973) 52–60.
[34] N.K. Jerne, Towards a network theory of the immune system, Ann. Immunology (Inst. Pasteur) 125 C (1974) 373–389.
[35] R. Jones, C. Barnes, Y.C. Lee and K. Lee, Fast algorithm for localized prediction, private communication (1989).
[36] S.A. Kauffman, Autocatalytic sets of proteins, J. Theor. Biol. 119 (1986) 1–24.
[37] S.A. Kauffman, Emergent properties in random complex automata, Physica D 10 (1984) 145–156.
[38] S.A. Kauffman, Metabolic stability and epigenesis in randomly constructed genetic nets, J. Theor. Biol. 22 (1969) 437.
[39] S.A. Kauffman, Origins of Order, Self-Organization, and Selection in Evolution (Oxford Univ. Press, Oxford, 1990), in press.
[40] C.G. Langton, ed., Artificial Life (Addison-Wesley, Redwood City, CA 1989).
[41] R. Marimon, E. McGrattan and T.J. Sargeant, Money as a medium of exchange in an economy with artificially intelligent agents, Technical Report 89-004, Santa Fe Institute, Santa Fe, NM (1989).
[42] W.S. McCulloch and W. Pitts, A logical calculus of the ideas immanent in nervous activity, Bull. Math. Biophys. 5 (1943) 115–133.
[43] G.F. Miller, P.M. Todd and S.U. Hegde, Designing neural networks using genetic algorithms, in: Proceedings of the Third International Conference on Genetic Algorithms, ed. J.D. Schaffer (Kaufmann, Los Altos, CA, 1989).
[44] J. Miller and S. Forrest, The dynamical behavior of classifier systems, in: Proceedings of the Third International Conference on Genetic Algorithms, ed. J.D. Schaffer (Kaufmann, Los Altos, CA), in press.
[45] S.L. Miller and H.C. Urey, Organic compound synthesis on the primitive earth, Science 130 (1959) 245–251.
[46] D. Montana and L. Davis, Training feedforward neural networks using genetic algorithms, in: Proceedings of the Eleventh International Joint Conference on Artificial Intelligence, 1989.
[47] J. Moody and C. Darken, Learning with localized receptive fields, Technical Report, Department of Computer Science, Yale University (1988).
[48] A. Newell, Production systems: models of control structures, in: Visual Information Processing, ed. W.G. Chase (Academic Press, New York, 1973).
[49] J. Pearl, Probabilistic Reasoning in Intelligent Systems: Networks of Plausible Inference (Kaufmann, Los Altos, CA, 1988).
[50] A.S. Perelson, Immune network theory, Immunol. Rev. 110 (1989) 5–36.
[51] A.S. Perelson, Toward a realistic model of the immune system, in: Theoretical Immunology, Part II, ed. A.S. Perelson (Santa Fe Institute/Addison-Wesley, Redwood City, CA, 1988).
[52] E.S. Pichler, J.D. Keeler and J. Ross, Comparison of self-organization and optimization in evolution and neural network models, Technical Report, Chemistry Department, Stanford University (1989).
[53] F.J. Pineda, Generalization of backpropagation to recurrent and higher order neural networks, in: Neural Information Processing Systems, ed. D.Z. Anderson (AIP, New York, 1988).
[54] T. Poggio and F. Girosi, A theory of networks for approximation and learning, MIT preprint (1989).
[55] S. Pope, unpublished research.
[56] P.H. Richter, A network theory of the immune system, European J. Immunol. 5 (1975) 350–354.
[57] R. Riolo, CFS-C: a package of domain independent subroutines for implementing classifier systems in arbitrary, user-defined environments, Technical Report, Logic of Computers Group, University of Michigan (1986).
[58] F. Rosenblatt, The perceptron: a probabilistic model for information storage and organization in the brain, Psychological Rev. 65 (1958) 386.
[59] D. Rummelhart and J. McClelland, Parallel Distributed Processing, Vol 1 (MIT Press, Cambridge, MA, 1986).
[60] A.C. Scott, Neurophysics (Wiley, New York, 1977).
[61] L.A. Segel and A.S. Perelson, Shape space analysis of immune networks, in: Cell to Cell Signalling: From Experiments to Theoretical Models, ed. A. Goldbeter (Academic Press, New York, 1989).
[62] T.J. Sejnowski and C.R. Rosenberg, Parallel networks that learn to pronounce English text, Complex Systems 1 (1987) 145–168.
[63] J. Maynard Smith, Evolutionary game theory, Physica D 22 (1986) 43–49.
[64] F.J. Varela, A. Coutinho, B. Dupire and N.N. Vaz, Cognitive networks: immune neural and otherwise, in: Theoreti-

cal Immunology, Part II, ed. A.S. Perelson (Santa Fe Institute/Addison-Wesley, Redwood City, CA, 1988).

[65] D. Whitley and T. Hanson, Optimizing neural networks using faster, more accurate genetic search, in: Proceedings of the Third International Conference on Genetic Algorithms, ed. J.D. Schaffer (Kaufmann, Los Altos, CA, 1989).

[66] S.W. Wilson, Bid competition and specificity reconsidered, Complex Systems 2 (1989) 705–723.

[67] S.W. Wilson, Perceptron redux: emergence of structure, Physica D 42 (1990) 249–256, these Proceedings.

[68] D.H. Wolpert, A benchmark for how well neural nets generalize, Biol. Cybern. 61 (1989) 303–313.

CONCERNING THE EMERGENCE OF TAG-MEDIATED LOOKAHEAD IN CLASSIFIER SYSTEMS

John H. HOLLAND

Computer Science and Engineering, Psychology, The University of Michigan, Ann Arbor, MI 48109, USA

This paper, after a general introduction to the area, discusses the architecture and learning algorithms that permit automatic parallel, distributed lookahead to emerge in classifier systems. Simple additions to a "standard" classifier system suffice, principally a new register called the *virtual strength register*, and a provision to use the bucket brigade credit assignment algorithm in "virtual" mode to modify values in this register. With these additions, current actions are decided on the basis of the expected values associated with the "lookahead cones" of possible alternatives.

1. Introduction

Whenever one studies adaptation or machine learning in realistic contexts one constraint soon comes to occupy a central position: Feedback about performance is intermittent and lacks detail. Samuel [1], at the very start of the modern study of emergent computation, realized that games – checkers is the example he used – provide a good paradigmatic example of the problem. During the play of a game there is a great flow of information, but only at the end of the game is there any feedback about performance, the game's payoff, and that is only a few bits of information. In more complex environments, such as ecological, economic or social settings, long sequences of actions are typically required before some reward or reinforcement occurs (e.g. reduction of a "drive" like hunger, or the filling of some "reservoir"). Somehow the system must utilize the "sparse" information about performance, and the large flow of other kinds of information, to improve its performance over time.

Samuel offered lookahead as a (perhaps, the) solution to this problem, and provided a convincing demonstration of its efficacy. To utilize lookahead a system must generate an internal model of its environment. This model enables the system to "look ahead", allowing it to make predictions about the expected consequences of different sequences of action. These predictions can be checked as experience accumulates, providing feedback that can be used directly in improving the model. Note that the resulting procedure for modifying the emergent model depends not at all on environmental measures of performance. Of course, if the model is to be the basis for improved performance, some of the predictions at least must be concerned with expected rewards or reinforcements. However, model-based lookahead neatly steps around a requirement for continual detailed information about performance, and it makes good use of the large flow of (non-performance) information supplied by the environment.

This paper is concerned with the emergence, in classifier systems, of organized, rule-based models that permit "lookahead". (The paper is a continuation and elaboration of the paper on classifier systems that appeared in 1986 in Physica D 22 [2]. Section 1 of that paper details the definitions and notations used here; however, most of the relevant ideas are sketched below in figs. 1, 2, 5 and 6.)

2. Symbols and internal models

An internal model is above all a "symbolic" entity and its predictions depend upon an appropriate manipulation of those symbols. Most current computation-based approaches to symbols and symbol-processing can be assigned to one of two broad classes: The "language-based" systems, such as those implementing the physical symbol system hypothesis [3], and the "stimulus-oriented" systems, such as those investigated by the connectionists (see, for example, ref. [4]). In general, physical symbol systems are good at lookahead, anticipation, and means-ends analysis, when supplied with an appropriate internal model of the environment (usually given as a problem space), but they generally lack procedures for the autonomous construction of experience-based internal models (new problem spaces). Connectionist systems are good at the autonomous construction of categories on the basis of knowledge acquired while exploring an environment, but they lack procedures for organizing that knowledge into models that guide the system by lookahead and anticipation.

It is important that both kinds of system, different as they are in most respects, share a common characteristic: Information about the environment, as supplied by the input interface, always comes with "labels" of some kind. These labels may be quite sophisticated (such as labeling a given input image a "chair") or they may be quite primitive (such as the retinal coordinates of an input neuron). The question, in generating internal models, is not whether or not input is labeled, but rather how sophisticated the labels are. Stated another way, it is a question of how much "intelligence" the input interface uses in translating the environment into the input messages processed by the system.

Both kinds of system thus share a common limitation on their ability to categorize the external world: Environmental states that cause the input interface to generate the same input "message" are indistinguishable, and further processing, however implemented, can only categorize the distinguishable. Indeed, this limitation is shared by any system that acquires all its environmental information via an input interface. If such a system is computationally complete with respect to sorting input "messages" into categories, then it has reached the limits of what categorization can do for it.

Clearly, when it comes to building models, there is a great difference between a system that has a selection of higher-level categories "wired" into its input interface and a system that uses only primitive, coordinate-like labels to formulate higher-level categories. In the latter case, categories and symbols, assuming they emerge, tend to be constructed of "building blocks" – new categories and symbols are constructed by using "good" building blocks, and experience is thereby transferred to new situations. In the former case, categories and symbols tend to be monolithic and experience must be transferred from one domain to another by other means.

Taking this into account, there are reasons that both the stimulus-oriented and language-based approaches should pay close attention to Edelman's [5] points about "re-entrant connections": A system can only generate autonomous internal activity – and lookahead is an example par excellance of such an activity – if it has "re-entrant connections". This point is closely allied to the one Hebb [6] makes in his magnum opus *The Organization of Behavior*: Re-entrant connections provide a recirculation of pulses that allows parts of the network to act independently of recent inputs. As the system learns, clusters of neurons use some of the re-entrant connections to form reverberating "cell assemblies", and these in turn become building blocks (a kind of flexible "compositionality", à la Pylyshyn [7]) for sophisticated sub-routines called "phase sequences". Several nerve net simulations of the 50's, now largely forgotten (e.g. Rochester et al. [8]), exhibited the emergence of "cell assemblies", under Hebb's learning rule, when repeating input patterns were applied to randomly connected networks with re-entrant connections.

For the stimulus-oriented connectionists this point bears on the construction of internal models in another way: It is a long-established theorem of automata theory (going back to McCulloch and Pitts [9]) that a system constructed of interconnected "logical" elements (such as formal neurons), without internal feedback loops, can attain only a very limited subclass of the class of finite automaton behaviors. For example, networks without internal feedback loops cannot exhibit indefinite memory ("at some time in the indefinite past, event x occurred"). Accordingly, without such loops, it is impossible to construct an internal storage for pulses or a counter for pulses. It follows that most computational routines are impossible for nets without loops. In particular, internal feedbacks are necessary if the networks are to be able to produce emergent, semi-autonomous internal models that provide predictions and anticipations.

At the other end of the scale, language-based systems are almost always computationally complete because they directly employ some "universal" language such as LISP. However, they have little to say about the emergence of categories and internal models under the impetus of experience. This is partly the result of using symbols that are pre-defined and close to monolithic, and partly the result of designing systems that require inputs ("symbols") that activate appropriate sections of a high-level interpreter. It is difficult to design physical symbol systems that can learn using the "low-level" data supplied by natural environments. The learning mechanisms used for language-based systems (such as the "chunking" mechanism used by Laird et al. [3] in Soar) look much more like compilation than like the origination of new categories.

Classifier systems occupy a middle ground between these two approaches. They construct models by using experience to extract simple substructures (building blocks) that can be combined in a variety of ways to yield plausible models. (Hebb makes allowance for similar possibilities by providing for the recombination of parts of cell assemblies via processes he calls "fractionation" and "recruitment".) It is important that these building blocks must be used in a fluid, context-dependent way. If we think of the resulting models as complexes of symbols, then, in the sense so well described by Hofstadter (in chapters XI and XII of ref. [10]), the symbols must be *active*. In a later discussion on the topic of active symbols [11], Hofstadter quotes E.O. Wilson:

"'Mass communication is defined as the transfer, among groups, of information that a single individual could not pass to another.'"

and then goes on to say:

"One has to imagine teams of ants [read "neural firings"] cooperating on tasks, and information passing from team to team that no ant ["neuron"] is aware of...

...[W]hat guarantee is there that we can skim off the full fluidity of the top-level activity of a brain and encapsulate it – without any lower substrate – in the form of some computational rules. To ask an analogous question, what guarantee is there that there are rules at the "cloud level" (more properly speaking, the level of cold fronts, isobars, trade winds, and so on) that will allow you to say accurately how the atmosphere is going to behave on a large scale?...

The difference between my active symbols ("teams") and the passive symbols (ants, tokens) of the information-processing school of AI is that the active symbols flow and act on their own. In other words, there is no higher-level agent (read "program") that reaches down and shoves them around."

It is the notion that symbols are composed of building blocks that can be recombined fluidly in response to context that makes emergence of symbols and internal models a natural, almost inevitable, process.

Different combinations of building blocks yield different internal models that compete and gain varying degrees of confirmation as experience accumulates. Parts of highly confirmed models, and sometimes whole models, serve as building blocks for still more sophisticated models and competitions. (As one implementation of this notion, see the parallel, rule-based, message-passing systems discussed in Holland et al. [12].) In principle, such a system could yield an "upper" layer that behaves much as described by the physical symbol system hypothesis. However, when it comes to the origination of new hypotheses and models, the upper layer is the servant of the lower layers. Whether one prefers the stimulus-oriented or the language-based approach, it seems to me a great risk to ignore processes that construct models by extracting and combining building blocks.

3. Classifier systems and marker-passing lookahead

The remainder of this paper is devoted to an outline of a classifier system (a distributed, parallel system, hence essentially connectionist) that is rule-based and lookahead-oriented (hence akin to physical symbol systems that use means–ends analysis). It continually augments its models by adding rules and proto-symbols (tags) as it gains experience in its environment. The objectives of the design are to (i) use a small set of domain-independent "local" mechanisms to (ii) provide for the emergence of a hierarchical, epoch-guided lookahead based on experience.

A standard classifier system involves a set of message-passing rules in condition/action form. The action part of a rule specifies a message that is to be posted when it is executed. A rule is only executed when there are messages present that satisfy its condition part. Many rules can be active simultaneously. Overt actions (affecting the environment) are the result of messages directed to the output (effector) interface. (See figs. 1, 2 and 5, and, for more detail, Holland [2].)

Lookahead amounts to an extension of this system wherein the system attempts to predict the effect of a sequence of actions so as to base its current overt action on expected long-term consequences. (For example, in the game of checkers, the program decides to move a checker to the edge

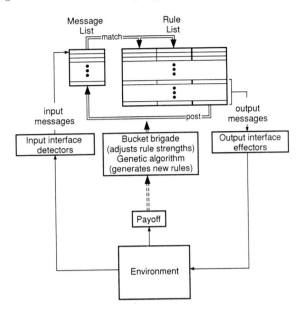

Fig. 1. General organization of a classifier system. Each *rule* is in condition/action form and has an assigned strength that reflects its past usefulness. On each time step, all conditions are checked against the *message list* for matching messages. If all the conditions of a rule are matched then it competes in terms of its strength to post the message specified by its action part. Many rules may win the competition, with many messages being posted for processing on the next round. The *input interface* provides one or more messages that describe the current state of the environment, and the *output* interface translates some messages into actions that affect that environment. Learning takes place by (i) modifying the strengths of rules to reflect experienced usefulness, and by (ii) generating new rules to replace weak rules. The overall performance of the system is measured in terms of *payoff* it receives from the environment.

of the board, anticipating that the move will make possible a "double-jump" four moves hence.)

Because classifier systems are parallel systems it is natural to think of a lookahead process that traces many possible courses of action simultaneously. *Marker propagation* over a network provides a useful metaphor for exploring this possibility. (Fahlman's [13] treatise provides a detailed description of parallel marker propagation.) In the present context we can describe the network, and marker propagation over it, as follows (see figs. 3 and 4):

(1) Each node in the marker propagation network corresponds to some equivalence class (category) over the environmental states. The directed

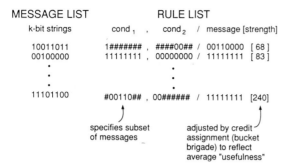

Fig. 2. Specification of messages and conditions in a classifier system. "#"'s in the condition part of a rule act as "wild cards" or "don't cares", accepting any value at that position in a message. A condition with more #'s is more *general* in the sense that it accepts a broader range of messages. Rules with matched conditions *bid* to post the message specified by their action part. The bid is equal to c(rule strength)(rule specificity) where c is a fraction (say $c = 0.1$) and rule specificity is given by $k -$ (no. of #'s). Winners are chosen with a probability proportional to the size of their bids.

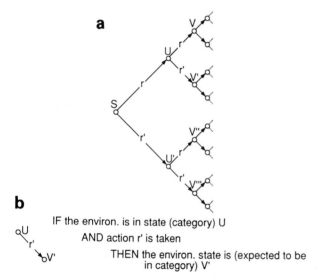

Fig. 3. An example of a causal network. (a) Fragment of a "causal net" version of an internal model. (b) Interpretation of an edge of the causal net in terms of rules.

edges connecting the nodes correspond to possible actions that will cause state transitions in the environment. (The directed edges amount to hypothesized causal relations.)

(2) The lookahead process is initiated by placing a marker on the node corresponding to the current state (more carefully, the equivalence class containing the current state).

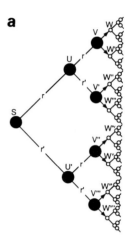

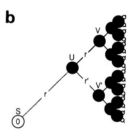

Fig. 4. An example of lookahead using marker propagation. (a) Marker propagation prior to an action decision at state S. (b) Marker propagation after action is carried out at state S.

(3) Copies of the marker are then propagated along each directed edge leading from that node, the result being that the initial node and all nodes that can be reached from it in one step are marked.

(4) The process is iterated T times, with the result that a "cone" of nodes is marked representing the states (equivalence classes) that can be attained from the current state via various action sequences of length T.

This elementary process can be made more sophisticated in several ways. Among the possible extensions, the following two play a key role in the extended classifier system:

(1) The system's experience will typically be indeterministic. That is, a given action applied to a given node (equivalence class) leads to different consequences (equivalence classes) at different

times. (This happens because the equivalence classes are too "coarse" to capture some distinctions necessary for a fully causal description.) Under such circumstances the edges can be assigned probabilities corresponding to the relative frequency with which the alternatives occur. Marker propagation still proceeds in parallel, but probabilistically. The lookahead "cone" now contains action sequences of various lengths, lookahead proceeding to greater depth along more probable paths.

(2) Equivalence classes often have "values" attached to them. (The evaluation functions used in game-playing programs provide a concrete example – the evaluation function provides an estimate of the general value of attaining a given equivalence class.) To induce the system to explore more valuable paths, the probability of propagation can be made proportional to the product of the probability assigned to the edge and the value assigned to the node at the tip of the edge. This number is akin to the expected value of the transition. With this addition, the length of each path in the lookahead "cone" varies roughly as the average expected value assigned to that path.

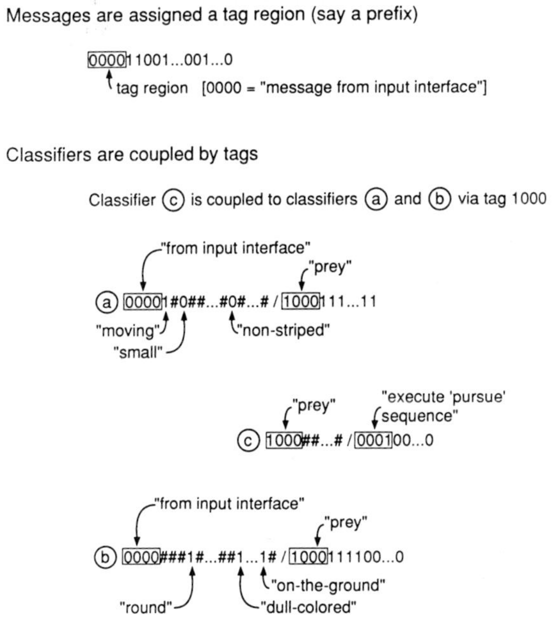

Note that the purpose of lookahead is to influence the decision as to what action is to be taken next. That is, the current action is predicated upon an estimate of the future value of moving in that general "direction". (More carefully, the first step away from the current state is chosen on the basis of the estimated values of the alternatives attainable in the part of the lookahead cone that includes that first step.) Note also that, once that first step has been taken, the parts of the lookahead cone involving other possible first steps are now largely irrelevant. However, the part of the lookahead cone involving the chosen first step applies and can be extended – marker propagation can continue from the *end-points* of the sub-cone, rather than starting all over at the new "current" state.

4. Tag-mediated lookahead

The first step in providing a classifier system with lookahead is implementation of the "causal" network that constitutes the system's internal model of its environment. For each transition to be modeled, this is a matter of implementing the rule "IF the environment is in state S AND action A is taken THEN (the system expects) state S' will occur". This can be accomplished by coupling a rule that is active when state (category) S is detected to a rule that is active when state (category) S' is detected. In classifier systems, tags (see fig. 5) typically provide the couplings that implement pointers, action sequences, and the like. Thus, tags play a central role in the construction of the causal network, and they constitute natural building blocks for emergent models.

Tags are implemented by setting aside certain regions in internal messages, typically a prefix or a suffix, though any region will do. For example, any message with the prefix 1101 will satisfy a condition of the form 1101#...#. Stated another way, a classifier with the condition 1101#...#

Messages are assigned a tag region (say a prefix)

0000 1 1001...001...0
↑ tag region [0000 = "message from input interface"]

Classifiers are coupled by tags

Classifier ⓒ is coupled to classifiers ⓐ and ⓑ via tag 1000

ⓐ 0000 1#0##...#0#...# / 1000 111...11
"from input interface"   "prey"
"moving"   "non-striped"
"small"

ⓒ 1000 ##...# / 0001 00...0
"prey"   "execute 'pursue' sequence"

ⓑ 0000 ###1#...##1# / 1000 111100...0
"from input interface"   "prey"
"round"   "on-the-ground"
                         "dull-colored"

Fig. 5. Tags and rule coupling.

has an "address". To send a message to it, simply put the prefix 1101 on the message. There are many variations on this theme. For example, consider a pair of classifiers C1 and C2 that send messages tagged with 1101 and 1001, respectively. A classifier with the condition $1101\# \ldots \#$ will attend only to C1, but a classifier with condition $1\#01\# \ldots \#$ will attend to both C1 and C2. Moreover, a message with a given tag can activate a whole cluster of classifiers, if all the classifiers in the cluster have conditions that are sensitive to that tag. Fig. 5 provides an example of this use of tags, and the interested reader will find a detailed description of the use of tags to implement a semantic net (KL-One) in Stephanie Forrest's paper [14].

Tags supply the "glue" for models in classifier systems and, as with any other part of a classifier, they are subject to modification and elaboration under the recombinations induced by the genetic algorithm. As the genetic algorithm supplies the system with additional coupled rules, the tags acquire meaning in terms of the model-based effects they mediate (actions, anticipations, predictions, and the like). In effect, tags serve as active proto-symbols providing context dependent associations: Many classifiers can be activated by a message with a single tag – the particular cluster activated being dependent on the other messages present.

As the system evolves, it seems reasonable to expect that these proto-symbols will become associated with external, manipulatable patterns (physical symbols). These external patterns, feeding back through the input interface, would "close the loop", moving the proto-symbols to full-fledged symbols with distal access.

5. Hierarchical models

The internal models that arise naturally in the classifier system format are best described as *default hierarchies* (see fig. 6). The (useful) rules that are easiest for the system to discover are those with many #'s (don't cares) in the condition part. This is true both because such rules are easy to formulate and because they are tested often. Any such rule that gives the system a slight statistical advantage will be quickly strengthened. These rules act as *default* rules. More specific rules that contradict the default rules in specific situations are tested and established less quickly. These rules act as *exception* rules. There can of course be excep-

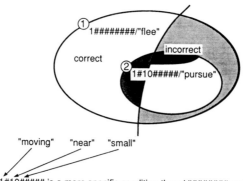

1#10##### is a more specific condition than 1####### and hence tends to win competitions when both conditions are satisfied.
The emerging default hierarchy is "symbiotic"

② prevents ① from making mistakes, therefore increasing ①'s net payoff rate, while ② increases the overall pay off rate.

Fig. 6. An example of a simple internal model in a classifier system.

tions to the exceptions, and so on, whence comes the *default hierarchy*. Fig. 6 illustrates the point that, in classifier systems, these contradictory rules, rather than competing to displace each other, can act in a symbiotic fashion. Because of this, default hierarchies form a "natural" emergent structure in classifier systems.

The elements (rules) in the hierarchy can also respond to events of different duration. For example (see fig. 7), there could be a rule that has a condition that is satisfied so long as it is daytime and the "food reservoir" (stomach) is unfilled. Such a rule could continue to post its message over a considerable time interval, which could be called the HUNT epoch. There could also be a more specific rule that has its condition satisfied under the same conditions but only so long as there is no "prey" in sight. This rule would be active for only a part of the HUNT epoch, a sub-epoch that could be called the SEARCH epoch. Again a hierarchy forms, an *epoch* hierarchy, involving conditions that are increasingly specific (as in the earlier default hierarchy) and activities of progressively shorter duration with more detailed specification.

In the figures and discussions that follow, it is useful to keep in mind a simple system–environment configuration, wherein the state of the configuration can be described as a vector over four properties:

$\{day, night\} \times \{hungry, not\ hungry\}$

$\times \{prey\ (in\ sight), no\ prey\ (in\ sight)\}$

$\times \{(prey)\ off\text{-}center, (prey)\ centered\}$.

Actions (the edges of the transition graph) can be restricted to the set:

$\{"run", "twiddle", center, forward\}$.

The intended action sequences of the system could then be collected into equivalence classes with various levels of refinement and duration (see fig. 7). At the coarsest level, the system's action, "hunting", is an activity that could last the better part of a day and would consist of admixtures of the elementary actions "run", "twiddle", etc. Early in the system's experience the epoch hierarchy would consist of a single level and this admixture of elementary actions would be more or less randomly determined. With experience, the epoch hierarchy supplies additional levels of specificity and the admixture becomes more context dependent. That is, the coarsest equivalence class, which extends over both "space" (different instantaneous states) and "time", is progressively refined into classes of shorter duration and fewer states, yielding an epoch hierarchy.

6. The lookahead process in classifier systems

We are now in a position to look at an outline of the mechanisms necessary for a classifier system to produce emergent experience-based internal models. *The basic objective is to add lookahead to a standard classifier system without adding new rule types or changing the operators that generate the rules.* In particular, this means that the lookahead process should use the same coupled rules that are generated by the "triggering" processes in the standard classifier system (see section 3.2 of ref. [12] or section 9.3 of ref. [15]).

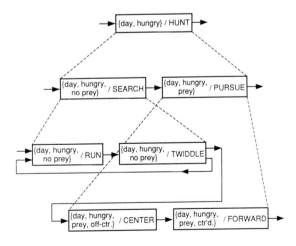

Fig. 7. An example of an epoch-directed hierarchical model (see text).

The main triggering mechanism is that for producing coupled rules (see p. 90 of ref. [12]). The paradigm case occurs when:

(1) Rule R is executed at time t,

(2) Rule R′, not coupled to R, is executed at time $t + 1$,

(3) R′ is substantially strengthened (via the bucket brigade credit assignment algorithm or by direct payoff from the environment).

This condition triggers the generation of a pair of coupled offspring rules, R1 and R1′, that are activated under the same conditions as R and R′ but are coupled by a (newly generated) tag. The pair so generated does *not* replace the parents, but simply enters the system as a competing hypothesis. If the coupled pair captures a causal relation in the environment, it will be strengthened under the bucket brigade (because of the profit made by R1′) and it will persist. If not, it will quickly lose strength (because R1 fails to set the stage for the activation of R1′), becoming a candidate for replacement by other newly generated hypotheses.

There are three types of rules that serve as the elements of a causal network in a standard classifier system:

(1) *Node* rules. The condition part of a node rule is satisfied by a message designating a particular state (category). The action part of the rule sends a message indicating that the state has been "marked".

(2) *Transition* rules. The condition part of a transition rule is satisfied by a message designating a particular state (category). The action part of the rule sends a message designating a response (the label of the corresponding edge in the causal network) and the state (category) expected when that response is made to the state designated in the condition part.

(3) *Action* rules. An action rule has two conditions in its condition part: The first condition is satisfied by a message designating a particular state (category), and the second condition requires the presence of a special *action* message. The action part of the rule sends a message to the output interface that causes some overt action in the environment.

The messages produced by these rules have three parts:

(1) a prefix part allocated to tags,

(2) a middle part typically allocated to response specification, and

(3) a final part typically allocated to state specification.

Where helpful in the discussions and figures that follow the three parts will be indicated by a sequence of three pairs of parentheses corresponding to the three parts: ( )( )( ).

Three different tags are used to distinguish different interactions between the coupled rules representing the causal network:

(1) An "i" ("input") tag indicates that the message originates from the input interface.

(2) An "a" ("action") tag indicates that the message is directed to the output interface.

(3) A "v" ("virtual") tag indicates a message that is involved in node marking. There are two subtypes to the v tag:

(i) A "v0" tag is used on messages that initiate the marking process.

(ii) A "v1" tag is used on messages involved in an ongoing marking process.

The lookahead process depends upon the fact that the node and transition rules are coupled so that messages tagged with (v) are propagated just as messages tagged with an (a). However, when the message is tagged with a (v), actions upon the environment are not carried out, and the "next states" are those anticipated by the transition rules under the response (r) specified by the message.

Fig. 8 uses these conventions to show in detail how a classifier system would carry out the marking process illustrated in fig. 4. The caption for the figure gives an overview of the process.

7. The virtual bucket brigade

We emphasized at the start that the purpose of lookahead is to allow a system to make current action decisions on the basis of anticipated future consequences of those decisions. So far we have described a classifier system that can use causal

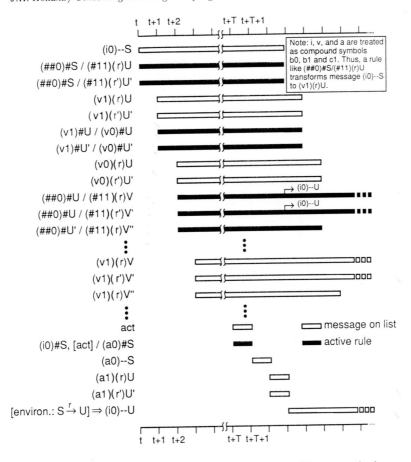

Fig. 8. Timing diagram for classifiers implementing the lookahead process of fig. 4. The process begins, on the top line of the diagram, when a message (i0)--S comes from the input interface, indicating that the environment is in state (category) S. This message starts the lookahead process by activating the transition rules on lines 2 and 3 of the diagram. The "pass through" # in the action part of these rules transforms the (i0) tag to a (v1) tag producing the messages (v1)(r)U and (v1)(r')U', at time $t+1$ on lines 4 and 5, indicating that markers are passing over the corresponding edges in the causal network. These messages activate the node rules on lines 6 and 7, issuing the messages (v0)(r)U and v(0)(r')U', at time $t+2$ on lines 8 and 9. These messages initiate further marker passing from the nodes corresponding to U and U'. This basic process is continued until an "act" signal appears at time $t+T+3$, indicating that the system must take some overt action. This causes the action rule associated with state S, on the fifth line from the bottom of the diagram, to be activated yielding the message (a0)--S. This message now causes the transition rules on lines 2 and 3 to issue messages (a1)(r)U and (a1)(r')U', which, because of their tags, are directed to the output interface. There they compete since they specify different responses. (Lookahead affects this competition through the *virtual bucket brigade* discussed below.) In this example, the response r wins out, causing the state transition from S to U to occur in the environment. The new environmental message (i0)--U keeps the transition rules for V and V' (lines 10 and 11) active, but the transition rules for V'' and V''' (line 12 and ...) go inactive, in effect dropping the markers from the latter nodes because they are not accessible from the new state U.

couplings (acquired through triggering) as the basis for marker-passing lookahead. We have yet to indicate the way in which this lookahead can influence current decisions. Once again Samuel [1] leads the way. In the checkersplayer, he associates a "cone" of future possibilities (cf. step 4 of the description of marker-passing lookahead in section 3) with each currently possible response (move). That is, for each currently possible move, his program looks down all sequences of moves (to some feasible depth) that begin with that move. He then weighs each first move by applying an evaluation function to the far ends (anticipated future states) of the associated cone. In effect, he "backs up" the values of future possibilities to the current decision point.

The bucket brigade algorithm does a similar kind of "backing up" over coupled classifiers. Classifiers at the end of a chain of coupled classifiers "pass their strength back", via bids, to earlier classifiers in the sequence (see p. 309 of ref. [2], section 3.1.2 of ref. [12], or section 5 of ref. [15]). This occurs only over an overt sequence of actions (and over several trials of the action sequence), but it does suggest that the bucket brigade might be used for a similar purpose during the "virtual" actions.

To this end, we add a *virtual strength register* to each rule in the system, to supplement the *strength register* already possessed by each rule. (This is the only substantial modification we have to make to a standard classifier system to implement the overall lookahead process.) Now, a rule activated by a message with an (a) tag passes its bid back to the *strength register* of the sender, in the usual way, but a rule activated by a message with a (v) tag passes its bid back to the sender's *virtual strength register*. Each time a rule is *first* activated in a marker-passing (v) process, its virtual strength register is set to the value of its strength register. Thereafter, as long as it is involved in the ongoing marker-passing process, the virtual strength is repeatedly modified by the bucket brigade. Moreover, the bids it makes are based on the virtual strength rather than the actual strength. Under this regime, the associated rules in each cone, once active, stay active as long as they are part of the cone (i.e., so long as the corresponding nodes are "marked"). This means that the bucket brigade is iterated over and over again through those rules. In effect, the virtual strengths come to reflect the value that would have been passed back over *many* overt trials of the action sequence (assuming that sequence accurately predicts the state changes in the environment).

To have the lookahead process influence the current action decision, we need only have the bids of the rules sending action messages be influenced by values in their virtual strength registers. As stated earlier, overt acts occur only when an "act" message is posted. Then, each rule at the start of a lookahead cone (1) posts a message with an (a) tag and a mid-part that specifies a response (see fig. 8), and (2) makes a bid that determines how it fares in the competition to control the output interface. When this bid is influenced by the virtual strength register, the competition at the output interface is affected by the cumulative "backing up" of future values under the virtual bucket brigade. (It is probably sensible to let the actual strength also influence the bid; the relative influence of the two strengths could be determined by a "daringness" coefficient, which could be "wired in" or could be set adaptively by other rules.)

Note that, if the competition between rules is probabilistic, based on the relative sizes of their bids, then lookahead proceeds more rapidly along paths wherein the corresponding transition rules make high bids. (Even though the competition is probabilistic, many winners are allowed at any given time, thereby exploiting the inherent parallelism of the system. See (3) in section 8.) Moreover, under this regime, the virtual bucket brigade produces a sophisticated estimate of the future: The value returned to each rule amounts to the expected value of its lookahead cone (the values of the endpoints weighted by the probabilities of reaching them).

8. Further refinements

(1) Because epoch hierarchies arise naturally when the genetic algorithm is applied to classifier systems, it is worth while to try to exploit them under marker-passing lookahead. One way to do this is to modify the triggered coupling. Under the modification, *node* rules would be formed with two conditions in the condition part. One condition would be the same as before, being satisfied by a message designating the state (category) of the corresponding node. The other condition would be satisfied by the message of some *more*

general classifier active at the same time. That is, the second condition would (often) couple the node rule to high level, "epoch-marking" rules.

Fig. 9 sketches the use of rules to implement this kind of epoch hierarchy. *Support* (see sections 2.3.2 and 4.1.5 of ref. [12]) enables rules (and anticipations) belonging to the coarser epochs to influence the more detailed actions at deeper levels of the hierarchy. In other words, the "desirability" of a given epoch, as measured by the strength of the corresponding rule and the virtual strength supplied by lookahead at that level, adds support (as defined in section 2.3.2 of ref. [12]) to various branches at lower levels, influencing lookahead and decisions at that level. The resulting structure can be interpreted as a quasi-homomorphic image (see section 2.1 and appendices 2A and 2B of ref. [12]) of the environmental dynamics.

(2) For rules dealing with coarse equivalence classes, as in the upper levels of an epoch hierarchy, a given response may at different times lead to states in different categories. That is, the model at that level is ambiguous as to outcomes because it inadequately distinguishes external states. As a result, the triggering operators generate more than one rule for a given category–response pair (S, r). Under such conditions, it is important that probabilities be assigned to each transition rule that leads from S and predicts a different outcome under response r. This probability can be treated

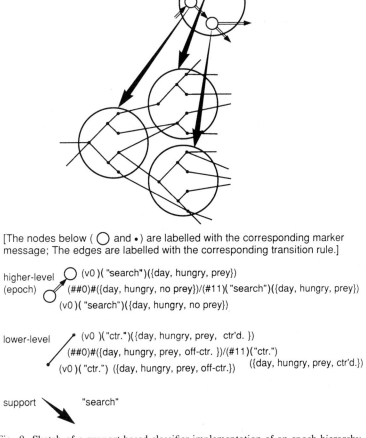

Fig. 9. Sketch of a support-based classifier implementation of an epoch hierarchy.

as a frequency count that is updated each time the response r is overtly executed. The frequency count is increased for the rule R predicting the state category S' that actually follows the response, and it is decreased for all other rules associated with the pair (S, r). This can all be normalized to a fraction between 0 and 1 by treating the count as an average: Let $P(R, t)$ be the normalized frequency associated with rule R at time t, and let $d(S', t) = 1$ just in case S' occurs at time t. Then

$$P(R, t+1) = [1 - (1/n)] P(R, t) + d(S', t)/n$$

provides a reasonable updating formula when a constant $n \gg 1$ is used. For example, if $P(R, T) = 0.5$ and $n = 4$, and the corresponding $d(S', t) = 1$ at T, $T + 3$, and $T + 4$, then $P(R, T + 4) = 0.70$, a reasonable approximation to the 0.75 experienced in the last 4 steps. If S'' corresponds to the category S', and R' is the transition rule from (S, r) to S'', then $P(R', T + 4) = 0.30$, as is appropriate.

Note that, when a prediction is verified, the corresponding rule is also strengthened. Consider, then, two *pairs* of rules:

(i) {(#)(#)(no prey)/(#)("twiddle")(no prey); (#)(#)(no prey)/(#)("twiddle")(prey)}

(ii) {(#)(#)(no prey)/(#)("run")(no prey); (#)(#)(no prey)/(#)("run")(prey)}.

Under either "twiddle" or "run" the result of a search at any instant may be either "no sighting of prey", or "prey comes into view". Past probabilities will determine the relative frequency with which the second rule in each pair wins, which is tantamount to determining the expected length of a "search" before prey is encountered. This, in turn, under the virtual bucket brigade and competition (see (3) below), determines the probability that "twiddle" will be employed over "run".

(3) The list of satisfied rules becomes a sample space, once the associated probabilities of winning are available. This observation provides a useful way for determining the number of rules that will be allowed to win the competition at any given time: The list of satisfied rules is sampled repeatedly until some rule is selected for a second time, at which point the sampling process terminates. The rules so selected post their messages. This process has the advantage that the list will be short if there are a few high-probability rules (the system has well-established means of acting upon the situation), and long otherwise (allowing extensive alternatives, to be resolved in terms of mutual exclusions at the output interface).

(4) In order that competition and limitation of the size of the message list not cause "marked" nodes to be deactivated, the system could be supplied with a special message list for messages from nodes. Note that a special list is only necessary if the node rules are weak. Otherwise the method in (3) provides for an expanding message list with occasional losses – something closer to human lookahead with its difficulty of retaining a conscious picture of all the branches in a "bushy" structure.

9. Commentary

The emergent structure discussed in this paper is an architecture that provides for parallel lookahead under distributed control. Though we have, by now, accumulated considerable experience with learning procedures and emergent structures in "standard" classifier systems (see ref. [15]), we have no experience with classifier systems exhibiting lookahead. For this reason, the new system is organized to exploit structures that are known to emerge under the learning algorithms (the bucket brigade algorithm and triggered genetic algorithms) of the standard systems. It is interesting that a small change in the standard system – addition of a *virtual strength register* coordinated with an additional use of the bucket brigade algorithm in *virtual* mode – provides a sophisticated way for future anticipations to influence current action. In effect, the decisions are based on the "expected" value of the cone of future possibilities associated with each perceived action possibility. It should be emphasized that these ideas have yet to be tested in a complete system – we have experience only with fragments.

Acknowledgements

Many of the ideas presented here have been hammered out over the past couple of years in meetings of the BACH group at the University of Michigan. Dr. Rick Riolo is currently preparing a simulation to test his own version of these ideas in the context of latent learning. The research reported has been supported, in part, by the National Science Foundation under grant IRI-8904203 and its predecessors, and a substantial part of the work was done during visits to the Santa Fe Institute.

References

[1] A.L. Samuel, Some studies in machine learning using the game of checkers, IBM J. Res. Dev. 3 (1959) 210–229.
[2] J.H. Holland, A mathematical framework for studying learning in classifier systems, Physica D 22 (1986) 307–317.
[3] J.E. Laird, P.S. Rosenbloom and A. Newell, Chunking in Soar: The anatomy of a general learning mechanism, Machine Learning 1 (1986) 11–46.
[4] D.E. Rumelhart and J.L. McClelland, eds., Parallel Distributed Processing. (MIT Press, Cambridge, MA, 1986).
[5] G. Edelman, Neural Darwinism: The Theory of Neuronal Group Selection (Basic Books, New York, 1987).
[6] D.O. Hebb, The Organization of Behavior (Wiley, New York, 1949).
[7] Z.W. Pylyshyn, Computation and Cognition (MIT Press, Cambridge, MA, 1986).
[8] N. Rochester, J.H. Holland, L.H. Haibt and W.L. Duda, Tests on a cell assembly theory of the action of the brain, using a large digital computer, IRE Trans. Information Theory IT2 (1956) 80–93.
[9] W.S. McCulloch and W. Pitts, A logical calculus of the ideas immanent in nervous activity, Bull. Math. Biophys. 5 (1943) 115–133.
[10] D.R. Hofstadter, Gödel, Escher, Bach: An Eternal Golden Braid (Basic Books, New York, 1979).
[11] D.R. Hofstadter, Metamagical Themas: Questing For The Essence of Mind and Pattern (Basic Books, New York, 1985).
[12] J.H. Holland, K.J. Holyoak, R.E. Nisbett and P.R. Thagard, Induction: Processes of Inference, Learning, and Discovery (MIT Press, Cambridge, MA, 1986).
[13] S.E. Fahlman, NETL: A System for Representing and Using Real-World Knowledge (MIT Press, Cambridge, MA, 1979).
[14] S. Forrest, Implementing semantic network structures using the classifier system, in: Proceedings of an International Conference on Genetic Algorithms and Their Applications (Erlbaum, London, 1985).
[15] L.B. Booker, D.E. Goldberg and J.H. Holland, Classifier systems and genetic algorithms, Artificial Intelligence 40 (1989) 235–282.

LEARNING AND BUCKET BRIGADE DYNAMICS IN CLASSIFIER SYSTEMS

M. COMPIANI[1], D. MONTANARI[2] and R. SERRA[3]

ENIDATA (ENI Group), Viale Aldo Moro 38, 40127 Bologna, Italy

Classifier systems are rule-based adaptive systems whose learning capabilities emerge from processes of selection and competition within a population of rules (classifiers). These processes are ruled by the values of numerical variables which measure the fitness of each rule. The system's adaptivity is ensured by a fitness reallocation mechanism (the bucket brigade algorithm) and by genetic algorithms which are responsible for the internal dynamics of the system. In this paper we discuss classifier systems as dynamical systems, the main focus being on the asymptotic dynamics due to the bucket brigade, abstracting from the action of the genetics. This topic is discussed with reference to a specific task domain, in which the system is used as a detector of statistical properties of periodic or fluctuating external environments. We also describe a major consequence of the genetics on the bucket brigade dynamics, namely the proliferation of individual rules into subpopulations of equivalent classifiers; we then show that this can eventually lead to undesired stochastic behavior or to the destabilization of correct solutions devised by the system.

1. Introduction

Classifier systems are rule-based message-passing systems (see Holland [1, 2], Riolo [3] and Holland's paper in this volume [4]) which are particularly interesting since they combine features typical of classical AI systems with those peculiar to the "dynamical systems approach" to AI (Serra [5], Serra and Zanarini [6]) (exemplified by neural networks and reaction–diffusion models (Steels [7])).

In some respects, the overall framework of classifier systems is very similar to that of production systems[#1], but classifier systems depart from the approach of classical AI, which is based exclusively upon inference chaining, since they use evolution equations of numerical variables (strengths of classifiers and message intensities) as a basic tool for learning. These variables provide a measure of fitness for the "logical" components (the rules) to which they are attached, and change in time according to definite evolution equations, thus making classifier systems true dynamical systems[#2].

Moreover, like neural networks, classifier systems are able to learn starting from a condition of tabula rasa through a process of self-organization. The relationship between classifier systems and neural networks has been recently studied in several papers (Compiani et al. [10], Davis [11], Miller and Forrest [12]). The relevance of self-organizational processes is discussed in Serra [5] and Serra and Zanarini [6].

The learning capabilities of classifier systems are intimately linked to the dynamical evolution

---

[1]Present address: ARS, Viale Aldo Moro 38, 40127 Bologna, Italy.
[2]Present address: TEMA, Viale Aldo Moro 38, 40127 Bologna, Italy.
[3]Present address: DIDA*LAB, Via Lamarmora 3, 20122 Milan, Italy.
[#1]Production systems are rule-based systems relying on condition–action rules, named productions. Condition–action rules are said to be triggered when all the condition parts are satisfied; this is a necessary (but not always sufficient) condition for the rules to perform their own actions. Rules may be triggered by other rules thus forming chains of concatenated rules which mimic the process of inferring conclusions from premises.

[#2]Classifier systems also provide an example of a computational ecology (Huberman [8], Kephart et al. [9]), since they can be viewed as open systems of cooperating computational agents which operate with no central control and compete for resource utilization.

of the system's components, and the rules emerge under the combined action of two different learning mechanisms: the *bucket brigade* is a strength reallocation algorithm which modifies the fitness of the rules in the system, whereas the *genetic operators* induce structural changes in the population of classifiers, creating new individuals, replicating or deleting old ones depending on their fitness.

These mechanisms make classifier systems extremely powerful, but they also make their behavior difficult to analyze and predict. Achieving a good insight into some of the fundamental aspects of classifier systems will allow us to appreciate their possibilities and limitations and to devise meaningful extensions.

In this paper we will elucidate some properties of classifier systems that arise from their complex dynamics. For the sake of simplicity, we restrict our analysis to a fixed population of rules and consider the dynamics of the bucket brigade alone, assuming that the genetic operators have been "frozen". Such an "adiabatic" approximation is acceptable provided that the time scales of the two processes are different; this is the case in our applications in which the genetic operators act upon a population of rules which have already been correctly ranked by the bucket brigade according to their usefulness. The only exception occurs in section 3, where we describe a phenomenon which is a direct consequence of the interaction of the two mechanisms.

Our results are based on a theoretical analysis integrated with numerical work. The latter consisted of computer simulations of a classifier system acting in a letter sequence learning task domain (see Robertson and Riolo [13]). A fixed letter sequence is cyclically presented to the classifier system which, given as input one letter (the "current" letter), is expected to predict the next coming letter. Regardless of their length, finite cyclic letter sequences may be divided in two major classes. A sequence is called *non-ambiguous* if every letter has a unique successor (in the sequence "neural" "n" is followed only by "e", "e"

only by "u" etc.), while an *ambiguous* sequence contains letters with different successors (in the sequence "systems" the letter "s" is followed by either "y" or "t"). In order to forecast correctly the next letter in "ambiguous" sequences the system must learn to build up an internal memory, whose size varies with the degree of ambiguity of the sequence[#3].

The internal dynamics of classifier systems are intimately related to their learning properties. In section 2 it is shown that different kinds of solutions to the letter prediction task can be developed by the system, based upon different bucket brigade dynamics.

As a side effect of the genetic operators, good rules are replicated and subpopulations of similar classifiers are generated. The combined action of bucket brigade and genetics leads to performance instabilities in the long term, which are described in section 3; a quantitative description of the dynamics within a given subpopulation, under the action of the bucket brigade alone, is given in section 4.

In many applications it is realistic to admit a finite error probability in the replication of the sequence, which thus becomes a stochastic process. The effects of environmental fluctuations upon the internal dynamics are addressed in section 5.

Finally, the value of the relevant parameters used in the simulations are listed in the appendix.

[#3] Limiting the size of the input window makes the task harder to be solved since the adaptive mechanisms of the system are required to build more complex structures (e.g. chains) of classifiers. In turn, allowing for an input window of suitable size, each ambiguous sequence can be transformed into a non-ambiguous sequence. It should be noted that feed-forward neural networks cannot handle this task, unless they are augmented with recurrent interactions from the output or hidden layers to the input layer (Elman [14]). It is also noteworthy that the maximum length of such a built-in memory must be specified in advance in neural networks, while classifier systems can adjust it to the sequence in an automatic manner.

```
C(1):   D: a, X: c / E: b

C(2):   D: b, D: b / E: a

C(3):   D: b, D: b / X: b

C(4):   D: a, X: b / E: c

C(5):   D: c, D: c / E: a

C(6):   D: c, D: c / X: c
```

Fig. 1. Static solution to the ambiguous sequence "abac". In CFS-C each classifier has two conditions and one action. The semantics of each classifier is specified by means of labels which describe the type of message read by the conditions and produced by the action. "D:" indicates that the condition reads messages coming from the detectors (the input channels), "E:" that the action produces an output message to the effectors (the output channels). The internal messages have label "X:" and are produced and read exclusively by the system's classifiers. Classifier C(1) is activated by a message coming from the detectors and coding for the letter "a" and an internal message coding for "c", and outputs a message towards the effectors which codes for "b". Note that C(1) needs the internal message which is produced by C(6) for getting activated. Internal messages provide classifiers C(1) and C(4) with a kind of tag which allows the selective activation of only one of them at a time.

2. Dynamical solutions

In this section we will briefly recall the salient points of the taxonomy of experimental solutions to the letter prediction task, in the case of ambiguous sequences. A more detailed description of the solutions to non-ambiguous and ambiguous sequences can be found in Compiani et al. [15–17].

The solutions to the task of predicting an ambiguous letter sequence can be grouped in two classes, i.e. the "static" and the "dynamical" solutions. Static solutions[#4] accomplish the prediction task on each occurrence of the ambiguous letter by using classifiers that need a message from the detectors and simultaneously an internal message for being activated. This enables the system to selectively activate only the correct classifier (see fig. 1). These solutions admit variations (see

[#4] These solutions were referred to as "anthropomorphic" solutions in Compiani et al. [15–17] since they look like the most intuitive solution that a human would devise to manage this task.

```
C(1):   D: a, D: a / E: b

C(2):   D: b, D: b / E: a

C(3):   D: b, D: b / X: b

C(4):   D: a, X: b / E: c

C(5):   D: c, D: c / E: a
```

Fig. 2. Dynamical solution to the sequence "abac". Note that C(1) and C(4) can be activated simultaneously and try to post contradictory messages; the success of this solution is due to the fact that whenever C(1) and C(4) compete the bid of C(4) always exceeds that of C(1). The conflict resolution mechanism then selects the classifier which is making the highest bid. In the absence of the internal message posted by C(3) no competition occurs since only C(1) is activated.

Compiani et al. [16]) in which efficient tagging (Belew and Forrest [18]) of the messages is achieved by means of markers located in the non-coding loci (i.e. in the non-interpreted part of the messages). In the dynamical solutions no selective activation of the different alternatives takes place; this implies that more than one classifier at a time might be active and compete to forward their respective guesses to the effectors. The discrimination among the alternatives and the final choice of the winner is achieved by the conflict resolution mechanism on the ground of a highest-bid-wins criterion. As a consequence, the bids of classifiers which enter the competition must be properly modulated to ensure the success of the one or the other on the proper time step. The frequency of occurrence of the ambiguous letter is eventually mirrored in the suitable temporal modulation and proper staggering of the bids of the attendant classifiers[#5]. This is the case of the solution reported in fig. 2.

3. Instabilities in classifier systems

In this section we discuss some effects derived from the combined action of the bucket brigade

[#5] Adjusting the frequency of oscillation of some internal parameters to the frequency of the external stimulus (the reward) recalls the entrainment of non-linear oscillators driven by an external force.

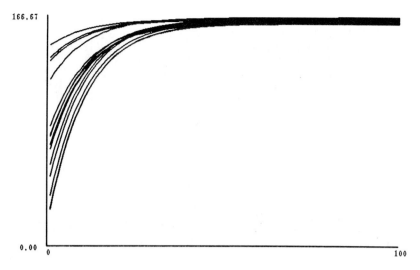

Fig. 3. This plot shows the behavior in time (measured in number of cycles) of the strengths of 15 classifiers starting with a 90% spread around the initial average strength and in the absence of genetics. The classifiers are activated by the same detector message and are supposed to send directly the same output message to the effectors (as far as the interpreted parts of the classifiers are concerned, they are coincident with C(2) in fig. 1). In the absence of competition for entering the message list, the strengths undergo a process of fast convergence towards a common value in about 50 steps.

and genetic algorithm with the aim of clarifying the relationship between some aspects of the internal dynamics and overall performance in classifier systems. Let us start by remarking that genetic operators are "blind", in the sense that their action may lead to a decline in the system performance in some stages of its evolution. This is the case when good rules are occasionally removed and replaced by individuals which perform the task in a less satisfactory way. While these undesired effects could be avoided by protecting the good rules prohibiting their replacement by genetic operators (Zhou [19]), it should be noted that restricting the action of genetics may prevent the discovery of new and possibly better rules. Therefore, our approach has been to handle all of the rules on an equal footing in order to analyze the resulting global dynamics.

Let $\mathscr{C}$ be a classifier system which involves a number of rules, partitioned into disjoint subpopulations: a subpopulation is defined here as a subset Ω of $\mathscr{C}$, whose members are the classifiers which, in the given task domain, match the same messages. According to this definition (which does not coincide with the one given in Wilson [20]) the members of Ω need not be equal to each other. Indeed, they may have different "genotypes" (e.g. due to a redundant encoding) or they may differ in the action part.

We will assume that, as is usually the case in our experiments, the time constants of genetic operators are much slower than that of the bucket brigade, so that the classifiers attain quasi-asymptotic values for the strengths between different activations of the genetics. From a formal point of view the activity of the genetic operators introduces discontinuous changes into the set of equations ruling the process of strength reallocation within the population.

Let us now focus on a typical effect of the genetics: the fittest rules are more often chosen by the genetic operators as a template so that eventually they proliferate and evolve into subpopulations of rules. Let us consider a particular case of subpopulation made of classifiers (synonyms, for brevity) that, in the given task, produce non-contradictory messages on the output channel (i.e. messages labelled "E:"). This implies that as long as the number of classifiers in the subpopulation is lower than the message list size m, they get re-

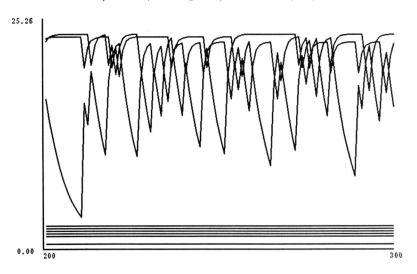

Fig. 4. This figure shows the steady state dynamics of the strengths of a population of ten classifiers (of the type considered in fig. 3) competing to access a message list of size 2. Seven classifiers have undergone complete decay to zero strength, and do not take part in the asymptotic activity of the subpopulation. Their strengths, although they are exactly zero, have been slightly shifted above the x-axis for major clarity. For the three active classifiers the time lapse between successive rewards is a stochastic variable depending on the probabilistic mechanism selecting the two winners on each time step. However, the strength of the classifier which from time to time is temporarily decaying is never subject to dramatic decrease; rather, on getting the next reward it recovers a strength comparable in value to its full asymptotic value so that it can successfully compete with the other two classifiers. Accordingly, the strengths oscillate irregularly around a mean value.

warded simultaneously and all the strengths converge towards a common steady state value (strength equalization – see fig. 3).

When the size μ of the subpopulation exceeds m, the equalized regime breaks down and fluctuations begin to appear, due to the probabilistic choice of posting classifiers. While a quantitative description of this phenomenon is deferred to section 4, let us remark that, at least for some values of some critical parameters, a kind of strength equalization in the average takes place, where all the classifiers which survived the competition up to that point, exhibit fluctuating strengths around the same average value (see fig. 4).

Typically, in our task domain strength equalization occurs at about the same time for all the subpopulations performing independent subtasks and the global result is that the whole system evolves into a configuration where almost all the rules are good and they all have approximately the same strength.

This implies that when new members are introduced into the population, even if their initial strength is slightly lower than the strength of the older members, the new rules have a chance of posting their message which is comparable with the chances of the older rules.

Far reaching effects occur when the genetics produce rules that belong to a subpopulation but are not synonymous (e.g. "D: b, D: b / E: z" in a subpopulation of classifiers "D: b, D: b / E: a"). Such a new rule is given an average strength which, owing to the equalization process described above, is comparable with the strength of all the other classifiers. Then, all rules have about the same chance to post their messages. One of the wrong rules will eventually post its message, which will be selected by the conflict resolution mechanism, leading to a wrong answer. This will have two consequences: the good rules will miss one of their "feeding times", thus decaying instead of recovering their strength; more importantly, error-triggered genetics (the so-called cover effector operator used by Riolo [3]) may be activated, possibly generating more members of the subpopulation *with random actions*, and further

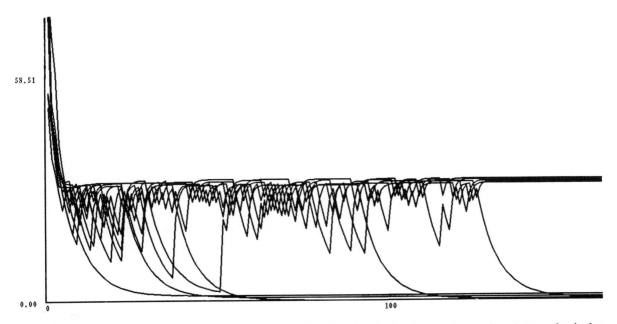

Fig. 5. Ten classifiers were made to compete to enter a message list of size 5 under the same conditions of fig. 4, except for the fact that the competition was biased towards the strongest classifiers by imposing $b=2$ in the probability function of eq. (6). As a result, small differences in the strengths, due to the random access to the message list, are strongly amplified and the classifiers which first happen to have below-average strengths are almost irreversibly doomed to decay to zero strength (as in fig. 4, lines close to the time axis denote zero strengths). Very rarely, a weak classifier still succeeds in being rewarded and taking part anew in the competition with a strength comparable with that of the strong classifiers. Consequently, within about 150 steps only five classifiers have survived the hard competition and the subpopulation reaches a condition of full equalization.

reducing the chance of winning for the good rules in Ω. The cover effector operator thus introduces a positive feedback which may transform an occasional mistake into a serious problem. In fact it is frequently observed that the subpopulation considered is overcrowded by incorrect classifiers with random actions. Such effects have been observed in several runs. The number and severity of the errors depend on several factors, the most significant being the degree of equalization of the strength within the subpopulation, the size of the message list and the policy adopted to select from the pool of bidders the classifiers that will post their messages.

A detailed discussion of the effects of these parameters is given in Compiani et al. [16]. Here we simply mention the fact that a completely different situation has also been observed, when the probability distribution for the choice of the posting classifiers is more sharply dependent upon the classifiers' strengths[#6]. In this case, one frequently observes that just as many classifiers as can be contained in the message list survive, while the other synonyms have a vanishingly small probability of posting and therefore undergo irreversible decay (see fig. 5).

Another effect of the action of the genetic operators is the generation of entirely new subpopulations that solve the same subtask. The case of the non-ambiguous sequence offers a clear example of this feature. It has been observed that in the long run new classifiers like "D: a, E: a / E: b" would appear and quickly reproduce themselves, replacing the rules of type "D: a, D: a / E: b". Solutions of this type are harder to find for the

---

[#6] This option available in the CFS-C implementation (Riolo [3]) corresponds to choosing values larger than 1 for the parameter b in eq. (6) below.

system, but they are favored by the bucket brigade (i.e. attain higher steady state strengths) if the relative support (see Holland et al. [2] and Riolo [3]) is taken into account in computing the bid (see eq. (2) below). In the long run, therefore, the previous solution will be superseded by the new one (Compiani et al. [16]).

4. Subpopulation dynamics

In this section we investigate, under some simplifying assumptions, the internal dynamics of the bucket brigade algorithm in classifier systems. Namely, we discuss the evolution of the strengths of a single subpopulation Ω when no genetics take place, i.e. we will assume that the number of classifiers is constant. Furthermore, while the analysis described in this section could be extended in a straightforward way to cover different cases, we restrict ourselves to subpopulations of synonymous classifiers matching detectors messages and producing output messages to the effectors[#7].

The time evolution equation for the strength $s_i(t)$ of a single classifier $C_i \in \Omega$ is given by the fundamental equation of the bucket brigade

$$s_i(t+1) = s_i(t)(1-\alpha) - b_i(t) + R_i(t), \quad (1)$$

where α is the decay constant[#8], $b_i(t)$ is the bid offered by classifier C_i to post a message at time t (it is zero if the classifier does not post any message), and $R_i(t)$ is the reward (if any) which the classifier receives at time t. Under share-reward conditions the system receives from the external environment a fixed reward R to be divided among

---

[#7] Results concerning the bucket brigade of long chains of coupled classifiers are reported in Riolo [21].

[#8] We limit ourselves to the linear decay term ("headtax", in CFS-C jargon (see Riolo [3])) in order to make some progress with the analytical treatment; other "taxes" are sometimes introduced whereby $1-\alpha$ is generalized to a non-linear function of the strengths.

all of the N classifiers posting output messages; hence, each of them gets a fraction $R_i = R/N$.

The bid is itself a function of the strength s_i,

$$b_i(t) = ks_i(t). \quad (2)$$

Eq. (2), with $k = $ constant < 1, is the one originally proposed by Holland [1]. We have omitted for simplicity the so-called "relative support" of a classifier (see Holland et al. [2]), which introduces a non-linear coupling with delay among the strengths of different classifiers. Note that in eq. (2) we do not consider the specificity of the classifier; the reason for this is explained in Compiani et al. [15] and in Serra and Zanarini [6].

In order to fix the ideas, let us consider a classifier of type "$a \Rightarrow b$" (like C(1) in fig. 2), which reads a given letter "a" from the detectors and makes a correct guess "b" concerning the following letter. If the sequence is cyclically presented to the system and it is T letters long, then this classifier will become active and be rewarded once every T steps and the asymptotic behavior of $s_i(t)$ is periodic with period T. If we suppose that C_i becomes active at nT ($n = 0, 1, 2, \ldots$) and that $\alpha T \ll 1$ and $k \ll 1$ (these restrictions could easily be abandoned, providing an exact solution, like in Compiani et al. [17]), we obtain

$$s_i(t+T) \approx s_i(t)(1-\alpha T) - ks_i(t) + R_i, \quad (3)$$

where R_i (see eq. (1)) is the amount of the external reward to C_i, paid once per cycle. The strength will then decay, until a new activation occurs at time $t + 2T$, etc. The maximum value of the asymptotic cycle, $s_i(\infty)$, can be obtained by imposing $s_i(t+T) = s_i(t)$. This gives

$$s_i(\infty) = \frac{R_i}{k + \alpha T}, \quad (4)$$

which shows how the periodicity of the ambiguous letter affects the asymptotic strength of C_i.

The previous equations hold as long as every classifier in Ω posts once every T time steps, that is if $\mu \leq m$; as soon as μ exceeds m, each classifier in Ω is no longer sure to post every time it matches. Two basically different behaviors have been observed in such a case, depending on the values of some critical variables:

(i) a situation of *strength equalization*, where the different classifiers have the same average strength, while their actual strengths oscillate randomly around it;

(ii) a situation of *strong polarization*, where m classifiers have high strengths, while the others decay without posting (in this case, eqs. (3) and (4) still hold for the m strong classifiers).

Which one of these two regimes will prevail depends upon the value of some parameters: a very important factor is the shape of the probability distribution (see eq. (6) below) for choosing the m winners of the competition to post a new message in the message list among all the μ classifiers active at a given time step. The role of b in eq. (6) in determining the final behavior is apparent in figs. 4 and 5, and it is discussed in Compiani et al. [15, 17].

In order to discuss the case where $\mu < m$ let us consider a subpopulation Ω having μ classifiers. The time evolution equation for a single rule is

$$s_i(t+1) = s_i(t)(1-\alpha) + [R_i - ks_i(t)]\phi_i(t), \tag{5}$$

where

$\phi_i(t) \equiv 1$ if C_i posts at time t

$\equiv 0$ otherwise.

$\phi_i(t)$ depends upon $s_i(t)$ through the probability distribution for choosing the posting classifiers. In the CFS-C implementation, this probability is proportional to a power b of the classifier's bid: the probability $p(i, t)$ that an active classifier C_i wins at time t is given by

$$p(i,t) = m\frac{s_i^b}{\sum_{j=1}^{\mu}s_j^b}. \tag{6}$$

The general features of this dynamical regime can be properly represented with the aid of the map of fig. 6. $s_i(t)$ evolves in time following eq. (5); the iterates have to be calculated on two different straight lines according to whether C_i is simply decaying or is also being rewarded. The intersections with the bisecting line individuate the two attractors between which the representative point oscillates. The stochastic outcome of the competition simply reflects the fact that the number of steps of decay and reward is a random variable. As a consequence, the representative point undergoes an erratic motion along open trajectories lying inside the region ABCO. In the absence of competition the system visits the two branches of the map following closed and periodic trajectories according to the periodic function $\phi_i(t)$.

Let us now average eq. (5), by making the assumption that the probability that C_i posts is independent of its strength (let us call it ϕ). This assumption is strictly valid in the case of strength equalization, or with $b = 0$ in eq. (6) (so that $\phi = m/\mu$); it is also acceptable in the case of strength equalization in the average, if the spread of the strengths is not too wide. We obtain

$$\langle s_i(t+T)\rangle = \langle s_i(t)\rangle(1-\alpha T) + [R_i - k\langle s_i(t)\rangle]\phi. \tag{7}$$

Its asymptotic limit is

$$\langle s_i\rangle_\infty = \frac{R_i\phi}{\alpha T + k\phi} = \frac{R_i}{k + \alpha T\phi^{-1}}. \tag{8}$$

Eq. (8) has the same form as eq. (4), save for the fact that the average interval between two successive posting of the same classifier is, under the assumptions made here, equal to $T\phi^{-1}$. With the parameter values specified in the appendix, the accuracy of the above results is within a few percent of the measured values.

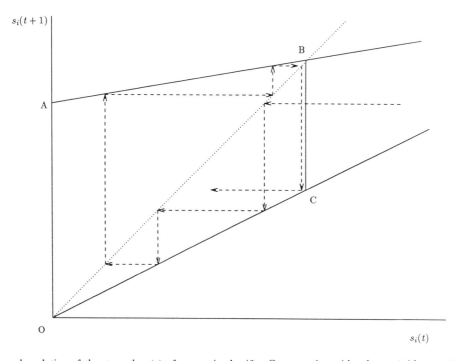

Fig. 6. Temporal evolution of the strength $s_i(t)$ of a generic classifier C_i competing with other μ (with $m < \mu$) for entering the message list. The equations of the two lines defining the map follow from eq. (5) by posing $\phi = 1$ and $\phi = 0$ respectively. $s_i(t)$ decays along line OC approaching the stable fixed point O, and makes a transition to the line AB whenever C_i gets rewarded. In this case $s_i(t)$ points towards the new stable attractor located in B. Note that $s_i(t)$ cannot escape the region enclosed by the polygonal ABCO; it wanders inside this trapping region in a random way being alternatively attracted by the fixed points O and B. The location of the attractors and the geometry of the basins of attraction may change in time due to the action of the genetics. By way of example, the magnitude of R_i in eq. (5) can be influenced by the possible variations of the current number N of classifiers in the subpopulation whenever the system is operated under share-reward conditions.

5. Environmental fluctuations

The preceding calculations referred to the case of a letter sequence which was continuously presented to the system, without any error or fluctuations. Under these conditions the randomness which can be observed in the system's strengths and – under some conditions – in its answers, is entirely due to the internal random mechanisms which rule the competition processes.

We will now consider a different situation where the system senses a stochastic environment and the classifiers are expected to learn the statistical regularities of the input signal. Let us think of a sequence which is produced by a stochastic process, and therefore is subject to possible error of replication. To study the effects of unbiased environmental fluctuations, the spurious random behavior due to the subpopulation dynamics will be prevented from occurring by taking $\mu \leq m$ (see section 4).

Environmental fluctuations can be modelled by generating a stochastic sequence of symbols, where the probabilities of obtaining each letter x at time $t + 1$ are specified. These probabilities will in general depend upon the sequence of all the previous letters. To simplify our analysis we will restrict ourselves to the case (analogous to the "non-ambiguous" sequence in the deterministic case) of a Markovian sequence, where the transition probabilities depend only upon the last letter, regardless of the preceding ones. Moreover, if we assume that the transition densities do not change in time (i.e. that the letter sequence is a homogeneous stochastic process), the conditional probability of reading x at $t+1$ if letter y is read at time t can be written as $w(x|y)$, where the dependence on time can be dropped and a unit time increment between x and y is understood. The time evolution equation for the strength of a classifier of the type $b \Rightarrow a$ is

$$s_i(t+1) = s_i(t)(1-\alpha) + R_i \phi_b(t) \phi_a(t+1) - k s_i(t) \phi_b(t), \qquad (9)$$

where

$$\phi_x(t) = 1 \quad \text{if letter } x \text{ is read at time } t,$$
$$= 0 \quad \text{otherwise.} \qquad (10)$$

Let us now average eq. (9) (considering e.g. ensemble averages); by taking the infinite time limit, and assuming that a stationary average is obtained, we obtain

$$\langle s_i \rangle_\infty = \frac{R_i p(b) w(a|b)}{\alpha + k p(b)}, \qquad (11)$$

where $p(x)$ is the infinite time limit of the probability $p(x, t)$ of finding letter x at time t. In the simple case of a sequence of two letters a, b, denoting

$$w(a|b) = \rho, \quad w(b|a) = \eta \qquad (12)$$

and expressing the asymptotic single letter probabilities as functions of ρ and η, eq. (11) can be reduced to

$$\langle s_i \rangle_\infty = \frac{R_i \rho \eta}{k\eta + \alpha(\rho + \eta)}. \qquad (13)$$

Note that eq. (13) has the same structure as eq. (4); it can be read as the asymptotic strength of a classifier which has a period of activation $T=1$ and R_i and k renormalized by factors which measure the average number of rewards received and the bids paid by the classifier per unit time.

Eq. (13) has been numerically verified in a range of parameter values. With the usual parameters used in our experiments (given in the appendix) its accuracy is typically within 1%.

6. Conclusions

The bucket brigade dynamics is responsible for some peculiar features of classifier systems, compared with traditional AI and connectionist systems, which manifest themselves both in the nature of the solutions devised by the system, and in their time evolution. In particular, the comparison of the static and dynamical solutions points to the fact that classifier systems may exhibit dynamical properties which enable them to devise solutions to cognitive tasks which go beyond the logical paradigm of classical AI. In this spirit we have examined the dynamics triggered by the bucket brigade in the case of a periodic modulation of the reward (this is the case when the system is confronted with a periodic environment). In our analysis the genetics have been considered only as a potentially disturbing factor of the bucket brigade dynamics. More precisely, the proliferation of equivalent classifiers has been shown to be deceptive for the bucket brigade; in fact this can result in random regimes in the dynamical evolution of the strengths which, in turn, may lead to a decline of the system performance.

Finally, we have studied the bucket brigade dynamics in the presence of external sources of randomicity (the environment properties are no longer periodic but obey a statistical distribution). Estimates of the average values of the strengths are carried out in view of the possible use of classifiers as detectors of statistical properties of noisy environments.

Appendix

The experiments described in this paper have been carried out using the simulation program LETSEQ implemented by R. Riolo using his software package CFS-C (Riolo [3, 22]). Additional software has also been developed to perform specific experiments on various forms of the bucket brigade algorithm and to generate some of the figures in this paper.

The experiment of fig. 3 was performed under the following conditions; 15 classifiers, $\alpha = 0.01$, $k = 0.08$, $b = 1$, $m = 15$, $R_i = 15$. The experiment of fig. 4 was performed with 10 classifiers, $\alpha = 0.15$, $k = 0.5$, $b = 1$, $m = 2$, $R_i = 15$. The experiment of fig. 5 was performed with the same parameters of fig. 4 except for $b = 2$, $m = 5$, $R_i = 20$.

The experiments concerning the competition between members of the same subpopulation have been performed with $R_i = 30$, $k = 0.1$. The cases $\mu = 20$ and $\mu = 100$ and the cases $m = 1$ and $m = 5$ have been considered, while α has been varied between 0.001 and 0.01 and b has been varied between 1 and 4.

Experiments with the stochastic sequence have been performed with $\mu = m = 20$, $R_i = 30$, $k = 0.1$, $b = 1$ and $\alpha = 0.001$, with ρ and η ranging from 0.5 to 0.9.

References

[1] J.H. Holland, Escaping brittleness, in: Machine Learning. An Artificial Intelligence Approach, Vol. II, eds. R.S. Michalski, J.G. Carbonell and T.M. Mitchell (Kaufmann, Los Altos, CA, 1986) p. 592.
[2] J.H. Holland, K.J. Holyoak, R.E. Nisbett and P.R. Thagard, Induction. Processes of Inference, Learning and Discovery (MIT Press, Cambridge, MA, 1986) ch. 4.
[3] R. Riolo, CFS-C: A Package of Domain Independent Subroutines for Implementing Classifier Systems in Arbitrary, User-defined Environments, Logic of Computers Group, University of Michigan, Ann Arbor, MI (1986).
[4] J.H. Holland, Physica D 42 (1990) 188–201, these Proceedings.
[5] R. Serra, Dynamical systems and expert systems, in: Connectionism in Perspective, eds. R. Pfeifer, Z. Schreter, F. Fogelman-Soulié and L. Steels (North-Holland, Amsterdam, 1989) p. 331.
[6] R. Serra and G. Zanarini, Complex Systems and Cognitive Processes (Springer, Berlin), in press.
[7] L. Steels, Steps towards common sense, in: Proceedings of ECAI 1988, ed. Y. Kodratoff (Pitman, London, 1988) p. 49.
[8] B.A. Huberman and T. Hogg, The behavior of computational ecologies, in: The Ecology of Computation, ed. B.A. Huberman (North-Holland, Amsterdam, 1988).
[9] J.O. Kephart, T. Hogg and B.A. Huberman, Dynamics of computational ecosystems, in: Distributed Artificial Intelligence, Vol. 2 (Kaufmann, Los Altos, CA, 1989), in press.
[10] M. Compiani, D. Montanari, R. Serra and G. Valastro, Classifier systems and neural networks, in: Parallel Architectures and Neural Networks, ed. E. Caianiello (World Scientific, Singapore, 1988) p. 105.
[11] L. Davis, Mapping classifier systems into neural networks, in: Proceedings of the 1988 Conference on Neural Information Processing Systems (Kaufmann, Los Altos, CA, 1988).
[12] J.H. Miller and S. Forrest, The dynamical behavior of classifier systems, in: Proceedings of the Third International Conference on Genetic Algorithms (Kaufmann, Los Altos, CA, 1989) p. 304.
[13] G.G. Robertson and R.K. Riolo, A tale of two classifier systems, Machine Learning 2 (1988) 139.
[14] J.L. Elman, Finding Structure in Time, Technical Report 8801, Center for Research in Language (1988).
[15] M. Compiani, D. Montanari, R. Serra and P. Simonini, Asymptotic dynamics of classifier systems, in: Proceedings of the Third International Conference on Genetic Algorithms (Kaufmann, Los Altos, CA, 1989) p. 298.
[16] M. Compiani, D. Montanari, R. Serra and P. Simonini, Dynamical systems in artificial intelligence: The case of classifier systems, in: Connectionism in Perspective, eds. R. Pfeifer, Z. Schreter, F. Fogelman-Soulié and L. Steels (North-Holland, Amsterdam, 1989) p. 331.
[17] M. Compiani, D. Montanari, R. Serra and P. Simonini, Dynamics of classifier systems, in: Proceedings of the Second Workshop on Parallel Architectures and Neural Networks, ed. E. Caianiello (World Scientific, Singapore), to be published.
[18] R.K. Belew and S. Forrest, Learning and programming in classifier systems, Machine Learning 3 (1988) 193.
[19] H.H. Zhou, A prototype of long-lived, rule-based learning system, in: Proceedings of Computational Intelligence (Milan, 1988) p. 79.
[20] S.W. Wilson, Classifier systems and the animat problem, Machine Learning 2 (1988) 199.
[21] R. Riolo, Bucket brigade performance I. Long sequences of classifiers, in: Genetic Algorithms and their Applications, Proceedings of the Second International Conference on Genetic Algorithms, ed. J.J. Grefenstette (Erlbaum, Hillsdale, NJ, 1987) p. 184.
[22] R. Riolo, LETSEQ Logic of Computers Group, University of Michigan, Ann Arbor, MI (1986).

EMERGENT BEHAVIOR IN CLASSIFIER SYSTEMS

Stephanie FORREST
Center for Nonlinear Studies and Computing Division, CNLS-MS-B258, Los Alamos National Laboratory, Los Alamos, NM 87545, USA

and

John H. MILLER
Santa Fe Institute and Carnegie-Mellon University, SFI, 1120 Canyon Road, Santa Fe, NM 87501, USA

The paper presents examples of emergent behavior in classifier systems, focusing on symbolic reasoning and learning. These behaviors are related to global dynamical properties such as state cycles, basins of attraction, and phase transitions. A mapping is defined between classifier systems and an equivalent dynamical system (Boolean networks). The mapping provides a way to understand and predict emergent classifier system behaviors by observing the dynamical behavior of the Boolean networks. The paper reports initial results and discusses the implications of this approach for classifier systems.

1. Introduction

Classifier systems are computational systems that model cognitive behavior. Like connectionist networks, classifier systems consist of a parallel machine (most often implemented in software) and learning algorithms that adjust the configuration of the underlying machine over time. Classifier systems differ from connectionist networks in the details of both the parallel machine and the learning algorithms. Specifically, classifier systems compute with patterns called messages instead of real-valued weights, and they control their state with IF/THEN rules that specify patterns of messages. Classifier systems incorporate an additional mechanism, called the genetic algorithm [15], which synthesizes connectivity patterns (from inputs through intermediate states to outputs) from initially random configurations. More detailed descriptions of classifier systems and genetic algorithms are available in refs. [16, 15, 12, 17, 4].

In the context of classifier systems, emergent computation arises when co-adapted sets of rules evolve which together perform a coherent function. This includes both the case of a set of classifiers interacting with an external environment and the case in which different groups of rules within one classifier set interact with one another at the group level. A set of classifier rules forms an ecology in which each individual rule evolves in the context of the external environment and the other rules in the classifier system. Competition forces individuals into uncrowded and productive niches, and over time, the rules in the system can learn to act cooperatively.

Classifier systems are interesting for several reasons. They provide a computational theory of cognition based on learning, intermittent feedback from the environment, and hierarchies of internal models that represent the environment [17]. As a theory of cognitive activity, classifier systems can be used to model other "intelligent" processes, such as how people behave in economic and social situations (trading goods in a simple market, playing the stock market, or obeying social norms). The classifier system model also provides an example of a programming language in which correct programs can be either learned or programmed [2]. Any particular configuration of a classifier system can be viewed as a program in which the individual rules (called classifiers) correspond to instructions. Previous attempts to apply machine learning techniques to the problem of generating or debugging computer programs have been largely unsuccessful, due to the brittle nature of most programming languages, in which a program's behavior can be changed dramatically by one misplaced character. In a classifier system, however, the restricted syntax of each instruction allows almost any combination of instructions to form a legal program. Additionally, the relative position of a single instruction does not determine its effect on the program. These two properties of classifier systems support the notion of a program as an "ecology" of individual instructions, each instruction filling some useful niche in the overall program and evolving in the context of the other instructions.

Classifier systems have various global properties that arise from interactions among their components. These include interactions among individual rules (as mentioned above) and interactions at a coarser level among the rule system, the interface to the environment, and the learning algorithms (bucket brigade and genetic algorithm). These global properties are poorly understood even though they have a profound effect on the overall performance of the system. In the absence of analytical results that characterize the aggregate behavior of classifier systems, most of the current understanding of their behavior has been obtained through trial and error, or through careful experimentation on small sets of classifiers in artificial environments [23]. Recently, however, there have been several attempts to study the emergent properties of classifier systems using techniques from nonlinear dynamical systems [25, 1, 26, 21].

This paper first discusses several examples of emergent behavior in classifier systems. It then describes a recently developed method for studying them based on nonlinear dynamical systems and presents some initial results produced by the methodology. Some of the described emergent behaviors have actually been observed in learning classifier systems while others are predicted even though they have not yet been demonstrated. The goal of this work is to find techniques for noticing if and when interesting emergent behaviors arise, to study how such behaviors evolve over time, and to make suggestions for designing classifier systems to exhibit preferred behaviors.

2. Emergent behaviors in classifier systems

This section discusses three related aspects of emergent behavior in classifier systems and their connection to emergent computation: (1) emergent symbolic reasoning, (2) the role of learning, and (3) global dynamical properties. Emergent symbolic reasoning has been a goal for classifier systems since their inception, but in practice has been difficult to achieve. It is hoped that the dynamical systems perspective presented in this paper will provide insight about how to elicit emergent symbolic reasoning behavior from classifier systems. Learning adds another dimension of emergent behavior to classifier systems by providing a mechanism for the "emergence of emergence" – that is, over time the learning algorithms cause structures to be formed that exhibit emergent computation. Finally, we discuss the relevance of global dynamical behaviors (basins of attraction, phase transitions, etc.) to emergent computation in classifier systems.

2.1. Symbolic reasoning

When co-adapted sets of rules act together as a unit, that unit may be viewed as a higher-level structure. Of central importance to the theme of emergent computation is understanding how high-level structures can emerge in learning classifier systems and come to have a selective advantage over direct input/output maps. For example, if a group of rules were to evolve that implemented the system's understanding of the concept "food" and that group functioned as a unit with respect to the rest of the system, we would say that the concept "food" was operating at a higher level than that of the system description (individual rules). Suppose that the "food" concept evolved as a default hierarchy of rules, ranging from very general descriptions of food (If Object-Is-Moving THEN Object-Is-Food) to slightly more specific (If Object-Is-Moving AND Object-Is-Alive THEN Object-Is-Food) to highly detailed descriptions (If Object-Is-On-Plate And Object-Is-Leaf THEN Object-Is-Food). Note that the most specific rule in this three-rule hierarchy contradicts the most general rule, since a leaf on a plate would not usually be moving. The idea here is that the higher-level defaults operate most of the time, and the more specific rules handle exceptions to and refinements of the defaults. If other high-level concepts evolved (the concepts "Leaf" and "Alive" might themselves be default hierarchies) and the system was able to use these concepts in its reasoning process (e.g. first recognizing an object as a leaf and then using the food hierarchy to determine whether it was edible), the classifier system would be exhibiting emergent computation. This form of emergent computation in classifier systems can be contrasted with direct mappings from inputs to outputs in which there is no internal representation or processing. We refer to a direct mapping which does not use any intermediate reasoning steps or representations as a stimulus/response system.

The question of how high-level concepts can come to have a life of their own in a classifier system is an example of the more general problem of how symbolic computations can arise in a low-level learning system that is closely tied to its input/output interface [2]. By symbolic computation we mean both the formation of representations that are somewhat removed (abstracted) from the sensory interface to the system, which we call symbols, and procedures for manipulating symbols. High-level symbolic representations have several advantages over lower-level direct mappings. They can express sophisticated reasoning strategies, for example, allocating appropriate amounts of processing time to different aspects of a problem. Further, such representations are potentially more comprehensible to an outside observer trying to understand what the system has learned (or to an external agent trying to communicate with the system). More importantly, in any realistic problem space, the number of explicit input/output mappings will exceed the capacity of the system, so higher-level abstractions of the input/output map are advantageous for reasons of efficiency. Thus, a classifier system exhibiting emergent computation in the form of high-level symbolic knowledge is desirable. However, there must be some competitive advantage or utility for these higher-level constructs if they are to be maintained by learning. Additionally, the concepts need to evolve in such a way that they will be stable under variable inputs from the environment and stochastic learning algorithms.

What differences do we expect between stimulus/response classifier systems and those that manipulate internal representations and symbols? In a stimulus/response system, there is no pressure for rules to activate one another to form *chains* of rule activation. A chain is formed when the output of one rule satisfies the condition of another rule (e.g. in the rules IF A THEN B, and IF B THEN C, activation of the first rule would cause the second rule to become active). Thus, we expect that there would be little communication among rules in a stimulus/response system and a high amount of communication in a symbol-processing system.

A crucial aspect of emergent concepts is that their representation is "distributed" over many different classifiers. In a distributed representation, any one concept is comprised of several classifiers, and each individual classifier may be part of many different concepts simultaneously. For example, in a default hierarchy [17] of concepts built from classifiers the complete meaning of a concept would be distributed across all of the higher-level classifiers in the hierarchy. Likewise, any one classifier in the hierarchy could be participating in many different hierarchies simultaneously. Booker [3] has identified three different levels at which classifier systems use distributed representations: (1) bit-level encodings, (2) individual rules, and (3) groups of rules. While there are some theories of distributed and associative memories [18, 27, 14], distributed representations are difficult to recognize except in carefully controlled settings. In classifier systems, we expect that distributed representations will result in sets of rules with a large amount of overlap such as shared conditions and shared output messages.

Thus, an important question is how distributed, high-level representations can emerge in learning classifier systems and come to have a selective advantage over direct input/output maps. Although classifier systems were designed with this goal in mind and there are many proposals for how classifier systems can exhibit this kind of behavior, we do not know of any learning classifier systems that demonstrably exhibit high-level symbolic manipulations in any but the most artificial situations. Riolo [23] has studied this question in the most detail using an environment built from finite Markov processes in which some states are assigned nonzero payoffs.

2.2. Learning

A classifier system consists of three layers, with the rule and message passing system forming the lowest level. The rule system is the fundamental computational component of the system; the remaining layers are algorithms for modifying its structures. At the second level is the bucket brigade learning algorithm which manages credit assignment among competing classifiers (or rules). It plays a role similar to that of back-propagation in neural networks. Finally, at the highest level are genetic operators that create new classifiers.

Both the genetic algorithm and the bucket brigade have interesting emergent properties. The genetic algorithm raises questions about how rule ecologies develop, how stable they are, and the effects of competition and randomness on the rule base. The bucket brigade algorithm raises questions about how the strengths evolve through time: do they converge on a stable set of strengths, oscillate, or vary indeterminately? Because these systems are nonlinear, the conventional tools of computer science are of limited use in answering these questions. (See Introduction to these Proceedings [10].)

It is the interactions between the environment and a learning system that are at the heart of learning. This makes learning systems especially difficult to analyze because both the learning system and its external environment (reward function, test cases, etc.) must be taken into account. In many cases the environment is so complex that no reasonable analytical solutions exist. In fact, problem domains that are not understood analytically are often considered to be good candidates for learning, since if we had a good understanding of the environment we could simply program a solution to it.

To summarize, the learning algorithms of classifier systems specifically, and more generally those of other low-level learning systems, have interesting emergent properties. These are difficult enough to analyze in isolation, but the problem is significantly more complicated when other learning systems, the environment, and different encoding strategies are taken into account.

2.3. Emergent dynamical properties of classifier systems

Since it changes its structure and state over time, a classifier system can be viewed as a dynamical system and studied from that perspective. In this subsection, we discuss why we believe that the emergent dynamical properties of classifier systems are interesting and relevant to the questions of emergent computation.

The computational properties of the individual components of classifier systems are quite well understood. In addition to outlining the classifier system architecture [17], Holland [15] described both the mechanics and underlying theory of genetic algorithms. Forrest [9] proved one form of computational completeness for the production rule system and in related work demonstrated that the classifier system architecture is suitable for problems requiring deep reasoning and sophisticated data structures[#1]. Riolo [23] investigated the ability of the bucket brigade to maintain hierarchies and sequences of rules, and several other researchers have recently explored the mathematics of bucket brigades.

None of this work adequately accounts for the aggregate behavior of a classifier system that combines the production rule system, the bucket brigade, and a genetic algorithm. Specifically, there are no results that guarantee that a classifier system will converge on a solution or that once it has found a solution that it will be stable (see ref. [23] for empirical studies of this question). There are also no reachability results that discuss which sorts of problems classifiers systems can and cannot solve (see ref. [11] for similar results for genetic algorithms). Empirical results indicate that classifier systems are significantly more complex and difficult to get working correctly than any of the individual components. It has proved difficult to design learning classifier systems with the ability to follow long chains of reasoning (long sequences of computation), and there is little solid understanding of why some classifier system designs are successful at solving a given problem and others are not.

Basic concepts from nonlinear dynamical systems theory provide a natural vehicle for answering these questions. By viewing a classifier system as a dynamical system, different regimes of behavior (periodic, chaotic, etc.) can be associated with certain aspects of classifier systems. For example, the concept of basins of attraction and state cycles can be used to discuss how robust a classifier system is to noisy data from the environment or random perturbations of its rule base. A robust system should remain in its basin of attraction under small perturbations and similar initial conditions. Likewise, a responsive system is one in which significantly different inputs will cause it to converge on different attractors. A second example is provided by classifier systems that suddenly change their behavior as a result of learning, adding more rules, or varying parameters. These systems can be understood in terms of phase transitions. The nonlinear dynamics perspective suggests that there are likely to be narrow ranges in which classifier system performance is particularly good or bad. The dynamical systems perspective is also useful for studying the formation of chains (see section 3.1 for details).

Thus, the following kinds of questions, which are highly relevant to understanding classifier systems, have natural formulations as emergent dynamical behaviors:

(i) How many classifiers must be in the system for interesting behavior to occur?

(ii) Under what conditions will chains of classifiers form?

(iii) How dense will these chains be (what proportion of classifiers participate in some chain, how long is the average chain)?

(iv) What is the effect of various classifier parameter settings (number of classifiers, size of the message list, percentage of #s, percentage of conditions that are negated, whether or not pass-through is used, etc.) on performance?

(v) What is the impact of learning and different representations on aggregate behavior?

(vi) How stable are these systems to random perturbations such as those introduced by learning algorithms or by noisy data from the environment?

3. Boolean network models of classifier systems

In section 2, we discussed several kinds of emergent behavior and argued that they have natural

---

[#1] Here "suitable" means that the architecture can implement solutions to these problems efficiently (in terms of number of classifiers and computation time).

interpretations as emergent dynamical systems. In this section, we illustrate these ideas by showing how Boolean network models of classifier systems can be used to answer the questions of the previous section.

Nonlinear dynamical systems theory provides a number of tools for studying complex phenomena that change in time, such as learning systems. There are several problems, however, with trying to apply the standard techniques directly. First, most nonlinear dynamical systems analysis emphasizes asymptotic behavior. Thus, a standard approach is to run the system under study for a long time before observing its behavior. After a sufficiently long time, most systems settle into some sort of steady state behavior, possibly converging on one fixed point, oscillating among a small set of states, or wandering around a strange attractor. Analysis of complex systems focuses on this steady state behavior that arises "in the limit" and generally ignores the transient behavior of the system as it approaches the steady state. Any learning system that must interact with a dynamic environment is highly unlikely to reach a meaningful asymptote in any space of reasonable dimensions. Further, even if it did we would be at least as interested in understanding how it approached asymptotic behavior as in understanding what particular asymptote it reached.

A second problem with applying dynamical systems techniques to learning systems is that most of the mathematics has been designed to describe continuous physical systems. Most learning systems have at least some aspects that are discrete. It is well known that dynamical systems can be discrete or continuous along several dimensions, including time, state spaces, and internal variable values. If we are to apply the techniques of dynamical systems theory to the analysis of learning systems this distinction between discrete and continuous must be modeled appropriately.

In the case of classifier systems, most of the relevant dimensions are discrete. The rule system runs synchronously and time is discrete. The messages on the message list are discrete and classifiers are either activated or not activated with no intermediate states. Continuous values do appear, however, both for the strengths associated with classifiers that are adjusted by the bucket brigade and the intensities associated with messages[#2].

3.1. Boolean networks

Boolean networks are well suited for studying classifier systems. They are discrete along the same dimensions as classifier systems, although a slight modification to the conventional model is required to accommodate classifier strengths. A Boolean network consists of a set of nodes, each of which has two possible states, 0 or 1. The state of each node at time $t + 1$ is determined by its own Boolean function, which takes as input the states of other nodes in the network (indicated by a directed arc between the nodes) at time t. The variables of a node's Boolean function correspond to the states of the connected nodes. The Boolean functions can vary for different nodes, as can the number and location of the input nodes. A more detailed description of Boolean networks is provided by ref. [19].

The properties of Boolean networks have been studied extensively [20, 7, 8], and general techniques have been developed for determining the dynamical properties of specific networks. These studies show that the dominant dynamical behaviors of Boolean networks can be characterized by a small set of emergent properties such as state cycles (size and number), frozen components, and stability to perturbation. Further, these emergent properties depend directly on structural properties of the Boolean network, notably the number of input arcs at each node and the form of the Boolean function stored at each node. Different classes of Boolean networks can be studied by specifying the number of input arcs, the form of the Boolean functions, and within these con-

---

[#2] In some classifier systems a continuous quantity called "intensity" is associated with each message. A high intensity message increases the corresponding classifier's bid.

straints assigning the connections and functions randomly. The dynamics of a particular network are studied by initializing each node to state 0 or state 1 (by random or uniform assignment, or to an initial condition that corresponds to a specific input pattern), iterating the network, and watching how the patterns of state variables change. In the following paragraphs we review briefly some of the more relevant results from this body of work [19].

A given Boolean network with n nodes has 2^n possible states ($\{0,1\}^n$). The Boolean functions and connections among nodes imply a deterministic state transition function. Given that the network is deterministic and finite, it must eventually fall into some state cycle during the course of its iteration. Different initial conditions may, however, cause the network to enter different state cycles. All points in the state space are either part of some state cycle, or they lie on a trajectory that leads to a cycle.

For networks with two inputs at each node and unbiased random Boolean functions[#3], the actual number of distinct state cycles is on the order of the square root of the number of nodes. For a network with 10 000 nodes, this implies that there are only 100 distinct state cycles. Cycle lengths are also typically small, with median cycle length again on the order of $\sqrt{n}$, where the theoretical maximum is 2^n (in the 10 000-node case this is the difference between 100 and $2^{10\,000}$). A large fraction of the nodes (60–80%) tend to "freeze"[#4]. Different cycles tend to have similar states, with hamming distances[#5] between 1 and 10%. State

---

[#3] By unbiased we mean that the corresponding truth table has an equal number of 1's and 0's.

[#4] Boolean networks are subject to a condition called "freezing" in which a region of the network becomes locked into one state (either 1 or 0) and is impervious to fluctuating states in the rest of the network).

[#5] Two global states of a Boolean network can be compared by assigning each node in the network one bit position in a binary string and setting the bit position according to the current state of the corresponding node. The hamming distance between two such strings then provides a measure of similarity between the two states.

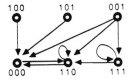

Classifier Set

1#0, #01;000
00#,~111;11#
001, 110;110

Node	Boolean Function
000	(100 v 110) ∧ (001 v 101)
110	(000 ∧ ~111) v (001 ∧ 110)
111	(001 ∧ ~111)

Equivalent Boolean Network

Fig. 1. Mapping from classifier systems to Boolean networks. This figure illustrates the mapping from classifier systems to Boolean networks for three sample classifiers. Each node in the Boolean network represents one possible output message from a classifier. The 3-bit 2-condition classifier 1#0, #01; 000 corresponds to the Boolean function ((100 ∨ 110) ∧ (001 ∨ 101)) on the 000 node. The classifier 00#, ~111;11# is distributed across the nodes 110 and 111, which correspond to the possible passed-through messages that it could produce. The classifier 001,110;110 shares an output node with the second classifier; thus, the Boolean expressions for each classifier are combined disjunctively.

cycles tend to be stable to most one-node perturbations, i.e. if one node in the network randomly changes state, the network usually does not enter a new cycle. As discussed in section 3.2, these emergent properties have direct implications for the behavior of classifier systems.

3.2. Mapping classifier systems to Boolean networks

By defining a mapping between classifier systems and Boolean networks [22] the techniques that have already been developed for studying Boolean networks can be applied to classifier systems. Fig. 1 illustrates a mapping in which classifier messages are represented by nodes in the network. The set of classifier rules is mapped onto a set of Boolean functions stored at the output message nodes corresponding to the classifier. More precisely, the mapping takes the node associated with the given message(s) a rule may post, and connects it to those nodes that can actually fire the rule with the appropriate Boolean function. A formal definition of the mapping appears in ref. [22].

The mapping shown in fig. 1 between classifier systems and Boolean networks preserves the functional behavior of classifier systems. For each possible state of the message list at time t (corresponding to a set of nodes in the Boolean network that are in state 1), the message list produced by the classifier system at time $t + 1$ will be equivalent to the set of active Boolean net nodes (those in state 1) at time $t + 1$. By mapping a particular classifier system to its corresponding Boolean network, one can easily follow the dynamics of the message system by iterating the Boolean network. Moreover, by considering generic classes of classifier systems a link between such systems and specific classes of randomly connected Boolean networks can be formed.

Since the emergent properties of a Boolean network depend on a relatively small set of defining structural properties, it is natural to ask how different parameter choices in standard classifier systems influence these properties and how these properties affect the dynamic behavior of the corresponding network. The results reported here are preliminary, but they focus on both of these issues. The types of properties that can be explained by the Boolean network mapping of classifier systems include the degree of internal connectivity among rules (see section 2.1), freezing, the effect of learning (adding new rules to the system), and the impact of noise on the classifier system. These properties can be linked to different parameter settings (e.g. the number of rules in the system), strength-assignment rules, and learning mechanisms for generating new rules.

The experiments reported in the following subsections were conducted using 2-condition 8-bit classifier systems without negative conditions (in other runs we have discovered that the results are not significantly altered when typical numbers of negative conditions are included). The classifier systems were randomly generated and mapped to their corresponding Boolean networks. An 8-bit system, rather than more common 16- or 32-bit systems, was used in order to simplify computation. From the Boolean network perspective, an 8-bit classifier system with 20 classifiers is roughly equivalent to a 16-bit classifier system with over 1000 classifiers. We expect that relatively simple scaling relations exist that will allow the results to be applied to arbitrarily-sized systems. Each reported data point is a mean computed from 30 randomly constructed classifier systems. We report data for classifier systems ranging from 5 to 80 classifiers. Results are shown for classifier systems with either 25% or 50% do not cares (#s) in the population, both with and without pass-through (P-T).

3.3. Connectivity

The earlier discussion of Boolean networks suggests that the average number of input arcs per node is an important determinant of a network's dynamical properties. Figs. 2 and 3 show the average number of input arcs per internal node[6].

The first figure shows the average for both internal and external arcs, while the second one considers only those arcs coming from other internal nodes. As expected, the use of more #s in the conditions significantly increases the amount of connectivity. The average number of arcs also increases with the number of classifiers in the system. For small numbers of classifiers – where small is relative to the number of potential messages – the average number of inputs should not change with additional classifiers. However, as the number of classifiers relative to potential messages increases or the learning algorithms concentrate classifiers around certain nodes, multiple classifiers sharing the same internal node result in greatly increased connectivity. Fig. 2 indicates that pass-through inhibits connectivity. This occurs because a higher number of internal nodes are created from pass-through for the different possible

---

[6] We distinguish between external and internal nodes. External nodes (100, 101, and 001 in fig. 1) have empty Boolean functions, and correspond to messages that might be generated by the input interface of the classifier system. Internal nodes 000, 110, and 111 in fig. 1 have nonempty Boolean functions and correspond to messages that can be posted by a classifier.

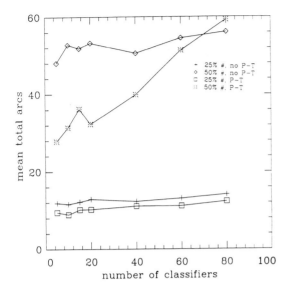

Fig. 2. Mean total input arcs per internal node.

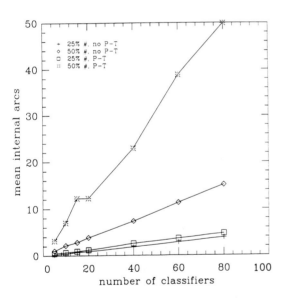

Fig. 3. Mean internal input arcs per internal node.

output messages, and they share the same number of connections. One unexpected result of pass-through is an increase in the internal-to-internal node connectivity (see fig. 3). With pass-through the ratio of internal to external nodes in the system increases, and thus a random set of classifiers will have a higher rate of internal connectivity.

In section 2.1 the importance of classifier chains was discussed. Chains support internal reasoning processes that allow a classifier system to go beyond simple stimulus/response behaviors. In particular, cyclic chains (loops) are one of the most powerful computational constructs. State and substate cycles that emerge in Boolean networks are closely related to chains in a classifier system. Based on the connection between the structure of a Boolean network and its state cycles, we can predict that the likelihood of chain formation in classifier systems is closely tied to the classifier system's configuration (number of classifiers, percentage of do not cares in the population, etc.). Critical values probably exist that catalyze the formation and survival of chains. The earlier discussion of Boolean networks suggests that the number of cyclic chains that form for typical classifier system configurations will be relatively small and that such chains will have few members. The letter-prediction classifiers studied in section 2.2 are an example of this.

Cyclic chains allow classifier systems to exhibit self-sustaining activity, that is, to generate internal activity in the absence of sustained external input, while acyclic chains allow classifier systems to exhibit transient internal activity. Some level of internal activity is required if a classifier system is to form large internal representations and operate with intermittent environmental input. Without any internal activity, these important characteristics of classifier systems would be lost. Too much activity, however, would be likely to hurt performance. We conjecture that "interesting" classifier system behavior occurs at or near the boundary between these two extremes [20].

To test the property of self-sustaining activity all internal nodes were initialized to state 1 and the network was iterated until it reached a state cycle or all activity died out (all nodes in state 0). In the latter case, the length of the state cycle is 0. For state cycles with length 0, the number of time steps it takes for the activity to die out (the length of the transient) indicates the length of the longest acyclic chain. Self-sustaining activity indicates

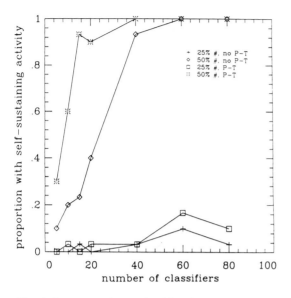

Fig. 4. Percentage of networks with self-sustaining activity.

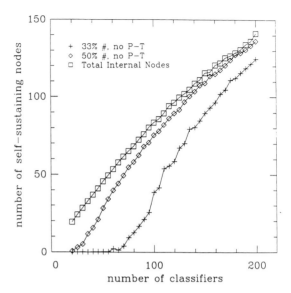

Fig. 5. Mean number of nodes in the self-sustaining cycle.

chaining in randomly constructed networks, since networks with long acyclic chains are likely to have at least some cyclic chains.

Fig. 4 shows the proportion of networks from different classifier system configurations that have self-sustaining components. Fig. 4 indicates that a rapid transition occurs from systems without self-sustaining components to those that do have them. The average size of a self-sustaining cycle (the number of nodes active in the self-sustaining state cycle) is shown in fig. 5. This figure shows a potential for rapid growth and saturation of internal networks.

Fig. 5 shows the average length of self-sustaining cycles (that is, the number of different nodes in the self-sustaining cycle) for classifier systems with two different percentages of #s[7]. The third line shows the total number of nodes in the network for variously sized classifier systems. Fig. 5 shows the possibility for rapid growth and saturation of internal networks in some configurations,

[7] Since these results are for 8-bit randomly generated classifier systems, the actual number of classifiers needed to generate self-sustaining activity is substantially lower than what we expect for the 16-bit case. Because it is so easy to saturate the 8-bit system with connections, the sharp transition of interest is only apparent in the case with 33% #s.

where small changes in the number of classifiers cause a rapid transition from no self-sustaining activity to a large amount.

3.4. Freezing

The potential for messages to become frozen either on or off depends on the percentage of ones in the Boolean function associated with that message node [20]. As the measure of "internal homogeneity" (percentage of ones) moves away from 50%, networks tend to exhibit large "frozen components" in which the states of nodes become locked. For two-condition classifiers without negation that do not have pass-through, the percentage of ones in the truth table is given by

$$p_1 = \frac{(2^m - 1)(2^n - 1)}{2^{m+n}}, \qquad (1)$$

where m and n denote the number of messages that could possibly match the first and second conditions respectively. This equation assumes that the set of matching messages for the two condi-

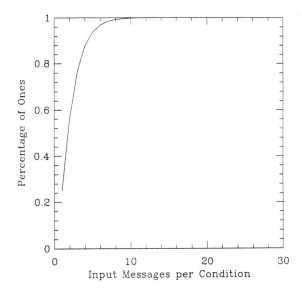

Fig. 6. Percentage of ones in the truth table using eq. (1) ($n = m$).

Table 1
The number of #s in each condition that will prevent freezing (f implies "freeze", * implies "not freeze").

	0	1	2	≥ 3
0	f	*	*	*
1	*	*	*	f
2	*	*	f	f
≥ 3	*	f	f	f

tions is disjoint[#8]. If the second condition is negated, then

$$p_2 = \frac{(2^m - 1)(1)}{2^{m+n}}. \qquad (2)$$

Similar expressions can be written for Boolean functions that contain clauses for multiple classifiers (i.e. when two or more classifiers share an output node), multiple classifiers that have identical conditions, etc. More complicated expressions can be derived for classifiers that violate the assumption of disjointness among conditions.

Fig. 6 graphs the proportion of ones using eq. (1) and assuming that $m = n$. Under these conditions a very narrow range of values exists that avoids frozen components. If the percentage of ones in a node's truth table is between 30 and 70% then the probability of that node freezing is low. Assuming that eq. (1) is a reasonable approximation, the number of #s in each condition that will prevent freezing is as given in table 1 (f, * imply freeze, not freeze respectively). Thus, if the first condition has no #s the node will be active when the second condition has at least one #. If the first condition has two #s then freezing will occur when the second condition has more than one #. The actual distribution of #s in each condition can be controlled when the system is initialized or through biases in the learning operators. These results suggest that classifier systems are highly sensitive to the proportion of #s in the population and that the nature of this sensitivity should be studied carefully. Preliminary analysis indicates that if #s are chosen randomly with probability 0.25 then about 53% of 8-bit system nodes and 95% of 16-bit system nodes will be frozen. With a 0.50 probability almost all of the nodes will be frozen in either system.

Note, however, that these results assume that every row in the truth table has an equal probability of being accessed. Since most of the variables of most of the Boolean functions are for external messages being posted from the environment, there is not necessarily a 0.5 probability that each variable will be true at any time. In fact, most environmental messages will have a very small probability of being true. This implies that each row of the truth table will not necessarily be accessed with equal probability. Determining how this bias in environmental messages affects the likelihood of frozen components is an area of future investigation.

[#8] The logic of the equation is as follows: the first condition will not be true iff all of the m conditions are false (since it is the disjunction of possible matching messages) and therefore it will be true for $2^m - 1$ of its possible 2^m states. Similarly, the second condition will hold for $2^n - 1$ of its states. Thus they will both hold (and the full Boolean function will be true) for $(2^m - 1)(2^n - 1)$ out of the 2^{m+n} states.

3.5. Learning

An important question is how the structural properties of networks differ for networks corresponding to randomly generated classifier systems and for classifier systems that have evolved under learning. We have begun to investigate this question by looking for cyclic and acyclic chains in classifier sets before and after learning. To date we have looked at classifier sets that learn letter prediction. In the letter prediction problem, a sequence of letters is presented repeatedly (i.e. as a periodic time series), one letter being posted to the message list at a time. In the case of ambiguous sequences (e.g. mississippi), the classifier system must learn to use some form of memory, so it can correctly predict what letter will follow an "i".

When we compared internal connectivity (both self-sustaining activity, and how long activity takes to die out) for classifier sets[#9] before and after learning, we found no significant differences. For example, in a classifier set with 100 classifiers, activity died out after two time steps for the initial (randomly generated) set of classifiers. When we looked at the classifier set after learning for 20 000 time steps, activity also died out after two time steps. Finally, after 60 000 time steps activity died out after three time steps. Even though we have a limited sample of classifier systems performing one task, these results add weight to our hypothesis that most current classifier systems are not performing significant amounts of internal computation.

Another aspect of learning is how stable a classifier system is to the operations of the genetic algorithm. It is important, for example, that a random mutation or cross-over be capable of having a measurable effect on the overall system, but that it not completely disrupt all ongoing activity. In Boolean network terminology, it would be interesting to know under what conditions one genetic algorithm operation (e.g. a cross-over) is capable of unfreezing a set of frozen components (or freezing a set of unfrozen components). More generally, we are interested in the expected amount of perturbation caused by the application of the learning operators. This could be measured by comparing the dynamics of networks before and after the application of genetic operators, although we have not yet conducted this experiment. Compiani et al. [6] have also investigated the effect of genetic operators on a set of classifiers. They report scenarios in which one incorrect rule added to a population of correct rules eventually disrupts a chain of correct rules.

The mapping from classifier systems to random Boolean networks provides a framework for studying classifier system components and their interactions. To date, we have preliminary results about the structure (input arcs, Boolean functions) and the dynamics (self-sustaining activity). The general methodology presented here provides an example of how important questions about classifier systems can be addressed by viewing a classifier system as a dynamical system.

4. Related work

A number of other researchers are currently investigating classifier systems from the dynamical systems perspective. The Boolean network method differs from these efforts in that it can potentially model the aggregate behavior of the entire classifier system, rather than concentrating on one component at a time. Arthur and Simon are studying how rule strengths change under the bucket brigade [1, 25]; Smith and Valenzuela-Rendón are analyzing how the recombination operators of the genetic algorithm affect an existing rule set [26]; Compiani et al. are studying several aspects of classifier systems such as oscillation of strengths between rewards in correct classifier sets and the effect of introducing a wrong rule into a population [5]. In this work they treat the bucket brigade and genetic algorithm as acting on different time

---

[#9] 16-bit classifier sets with 40% #s and 0.45% ~ s.

scales. Each of these approaches is contributing important insights about some aspect of classifier systems and is accordingly reviewed in the following paragraphs.

4.1. Bucket brigades

A classifier system has a set of rules each of which can be fired by the presence of appropriate messages posted on a centralized message list. Once the appropriate messages are on the message list, the probability that a rule will be fired is proportional to a measure of the rule's own strength. The bucket brigade is the credit allocation algorithm [24] that distributes *strength* among classifiers.

A rule's strength is determined by the impact of the rule's output message. If the rule's message is posted when external payoffs are received from the environment then strength is increased. If the message is instrumental in allowing other rules to fire and receive payoff, then its strength will be increased over time by an amount proportional to the downstream rule's strength. Reward propagates back through the chain of contributing classifiers indirectly; on every iteration of the system, each active classifier$_i$ passes some of its strength to the classifier$_j$, immediately "upstream" of it, that is, the classifier whose output message made it possible for classifier$_i$ to become active. Over time, classifiers that help "set up" other classifiers to be rewarded will have their strength increased through this indirect payoff method. Since the probability that any rule will fire is closely tied to its strength (as well as the set of currently posted messages matching the rule's conditions), an understanding of the dynamics of strength allocation provides insight to the set of rules that will emerge over time.

Arthur [1] and Simon [25] have both begun to analyze the dynamics of strengths by mapping variations of the standard credit allocation algorithms for simple classifier systems into an equivalent set of stochastic differential equations. They have focused on N-armed bandit problems, studying what kinds of strategies evolve under different algorithms. This work has shown that in classifier systems the choice of a particular allocation strategy within the broad guidelines of the bucket brigade can have a large impact on the set of final strengths. As a result it is possible to tune the strategy so that different properties can be emphasized. Specifically, they have found that one version of the standard classifier system bidding algorithm (in the standard algorithm a classifier's bid is subtracted from its strength) leads to a probability matching strategy in which the classifier system selects the N arms with probabilities that match their payoffs, while a bidding mechanism in which individual bids are not subtracted from a classifier's strength leads to an optimizing strategy in which the classifier system learns to always select the arm with the highest payoff. These criteria clearly show the trade-off between maintaining high levels of adaptability (probability matching) and the ability to exploit the current environment (optimization). Goldberg [13] has obtained similar results.

4.2. Karnaugh maps

Smith and Valenzuela-Rendón [26] have studied the dynamics of rules from an ecological perspective to determine how recombination of old rules affects the existing ecosystem of rules. Through the use of a simplified environment and classifier system (2-bit, single-condition classifiers), they have explicitly modeled the dynamics of the recombination process. Because the environment is so simple (2-bit Boolean functions), it is possible to assign each classifier a fitness a priori without running the classifier system and allocating strength via the bucket brigade. Their analysis indicates that the coevolution of rule sets can follow complicated paths, and is modified by the introduction of operators that encourage speciation and the formation of niches. Rule sets may have long metastable states, which eventually converge on small sets of inadequate rules dominating the population.

4.3. Asymptotic dynamics

Compiani et al. [5] have used dynamical systems techniques to study the behavior of classifier systems on the letter prediction problem. They view the genetic algorithm and bucket brigade as operating on two completely different time scales, so they can be treated independently. They are concerned with questions such as the stability of a learned solution under the operations of the genetic algorithm, and the influence of system parameters on performance. They have found that the two major influences on classifier system performance are the size of the message list and the maximum number of copies that are allowed of any one classifier in the reproductive phase of the genetic algorithm.

5. Conclusions

Classifier systems exhibit many interesting emergent behaviors. Some of these are desirable and others are not. Nevertheless, understanding them is important. The above work is attempting to understand the emergent properties of classifier systems by focusing on their dynamics. By concentrating on the dynamics of classifier strengths or on the effects of genetic operations it is possible to study the performance of different learning algorithms in the context of classifier systems. The Boolean network model of classifier systems makes it possible to study the integrated behavior of classifier systems by viewing the patterns of messages on the message list as a dynamical system. The message list reflects all of the interacting components in that the rule set determines which messages could be posted, the bucket brigade and the bidding mechanism control the strengths (and hence the probability that an activated rule will actually be allowed to post its message), and the genetic algorithm controls which rules are in the current rule set. We hope that this work will eventually make it possible to build classifier systems that tend toward the desired properties and away from the undesirable ones.

There are several areas of future research that are necessary in order to achieve this goal. Noise tolerance is an important property for classifier systems because they use stochastic learning procedures and operate in potentially noisy environments. The effect of noise on network dynamics has been studied in random Boolean networks, but we have not yet related these results to classifier systems. A second direction is to extend the results to classifier systems undergoing learning. Randomly generated classifier systems provide an important baseline, but clearly the systems of interest are those that incorporate both bucket brigade and genetic algorithm learning procedures. Finally, a more detailed investigation of the dynamics for classifier networks is needed. The existing literature focuses on only a few special classes of Boolean networks. The networks corresponding to classifiers differ from these previously studied classes. The most notable difference is that for classifier systems the number of input arcs at a nodes varies within a network, while that number is constant in the previously-studied networks.

Previously unexplored properties of Boolean networks are likely to be important in the analysis of classifier systems. The impact on the dynamics of randomly firing external nodes may be useful, since classifier systems typically operate in rapidly changing environments. The analysis of network propagation paths (the Boolean network analog of an execution trace) caused by the activation of particular subsets of nodes as well as the length of transients will provide insights into acyclic classifier chains. Extended investigation of the attractor basin structure for subsets of firing external nodes will provide useful information about the ability of the systems to differentiate between different inputs. A closer focus on the state cycles in the subnetworks (global states are likely to be less important than functional subpieces of a network) may be more appropriate for classifier system analysis.

Acknowledgements

The work reported here has benefited from discussions with John Holland and Stuart Kauffman. We are grateful to Rick Riolo for providing us with sample classifier sets. Lashon Booker and Erica Jen both provided helpful comments on the manuscript.

References

[1] B. Arthur, On classifier systems and models of learning (1989), unpublished.
[2] R.K. Belew and S. Forrest, Learning and programming in classifier systems, Machine Learning 3 (1988) 192–223.
[3] L.B. Booker, Using classifier systems to implement distributed representations, in: Advances in Neural Information Processing Systems II (Kaufmann, Los Altos, CA), in press.
[4] L.B. Booker, D.E. Goldberg and J.H. Holland, Classifier systems and genetic algorithms, Artificial Intelligence 40 (1989) 235–282.
[5] M. Compiani, D. Montanari and R. Serra, Asymptotic dynamics of classifier systems, in: Proceedings of the Third International Conference on Genetic Algorithms, ed. J.D. Schaffer (Kaufmann, Los Altos, CA, 1989) pp. 298–303.
[6] M. Compiani, D. Montanari and R. Serra, Learning and bucket brigade dynamics in classifier systems, Physica D 42 (1990) 202–212, these Proceedings.
[7] B. Derrida and D. Stauffer, Phase-transitions in two-dimensional Kauffman cellular automata, Europhys. Lett. 2 (1986) 739–745.
[8] B. Derrida and G. Weisbuch, Evolution of overlaps between configurations in random boolean networks, J. Phys. (Paris) 47 (1986) 1297–1303.
[9] S. Forrest, A study of parallelism in the classifier system and its application to classification in KL-ONE semantic networks, Ph.D. Thesis, The University of Michigan, Ann Arbor, MI (1985).
[10] S. Forrest, Emergent computation: self-organizing, collective, and cooperative phenomena in natural and artificial computing networks, Introduction to the Proceedings of the Ninth Annual CNLS Conference, 1989, Physica D 42 (1990) 1–11, these Proceedings.
[11] D.E. Goldberg, Genetic algorithms and Walsh functions, Part II, deception and its analysis, TCGA Report 89001, The University of Alabama, Department of Engineering Mechanics, Tuscaloosa, AL (1988).
[12] D.E. Goldberg, Genetic Algorithms in Search, Optimization, and Machine Learning (Addison–Wesley, New York, 1989).
[13] D.E. Goldberg, Probability matching, the magnitude of reinforcement, and classifier system bidding, TCGA Report 88002, The University of Alabama, Department of Engineering Mechanics, Tuscaloosa, AL (1988).
[14] G.E. Hinton and J.A. Anderson, eds., Parallel Models of Associative Memory (Erlbaum, Hillsdale, NJ, 1981).
[15] J.H. Holland, Adaption in Natural and Artificial Systems (University of Michigan Press, Ann Arbor, MI, 1975).
[16] J.H. Holland, Concerning the emergence of tag-mediated lookahead in classifier systems, Physica D 42 (1990) 188–201, these Proceedings.
[17] J.H. Holland, K.J. Holyoak, R.E. Nisbett and P.R. Thagard, Induction: Processes of Inference, Learning, and Discovery (MIT Press, Cambridge, MA, 1986).
[18] P. Kanerva, Self propagating search: a unified theory of memory, Ph.D. Thesis, Stanford University (1984).
[19] S.A. Kauffman, Requirements for evolvability in complex systems: orderly dynamics and frozen components, Physica D 42 (1990) 135–152, these Proceedings.
[20] S.A. Kauffman, Emergent properties in randomly complex automata, Physica D 10 (1984) 145–156.
[21] C.G. Langton, Computation on the edge of chaos: phase transitions and emergent computation, Physica D 42 (1990) 12–37, these Proceedings.
[22] J.H. Miller and S. Forrest, The dynamical behavior of classifier systems, In: proceedings of the Third International Conference on Genetic Algorithms, ed. J.D. Schaffer (Kaufmann, Los Altos, CA, 1989) pp. 304–310.
[23] R.L. Riolo, Empirical studies of default hierarchies and sequences of rules in learning classifier systems, Ph.D. Thesis, The University of Michigan, Ann Arbor, MI (1988).
[24] A.L. Samuel, Some studies in machine learning using the game of checkers, in: Computers and Thought (McGraw-Hill, New York, 1963) pp. 71–108.
[25] C. Simon, personal communication (1989).
[26] R.E. Smith and M. Valenzuela-Rendón, A study of rule set development in a learning classifier system, in: Proceedings of the Third International Conference on Genetic Algorithms, ed. J.D. Schaffer (Kaufmann, Los Altos, CA, 1989) pp. 340–346.
[27] D. Touretzky and G.E. Hinton, Symbols among the neurons: details of a connectionist inference architecture, in: Proceedings of the Ninth International Joint Conference on Artificial Intelligence (AAAI, 1985).

CO-EVOLVING PARASITES IMPROVE SIMULATED EVOLUTION AS AN OPTIMIZATION PROCEDURE

W. Daniel HILLIS

Thinking Machines Corporation, 245 First Street, Cambridge, MA 02142-1214, USA

This paper shows an example of how simulated evolution can be applied to a practical optimization problem, and more specifically, how the addition of co-evolving parasites can improve the procedure by preventing the system from sticking at local maxima. Firstly an optimization procedure based on simulated evolution and its implementation on a parallel computer are described. Then an application of this system to the problem of generating minimal sorting networks is described. Finally it is shown how the introduction of a species of co-evolving parasites improves the efficiency and effectiveness of the procedure.

1. Introduction

The process of biological evolution by natural selection [5] can be viewed as a procedure for finding better solutions to some externally imposed problem of fitness. Given a set of solutions (the initial population of individuals), selection reduces that set according to fitness, so that solutions with higher fitness are over-represented. A new population of solutions is then generated based on variations (mutation) and combinations (recombination) of the reduced population. Sometimes the new population will contain better solutions than the original. When this sequence of evaluation, selection, and recombination is repeated many times, the set of solutions (the population) will generally evolve toward greater fitness.

A similar sequence of steps can be used to produce *simulated evolution* within a computer [3, 4, 12, 17–19]. In simulated evolution the set of solutions is represented by data structures on the computer and the procedures for selection, mutation, and recombination are implemented by algorithms that manipulate the data structures. Although the term "simulated evolution" deliberately suggests an analogy to biological evolution, it is understood that the real biological processes are far more complex than the simulation; simulated evolution represents only an idealization of certain aspects of a biological system. Such simulations are sometimes used as tools for understanding biological evolution [15], but this paper will concentrate on the use of simulated evolution for optimization; that is, as a practical method of generating better solutions to problems. Biological systems will serve as a source of metaphor and inspiration, but no attempt will be made to apply the lessons learned to biological phenomena.

As an optimization procedure, the goal of simulated evolution is very similar to that of other domain-independent search procedures such as *generate and test*, *gradient descent*, and *simulated annealing* [13, 16]. Like most such procedures, simulated evolution searches for a good solution, although not necessarily the optimal one. Whether or not it will find a good solution will depend on the distribution of solutions within the space.

These methods are all useful in searching solution spaces that are too large for exhaustive search. As in gradient descent and simulated annealing procedures, simulated evolution depends on information gathered in exploring some regions of the solution space to indicate which other regions of the space should be explored. How well this works obviously depends on the distribution of solutions in the space. The types of fitness spaces for which

0167-2789/90/$03.50 © Elsevier Science Publishers B.V.
(North-Holland)

simulated evolution produces good results are not well understood, but one important type of space for which it works is a space that is independently a good domain for hill climbing in each dimension.

Another attractive property of simulated evolution is that it can be implemented very naturally on a massively parallel computer. During the selection step, for example, the fitness function can be evaluated for every member of the population simultaneously. The same is true for mutation, recombination, and a computation of statistics and graphics for monitoring the progress of the system. In the system described below, we routinely simulate the evolution of populations of a million individuals over tens of thousands of generations. Since these simulations take place on several generations per second, such experiments take only a few hours.

In these simulations, individuals are represented within the computer's memory as pairs of number strings that are analogous to the chromosome pairs of diploid organisms. The population evolves in discrete generations. At the beginning of each generation the computer begins by constructing a phenotype for each individual, using the set of number strings corresponding to an individual (the "genome") as a specification. The function used for the interpretation is dependent upon the experiment, but typically a fixed region within each of the chromosomes is used to determine each phenotypic trait of the individual. Discrepancies between the two bit strings of the pair are resolved according to some specified rule of dominance. This is similar to the diploid "genetic algorithms" studied by Smith and Goldberg [18].

To simulate selection, the phenotypes are scored according to a set of fitness criteria. When the system is being used to solve an optimization problem, the traits are interpreted as solution parameters and the individuals are scored according to the function being optimized. This score is then used to cull the population in a way that gives higher scoring individuals a greater chance of survival.

After the selection step, the surviving gene pool is used to produce the next generation by a process analogous to mating. Mating pairs are selected by either random mating from the entire population, some form of inbred mating, or assortive mating in which individuals with similar traits are more likely to mate. The pairs are used to produce genetic material for the next generation by a process analogous to sexual reproduction. First, each individual's diploid genome is used to produce a haploid by combining each pair of number strings into a single string by randomly choosing substrings from one or the other. At this point, randomized point mutations or transpositions may also be introduced. The two haploids from each mating pair are combined to produce the genetic specification for each individual in the next generation. Each mating pair is used to produce several siblings, according to a distribution normalized to ensure a constant total population size. The entire process is repeated for each generation, using the gene pool produced by one generation as a specification for the next.

The experiments that we have conducted have simulated populations ranging in size from 512 to $\sim 10^6$ individuals, with between 1 and 256 chromosomes per individual. Chromosome lengths have ranged from 10 to 128 bits per chromosome, mutation rates from 0 to 25% probability of mutation per bit per generation, and crossover frequencies ranged from 0 to an average of 4 per chromosome. Using a Connection Machine® [#1] with 65 536 processors, a typical experiment progresses at about 100 to 1000 generations per minute, depending on population size and on the complexity of the fitness function.

2. Sorting networks

As an example of how simulated evolution can be applied to a complex optimization problem, we consider the problem of finding minimal *sorting*

[#1] Connection Machine is a registered trademark of Thinking Machines Corporation.

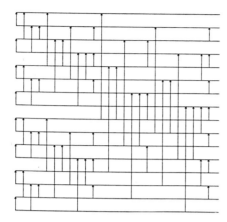

Fig. 1. Sorting network.

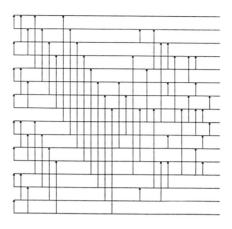

Fig. 2. Green's 60-comparison sorter.

networks for a given number of elements. A sorting network [14] is a sorting algorithm in which the sequence of comparisons and exchanges of data take place in a predetermined order. Finding good networks is a problem of considerable practical importance, since it bears directly on the construction of optimal sorting programs, switching circuits, and routing algorithms in interconnection networks. Because of this, the problem has been well studied, particularly for networks that sort numbers of elements that are exact powers of two.

Sorting networks are typically implemented as computer programs, but they have a convenient graphical representation, as shown in fig. 1. The drawing contains n horizontal lines, in this case 16, corresponding to the n elements to be sorted. The unsorted input is on the left, and the sorted output is on the right. A comparison–exchange of the ith and jth elements is indicated by an arrow from the ith to the jth line. Two specified elements are compared and they are exchanged if and only if the element at the head of the arrow is less than the element at the tail; the smallest element will always end up at the tail. The sorting network pattern shown in fig. 1 is a Batcher sort [1], which requires $n \log^2 n - 1$ exchanges to sort n elements.

A useful property of sorting networks is that they are relatively easy to test. A sorting network that correctly sorts all sequences of 1 and 0 will correctly sort any sequence, so it is possible to test an n-input sorting network exhaustively with 2^n tests.

In this section we describe how simulated evolution is used to search for networks that require a small number of exchanges for a given number of inputs. In particular, the case $n = 16$ is of particular interest, and has a long history of successive surprises. In 1962, Bose and Nelson [2] showed a general method of sorting networks which required 65 exchanges for a network of 16 inputs. They conjectured that this was the best possible. In 1964, Batcher [1], and independently, Floyd and Knuth [6], discovered the network shown in fig. 1, which requires only 63 exchanges. Again, it was thought by many to be the best possible, but in 1969, Shapiro [14] discovered a network using only 62 exchanges. Later that year, Green [14] discovered a 60-comparison sorter, shown in fig. 2, which stands as the best known. These results are summarized in table 1. For a lively and more detailed account of these developments, the reader is referred to the book by Knuth [14, pp. 227–229].

There are two ways to cast the search for minimal sorting networks as an optimization problem. The first is to search the space of functional sorting networks for one of minimal length. The second is to search the space of short sequences of comparison/exchanges for ones that sort best. The difficulty with the first approach is that there is no obvious way of mutating a working sorting

Table 1
Summary of number of exchanges required for best known sorting networks with 16 inputs.

Best known networks		
1962	Bose and Nelson	65
1964	Batcher, Knuth	63
1969	Shapiro	62
1969	Green	60

Networks found by simulated evolution	
without parasites	65
with parasites	61

network into another one that is guaranteed to work, so almost all mutations and recombinations will create a network that is outside of the search space. It is much easier in the second approach to produce mutations and variations of a small program that stay within the space of small programs. Mutation can be implemented by changing the position of one of the exchanges, and recombination by splicing the first part of one sorting network with the last part of another. This is essentially the approach we have adopted.

One difficulty with this approach is that even if the solution is in the space of small networks, the easiest paths to the solution may not be. It may be easier, for example, to produce a short correct network by optimizing a slightly longer correct network than by fixing a bug in a short uncorrect network. For this reason, we have taken advantage of the diploid representation of a genotype to allow longer networks to be generated as intermediate solutions.

The genotype of each individual consists of 15 pairs of chromosomes, each consisting of 8 codons, representing the digits of 4 chromosome pairs. Each codon is a 4-bit number, representing an index into the elements, so the genotype of an individual is represented as 30 strings of 32 bits each. The phenotype of each individual (an instance of a sorting network) is represented as an ordered sequence of ordered pairs of integers. There is one pair for each exchange within the network. The elements of the pair indicate which elements are to be compared and optionally exchanged. Each individual has between 60 and 120 pairs in its phenotype, corresponding to sorting networks with 60 to 120 exchanges.

The phenotype is generated from the genotype by traversing the chromosomes of the genotype in fixed order, reading off the pairs to appear in the phenotype. If a pair of chromosomes is homozygous at a given position (if the same pair is specified in both chromosomes), then only a single pair is generated in the phenotype. If the site is heterozygous, then both pairs are generated. Thus the phenotype will contain between 60 and 120 exchanges, depending on the heterozygosity of the genotype. Sixty was chosen as the minimum size so that a completely homozygous genotype would produce a sorting network that matches the best known solution. Because most of the known minimal 16-input networks begin with the same pattern of 32 exchanges, the gene pool is initialized to be homozygous for these exchanges. The rest of the sites are initialized randomly.

Once a phenotype is produced, it is scored according to how well it sorts. One measure of ability to sort is the percentage of input cases for which the network produces the correct output. This measure is convenient for two reasons. First, it offers partial credit for partial solutions. Second, it can be conveniently approximated by trying out the network on a random sample of test cases. After scoring, the population is culled by truncation selection at the 50% level; only the best scoring half of the population is allowed to contribute to the gene pool of the next generation.

To implement recombination, the gamete pool is generated by crossover among pairs of chromosomes. For each chromosome pair in the surviving population, a crossover point is randomly and independently chosen, and a haploid gamete is produced by taking the codons before the crossover point from the first member of each chromosome pair, and the codons after the crossover point from the second member. Thus, there is exactly one crossover per chromosome pair per generation. Point mutations are then introduced in the gamete pool at a rate of one mutation per one thousand sites per generation.

The next stage is the selection of mates. One way to do this would be to choose pairs randomly, but our experience suggests that it is better to use a mating program with some type of spatial locality. This increases the average inbreeding coefficient and allows the population to divide into locally mating demes. The sorting networks evolve on a two-dimensional grid with torroidal boundary conditions. Mating pairs are chosen to be nearby in the grid. Specifically, the x and y displacement of an individual from its mate is a binomial approximation of a Gaussian distribution. Mating consists of the exchange of haploid gametes. After a pair mates, they are replaced by their offspring in the same spatial location, so the genetic material remains spatially local.

Simulations were performed using the procedure on populations of 65 536 individuals for up to 5000 generations. Typically, one solution, or a few equal scoring solutions, were discovered relatively early in the run. These solutions and their variants then spread until they accounted for most of the genetic material in the population. In cases where there was more than one equally good solution, each "species" dominated one area of the grid. The areas were separated by a boundary layer of non-viable crosses. Once these boundaries were established, the population would usually make no further progress. The successful networks tend to be short because the descendant of heterozygotes tended to be missing crucial exchanges (recessive lethals). The best sorting networks found by this procedure contained 65 exchanges.

3. The co-evolution of parasites

While the evolution of the sorting networks produced respectable results, it was evident on detailed examination of the runs that a great deal of computation was being wasted. There were two major sources of inefficiency. One was a classical problem of local optima: once the system found a reasonable solution, it was difficult to make progress without temporarily making things worse. The second problem was an inefficiency in the testing process. After the first few generations, most of the tests performed were sorted successfully by almost all viable networks, so they provided little information about differential fitness. Many of the tests were too "easy." Unfortunately, the discriminative value of a test depends on the solutions that initially evolve, and in the case where several solutions evolve, the value of a given test varies from one sub-population to another.

To overcome these two difficulties, various methods were implemented for accelerating progress by encouraging a wider diversity of solutions and limiting the number of redundant test cases. Three general methods were investigated: varying the test cases over time, varying the test cases spatially, and varying the test cases automatically by independent evolution. Because the third case has yielded the most interesting results, only it will be described in detail.

The co-evolution of test cases is analogous to the biological evolution of a host parasite, or of prey and predator. Hamilton has used both computer simulation and mathematical/biological arguments to show that such co-evolution can be a generator of genetic diversity [7–11]. The improved optimization procedure uses this idea to increase the efficiency of the search.

In the improved procedure, there are two independent gene pools, each evolving according to the selection/mutation/recombination sequence outlined above. One population, the "hosts", represents sorting networks, while the other population, the "parasites", represents test cases. (These two populations might also be considered as "prey" and "predator", since their evolution rates are comparable.) Both populations evolve on the same grid, and their interaction is through their fitness functions. The sorting networks are scored according to the test cases provided by the parasites at the same grid location. The parasites are scored according to how well they find flaws in sorting networks. Specifically, the phenotype of

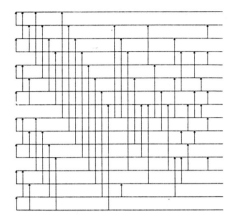

Fig. 3. 61 exchanges.

each parasite is a group of 10 to 20 test cases, and its score is the number of these tests that the corresponding sorting network fails to pass. The fitness functions of the host sorting networks and the parasitic sets of test patterns are complementary in the sense that a success of the sorting network represents a failure of the test pattern and vice versa.

The benefits of allowing the test cases to co-evolve are twofold. First, it helps prevent large portions of the population from becoming stuck in local optima. As soon as a large but imperfect sub-population evolves, it becomes an attractive target toward which the parasitic test cases are likely to evolve. The co-evolving test cases implement a frequency selective fitness function for the sorting networks that discourages large numbers of individuals from adopting the same non-optimal strategy. Successive waves of epidemic and immunity keep the population in a constant state of flux. While systems with a fixed fitness criteria tended to get stuck in a few non-optimal states after a few hundred generations, runs with co-evolving test cases showed no such tendency, even after tens of thousands of generations.

The second advantage of co-evolving the parasites is that testing becomes more efficient. Since only test-case sets that show up weaknesses are widely represented in the population, it is sufficient to apply only a few tests to an individual each generation. Thus, the computation time per generation is significantly less. These two factors taken together make it both more practical and more productive to allow the system to run for larger numbers of generations.

The runs with co-evolving parasites produced consistently better and faster results than those without. Fig. 3 shows the best result found to date, which requires 61 exchange elements. This is an improvement over Batcher's and Shapiro's solutions, and over the results of the simulation without parasites. It is still not the optimum network, since it requires one more sorting exchange than the construction of Green.

These preliminary results are encouraging. They demonstrate that simulated evolution of co-evolving parasites is a useful procedure for finding good solutions to a complex optimization problem. We are currently applying similar techniques to other applications in an attempt to understand the range of applicability. It is ironic, but perhaps not surprising, that our attempts to improve simulated evolution as an optimization procedure continue to take us closer to real biological systems.

Acknowledgements

The author would like to thank Chuck Taylor, Bill Hamilton, Stuart Kaufmann, Steve Smith, and the reviewers for helpful discussions and comments in the preparation of this paper.

References

[1] K.E. Batcher, A new internal sorting method, Goodyear Aerospace Report GER-11759 (1964).
[2] R.C. Bose and R.J. Nelson, A sorting problem, J. Assoc. Computing Machinery 9 (1962) 282–296.
[3] D.G. Bounds, New optimization methods from physics and biology, Nature 329 (1987) 215–219.
[4] H.J. Bremermann, Optimization through evolution and recombination, in: Self-Organizing Systems, eds. M.C.

Yovits, G.D. Goldstein and G.T. Jacobi (Spartan, Washington, DC, 1962) pp. 93–106.

[5] C. Darwin, The Origin of the Species by Means of Natural Selection; or, The Preservation of Favoured Races in the Struggle for Life (Murray, London, 1859).

[6] R.W. Floyd and D.E. Knuth, Improved constructions for the Bose–Nelson sorting problem, Notices Am. Math. Soc. 14 (1967) 283.

[7] W.D. Hamilton and J. Seger, Parasites and sex, in: The Evolution of Sex, eds. R.E. Micmod and B.R. Levin (Sinauer, Sunderland, MA, 1988) ch. 11.

[8] W.D. Hamilton, Pathogens as causes of genetic diversity in their host populations, in: Population Biology of Infectious Diseases, eds. R.M. Anderson and R.M. May, Dahlem Konferenzen (Springer, Berlin, 1982) pp. 269–296.

[9] W.D. Hamilton, Sex versus non-sex versus parasite, OIKOS (Copenhagen) 35 (1980) 282–290.

[10] W.D. Hamilton, P. Henderson and N. Moran, Fluctuation of environment and coevolved antagonist polymorphism as factors in the maintenance of sex, in: Natural Selection and Social Behavior: Recent Research and Theory, eds. R.D. Alexander and D.W. Tinkle (Chiron Press, New York, 1980) ch. 22.

[11] W.D. Hamilton, Gamblers since life began: barnacles, aphids, elms, Quart. Rev. Biol. 50 (2) (1975) 175–180.

[12] J.H. Holland, Adaption in Natural and Artificial Systems (University of Michigan, Ann Arbor, 1975).

[13] S. Kirkpatrick, C. Gelatt Jr. and M. Vecchi, Optimization by simulated annealing, Science 220 (1983) 671–680.

[14] D. Knuth, Sorting and Searching, Vol. 3. The Art of Computer Programming (Addison–Wesley, New York, 1973).

[15] R. Lewontin and I. Franklin, Is the gene the unit of selection?, Genetics 65 (1970) 707–734.

[16] N. Metropolis, A. Rosenbluth, M. Rosenbluth, A. Teller and E. Teller. J. Chem. Phys. 21 (1953) 1807.

[17] I. Rechenberg, Evolutionsstrategie: Optimierung Technischer Systeme nach Prinzipien der Biologischen Evolution (Frommann–Holzboog, Stuttgart, 1973).

[18] R. Smith and D. Goldberg, Nonstationary function optimization using genetic algorithms with dominance and diploidy, in: Genetic Algorithms and Their Applications, Proceedings of the Second International Conference on Genetic Algorithms, July 1987.

[19] Q. Wang, Optimization by simulating molecular evolution, Biol. Cybern. 57 (1987) 95–101.

COMPUTER SYMBIOSIS – EMERGENCE OF SYMBIOTIC BEHAVIOR THROUGH EVOLUTION

Takashi IKEGAMI[1] and Kunihiko KANEKO[2]

Center for Nonlinear Studies, Los Alamos National Laboratory, Los Alamos, NM 87545, USA

Symbiosis is cooperation between distinct species. It is one of the most effective evolutionary processes, but its dynamics are not well understood as yet. A simple model of symbiosis is introduced, in which we consider interactions between hosts and parasites and also mutations of hosts and parasites. The interactions and mutations form a dynamical system on the populations of hosts and parasites. It is found that a symbiotic state emerges for a suitable range of mutation rates. The symbiotic state is not static, but dynamically oscillates. Harmful parasites violating symbiosis appear periodically, but are rapidly extinguished by hosts and other parasites, and the symbiotic state is recovered. The relation between these phenomena and "TIT for TAT" strategy to maintain symbiosis is discussed.

1. Introduction

Symbiosis is a sophisticated evolutionary tactic. Symbiosis is a close relationship between species by which both species in the relationship benefit. Such "cooperative" relationships may actually begin as prey–predator relationships. The Lotka–Volterra equation [1], for example, shows an out-phased oscillating behavior of hosts and parasites. A prey and a predator in such model are not mutally beneficial. When we call symbiosis, it requires cooperative behavior between hosts and parasites. We argue that some qualitative changes in both hosts and parasites are requisites to symbiosis. Such changes are brought about by mutation, which may make the coexistence of hosts and parasites possible.

Evolution does not occur in a *fixed* environment. In nature, species can interact with many other species, and the term "environment" should include all interactions from other species. The strength of the interaction is dependent on the population of other species, and generally is time-dependent as well.

Here, the "environment" of hosts consists of parasites and vice versa. Both mutants of parasites and hosts can have more offspring if they gain larger benefits from the interaction with each other in the population at that time. Parasites have a benefit by mutating in a direction that increases harm to the hosts. We are interested in the question of how the population dynamics of hosts and parasites can reach a state in which they help each other, instead of attaining a state with growing harmful parasites.

Some of the symbiotic relationships are so strong that the species involved can no longer dissolve into free independent species. A eukaryotic cell is known to have a strong symbiotic relationship with Mitochondorous [2]. Such strong symbiosis is often encoded at the genetic level.

Another class of symbiosis is called loose symbiosis [2]. Such symbiotic relationships are loose enough to dissolve. A well-known example is lichen, which are the symbiosis between fungi and algae. Depending on environmental conditions, lichen may dissolve into two independent living

---

[1] On leave from Research Institute for Fundamental Physics, Kyoto University, Kyoto 606, Japan.
[2] On leave from Institute of Physics, College of Arts and Sciences, University of Tokyo, Tokyo 153, Japan.

species, or establish symbiosis from independent species [2]. Jeon's study of amoebae shows us that such symbiosis can be generated within a few days [3]. Another related example is the interaction between a virus and a host. Through evolution, the virus and the host may find a relationship in which the former attacks the latter less harmfully. The class of loose symbiosis is the focus of our study.

In the present paper we introduce a simple model for symbiosis. Through the simulation of the model, we try to understand the condition of the emergence of symbiosis and the dynamical nature of such symbiotic states.

The model consists of the population dynamics of two distinct genetic systems, called "host" and "parasite". Both hosts and parasites can change their genotypes by mutation. The genotypes are coded by bit strings. The difference between a host and a parasite lies in the interaction between the two. A host is uni-directionally attacked by all types of parasites but one. Such an exceptional parasite can be a symbiotic partner to the host. The more strongly a parasite exploits a host, the more offspring it has. Through mutations, hosts and parasites have a chance to generate a symbiotic relationship. A symbiotic state, however, is often unstable. There is no explicit reason to prevent the parasite from mutating into more harmful types, instead of remaining symbiotic. As will be seen, the symbiotic state exists only if the mutation rate of hosts is slightly larger than that of parasites. Furthermore, it will be shown that the attractor of the symbiotic state is not a fixed point but a limit cycle.

2. Model

We represent each species of host and parasite by a simple binary sequence of length L. Thus hosts and parasites can have 2^L different genotypes. Each genotype is represented as $0 = 0000000$, $1 = 0000001$, $2 = 0000010$, $3 =$ 0000011, ... The proportion[#1] of each genotype j in the population is represented by continuous variables on the 2^L-dimensional space, and is denoted by h_j and p_j corresponding to a host j and a parasite type j, respectively. To construct our dynamical model, we take into account the following constraints:

(1) Mutation of hosts and parasites: Hosts as well as parasites can mutate to other genotypes by only 1 bit. Hosts cannot become parasites or vice versa, that is, they mutate only among themselves. Since we restrict the gene space in a bit string of length L, each genotype has L neighbors by a 1-bit mutation. Hosts may mutate to a type which suffers less damage from parasites. Parasites can also mutate to a more harmful type. By denoting the mutation rates of hosts and parasites as μ_H and μ_P, we get the terms $\mu_H \Sigma_{j'}(h_{j'} - h_j)$ and $\mu_P \Sigma_{j'}(p_{j'} - p_j)$ for the population dynamics of hosts and parasites, respectively. Here, the summation over j' runs over all 1-bit neighbors of the binary sequence j.

(2) Interaction of hosts and parasites: For most parasites, hosts are uni-directionally exploited by parasites. The more advantage a parasite receives through the interaction with a host, the less advantage (or the more damage) the host receives by the interaction. If both a parasite and a host happen to take advantage of each other, such a pair of parasites and hosts is said to be in a symbiotic relation. This constraint leads to positive growth rates for parasites and mostly negative for hosts. Since a symbiotic host receives a benefit from a symbiotic parasite, a growth rate should be positive for the symbiotic host and parasite. The interaction term is written as $\Sigma_{k=1}^{2^L} a_{kj}^P p_j h_k$.

(3) Interaction among parasites: Since different parasites compete at the same host, we assume a mutual suppressing term in the equation of parasite.

[#1] "Proportion" here is not normalized by unity. It is normalized by the population at some (arbitrary chosen) time. Since the total population size can change in time, neither Σh_j nor Σp_j is constant.

No mutual of self suppression is assumed for the host system. The host can only increase or decrease its number by interacting with parasites. This constraint leads to the term $-\sum_{i \neq j} p_i p_j$ on the growth of parasite j.

Combining the constraints (1)–(3), we can write down the following set of equations on the population of the jth species of hosts $h_j(t)$ and that of parasites $p_j(t)$:

$$\frac{dp_j}{dt} = \sum_{k=1}^{2^L} a^P_{kj} p_j h_k - \sum_{i \neq j} p_i p_j + \mu_P \sum_{j'} (p_{j'} - p_j), \tag{1}$$

$$\frac{dh_j}{dt} = \sum_{k=1}^{2^L} a^H_{kj} h_j p_k + \mu_H \sum_{j'} (h_{j'} - h_j). \tag{2}$$

If $a^P_{kj} > 0$ and $a^H_{kj} < 0$, the interaction term takes the same form as the Lotka–Volterra equation with prey and predator. If both a^P_{kj} and a^H_{kj} are positive, the host and parasite help each other to increase the population through the interaction.

The next choice of our model is the dependence of interaction terms a^P_{kj} and a^H_{kj} on types k and j. From the constraint (2), the terms must have the following properties:

(i) There must be a pair of host and parasite types such that both a^P_{kj} and a^H_{kj} are positive, that is, the interaction results in "cooperation". We call such pairs of indices k, j a *symbiotic pair*. The interactions are constructed to have the symbiotic pairs.

(ii) If a host and a parasite do not form a symbiotic pair, then the parasite gets a positive gain from the interaction whereas the host receives a negative gain. This difference of the sign in the additional interaction term sets apart a parasite from a host.

We adopt the simplest form with these conditions as follows:

$$a^P_{kj} = f_P + \text{Ham}(k, j), \tag{3}$$

$$a^H_{kj} = f_H - \text{Ham}(k, j), \tag{4}$$

where the coefficients f_P and f_H are positive, and the function $\text{Ham}(k, j)$ denotes the Hamming distance between the binary sequences of k and j[#2]. In the present model, parasites have larger growth rates if their bit patterns *mismatch* those of hosts, while hosts have larger growth terms if their bit patterns *match* those of parasites.

We take $f_H < 1$, so that only a pair with a perfect matching develops a symbiotic state. Since all pairs (m, m) are symbiotic, there are 2^L possible symbiotic pairs. It can be shown by linear stability analysis that a state with only one symbiotic pair ($p_j = h_j = 0$ for $\forall j \neq m$) is unstable. As we see in section 3, an observed symbiotic state indeed contains small population ratio of non-symbiotic pairs.

In what follows, we set the bit length L equal to 7, and represent each genotype as $0 = 0000000$, $1 = 0000001, \ldots, 127 = 1111111$.

It may be natural to include a term γh_i which describes the growth of hosts in the r.h.s. of eq. (2). Indeed, we have studied such a system. The results obtained are qualitatively the same as those reported below, provided that γ is not too large.

3. Emergence of periodic symbiosis

To judge whether or not our system is in a symbiotic state, we compute the following average interaction between hosts and parasites:

$$\overline{a^\kappa} = \frac{\sum_{j,k} a^\kappa_{kj} p_j h_k}{\sum_j p_j \sum_k h_k}, \tag{5}$$

where κ is either P or H. Each interaction among hosts and parasites is averaged over all species. $\overline{a^\kappa}$ measures the ratio of symbiotic pairs in the total pairs of hosts and parasites. A positive $\overline{a^H}$ thus indicates that a system is in a symbiotic state.

[#2] The results in section 3 are not strongly dependent on the specific choice of a^P_{kj} and a^H_{kj}. Our conclusion is thought to be valid within the conditions of (i) and (ii) and for suitable parameter values.

Since the whole parasite receives positive interaction from hosts (see eq. (3)), $\overline{a^P}$ is always positive. On the other hand, $\overline{a^H}$ is positive only when symbiotic pairs are dominant. If the population is completely concentrated on a symbiotic pair, $\overline{a^P} = f_P > 0$ and $\overline{a^H} = f_H > 0$ would follow.

In wide parameter regimes, our system ends up with a state of extinction: Harmful parasites distribute widely in gene space, and hosts are exploited by the parasites[#3]. Escape from attack by the parasites is possible by mutation of hosts, but the escape is in vain if parasites are widely distributed. All hosts are thus extinguished by the parasites, which after all starve by loosing their hosts (preys).

If a parasite mutates to others whose bit sequence has a large Hamming distance from that of a host, the mutant can become more harmful to the host. On the other hand, hosts can escape from the attack of parasites by mutating to match the gene sequence of that parasite.

Parasites often spread further through gene space than hosts do. Hosts spread to follow the spread of parasites, but begin to shrink after a certain size is reached since parasites always outgrow hosts.

An example of successful symbiotic behavior is shown in figs. 1 and 2. Let us see how a symbiotic state is generated from the initial condition, by taking a case with the initial distribution of a parasite type 85 and a host type 86. Almost immediately, all hosts of type 86 change into 85 by mutation. Thus a symbiotic pair of type 85 is generated. The distribution in genotypes has a sharp peak around the symbiotic pair (see fig. 2). That symbiosis is accomplished is clearly seen in the effective interaction, as $\overline{a^H} > 0$ and $\overline{a^P} > 0$, as is shown in fig. 1. In this symbiotic state, both the populations of hosts and parasites increase, until the state becomes unstable and dissolves. Through the spread in gene space, parasites reach the most harmful species to the host type 85. Thus the

---

[#3] The extinction can be removed by introducing the growth term in our model. In order to focus on the emergence of symbiosis, we have not included this term. This stresses the harm of some parasites.

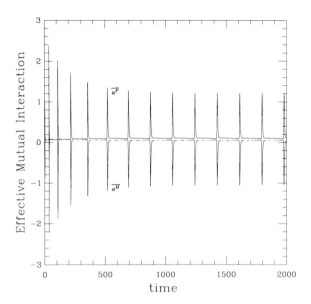

Fig. 1. Temporal evolution of order parameters $\overline{a^H}$ and $\overline{a^P}$. A symbiotic state occurs when both parameters have positive values. If the order parameter for hosts takes a negative value, it is out of the symbiotic state. A symbiotic state disintegrates periodically when the values of $\overline{a^H}$ and $\overline{a^P}$ are close together. The simulation is executed with parameter values $f_P = 0.05$, $f_H = 0.1$, $\mu_P = 0.001$ and $\mu_H = 0.002$.

parasite with the genotype 42 ($= 127 - 85$) appears. (Note that $85 = 1010101$ and $42 = 0101010$.) Since the parasite can exploit the host most strongly, its number increases. In order to recover the original symbiotic state, this harmful parasite must be eliminated. In the present case this elimination occurs through the decrease of host type 85 by mutation and the suppression of parasite type 42 by other parasites. A symbiotic pair of host and parasite drives down the parasite type 42, and a symbiotic state of type 85 has again been established.

In our present model, we have not found a fixed symbiotic state. After the system returns to a symbiotic state, the state lasts for a while but it again is destroyed. The present symbiotic state thus appears as a temporally periodic state. Two eras repeat periodically: a long symbiotic era (180 times steps in fig. 1) where the population increases slowly, and a shorter era (10 times steps in fig. 1) in which harmful parasites suddenly in-

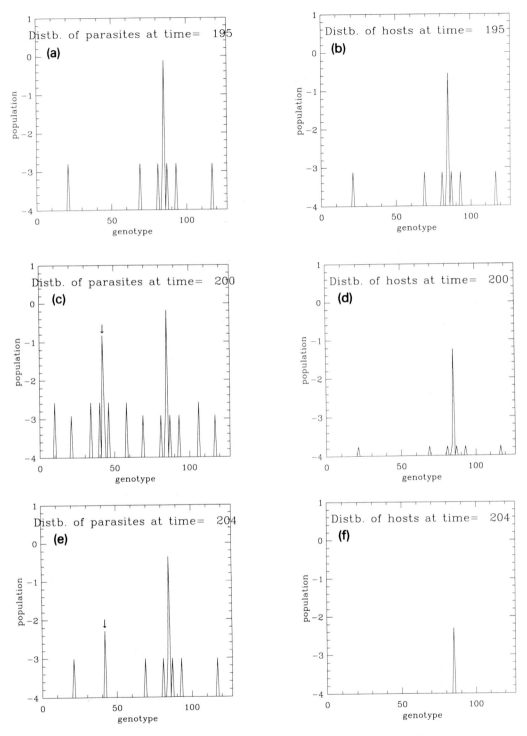

Fig. 2. Population distribution of hosts and parasites, which is simulated with the same parameter values as in fig. 1. Each horizontal line describes a genotype of species. Initially there exists one type of host with one type of parasite. A symbiotic pair of type 85 is accompanying 7 other types (a,b). At the disintegration of the symbiotic state, parasites spread in gene space but hosts begin to shrink (c,d). For symbiosis to be recovered, a host must fight with the most harmful parasite, which is indicated by a down arrow (e,f). If the host can beat the most harmful parasite, a symbiotic state is completely recovered.

crease, and then is eliminated accompanied by the decrease of populations of hosts. It should be noted that the duration of the latter era is much shorter (less than 1/10) than that of the symbiotic era. Neither a quasiperiodic nor a chaotic, but only a periodic temporal change has been observed.

Non-existence of fixed symbiotic states can be understood as follows: If our dynamics is in a symbiotic state, the population of hosts increases by the interaction term $a^H_{kj} p_j h_k$. As the population of hosts increases, the increase rate of harmful parasites gets larger by the interaction term $a^P_{kj} p_j h_k$. Since this increase rate is larger than that of other parasites and hosts, the population of harmful parasites eventually becomes large enough to destroy the symbiosis. Thus the symbiotic state cannot be temporally fixed.

The appearance of symbiosis is dependent on the initial distribution of hosts and parasites in gene space. We have studied some typical initial distributions as follows:

(a) One type of parasite versus a random or homogeneous distribution of host types.

(b) One type of host versus a random or homogeneous distribution of parasite types.

(c) Both types are randomly distributed.

(d) One type of parasite versus one type of host.

Of the above conditions, we have not found a stable symbiotic state for the conditions (b) and (c). Since various types of parasites already exist, hosts cannot escape from them, and are extinguished.

For suitable sets of parameters, a periodic symbiotic state is observed under the conditions (a) and (d). Hosts mutate so that they can form a symbiotic pair with the initially given type of parasite. Symbiosis lasts for a long time (say 200 time steps) and disintegrates, but it recovers through the punishment of harmful parasites, as has been discussed in the above. Even if the initial Hamming distance between hosts and parasites is large (say the types 85 and 42), the symbiotic pair is developed if the mutation rate of hosts is moderately large.

Next, let us discuss the condition of emergence of the symbiotic behavior in our model. By varying parameters in our model, we have found that the above dynamical symbiotic state exists only in a small parameter regime. The following conditions are found to be necessary for successful symbiosis:

(1) A mutation rate of hosts (μ_H) should be larger than that of parasites (μ_P). Too large mutation rate of hosts, however, brings the extinction again.

(2) The gain of hosts by the symbiotic pair (f_H) should be larger than that of parasites (f_P). Too large f_H again leads to the extinction.

The first restriction ($\mu_H > \mu_P$) is necessary for the escape of host from the attack of harmful parasites. The second condition ($f_H > f_P$) is understood as follows: Under the opposite condition ($f_H < f_P$), a larger growth rate of the symbiotic parasite generates more diverse parasites, which are harmful to the symbiotic host. As a result, the host is exploited by such parasites. If $f_H > f_P$, this tendency is suppressed.

The meaning of the restriction that neither f_H nor μ_H should be too large is more subtle. If these parameters are too large, the host population increases faster than that of parasites. The large population of hosts, however, leads to the growth of non-symbiotic parasites.

If the parameters do not satisfy the condition of symbiosis, a symbiotic state exists only for the initial stages and disintegrates by the spread of parasites (see figs. 3 and 4) for the extinction process. The spread of parasites in their gene space continues until it covers the whole gene space. Such a situation leads to the condition (b), and the hosts are completely exterminated by those parasites. Suppression of the spread of parasites is essential to the recovery of symbiotic behavior.

Emergence of TIT for TAT

The above mechanism of the suppression of harmful parasites is suggestive of *TIT for TAT* strategy [4] in the iterated prisoner's dilemma

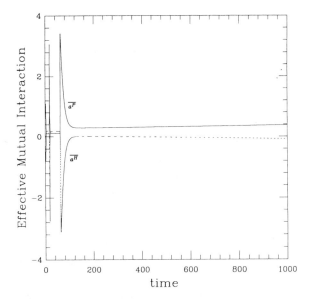

Fig. 3. A disintegration of a symbiotic state at time = 300 can be seen in the temporal evolution of order parameters. The simulation is executed with parameter values $f_P = 0.1$, $f_H = 0.2$, $\mu_P = 0.001$ and $\mu_H = 0.002$.

game. In the game, a community of programs has played a prisoner's dilemma with each other. Each player can defect or cooperate by some program. For just one iteration of game, the strategy "detect" may work, but for a long run of iterated games, it may not be a good strategy. Indeed, it is known that the robust program takes a strategy of TIT for TAT [4]. A player following the strategy of TIT for TAT trusts its component unless being defected. Once the opponent defects, it also defects.

A strategy similar to TIT for TAT is attained through a mutual supervision in a game of the evolution of norms [5].

In our model, harmful parasites (with large Hamming distance from hosts) are playing the role of "defect". If the proportion of such parasites increases, the number of parasites is eventually decreased by the decrease of the host population. Thus the existence of harmful parasites is also harmful to sustain the population of parasites, in the long run. To have sustained population size, we need a mechanism to suppress such harmful parasites. In our model, the interaction among parasites corresponds to mutual supervision. This mutual supervision inhibits the increase of the most harmful ("selfish") parasite like a type 42 in our example. Thus symbiosis is maintained by the mutual supervision, as in the case of the TIT for TAT strategy in the iterated prisoner's dilemma game.

We also note that the instant decrease of the harmful parasite is essential to the recovery of symbiosis in our model. Otherwise, the genotypes of parasites distribute widely, leading to the extinction of hosts. This instant decrease of harmful parasites is analogous to the instant punishment in TIT for TAT strategy in the iterated prisoner's dilemma game [4]. The instant "TIT" against a bad strategy (e.g. defect against any strategy) is necessary to the robustness of cooperative strategy in the game.

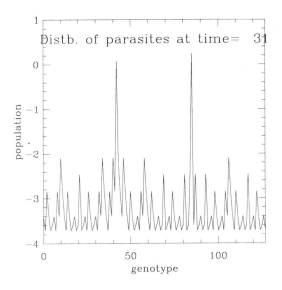

Fig. 4. If the host cannot beat the most harmful parasite, parasites spread out in gene space. After this event, all hosts are extinguished by parasites. This unsuccessful case is simulated with the same parameter values as in fig. 3.

Of course, neither host nor parasite plays a game with a strategy in our model. Both of them are just under random mutation without any purpose. As an emergent dynamical behavior, however, harmful parasites are extinguished instantly through supervision and mutation. If we observe our results from a macroscopic level, we can say that harmful parasites are "punished" by an "invisible hand". Thus we can conclude that the

"TIT for TAT" in our model appears as the *emergent* behavior to attain the symbiotic behavior.

4. Discussion

Let us consider the significance of our results from more general points of view, and discuss possible extensions.

First, our model adaptively changes the landscape through the interaction of hosts and parasites. Walks in a fixed landscape have recently been discussed especially in a rugged landscape [6]. In the evolution of species, the environment itself is generated by all species, and there is no a priori landscape. A model with a given landscape misses an essential quality of evolution. To study the evolutionary process, we need a model with a landscape which changes temporarily, depending upon the population of all species. The present model gives a simple example of this class of dynamics.

Second, our symbiosis appears only as a dynamical state. This class of symbiosis is predicted by Margulis [2] as "loose symbiosis". Symbiosis can be formed but it may dissolve again. It is a function of time. There are several reasons for the dissolution of a symbiotic relationship, as Margulis has pointed out. She also argues that the growth rate of partners should be approximately equal in order to keep a symbiotic relationship [2].

In the present model, the approximately same positive values for $\overline{a^P}$ and $\overline{a^H}$ lead to approximately equal growth rates for hosts and parasites. A dissolution of the symbiotic state of the present model is caused by the inbalance of these order parameters.

The importance of host–parasite interaction for the development of polymorphism has often been stressed (see e.g. ref. [7]). Polymorphism means a coexistence of different phenotypes of the same species. Let us consider a collection of ensembles of hosts and put parasites on each ensemble. Hosts in each ensemble may evolve to different genotypes (e.g. a different symbiotic pair). Thus polymorphism of hosts is induced by the interaction with parasites. If there exists a competition among different host types, this picture may not be justifiable.

Since our model is one of the simplest within the constraints in section 2, we expect that our results are rather general. Of course, other choices of interactions may be of importance to study specific models for symbiosis. Modification of parasite–parasite interaction or the inclusion of host–host interaction may be important. We have also studied a model in which parasites suppress themselves. It seems that the inclusion of this self-suppression term destroys the symbiotic state. Other choices of host–parasite interaction terms may be worth considering.

In our model we have assumed that all parasites and all hosts interact with each other. This assumption may be artificial. It may be better to use a model with the interaction only among restricted sets of hosts and parasites. A simple example is a model on a lattice (see also ref. [8]). It is expected that this kind of restricted interaction will enhance the stability of the symbiotic state, since the host which promotes a symbiotic relation with some parasite has a small chance to be attacked by other harmful parasites. In our long-ranged coupling model, however, such symbiotic hosts may be exploited by other parasites.

Another important future problem is the use of sex [9]. As stressed by Hamilton and others, recombination (i.e. cross-over) can be more effective than mere mutation. One of the reasons for this is that mere mutation cannot memorize the effective genetic sequences against the past parasites, but recombination does. Recombination is thought to be especially useful to protect from the attack of parasites, since it makes a large uncorrelated jump.

In our problem, recombination is useful for hosts for the same reason, to escape from the attack of parasites. Recombination, however, may in fact hinder the symbiotic relationship, since recombination may instantly create non-symbiotic hosts and parasites.

A creation of new species by recombination strongly depends on the population distribution.

Since recombination merely crosses over the already existing types, a symbiotic pair generates the symbiotic genotypes through recombination. Thus the above drawback may be removed. It is an open question, however, if our system attains such concentrated distribution under the existence of recombination.

If we regard a parasite in our model as a virus, we may ask why the mutation rate of hosts must be higher than that of viruses, to attain symbiosis. This would seem counter-intuitive unless one thinks of our "hosts" as antibodies in the host's immune system. The mutation rate of the immune system is known to be very fast to generate various types of antibodies, and our condition of "symbiosis" is satisfied. Through the mechanism we have discussed in the present paper, coexistence of less harmful viruses with the immune system is attained in the course of evolution. The loss of harm from viruses through evolution is frequently seen in nature.

Lastly, we have to point out that there is another form of symbiosis which we have not discussed in the present paper. This is the process which involves the sharing of information (gene sequence) and the creation of new species through it. For example, Margulis [2] has put forward a theory on the origin of eukaryotic cell through such joining of different species. For such a class of symbiosis, we need a model which includes the process of merging gene sequences and creating new species. This class of models will be discussed in the future[#4] [11].

In the present paper, we have discussed the emergence of symbiotic behavior. Without any supervising, our system attains a symbiotic relationship by suppressing harmful parasites. Emergence of such relationship seems relevant to cooperativity in computer community [12]. Since our model gives a simple example of symbiosis, it may be useful to extend our results to computer community, and study the emergent symbiotic computation.

Acknowledgements

The authors would like to thank Yoshitsugu Oono for critical comments and Howard Gutowitz for critical reading of the manuscripts. This work was partially supported by a Grant-in-Aid for Scientific Research from the Ministry of Education, Science, and Culture of Japan. One of the authors (T.I.) is indebted to the Japan Society for the Promotion of Science for financial support.

References

[1] M. Peschel and W. Mende, The Predator-Prey Model (Springer, Berlin, 1986).
[2] L. Margulis, Symbiosis in Cell Evolution (Freeman, San Francisco, 1981).
[3] K.W. Jeon, Science 202 (1978) 635.
[4] R. Axelrod, The Evolution of Cooperation (Basic Books, New York, 1984).
[5] R. Axelrod, Am. Political Sci. Rev. (1987) 44.
[6] S. Kauffman and S. Levin, J. Theor. Biol. 128 (1987) 11.
[7] R.M. Anderson and R.M. May, Parasitology 85 (1983) 411–426;
H. Ishikawa, Symbiosis and Evolution (Baifukan, Tokyo, 1988) (in Japanese).
[8] D. Hillis, Physica D 42 (1990) 228–234, these Proceedings.
[9] W.D. Hamilton, R. Axelrod and R. Tanase, Sexual reproduction as an adaptation to resist parasites, preprint;
W.D. Hamilton, presented at the Ninth Annual CNLS Conference on Emergent Computation, Los Alamos, 22–26 May 1989.
[10] T. Ikegami, Ph.D. Thesis (1989);
T. Ikegami and K. Kaneko, to be published.
[11] T. Ikegami and K. Kaneko, to be published.
[12] B.A. Huberman, Physica D 42 (1990) 38–47, these Proceedings.

[#4] A joint process of this type is also discussed in the "Urobors model" for the immune system [10].

USING GENETIC SEARCH TO EXPLOIT THE EMERGENT BEHAVIOR OF NEURAL NETWORKS

J. David SCHAFFER, Richard A. CARUANA and Larry J. ESHELMAN

Philips Laboratories, North American Philips Corporation, 345 Scarborough Road, Briarcliff Manor, NY 10510, USA

Neural networks are known to exhibit emergent behaviors, but it is often far from easy to exploit these properties for desired ends such as effective machine learning. We demonstrate that a genetic algorithm is capable of discovering how to exploit the abilities of one type of network learning, backpropagation in feedforward networks. Our results show that a network architecture evolved by the genetic algorithm performs better than a large network using backpropagation learning alone when the criterion is correct generalization from a set of examples. This is potentially a powerful method for design of neural networks – design by evolution.

1. Introduction

The class of feedforward neural networks trained with backpropagation [16] admits a large variety of specific architectures applicable to learning pattern recognition tasks. Unfortunately, not all network architectures can be expected to learn a given task successfully. For instance, some of these architectures exhibit a phenomenon called overlearning in which the network learns the exact characteristics of the training patterns rather than their general class properties, and subsequently performs poorly when shown novel patterns from the same classes. It is widely recognized that, in general, we do not know how to design a network capable of learning from a set of examples with the property that the knowledge will generalize successfully to other patterns from the same domain.

We propose an evolutionary approach to this problem. We present results from preliminary experiments suggesting a hybrid system consisting of a genetic algorithm (GA) and a neural network simulator (NNS) can learn more effectively than the network learning program alone.

2. Background

The evolutionary level in our system is a genetic algorithm [7], an exploratory procedure able to locate high performance structures in complex task domains. To do this, it maintains a set (called a population) of trial structures, represented as strings (called chromosomes). New test structures are produced by repeating a two-step cycle (called a generation) which includes a survival-of-the-fittest selection step and a recombination step. Recombination involves producing new strings (the offspring) by operations upon one or more previous strings (the parents). The principal recombination operator, abstracted from natural genetics, is called crossover. Holland has provided a theoretical explanation for the high performance of such algorithms [7] and this performance has been demonstrated on a number of complex problem domains such as function optimization [2, 4] and machine learning [8, 17, 18].

The action of the traditional crossover is illustrated in fig. 1. Starting with two strings from the population[#1], a point is selected between 1 and

For footnote see next page.

Fig. 1. The action of the traditional crossover operator. The two parent strings are shown as all zeros and ones for clarity.

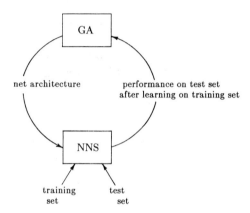

Fig. 2. A schematic of the linkage of the genetic algorithm and the neural network simulator.

$L - 1$, where L is the string length. Both strings are severed at this point and the segments to the right of this point are switched. The two starting strings are usually called the parents and the two resulting strings, the offspring. Taking the metaphor one step further, we will call this operation a mating with a single crossover event. In most previous work, the crossover point is chosen with a uniform probability distribution. Simple point mutation is also used, adding a local, though stochastic, hillclimbing capability and guarding against permanent loss of any bit patterns.

There are precedents for our evolutionary design paradigm for machine learning. Since the late 1970's, Holland and others have been using the GA to evolve classifiers (i.e. rules for parallel rule-based systems) [8, 11, 19]. When classifier systems themselves can learn, for example by using the bucket brigade algorithm [9, 10, 12], an analogous paradigm results. Complex emergent behavior has also been observed with classifier systems [14, 15]. More recently, work has begun on evolving neural networks with GAs [5, 13].

3. Method

The coupling of the GA with the NNS is shown schematically in fig. 2. The chromosomes in the GA's population represent architectures for feedforward networks. Specifically, they code for the number and layout of hidden neurons (i.e. neurons that are in neither input nor output levels) and three parameters that control the backpropagation algorithm (learning rate, momentum, and the range of the initial random weights, all using Gray coding). We allowed up to two hidden layers, either of which could be eliminated during evolution if this led to improved performance. The maximum number of neurons in each hidden layer was 16. The chromosome coding scheme is illustrated in fig. 3.

The task for this hybrid learning system was a pattern discrimination learning task, called the

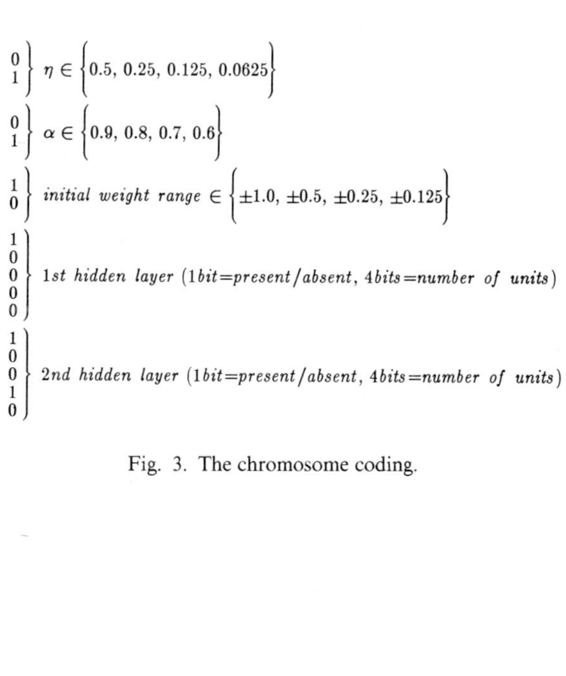

$$\left.\begin{matrix}0\\1\end{matrix}\right\} \eta \in \left\{0.5,\ 0.25,\ 0.125,\ 0.0625\right\}$$

$$\left.\begin{matrix}0\\1\end{matrix}\right\} \alpha \in \left\{0.9,\ 0.8,\ 0.7,\ 0.6\right\}$$

$$\left.\begin{matrix}1\\0\end{matrix}\right\} \text{initial weight range} \in \left\{\pm 1.0,\ \pm 0.5,\ \pm 0.25,\ \pm 0.125\right\}$$

$$\left.\begin{matrix}1\\0\\0\\0\\0\end{matrix}\right\} \text{1st hidden layer } (1bit=present/absent,\ 4bits=number\ of\ units)$$

$$\left.\begin{matrix}1\\0\\0\\1\\0\end{matrix}\right\} \text{2nd hidden layer } (1bit=present/absent,\ 4bits=number\ of\ units)$$

Fig. 3. The chromosome coding.

[#1] We use bit strings, but this is not a requirement imposed by the algorithm.

Table 1
The minimum interesting coding problem.

Input	Output	Input	Output
0000	00	0100	00
1100	00	1000	00
1001	01	0001	01
1101	01	0101	01
0010	11	1010	11
0110	11	1110	11
0011	10	0111	10
1011	10	1111	10

minimum interesting coding problem (MICP), illustrated in table 1. Among the input bits, the first two are noise, bearing no relation to the output pattern. The relationship between the two rightmost bits and the output is that of binary power-of-two-coded integers to their Gray codings. This task, contains elements of the "encoder problem" [1] in which a given network is forced to learn a compact encoding by being given an hourglass architecture and a task in which the input pattern must be reproduced at the output. The MICP complicates this by adding the elements of irrelevant inputs and a need to recode the meaningful inputs.

We assume that the learning system has access to only a subset of the possible patterns; in our experiment we used the first eight from table 1 (a stratified sample). The strategy for learning useful general knowledge was this: for each individual produced by the GA for evaluation, one of the eight sample patterns was chosen at random and set aside. The network was trained using the remaining seven patterns (the training set) halting either when the sum of squared errors reached a preset threshold (0.10) or at a prespecified number of epochs[#2]. Once a network had learned to criterion, it was evaluated by giving it the test case that had been set aside. The fitness of each individual

[#2] An epoch refers to one exposure to each pattern in the training set and one backpropagation of the accumulated errors for learning. Our threshold was 2000 epochs which we believe to be more than enough to achieve the learning criterion for most of the net architectures the chromosomes could express.

was its mean square error on this case, representing an estimate of the generality of the knowledge it had acquired during its learning phase. While this estimate has low reliability because the sample size is one, we depended on the inherent ability of the GA to deal with noisy evaluation information [3]. The randomization of the test case for each individual, gives the selection step an ensemble average to work with.

As a final test of the value of the best architecture produced by evolution, a runoff experiment was performed. This network was repeatedly trained on all eight of the available cases and then tested on the eight cases it had never seen (repeated 50 times). We recorded the total sum of squared errors on the test set as well as the number of times none of the eight test cases was incorrectly classified (error free tests). This procedure was also performed on the full architecture (i.e. one with two hidden layers of 16 neurons each) using backpropagation learning only. The comparison is of the value of the hybrid learning architecture versus backpropagation alone for learning the correct generalization.

4. Results

After producing and testing approximately 1000 individuals, the GA's population (population size 30) had converged on several network properties. All individuals but one had two hidden layers and the majority (19 of 30) had only a single neuron in the first hidden layer. The most prevalent single architecture was that shown in fig. 4[#3]. Furthermore, a strong majority of the population had the following values for the learning parameters: $\eta \in \{0.5, 0.25\}$ (29 of 30), $\alpha \in \{0.9, 0.8\}$ (24 of 30) and *initial weight range* $\in \{\pm 0.25, \pm 0.125\}$ (30 of 30).

At first, this design looks as though it could not possibly solve the problem. After all, there are

[#3] The evolution of this interesting architecture was replicated on independent genetic searches always starting from a random initial population of architectures.

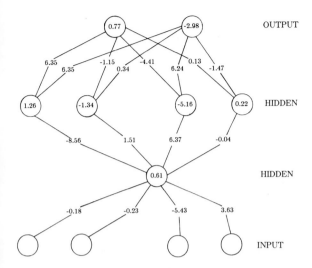

Fig. 4. An effective network architecture designed by evolution. The numbers beside the links are the weights and the numbers in the neurons are the firing thresholds set by training on eight cases.

Table 2
Results on 8 test cases for two network architectures trained on the same 8 training cases.

Criterion	Full network	Evolved network
total error		
(mean)	0.675	0.207
(standard error)	0.056	0.034
error free tests	19/50	48/50

four distinct classes of input patterns and the network channels all input information into a single neuron. The strategy, of course, is that the network discriminates the four classes by the activation level of this bottleneck neuron. The low weights on the two noise bits show the net has learned to ignore them and connections above the bottleneck perform the recoding of the input.

The results of the runoff experiment clearly demonstrate the superiority of this architecture over the full network. They are summarized in table 2. A t-test of the mean difference on total error is significant at the 0.001 level. The knowledge acquired by backpropagation alone, when given a generous supply of hidden units, exhibits overspecificity to the training set (overtraining). It generalizes poorly, while the severe restriction of the architecture evolved by the GA is right for this task.

5. Conclusion

The backpropagation learning algorithm is one approach to exploiting the emergent behavior of which feedforward networks are capable. However, it has known difficulties, including a risk of becoming trapped on local minima in weight space and a tendency to overlearning.

Our experiment illustrates how a GA can be used to exploit the properties of backpropagation to solve difficult tasks, something that humans are currently struggling to do. In the absence of prior information about the complexity of a task and knowledge of how to design networks that fit it, designers often resort to specifying what they hope are too many neurons and layers and trusting that backpropagation will be able to trim away the excess. This approach can fall victim to backpropagation's tendency to overlearn, as our experiment shows. The GA, on the other hand, is able to learn how to design effective networks by using a measure of the general utility of the knowledge learned by backpropagation (even a noisy estimate). Whether this behavior can be repeated on other problems remains to be shown.

We believe the novel architecture designed by the GA is a strong indicator of the ability of an evolutionary approach to exploit emergent properties of a process, even properties unknown to human designers. We, as designers, would probably never have thought of specifying so narrow a bottleneck and counting on backpropagation to learn to code the information in the activation level of a single neuron.

Hinton and Nowlan [6] have shown how learning can aid evolution. We have shown that the converse is also true: that evolution can discover novel and effective ways to exploit properties of complex learning processes.

There is still much to learn about the behavioral abilities of artificial networks and about the behavior of the genetic algorithm. The burgeoning research activity in both areas testifies to this. However, we are encouraged by the evidence that a synergistic hybrid of these technologies is possible.

References

[1] D.H. Ackley, G.E. Hinton and T.J. Sejnowski, A learning algorithm for Boltzmann machines, Cognitive Sci. 9 (1985) 147–169.

[2] K.A. De Jong, Adaptive system design: a genetic approach, IEEE Trans. Systems, Man Cybernetics SMC-10 (1980) 566–574.

[3] J.M. Fitzpatrick and J.J. Grefenstette, Genetic algorithms in noisy environments, Machine Learning (1988) 101–120.

[4] J.J. Grefenstette, Optimization of control parameters for genetic algorithms, IEEE Trans. Systems, Man Cybernetics SMC-16 (1986) 122–128.

[5] S.A. Harp, T. Samad and A. Guha, Towards the genetic synthesis of neural networks, in: Proceedings of the Third International Conference on Genetic Algorithms (Kaufmann, Los Altos, CA, 1989) pp. 360–369.

[6] G.E. Hinton and S.J. Nowlan, How learning can guide evolution, Complex Systems 1 (1987) 495–502.

[7] J.H. Holland, Adaptation in Natural and Artificial Systems (University of Michigan Press, Ann Arbor, MI, 1975).

[8] J.H. Holland and J.S. Reitman, Cognitive systems based on adaptive algorithms, in: Pattern-Directed Inference Systems, eds. D.A. Waterman and F. Hayes-Roth (Academic Press, New York, 1978).

[9] J.H. Holland, Escaping brittleness, in: Proceedings of the International Machine Learning Workshop, Monticello, IL (June 1983) pp. 92–95.

[10] J.H. Holland, Properties of the bucket brigade, in: Proceedings of an International Conference on Genetic Algorithms and Their Applications, (Lawrence Erlbaum, Hillsdale, NJ 1985) pp. 1–7.

[11] J.H. Holland, K.J. Holyoak, R.E. Nisbett and P.R. Thagard, Induction (MIT Press, Cambridge, MA, 1986).

[12] J.H. Holland, Escaping brittleness: The possibilities of general purpose learning algorithms applied to parallel rule-based systems, in: Machine Learning II, eds. R.S. Michalski, J. Carbonell and T. Mitchell (Kaufmann, Los Altos, CA, 1986) pp. 593–623.

[13] G.F. Miller, P.M. Todd and S.U. Hegde, Designing neural networks using genetic algorithms, in: Proceedings of the Third International Conference on Genetic Algorithms (Kaufmann, Los Altos, CA, 1989) pp. 379–384.

[14] R.L. Riolo, Bucket brigade performance II. Default hierarchies, in: Genetic Algorithms and Their Applications: Proceedings of the Second International Conference on Genetic Algorithms (Lawrence Erlbaum, Hillsdale, NJ, 1987) pp. 196–201.

[15] R.L. Riolo, Bucket brigade performance I. Long sequences of classifiers, in: Genetic Algorithms and Their Applications: Proceedings of the Second International Conference on Genetic Algorithms (Lawrence Erlbaum, Hillsdale, NJ, 1987) pp. 184–195.

[16] D.E. Rumelhart, G.E. Hinton and R.J. Williams, Learning internal representations by error propagation, in: Parallel Distributed Processing, Vol. I, eds. D.E. Rumelhart, J.L. McClelland and P.R. Group (MIT Press, Cambridge, MA, 1986) pp. 318–362.

[17] J.D. Schaffer, Some experiments in machine learning using vector evaluated genetic algorithms, Ph.D. Dissertation, Department of Electrical Engineering, Vanderbilt University, Nashville, TN (December 1984).

[18] S.F. Smith, Flexible learning of problem solving heuristics through adaptive search, in: 8th International Joint Conference on Artificial Intelligence, Karlsruhe, West Germany (August 1983).

[19] S.W. Wilson and D.E. Goldberg, A critical review of classifier systems, in: Proceedings of the Third International Conference on Genetic Algorithms (Kaufman, Los Altos, CA, 1989) pp. 244–255.

PERCEPTRON REDUX*: EMERGENCE OF STRUCTURE

Stewart W. WILSON
The Rowland Institute for Science, Cambridge, MA 02142, USA

Perceptrons were evolved that computed a rather difficult nonlinear Boolean function. The results with this early and basic form of emergent computation suggested that when genetic search is applied to its structure, a perceptron can learn more complex tasks than is sometimes supposed. The results also suggested, in the light of related work on classifier systems, that to hasten the *emergence* of an emergent computation it is desirable to provide evaluative feedback at a level as close as possible to that of the constituent local computations.

1. Introduction

This paper investigates the application of genetic search to the generation of effective emergent computation. We shall say that a system embodies an emergent computation if it has global information processing behavior that arises from the interaction of many small, local computations. The emphasis is on "local": no aspect of the system sees more than a fraction of the rest of the system or the environment, yet significant overall behavior results. Within this definition, many variations are possible, from cases where the local computations and their interaction are linear, e.g., the simplest linear associator, to human society, where both the component individuals and their interactions are highly nonlinear. Our focus will be on the three-layer perceptron [1, 2], an early and basic example of emergent computation in which a global decision is reached by weighing and combining the results of many small computations each looking at only a part of the input (fig. 1). Note that while the decision computation receives the results of the others, it, too, is local and limited in that it sees none of the system's environment. Related to the perceptron is the Pandemonium model [3], in which lower-level "demons" shout their computational results to higher-level demons. We shall apply genetic search to the problem of evolving perceptron structure, a process that itself forms an emergent computation.

In Minsky and Papert's [2] terminology, the perceptron computes a global predicate, i.e. a truth function with respect to the bits of its input. They showed that certain global predicates such as connectedness and parity could not be computed by a perceptron whose local computations, or *partial predicates*, were of restricted order, i.e. were based on fewer than the total number of input bits. However, there remain, as Minsky and Papert also showed, many useful and sometimes surprising global predicates that can be computed by order-restricted perceptrons.

The perceptron's attractiveness as a computing device is further enhanced by the *perceptron convergence theorem*, which states that if for a given perceptron there exists a choice of weights on its partial predicate outputs that would permit it to compute a certain global predicate correctly, then a weight adjustment procedure called the *perceptron learning algorithm* (PLA) will bring the system to perfect performance in a finite number of input trials. The drawback, of course, is that, for a given task, no such set of weights may exist, and

---

*"redux...2. Brought back, restored...", Oxford English Dictionary.

therefore the PLA will be of no use. The existence of appropriate weights is itself a function of the perceptron's partial predicates – that is, do the local computations or experiments on the input sufficiently recode it so that a linear weighting scheme can compute the global predicate? While the PLA could modify the partial predicates' output weights, at the time of *Perceptrons* no scheme was known for modifying the predicates themselves. One proposal was to modify weights placed *within* the partial predicates, but no technique for this was widely available until the PDP group [4] introduced back-propagation.

In this paper we investigate a quite different approach to the predicate modification problem, namely, genetic search over the space of possible perceptron structures. In proposing that the demons might "mutate" and undergo "fusion" to form new demons, Selfridge [3] made a related proposal for Pandemonium. Our approach will be to evolve a *population* of perceptrons under a version of the genetic algorithm [5] in the context of a specific task. A general characteristic of the GA is that partial solutions to a problem, called *schemata*, are distributed throughout the population, implicitly tested, and gradually combined to form the superior individuals that eventually emerge. Thus our approach has a two-fold emergence: (1) that of the GA in finding good structures through distributed testing and interaction of their constituent schemata, and (2) that of the perceptron computation itself.

Since the perceptron is a fundamental example of an artificial neural network, our investigation is related to several others in which network characteristics were evolved genetically. In some (e.g. refs. [6, 7]), the emphasis was on evolving connection weights. Others (e.g. refs. [8, 9]) addressed the question of evolving general network structure and are thus more related to the present work.

We shall offer some answers to the following questions: (1) Can genetic search over "perceptron space" indeed lead to higher performance? (2) Will the predicates in the evolved perceptrons have an interesting form? (3) How efficient is the evolutionary process in this case? (4) More generally, what do the results say for the generation of goal-oriented emergent computation? In the present example of evolution under the GA, the evaluated entity is an entire perceptron so that we are in effect designing by providing feedback solely at the top or system level. Such a regime puts "emergence" to the test, so we are interested in its efficiency relative to schemes in which evaluation is more local.

2. Chromosome representation

Genetic search takes place over a space of genotypes or chromosomes that encode relevant characteristics of the phenotypes or organisms whose worth or *fitness* is to be optimized. In the present problem, the phenotype is the perceptron itself, consisting of partial predicates, predicate output weights, and threshold decision unit. For the predicates (which he called *association cells*), Rosenblatt [1] generally used units that summed and thresholded inputs that could be excitatory or inhibitory – that is, positive or negative. We have simplified this somewhat by assuming that the inputs to the perceptron's input layer are binary variables, and that each predicate computes the logical conjunct (AND) of some subset of the set of those variables and their complements (given the right predicates, such a perceptron can compute any Boolean function of its inputs). However, there is no reason why threshold units, or any other fixed computation should not also be investigated. We assume that when a predicate is true, its output is $+1$; when false, its output is -1. The weights multiplying these values are allowed to be any real numbers. The threshold decision unit sums the weighted predicate outputs and itself outputs a 1 (true) if the sum is greater than or equal to 0, otherwise 0 (false).

This specification of the phenotype permits a straightforward genotype encoding. Notice that all such perceptrons have identical decision units. Notice also that the weights are not inherent to

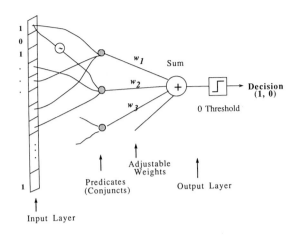

Fig. 1. A three-layer perceptron with binary input layer and predicates that compute the AND of subsets of the set of the inputs and their complements.

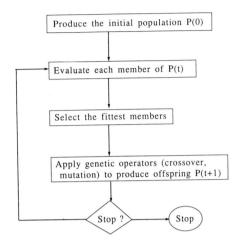

Fig. 2. A basic genetic algorithm (after a diagram in ref. [10]).

the perceptron structure, but result from training in a certain problem environment. Thus only the number and structure of the predicates need be encoded. We encode a particular predicate using an ordered string in which "1" means the predicate unit is connected to the corresponding input bit, a "0" means connected but through an inverter (the bit is complemented), and "#" means that bit is not involved in the predicate's computation. Thus if there were six input bits in all, the string "01#10#" would encode a predicate that computed the conjunct $x'_0 x_1 x_3 x'_4$, where the primes signify negation. As another example, the predicate second from the top in fig. 1 would have the encoding "0##1###1#...#" and computes $x'_0 x_3 x_7$ (reading input bits from the top down, the circled tilde indicating an inverter). To encode a whole perceptron, we simply concatenate, in any order, the encodings of its predicates.

3. Evolution under the genetic algorithm

The major steps of the standard genetic algorithm are shown in fig. 2. The initial population $P(0)$ usually consists of N fixed length chromosomes generated at random. In the evaluation step, each chromosome is decoded to its phenotype and the phenotype is subjected to a procedure that determines or estimates its fitness with respect to the task or other problem environment at hand. The same fitness value is then attached to the underlying chromosome. In the selection step, a new population of N members is formed by copying (reproducing) chromosomes by some procedure in which the fitter chromosomes receive more offspring. In the genetic operators step, some members of the new population exchange substrings (crossover) or undergo random changes at random string positions (mutation). Then the overall process iterates until the stop condition is satisfied, which may occur after a predetermined number of generations, or after some condition within the population has occurred, etc. For details on the genetic algorithm, see Goldberg [11]. Our experiments followed most of this pattern, with parameters and variations to be described later.

The GA's evaluation step is the task-dependent part of the algorithm. In the present case, we wish to determine "how good" is the perceptron encoded by a given chromosome. To do so, we construct the phenotype and train it in the problem environment, using some measure of performance as the fitness (fig. 3). "Training" means presenting the perceptron with random input strings, obtaining its decision, then adjusting the weights according to the degree to which the deci-

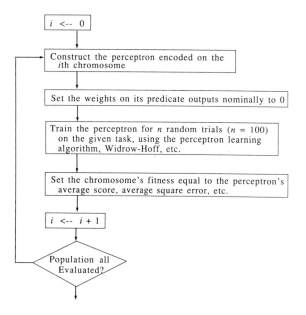

Fig. 3. Inside the GA's evaluate box for the perceptron problem.

sion matches the desired or correct decision for the global predicate that the trainer has in mind.

4. Task description

As a task or global predicate for the perceptrons to compute, we chose the six-bit Boolean multiplexer. This function is highly nonlinear, and has been used as a task with several different machine learning systems [12–17], providing a basis for comparing them. The six-multiplexer belongs to a family of progressively harder multiplexer functions, so that scale-up effects can be examined. In disjunctive normal form (DNF), the six-multiplexer is as follows:

$$F_6 = x'_0 x'_1 x_2 + x'_0 x_1 x_3 + x_0 x'_1 x_4 + x_0 x_1 x_5.$$

Bits x_0 and x_1 can be thought of as forming an address that selects one of the remaining bits as the function's value. In general, there is a multiplexer function for strings of length $L = k + 2^k$, with $k > 0$.

5. Results

After a series of experiments incorporating several changes that produced improvements, we obtained results of the sort illustrated in fig. 4. The graph plots population average score versus the total number of chromosome evaluations. The population contained 100 chromosomes, so that 100 evaluations constituted one generation. The score or individual performance of a perceptron was its percentage of correct decisions on the last 40 of 100 training trials, and these individual scores were averaged over the population. The quantity plotted is the average of three runs of the experiment. By about 40 generations, the population performance was close to 100%, meaning that essentially all of the 100 chromosomes encoded perceptrons that, once trained, easily solved the problem. Note that the performance of the initial population was about 64%, indicating that training a perceptron with random predicates produced better than random performance, but clearly did not solve the problem.

Besides performance, it was interesting to know what kinds of predicates were being evolved. Fig. 5 shows a typical set of predicates from one of the

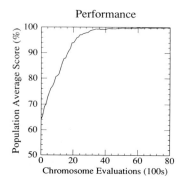

Fig. 4. Population performance. Average over the population of the score of each chromosome's perceptron on the last 40 of 100 training trials versus number of chromosome evaluations. Curve averages over three runs, each starting with a different random initial population.

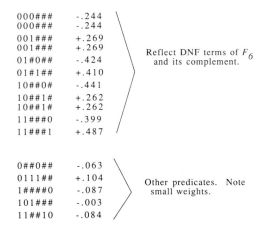

Fig. 5. Evolved predicates (left column) from a chromosome in one of the populations of fig. 4. Weights (right column) are after training.

chromosomes at the end of a run. For purposes of illustration, the predicates have been sorted into two groups. The upper group consists of predicates that turn out to correspond exactly to the DNF terms of F_6 and its complement. The lower group contains the remaining predicates from the chromosome. With each predicate is its weight after training. Note how strong are the weights of the upper group versus the lower – the latter predicates appear to be "chaff" resulting from the GA's stochastic nature. Ignoring the chaff and the duplicates in the upper group, it can be seen that the system discovered the most efficient possible solution in the sense that (1) when a predicate is true, the global predicate is either always true or always false, and (2) the predicates form a minimal non-overlapping cover of the input space.

One of the important experimental changes that helped lead to these results was to make the chromosome longer than needed. A chromosome eight predicates long has just enough room for the eight predicates corresponding to the DNF terms of F_6 and its complement. In fact, we used a chromosome that had room for 16 predicates, as is clear from fig. 5. With an eight-predicate chromosome, the system typically did not evolve more than about four of the DNF predicates before converging, and performance was lower. Of course,

in an arbitrary environment, one would not so conveniently know in advance how long to make the chromosome, but a more sophisticated system might employ chromosomes of variable length, along the lines of Smith's [18] work, so that the proper length could be found adaptively. But we are not sure just why extra length is needed. There was some informal evidence that evolving predicates compete for chromosomal "slots" during the early part of a run. With fewer slots, it is more likely that a proto-predicate that loses such a competition would later be unable to establish itself elsewhere. The effect deserves greater study, and may be related to Holland's [5] discussion of gene variation through intra-chromosomal duplication.

A second improvement in basic GA technique was to use so-called *reduced surrogate* crossover [19]. Instead of picking the crossover point at random, this technique picks at random but only among sites where the offspring are guaranteed to be different in at least one position from the parents. The result was to raise performance in the middle part of the evolution, so that the final level of performance was arrived at more quickly. The technique had no effect in the early stages, where totally random crossover is almost sure to be fruitful. It also did not raise final performance over that of standard crossover.

Two improvements resulted from changes in the evaluation procedure. In the first place, we were unsure just how to base *fitness* on a perceptron's performance during training. Initially, the fitness was the same as the performance measure: the average score on the last 40 of 100 training trials. It seemed reasonable to let the creature learn for a while before grading it. Eventually, however, we discovered that better results came from setting fitness equal to the average score over the entire training period, or all 100 trials. Perceptrons with superior predicates probably learn faster, and averaging from the very beginning may help detect them.

The largest single improvement came from changing the training algorithm from PLA to the

Widrow–Hoff algorithm [20]. The difference was to move final performance levels from approximately 95 to 100%. In PLA, no weight adjustment occurs if the perceptron is correct, whereas Widrow–Hoff adjusts whenever the output of the decision unit summer differs from either +1 or −1, and the adjustment is such as to make the error zero. Because WH is essentially always active, it may in some cases cause faster learning than PLA [21]. We did not observe this, but did observe the above final performance increase. Hinton [22] suggests that WH yields weight settings closer to optimal when the predicates are suboptimal. That would produce more accurate fitnesses in our case, and may explain the improvement.

Several more mundane aspects of the experiments should be included for completeness. The chromosomes were not encoded directly in the ternary $\{1, 0, \#\}$ vocabulary described in section 2, but were instead pure binary, 11 standing for the 1, 00 standing for 0, and both 01 and 10 standing for $\#$. This was based on the idea that binary encodings maximize the number of schemata processed by the GA [11]; empirical observations of the superiority of binary over ternary encodings were made by Schaffer [23], among others. The crossover rate in our experiments was 0.8, meaning that pairs of offspring were recombined with that probability. The mutation rate was 0.005 per allele (bit), the rate that gave the best results in an experiment with crossover turned off. The selection method used was a version of Wetzel's *tournament selection* [11] as modified by David Goldberg and the author. Tournament selection has a relatively simple implementation, and automatically eliminates problems of fitness scaling that can occur early and late in a run under more traditional selection methods. No direct comparison was made with traditional stochastic selection and its variants, but a version of Whitley and Kauth's [24] GENITOR algorithm was found to give performance similar to tournament selection along with somewhat less interesting final predicates.

6. Discussion

The basic impression gained from these experiments is that excellent performance and internal representation were achieved on a rather difficult problem, but the process was very lengthy. If, in fig. 4, we take 40 generations as the point at which a criterion performance was reached, then this required (40 generations) × (100 evaluations per generation) × (100 trials per evaluation) = 400 000 input trials. In comparison, *classifier systems* have been shown to solve the six-multiplexer problem much more quickly. A classifier system also employs genetic search, but the population consists of individual classifiers, somewhat analogous to a population of individual predicates. Results like the present ones have typically been achieved within approximately 5000 input trials or fewer [14, 17] and have been extended to as few as 500 trials [16]. A neural network system using back-propagation [13] solved the six-multiplexer in approximately 5000 input trials. Quinlan's [15] system learned the six-multiplexer rapidly, but that system cannot be described as emergent.

These comparisons are not quite fair to the present approach because the performance measure in fig. 4 is the population average performance, whereas high-performing individuals appeared in the population much earlier. In the runs of fig. 4, for example, the populations contained individuals scoring over 99% before 2000 evaluations, or about twice as quickly as the population average. Such individuals, however, did not yet have the clear DNF representation of fig. 5.

The question of evolution times looms especially large in connection with scale-up to bigger problems. Among other things, a "bigger" problem usually means a longer input string, and the size of the input space grows exponentially with string length. Consequentially, an exponentially growing number of trials per evaluation are required in order to sample a fixed fraction of the input space. Grefenstette and Fitzpatrick [25] presented evidence that the GA can sample a relatively small fraction of possible inputs and still get good results. We experimented with this concept on the six-multiplexer, trying as few as 16 random trials for evaluating fitness. Results held up quite well for 64 and 32 trials, but fell off sharply for 16. Since there are 64 possible input strings for the six-multiplexer, this may indicate that one should not sample less than about half of the input space. The input space of the 11-multiplexer has 2048 members, so this would indicate sampling at least 1000 points, which in itself would mean learning times for the 11-multiplexer an order of magnitude longer than for the six-multiplexer. Very long times were indeed found in preliminary experiments with the 11-multiplexer. In the best experiment, a population average score of 87% required (80 generations) × (200 evaluations per generation) × (300 trials per evaluation) = 4 800 000 trials. By contrast, in unpublished work of the author, a classifier system reached a 90% score on the 11-multiplexer within 3000 random trials. We do not know of any neural net investigation (using e.g. back-propagation) of the 11-multiplexer.

7. Conclusion

Perceptrons competent to solve quite difficult problems appear to be evolvable by genetic search – suggesting greater versatility for the perceptron architecture than is sometimes thought – but the process is long and does not scale up well. A clue to the shortcomings may lie in the comparison with classifier systems. There, each classifier receives evaluation whereas in the present work a single evaluation applies to a whole perceptron, and the individual predicates making up the perceptron are evaluated only indirectly. Since a predicate corresponds logically to a classifier, this would suggest the hypothesis that evaluation at the whole-system level, for systems as large as these perceptrons, is simply too diffuse – and is therefore less efficient than – evaluation at the level of the system's principal components. Of course, there may be situations where a component interaction more complex than simple linear summation requires that the whole system be evaluated as a unit. An example may be recent work on a control problem [26] in which each population member is a set of rules. But in that case, too, the evolution times were long, on the order of one million trials.

An emergent system is one that produces a global result through interaction of strictly local computations. Given an interaction scheme, design of such a system means design of the local computations. Such a process is suitable for genetic search, as has been illustrated by the perceptron example, because the GA is good at altering and recombining pieces of solutions. An emergent system designed by a GA may in fact be one of the best approaches to maximizing performance in an incompletely understood environment. However, the present results plus those on classifier systems suggest broadly that the "design time" of an emergent system will be significantly improved if evaluation occurs not at the system level but at the level of the system's component computations. Sic semper credit assignment!

Acknowledgements

The author would like to thank David Goldberg and Stephanie Forrest for valuable conversations. The suggestions of the anonymous reviewers are also appreciated.

References

[1] F. Rosenblatt, The perceptron: a probabilistic model for information storage and organization in the brain, Psych. Rev. 65 (1958) 386.
[2] M. Minsky and S. Papert, Perceptrons: An Introduction to Computational Geometry (MIT Press, Cambridge, MA, 1969).
[3] O.G. Selfridge, Pandemonium: a paradigm for learning, in: Mechanisation of Thought Processes: Proceedings of a Symposium Held at the National Physical Laboratory (HMSO, London, 1958).

[4] D.E. Rumelhart, J.L. McClelland and the PDP Research Group, Parallel Distributed Processing: Explorations in the Microstructure of Cognition (MIT Press, Cambridge, MA, 1986).

[5] J.H. Holland, Adaptation in Natural and Artificial Systems (University of Michigan Press, Ann Arbor, 1975).

[6] D. Whitley and T. Hanson, Optimizing neural networks using faster, more accurate genetic search, in: Proceedings of the Third International Conference on Genetic Algorithms, ed. J.D. Schaffer (Kaufmann, Los Altos, CA, 1989).

[7] D. Montana and L. Davis, Training feedforward neural networks using genetic algorithms, in: Proceedings of the International Joint Conference on Artificial Intelligence (1989).

[8] S.A. Harp, T. Samad and A. Guha, Towards the genetic synthesis of neural networks, in: Proceedings of the Third International Conference on Genetic Algorithms, ed. J.D. Schaffer (Kaufmann, Los Altos, CA, 1989).

[9] G.F. Miller, P.M. Todd and S.U. Hegde, Designing neural networks using genetic algorithms, in: Proceedings of the Third International Conference on Genetic Algorithms, ed. J.D. Schaffer (Kaufmann, Los Altos, CA, 1989).

[10] J.D. Schaffer and J.J. Grefenstette, A critical review of genetic algorithms, Report No. TR-88-009, Philips Laboratories, Briarcliff Manor, NY (1988).

[11] D.E. Goldberg, Genetic Algorithms in Search, Optimization, and Machine Learning (Addison–Wesley, Reading, MA, 1989).

[12] A.G. Barto, Learning by statistical cooperation of self-interested neuron-like computing elements, Human Neurobiol. 2 (1985) 229.

[13] C.W. Anderson, Learning and Problem Solving with Multilayer Connectionist Systems, Ph.D. Dissertation, Computer and Information Science, University of Massachusetts (1986).

[14] S.W. Wilson, Classifier systems and the animat problem, Machine Learning 2 (1987) 199.

[15] J.R. Quinlan, An empirical comparison of genetic and decision-tree classifiers, in: Proceedings of the Fifth International Conference on Machine Learning (Kaufmann, Los Altos, CA, 1988).

[16] S. Sen, Classifier system learning of multiplexer function, unpublished report available from Sandip Sen, Department of EECS, SCE Division, University of Michigan, Ann Arbor, MI 48197, USA.

[17] L.B. Booker, Triggered rule discovery in classifier systems, in: Proceedings of the Third International Conference on Genetic Algorithms, ed. J.D. Schaffer (Kaufmann, Los Altos, CA, 1989).

[18] S. Smith, A learning system based on genetic algorithms, Ph.D. Dissertation, Computer Science, University of Pittsburgh (1980).

[19] L.B. Booker, Improving search in genetic algorithms, in: Genetic Algorithms and Simulated Annealing, ed. L. Davis (Pitman, London, 1987).

[20] B. Widrow and M. Hoff, Adaptive switching circuits, in: IRE WESCON Convention Record (IRE, New York, 1960).

[21] J.A. Anderson and E. Rosenfeld, Neurocomputing: Foundations of Research (MIT Press, Cambridge, MA, 1988).

[22] G.E. Hinton, Connectionist learning procedures, Technical Report CMU-CS-87-115, Carnegie-Mellon University, Computer Science Department, Pittsburgh (1987).

[23] J.D. Schaffer, Learning multiclass pattern discrimination, in: Proceedings of the First International Conference on Genetic Algorithms, ed. J. Grefenstette (Lawrence Erilbaum, Hillsdale, NJ, 1985).

[24] D. Whitley and J. Kauth, GENITOR: a different genetic algorithm, in: Proceedings of the Rocky Mountain Conference on Artificial Intelligence, Denver, CO (1988).

[25] J.J. Grefenstette and J.M. Fitzpatrick, Genetic search with approximate function evaluations, in: Proceedings of the First International Conference on Genetic Algorithms, ed. J. Grefenstette (Lawrence Erlbaum, Hillsdale, NJ, 1985).

[26] J.J. Grefenstette, A system for learning control strategies with genetic algorithms, in: Proceedings of the Third International Conference on Genetic Algorithms, ed. J.D. Schaffer (Kaufmann, Los Altos, CA, 1989).

AN ENERGY FUNCTION FOR SPECIALIZATION

W. BANZHAF[1] and H. HAKEN
*Institut für Theoretische Physik und Synergetik, Universität Stuttgart, Pfaffenwaldring 57/IV,
D-7000 Stuttgart 80, Fed. Rep. Germany*

We present a model of unsupervised learning based on the minimization of an energy function. The minima of the energy function are related to the degree of specialization of a certain class of artificial neuronal cells – grandmother cells – in the neural network model proposed by Haken. The self-organizing properties of the system are demonstrated by feeding input into a network of such cells with originally randomized synaptic connections. The relation of this learning algorithm to other learning schemes, like e.g. Kohonen's feature maps, is outlined.

1. Introduction

In recent years dynamical systems have been used with considerable success for purposes of information processing. Especially in the field of pattern recognition they have proven to be useful. Processes modeled by differential or difference equations underlie all natural phenomena and are therefore candidates for successful implementations of natural information processing capabilities into computers of future generations.

In order to constrain the arbitrariness of dynamical processes, researchers are looking for dynamical laws which stand out in some sense, for instance those which are related to certain extremal or optimization principles. Thus, the derivation of a certain information processing dynamics from the maximization or (in physics) minimization of a particular scalar function increases its plausibility and gives a serious motivation to study this law in more detail. This may be one of the reasons for the recent success of the Hopfield model for associative memory [1].

Whereas the search for optimization principles has been successful in the case of a dissipative dynamics to recognize or classify patterns (see e.g. Haken [2]), the question of self-organized pattern learning has resisted such an approach for a long time. Again, a process observed in nature – adaptation of living organisms to their environment – provided a starting point to introduce different dynamical laws. Moreover, in the restricted case of "supervised learning" a particular scalar function, the error function, was a natural choice. In the more interesting case of unsupervised or self-organized learning, however, the formulation of an optimization principle is not so obvious. In the recent work of Linsker [3] we see one of the promising general approaches to this problem.

In the following, we shall propose another optimization principle for unsupervised learning, which may turn out to be equivalent if formulated sufficiently generally. For the moment, we shall restrict ourselves to a particular network architecture and study the consequences in that context in more detail. More specifically, we shall report here on recent progress made with the neural network architecture proposed in 1987 by Haken [2, 4]. In particular we use a local Hebb-like learning rule derived from what may be called a principle of cell specialization. This will be formulated in detail below.

To state the principle rather generally, a cell in the network competes with all the other cells to

---

[1] Now at: Central Research Laboratory, Mitsubishi Electric Corporation, 1-1, Tsukaguchi Honmachi 8-chome, Amagasaki, Hyogo, 661, Japan.

represent the patterns offered. The competition eventually settles when minimal overlap between patterns represented by different cells has been reached which accounts for a maximal specialization of the cells in the network. Thus the assumption is that a sort of "effectiveness" criterion is imposed on the cells due to the fact that it is costly to establish and supply any cell in a network. Though such a principle may not have a real justification in artificial systems, in natural systems at least it is reasonable. From this principle a dynamical law of connection modification is derived which will be demonstrated in the paragraphs to follow.

We only claim here that a principle of cell specialization can be applied to many neural network models and may lead to reasonable learning dynamics.

2. The network architecture

A few words are in order to give an overview of the system. The network consists of at least two layers of units (cf. fig. 1), the *input layer* (I) and the *processing layer* (II). Optional is a third layer (for output) or even more intermediate processing layers.

The input units, whose activity values q_i, $i = 1, \ldots, N$ vary continuously between -1 and $+1$,

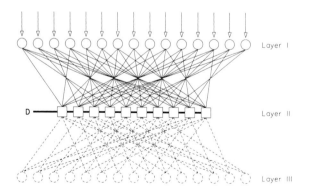

Fig. 1. Design of the overall system. Information flow is coming in through input cells q_i in layer I and is processed by layer II cells k with activity d_k. Output layer III is optional for generating patterns to implement associative recall.

communicate with the environment:

$q_i \in [-1, +1]$.

The processing units (layer II) receive their inputs from these units via synaptic connections A_{ki}, the sum of which prepares the initial conditions of their internal activity dynamics $d_k(t)$, $k = 1, \ldots, K$:

$$d_k(0) = \sum_{i=1}^{N} A_{ki} q_i.$$

The internal units are connected so as to implement a winner-take-all network by coupling every unit to a global field D,

$$D(t) = \sum_{k=1}^{K} d_k^2(t),$$

and a competitive dynamics described by

$$\dot{d}_k(t) = d_k(t)\left[1 - 2D(t) + d_k^2(t)\right]. \quad (1)$$

After relaxing the dynamics, one cell (let us call it k') wins the competition ($d_{k'} = 1$). The dynamics is constructed such that this will be the cell that had the maximal absolute activity value from the beginning.

This particular dynamics can be found in other natural systems, e.g. lasers [5], and it is therefore a good candidate for an implementation of winner-take-all networks. Moreover, the dynamics is derivable from a scalar (energy) function $V(d_k)$ of the cell activities:

$$V(d_k) = -\tfrac{1}{2}D + \tfrac{1}{2}D^2 - \tfrac{1}{4}\sum_k d_k^4 \quad (2)$$

by applying the gradient

$$\dot{d}_k(t) = -\nabla_{d_k} V.$$

As was shown elsewhere [2, 4, 6], this network has pattern recognition abilities if the synaptic couplings are suitably determined. The network and its dynamics were tested in detail in ref. [7] on a face recognition problem. A short review, however, will be given here to set the stage for the following discussion. For details, the interested reader may refer to refs. [2, 4, 6].

Suppose we have normalized prototype patterns v_l, $l = 1, \ldots, L$, that are to be recognized by the network. Then we use $K = L$ grandmother cells and either store pattern v_l in the respective synaptic filter A_k of the corresponding cell k (in the special case of orthogonal patterns) or we store the adjoint pattern v_l^+ in A_k (in the general case of non-orthogonal patterns). The adjoint vectors are defined as

$$v_l^+ = \sum_{l'} C_{l,l'} v_{l'} \tag{3}$$

with $C_{l,l'}$ being the inverse correlation matrix between patterns:

$$C_{l,l'} = \left(\sum_{i=1}^{N} v_{li} v_{l'i} \right)^{-1}. \tag{4}$$

The synaptic connections now act as filters decomposing arbitrary input patterns q which generally consist of superpositions of the known patterns v_l plus some noise n

$$q = \sum_l \alpha_l v_l + n.$$

They achieve this by translating the strength of any known pattern α_l into activities $d_k(0)$ of the corresponding grandmother cells k. Note that the noise n lies in a space orthogonal to all known patterns v_l. The highest activity $d_{k'}$, i.e. that of the cell k' responsible for the pattern with largest contribution α_l, is then amplified according to the network dynamics of eq. (1). If the network is equipped with the optional output layer connected to grandmother cells by another filter B_{ki} the patterns v_l can be stored in these filters. In this way, the network allows for associative recall observed at the output layer.

3. A specialization parameter

The purpose of this section is to identify a parameter which allows us to measure the degree of specialization of a certain cell k. Learning will then be derived from a maximization principle for such parameters.

Given M patterns to be learned by K cells with activities d_k, $k = 1, \ldots, K$, every cell may try to specialize on at least one of the patterns. The term specialization means – in the context of this network – the ability of a cell k to win a competition against the other $K - 1$ cells. On the basis of a pattern q this is provided for cell k by the following two criteria:

(a) an advantageous initial preparation $d_k(0) = A_k \cdot q$;

(b) a good position during the competition dynamics, as measured by

$$m_k = \langle d_k(t) \rangle_\tau \tag{5}$$

over transient times τ. Consequently, we claim that

$$s_k|_q = m_k d_k(0)|_q \tag{6}$$

measures the specialization of cell k on pattern q.

A general measure for the specialization state of a cell k (independent of the pattern q) is given by the ensemble average

$$\langle s_k \rangle_q,$$

whereas

$$s = \left\langle \frac{1}{K} \sum_k s_k \right\rangle_q$$

averages the specialization over all the K cells.

Measuring the specialization by this method is only one way of doing it in the context of this network. Alternative measures can be found and we do not state that there exists a unique method. We want to emphasize, however, that a learning rule based on a specialization measure is very

4. An energy functional and the learning dynamics

We now consider how to maximize the specialization of cells under the constraint that the length of vectors A_k should be equal to 1 or at least tend to 1. This constraint is necessary in order to implement a competitive learning rule which gives any cell equal opportunities. The maximization may be achieved by defining a specialization functional and deriving the learning dynamics as the gradient descent from it. We propose the following scalar functional:

$$E(A,q) = -\tfrac{1}{2}\sum_k s_k^2 l_k^2$$
$$= -\tfrac{1}{2}\sum_k m_k^2 d_k^2(0)\bigl(1 - \tfrac{1}{2}\|A_k\|^2\bigr), \qquad (7)$$

where l_k^2 was introduced for the dynamics to tend to normalized A_k vectors.

Note that the functional depends on q and thus results in different "landscapes" E for different q. The idea behind this is that a changing environment is able to modify the learning dynamics, at least its detailed trajectory. The system becomes history-dependent and – at the same time – adaptive.

The learning dynamics derives from this specialization functional as the gradient

$$\dot{A}_{ki} = -\frac{\partial E(A,q)}{\partial A_{ki}} = m_k s_k \bigl[l_k^2 q_i - \tfrac{1}{2} d_k(0) A_{ki} \bigr]. \qquad (8)$$

Under a given input q, the dynamics (8) for the synaptic connections A_{ki} tends to minimize E. One can easily verify that (8) tends to filters of equal length 1. Similar learning rules are studied in many models under the heading Hebbian rules, since the positive term proportional to q_i only needs local information about the cell's state as

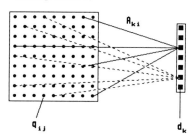

Fig. 2. The two-dimensional arrangement of cells in a sensory surface q_{ij} and a processing layer d_k.

well as the forgetting (or stabilizing) term proportional to A_{ki}. The global factor $m_k s_k$ modulates the learning velocity of individual cells according to their specialization state.

As mentioned before, changing q will result in another dynamics $\dot{A}_{ki}$ due to changes in the energy landscape. Different cells will specialize on different patterns and the average energy $\langle E \rangle_q$ stabilizes only if a suitable adaption of cells to the probability density of inputs $P(q)$ is achieved. This feature is particularly useful if more patterns than cells are present, a case we shall study in our simulations below. In that case, the system organizes itself to classify the presented patterns into (best-match) classes.

5. Simulations

For simulation purposes we have chosen a typical classification situation (see fig. 2): An arrangement of two-dimensional sensory cells (input units) with $N = 100$ sensors q_{ij} are connected to the processing layer of $K = 20, 16, 8, 4, 2$ grandmother cells k by initially randomized connections. There was no local constraint and the filters of every cell k covered at the beginning the entire input space. Fig. 3a displays the initial state of the filters for $K = 20$ grandmother cells.

A pattern is provided by a high stimulation q_{ij} at site i, j, together with lower stimulations in its

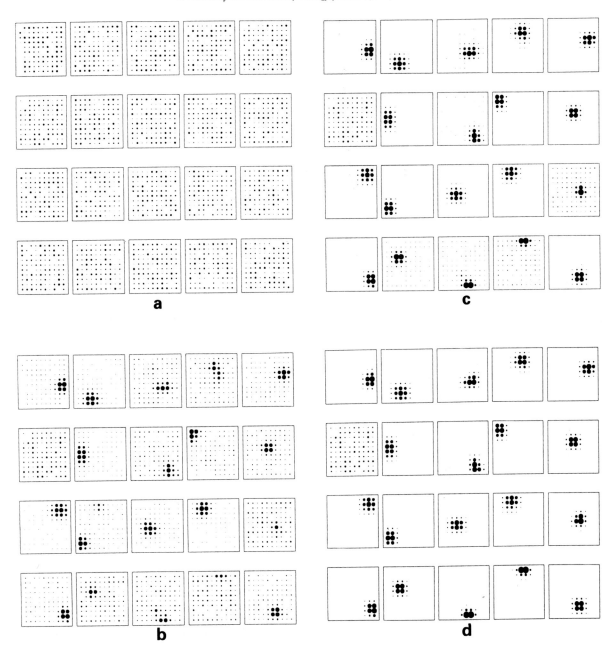

Fig. 3. Development of synaptic connections A_{ki} of grandmother cells $k = 1, \ldots, 20$. Synaptic strength proportional to the radius of black circles. (a) Before learning: All cells cover the whole surface. (b) After $r = 1000$ training steps. (c) After $r = 2000$ training steps. (d) After $r = 4000$ training steps. Cell 6 was not able to adapt to any pattern.

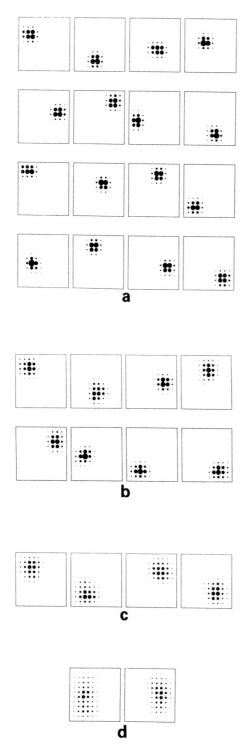

local neighborhood $q_{i'j'}$, $i' = i \pm l$, $j' = j \pm l$, $l = 1, 2, \ldots$. In other words, the patterns are chosen to be slightly correlated so as to allow a global ordering into neighborhoods. Here, the number of patterns, M, is 100 and we have many more patterns than cells for any of the cases studied.

Figs. 3b–3d show the synaptic filters (case $K = 20$) after $r = 1000, 2000, 4000$ training stimulations, respectively. The cells generally develop local sensitivities and respond after training to only a few patterns. One cell remained in its original state, a result which is not surprising since the gradient descent does not guarantee convergence to the global optimum. Quite evidently, after training the network, the cells are able to classify input patterns into different classes, the (maximal) number of which is given a priori by the number of grandmother cells participating in the competition.

Figs. 4a–4d show the results of runs with smaller numbers of grandmother cells. Clearly, the cells have to cover more and more sensory surface each.

A result of these simulations may be seen in the following.

(i) The system is able to classify patterns and thus to learn from noisy input data. Although no pattern of the kind seen in figs. 4a–4d was presented to the system, it nevertheless was able to develop a reasonable solution. The system reaches a stable state characterized by small fluctuations in the redistributed synaptic strength.

(ii) There is no built-in guarantee that the learning process will end up in the optimal solution, i.e. maximal specialization of all cells. Rather, the general result will be a nearly optimal solution.

(iii) A diffusion-like interaction in sensory cells is sufficient to generate local receptive fields, if a suitable competition between cells is implemented. The diffusion is able to correlate patterns which enables the cells in this case to generate neighborhood relations between patterns. The competition provides for learning of superpositions of presented patterns. After relaxation of the system, learning and competition may be turned off.

Fig. 4. Resulting connections in different runs with (a) $K = 16$, (b) $K = 8$, (c) $K = 4$, (d) $K = 2$ cells. The sensory surface is divided in nearly equal portions.

These sorts of patterns are by no means restricted to the simple point-like stimulations trained here. Those were merely chosen for purposes of demonstration. In another series of simulations we have shown what happens in cases $M = K$ and $M < K$ [8].

6. Discussion

This learning scheme has many similarities to Kohonen's learning algorithm and what he calls the formation of feature maps [9, 10]. Both learning algorithms could be termed non-equilibrium learning since the competitive systems are unrelaxed during learning. The adaptive abilities of both systems are comparable. We have hints on bound effects and on a shift in representation space towards regions with smaller probability density in our system, too.

The following differences from Kohonen's method are evident:

(i) The lateral inhibition between cells is uniform in our network. Neither time dependence of inhibition strength nor topological dependence of signals based on some notion of neighborhood in the network is introduced. The sharpening of signals during learning is due to an increase in specialization of cells; the topological mapping of signals is completely missing. Turning off competition results in a scattered map of input patterns, which is certainly useful for some applications.

(ii) The learning process automatically accelerates when specialization of cells proceeds. As a secondary effect of specialization and the forgetting term in eq. (8), the amount of redistributed synaptic strength decreases during training until it reaches minimal values if the optimal adaption of cells to the stationary probability distribution of inputs is reached.

(iii) The parameters which control the overall behavior are the three time constants: competitive activity dynamics, competitive learning dynamics and the training frequency.

In general, a multilayer system of cells is reasonable, as in the case of Linsker's network [11]. Accordingly, we should differentiate between a learning and a maturation state of the network, the latter without time-dependent connections and activity values in a layer. In this way, one layer after the other could process information and adapt to relevant signals in a self-organized manner. Our proposed learning dynamics leads to arbitrary synaptic strengths (with the constraint of being normalized for a cell), in contrast to Linker's connections, which saturate in extremal states. Carpenter and Grossberg [12] and Rumelhart's [13] competitive learning systems differ in the way they stabilize the network after learning.

Since the plasticity of the proposed algorithm is still present after the system has learned to classify input, and since it can readapt anew if the probability distribution of the input changes, experimental evidence [14, 15] concerning adaptive properties of living beings is at least not contradictory to the learning scheme presented here.

Acknowledgement

We wish to thank the "Stiftung Volkswagenwerk" for financial support.

References

[1] J.J. Hopfield, Proc. Natl. Acad. Sci. US 79 (1982) 2554.
[2] H. Haken, in: Computational Systems, Natural and Artificial, Proceedings of the Elmau International Symposion on Synergetics 1987, ed. H. Haken (Springer, Berlin, 1987).
[3] R. Linsker, IEEE Computer (March 1988) 105; presented at the Ninth Annual CNLS Conference on Emergent Computation, Los Alamos, 22–26 May 1989.
[4] H. Haken, Z. Phys. B 70 (1988) 121.
[5] H. Haken, Synergetics, An Introduction, 3rd Ed. (Springer, Berlin, 1983).
[6] H. Haken, in: Neural and Synergetic Computers, Proceedings of the Elmau International Symposion on Synergetics 1988, ed. H. Haken (Springer, Berlin, 1988).
[7] A. Fuchs and H. Haken, Biol. Cybern. 60 (1988) 17, 107.
[8] W. Banzhaf and H. Haken, Neural Networks, in press.

[9] T. Kohonen, Biol. Cybern. 43 (1982) 59.
[10] T. Kohonen, Selforganization and Associative Memory, 2nd. Ed. (Springer, Berlin, 1987).
[11] R. Linsker, Proc. Natl. Acad. Sci. US 83 (1986) 7508, 8390, 8779.
[12] G.A. Carpenter and S. Grossberg, Appl. Opt. 26 (1987) 4919.
[13] D.E. Rumelhart and D. Zipser, in: Parallel Distributed Processing, Vol. 1, eds. D.E. Rumelhart and J.L. McClelland (MIT Press, Cambridge, MA, 1986).
[14] M. Merzenich, presented at the Ninth Annual CNLS Conference on Emergent Computation, Los Alamos, 22–26 May 1989.
[15] W. Levy, Presented at the Ninth Annual CNLS Conference on Emergent Computation, Los Alamos, 22–26 May 1989.

A STOCHASTIC VERSION OF THE DELTA RULE

Stephen José HANSON[1]
Siemens Corporate Research, Princeton, NJ 08540, USA

Self-organizing networks of neuron-like elements naturally lead to high-dimensional, nonlinear parameter spaces which prove difficult to search. Back-propagation is one of the simplest neural network/connectionist models that uses a gradient descent (delta rule) in a high multi-dimensional parameter space. Search in such a space is subject to many difficulties including minima that are locally stable but do not at the same time provide solutions. Search time can also be shown under relatively good conditions to scale poorly. Although gradient descent in error is attractive as a "weak" learning method it also seems to suffer from many search inefficiencies. A principle is proposed about the relation between constrained local noise injections and global search. The delta rule is modified to include synaptic noise in the transmission of information and modification of the connection strength. This stochastic version of the delta rule seems to promote escape from poor locally stable minima, and can improve convergence speed and likelihood.

1. Introduction

Gradient descent methods in high multi-dimensional error spaces have become popular techniques in neural network/connectionist learning models. Boltzmann machines [1] and back-propagation [2] as well as many variations of these models such as those using different error metrics [3, 4], temporal feedback [5] or diverse architectures [6] all require a minimization of error in a very large, nonlinear parameter space. Such nonlinear parameter estimation approaches are known to be plagued by parameterizations that may be easily reached but at the same time seem to provide no solution to the given task – so-called local minima. These kinds of problems can lead to very long search times. In fact, even under good conditions (high learning rates, unbiased starting points) convergence time apparently can be a power function of the training set and exponential in terms of input dimension [7].

This sort of "scaling" problem[#1] could be a consequence of the inefficiency of the parameter search and inability of the method initially to detect a poor starting path. It is possible that noise introduced into the system at some point could be useful in improving both the starting point of the system and the stability of convergence to minima that represent solutions to the given problem. However, noise must be judiciously introduced and subsequently removed from the system as the solution is approached. Simulated annealing [8] is an attractive method for

---

[#1] "NP-Complete" and learning have gotten entangled recently. It is important to keep in mind that scaling over small ranges (say, in networks 10^4 or 10^5) of dimensionality may be biologically relevant and computationally tractable especially in *typical* or *best* case analysis. Remember that the statement that learning in networks is NP-Complete refers to a *worst* case analysis over an unbounded range. This rules out relatively trivial cases like "back-proping" through 10^{11} neurons while reading the encyclopedia. On the hand, notwithstanding NP-Complete results, traveling salemen still get around more than I prefer, and I have no fear of being in even really big grocery stores an exponential amount time. Computational complexity is not the same thing as biological complexity.

---

[1] Also a member of the Cognitive Science Laboratory, 221 Nassau Street, Princeton University, Princeton, NJ 08542, USA.

finding global minima. Nonetheless, the method can approach such global minima quite slowly. This is partly due to the fact that simulated annealing methods as implemented in Boltzmann machines, for example, introduce noise throughout the network independently of the learning process.

2. Modeling synaptic noise

Actual mammalian neural systems involve noise. Responses from the same individual unit (in isolated cortex) due to cyclically repeated identical stimuli will never result in identical bursts [9, 10]. Transmission of excitation through neural networks in living systems is essentially stochastic in nature. The activation function, typically a smooth monotonically increasing function used in neural network/connectionist models must reflect an integration over some time interval of what neurons actually produce: discrete randomly distributed pulses of finite duration. In fact, the typical activation function used in connectionist models must be assumed to be an average over many such intervals, since any particular neuronal pulse train appears quite random (in fact, Poisson; for example see ref. [9]).

This suggests that a particular neural signal in time may be modeled by a *distribution* of synaptic values rather than a single value. Further, this sort of representation provides a natural way to affect the synaptic efficacy in time. In order to introduce noise adaptively, we require that the synaptic modification be a function of a random increment or decrement proportional in size to the present error signal. Consequently, the weight delta or gradient itself becomes a random variable based on prediction performance. Thus, the noise that seems ubiquitous and apparently useless throughout the nervous system can be turned to at least three advantages in that it provides the system with mechanisms for (1) entertaining multiple response hypotheses given a single input (which can be useful for local search), (2) maintaining a coarse prediction history that is local, recent, and cheap,
thus providing more informed credit assignment opportunities and finally (3) revoking parameterizations[#2] that are easy to reach, locally stable, but distant from a solution. Thus, introducing noise locally, under constrained conditions, may provide the system with greater search efficiency.

3. Back-propagation

The delta rule or LMS rule [11] and its generalization [2] are easy to state. Given an output ($\hat{y}_i$) and a target (y_i) an error can be computed which can be accumulated over a sample of input-output tokens,

$$E = \tfrac{1}{2} \sum_s \sum_i (y_{is} - \hat{y}_{is})^2. \qquad (1)$$

A gradient can be found for the error as function of synaptic weights between units, assuming we start with a suitable fan-in function – say dot product ($x_i = \sum_j w_{ij} \hat{y}_j$) – and a reasonable fan-out function – smooth, differentiable and monotone. We wish to minimize these criteria, E, with respect to each weight in the network, so for example,

$$-\frac{\partial E}{\partial w_{ij}} \to 0. \qquad (2)$$

The weight update in the network is proportional to this gradient,

$$w_{ij}(n+1) = \alpha \frac{\partial E}{\partial w_{ij}} + w_{ij}(n). \qquad (3)$$

For back-propagation the weight updates for output distal parts of the network are based on the output gradient, which is recursively passed back through the network by weighting it with a connection strength most recently responsible for the

---

[#2] In the early 60s several methods for random search were introduced dubbed "random creep" which basically were hill-climbing variations which did parameter perturbation and subsequently saved improvements. These methods scale very poorly. Unlike these past methods, notice in the present approach that the noise is introduced within the constraints of a *directed* search process.

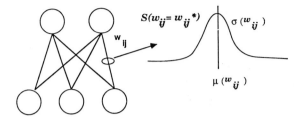

Fig. 1. Portion of a network showing the representation of a single connection in the present algorithm. Each weight is conceived of as a distribution of values with a finite mean and variance. Recognition passes are determined by randomly sampling from the distribution and learning consists of both updating the mean and variance based on prediction error.

error. Errors are precisely and deterministically sent back through the network affecting connections as a function of their prediction performance. Back-propagation is from the class of multivariate, nonparametric, nonlinear regression methods which simultaneously does dynamic feature selection and extraction[#3].

4. Stochastic delta rule

The present algorithm was implemented by assumming a connection strength to be represented as a distribution of weights with a finite mean and variance (see fig. 1). A forward activation or recognition pass consists of randomly sampling a weight from the existing distribution and then calculating the dot product producing an output for that pass,

$$x_i = \sum_j w_{ij}^* \hat{y}_j, \qquad (4)$$

where the sample is found from

$$S(w_{ij} = w_{ij}^*) = \mu_{w_{ij}} + \sigma_{w_{ij}} \phi(w_{ij}; 0, 1). \qquad (5)$$

[#3] Many seem to think that the fact that back-propagation is from a well-known class (albeit rarely explored) of algorithms also means it is a well-known algorithm. In fact, neural network approaches differ in that they *tend* to result from *non-normative* assumptions. In particular, back-propagation is unusual as compared to other pattern recognition approaches in that it is doing supervised learning and feature selection/extraction simultaneously.

Consequently $S(w_{ij} = w_{ij}^*)$ is a random variable constructed from a finite mean $\mu_{w_{ij}}$ and standard deviation $\sigma_{w_{ij}}$ based on a normal random variate (ϕ) with mean zero and standard deviation one. Forward recognition passes are therefore one-to-many mappings, each sampling producing a different weight depending on the mean and standard deviation of the particular connection while the system remains stochastic.

In the present implementation there are actually three separate equations for learning. First, the mean of the weight distribution is modified as a function of the usual gradient based upon the error. Note that the random sample point is retained for this gradient calculation and is used to update the mean of the distribution for that synapse,

$$\mu_{w_{ij}}(n+1) = \alpha\left(-\frac{\partial E}{\partial w_{ij}^*}\right) + \mu_{w_{ij}}(n). \qquad (6)$$

Second, the standard deviation of the weight distribution is modified as a function of the gradient; however, the sign of the gradient is ignored and the update can only increase the variance if an error results. Thus, errors immediately increase the variance of the synapse to which they may be attributed,

$$\sigma_{w_{ij}}(n+1) = \beta\left|-\frac{\partial E}{\partial w_{ij}^*}\right| + \sigma_{w_{ij}}(n). \qquad (7)$$

A third and final learning rule determines the decay of the variance of synapses in the network,

$$\sigma_{w_{ij}}(n+1) = \zeta \sigma_{w_{ij}}(n), \quad \zeta < 1. \qquad (8)$$

As the system evolves for ζ less than one, the last equation of this set guarantees that the variances of all synapses approach zero and that the system itself becomes deterministic prior to solution. For small ζ the system evolves very rapidly to being deterministic, while larger ζs allow the system to revisit chaotic states as needed during convergence. A simpler implementation of this algorithm involves just the gradient itself as a random variable (hence, the name "stochastic delta rule"); however, this approach confounds the growth in variance of the weight distribution with the decay and makes parametric studies more complicated to implement.

There are several properties of this algorithm that we have touched on implicitly in the above discussion that are worth making explicit:

(1) Adaptive noise injection early in learning is tantamount to starting in parallel many different networks and selecting one that reduces error the quickest, thus primarily dissociating initial random starting point from eventual search path.

(2) Noise injections are punctate and accumulate to parts of the network primarily responsible for poor predictions. This allows the network to maintain a cheap prediction history concerning the consequences of various response hypotheses.

(3) Given enough time in local minima (this is determined by both the gradient and ζ) a single synapse can accumulate enough noise to perpetuate variance increase throughout the entire network thus inducing a "chaotic episode".

(4) On average the network will follow the gradient as in a "drunkard's walk". This allows synapses of the network with greater certainty (lower variance) to follow the gradient exactly while synapses of less certainty to explore other parts of weight space.

(5) Finally, punctate (as opposed to global) synaptic noise injection introduces noise indirectly into the unit activation, output predictions and thus, errors, and on average implements a local, learning-dependent, simulated annealing process[#4].

[#4] These last two points are actually plausible assertions and it is possible that many of the beneficial effects seen in section 5 could be due to averaging effects from considering the algorithm as a distribution of networks in parallel that are smoothing ravines and gullys bounded by plateaus, rather than due to a sampling process that might "fortuituously jump" out of a local minimum. Experiments are being done in order to test these possible accounts. Although it is possible both sorts of effects are responsible for speed improvements. I thank Geoff Hinton for pointing this out to me.

5. Experiments

Parity. We have tested this algorithm in a number of domains and the following two are representative of its performance in a typical Boolean counting problem ("parity predicate") and a continuous problem possessing natural boundary variance and classification complexity (small sample, unconnected speech recognition where input and output are both continuous). In fig. 2 we show typical results using a parity predicate in which the input of the network is the number of bits to count and the output is a single bit signaling the even ("1") or odd ("0") determination. Shown are paired runs with random starting points from standard gradient descent back-propagation runs (solid lines) and from the stochastic delta rule runs (dashed lines).

As the input dimension (n) increases the sample size increases exponentially (2^n); so for example, for 2-bit parity there are 4 sample points (which take several minutes to run for 500 sweeps) while with 6-bit parity there are 65 (which take several days for 10k sweeps on a CONVEX C1 computer). All tests were run with the minimal number of hidden units to solve the problem (n-bit parity requires n hidden units), high learning rate ($\eta \approx$ 0.9–2.0) and slow variance decay ($\zeta \approx$ 0.9–0.999). The number of sweeps through the set increases with input dimension and the number of test cases are fewer with higher parity (10 @ 2-bit, 6 @ 3-bit and 2 @ 4-bit and 6-bit). The size of these networks ranges from 9 to 49 weights. In the 6-bit case the learning rate was lowered (0.4) to examine interaction of variance parameters on learning rate. Note that the main effect in small parity cases is to produce a good random starting point which seems uniformly to guarantee as well as speed up convergence. In the 4-bit and 6-bit cases, noise regeneration late in convergence is apparently associated with local minima. Examination of the paired back-propagation run seems to indicate that chaotic episodes co-occur with very long resolution phases of learning and are followed relatively soon by attaining global minimum.

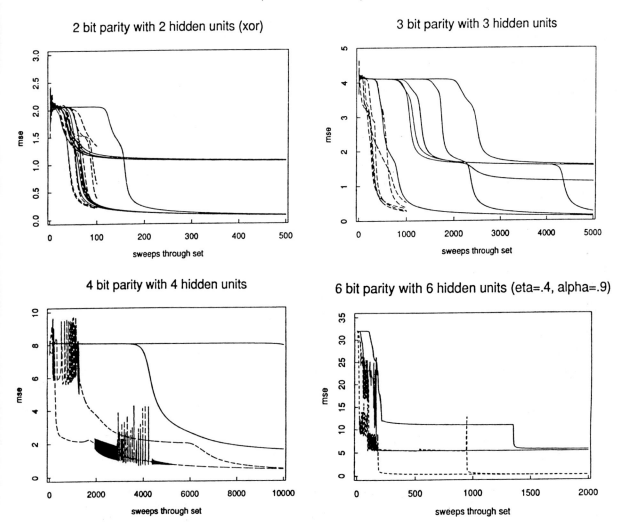

Fig. 2. Learning curves from four different cases of the parity predicate with back-propagation (solid lines) paired on random starting value with stochastic delta rule (dashed line). Note that in the xor (2-bit parity case) only 3 of the 10 runs of back-propagation converged with the 500 sweeps, while all but one of the stochastic delta rule runs converged within 100 sweeps. In the 3-bit parity case 3 of the 6 runs converged within 5000 sweeps, however all 6 runs of the stochastic delta rule converged within 1000 sweeps. In the 4-bit parity case, neither back-propagation run converged in 10 000 sweeps while noise is actually regenerated during late phases of convergence in one of the stochastic delta rule runs soon followed by attaining a global minimum. Finally, shown in the 6-bit parity run with a lowered learning rate are two back-propagation runs that do not converge in 2000 sweeps. While the stochastic delta rule runs both converge within 1000 sweeps, one producing a single chaotic episode near sweep 995, which apparently "shakes" it loose and convergence soon follows on sweep 998.

Speech recognition. The example involves a number of features different from the previous example. It is a continuous problem in which a larger network was employed ($\approx$ 200 weights) due to a larger input and output dimension. Spoken digits ("0–9") taken from the same speaker (D.J. Burr kindly supplied the data) were used to establish input to the network. Five samples of 10 digits were taken from the speaker for training and testing. The first 12 Cepstral coefficients were retained for the continuous input to the network and 10 output bits were assigned to the digits 0–9.

In this case, it is not possible to a priori assign the minimal number of hidden units for problem

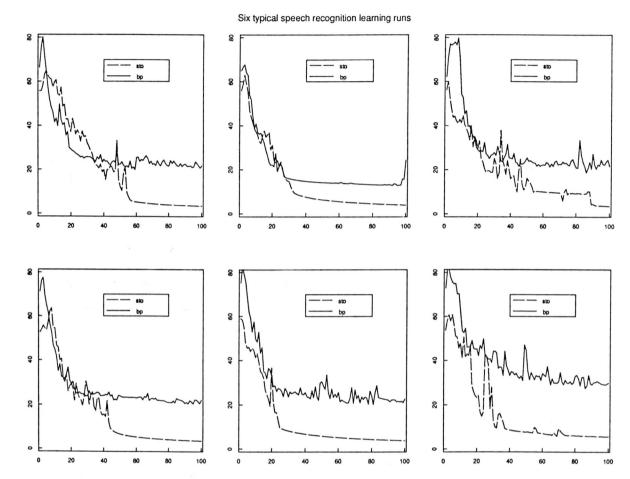

Fig. 3. Learning curves from six 50-digit speech recognition runs with back-propagation (solid lines) and paired stochastic delta rule runs (dashed line). A network with 12 inputs, 8 hidden units and 10 outputs ($\approx$ 200 weights) was trained on 50 tokens (5 samples each of 10 spoken digits) of unconnected single speaker speech. Note that all stochastic delta rule runs converge within 100 sweeps, none of the back-propagation runs converged within the criteria sweeps.

solution. However, exactly 4 hidden units will implement exactly 10 separate bounded (at least on two sides) regions in feature space. Smoother convex boundaries would require many more hidden units. In fact, prior experience with these data suggested that 10 hidden units was adequate for solution, consequently in these runs only 8 were used. The network was trained on all 50 digits for 100 sweeps. Shown in fig. 3 are six typical paired random starting point learning runs for the standard gradient descent back-propagation (solid line) and the stochastic delta rule (dashed line). Note in every case the stochastic delta rule converged to a solution within the 100 sweeps while back-propa-

gation never converged. A typical solution resulting from back-propagation and the stochastic delta rule are shown in fig. 4. These cluster dendrograms[5] represent the response of the network at the hidden layer after learning to all 50 digits. Notice that the back-propagation network after

---

[5] Clustering in networks was first suggested by me, and used by Sejnowski and Rosenberg in analyzing NETalk's hidden layer projections of the input features. This can be useful since the activations are not necessarily simple linear projections of input features. It would be especially useful to weight hidden unit activations with their output weights as "contributions" which scales the projections for output space. This was suggested first by Charlie Rosenberg. See also ref. [12] concerning more details of the use of this method.

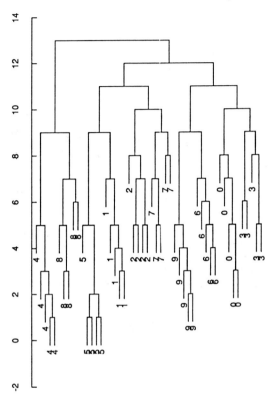

Fig. 4. Cluster dendrograms (rooted tree showing merge history of speech tokens based on Euclidean distance metric and centroid group membership rule) showing the representation of the 50 digits after learning for the back-propagation network (leftmost dendrogram) and the stochastic delta rule network (rightmost dendrogram). Note that the back-propagation network did not reach acceptable error levels and as indicated by the dendrograms could not separate first letter "ss" digits ("zero", "six", and "seven"), while the stochastic delta rule network correctly separates all tokens of all ten digits.

learning does not separate the 10 classes of digits while the stochastic delta rule representation causes the same digits presented in input to be close while different digits are well separated in hidden unit space.

Presently it is not known whether real synaptic modification is concomitant with a change in noise from single unit recording[#6], although algorithms like the stochastic delta rule and Boltzmann machine have provided some impetus for expecting beneficial effects of dynamic noise introduction. In the present case, the novel aspect of the algorithm is apparently due to local, transient noise injections on search. Further experiments and theoretical work must be done in order to confirm the specific desirable aspects of noise introduction. At a more general level the injection of noise into algorithms like back-propagation increases their biological plausibility, as the presence of noise can mitigate the reliance on architectural symmetry and precise target values introducing a natural way to utilize reinforcement signals [13].

Acknowledgements

I thank Carl Olson for conversations concerning possible neural transmission mechanisms, Mark Gluck for conversations about the delta rule, and

---

[#6] Indeed, neuroscientists tend to try to minimize the noise in the measurement by averaging many such observations.

the Bellcore connectionist group for support and interesting discussions.

References

[1] D. Ackely, G. Hinton and T. Sejnowski, A learning algorithm for Boltzmann machines, Cognitive Sci. 9 (1985) 147–169.

[2] D.E. Rumelhart, G.E. Hinton and R. Williams, Learning internal representations by error propagation. Nature 323 (1986) 533–536.

[3] S.J. Hanson and D.J. Burr, Minkowski back-propagation: learning in connectionist models with non-Euclidean error signals, in: Neural Information Processing Systems, ed. D. Anderson (AIP, New York, 1988).

[4] E. Baum and F. Wilczek, Supervised learning of probability distributions by neural networks, in: Neural Information Processing Systems, ed. D. Anderson (AIP, New York, 1988).

[5] M. Jordan, Sequential behavior, ICS Technical Report 8604 (May 1986).

[6] F. Pineda, Dynamics and architecture in neural computation, J. Complexity, in press.

[7] G. Tesauro and R. Janssens, Scaling relationships in back-propagation learning: dependence or predicate order, Technical Report, CCSR-88-1, Center for Complex Systems Research (1988).

[8] S. Kirkpartrick, C.D. Gelatt and M. Veechi, Optimization by simulated annealing, Science 220 (1983) 671–680.

[9] G.J. Tomko and D.R. Crapper, Neural variability: Nonstationary response to identical visual stimuli, Brain Research 79 (1974) 405–418.

[10] B.D. Burns, The Uncertain Nervous System (Arnold, Paris, 1968).

[11] G. Widrow and M.E. Hoff, Adaptive switching circuits. Institute of Radio Engineers, Western Electronic Show and Convention, Convention Record, Part 4 (1960) pp. 96–104.

[12] S.J. Hanson and D.J. Burr, What connectionist models learn; Learning and representation in connectionist networks, Behavioral Brain Sci., in press.

[13] R. Williams, Towards a theory of reinforcement learning connectionist systems, Technical Report NU-CCS-88-3, Northeastern University (1988).

SYNCHRONOUS OR ASYNCHRONOUS PARALLEL DYNAMICS. WHICH IS MORE EFFICIENT?

I. KANTER

*Joseph Henry Laboratories of Physics, Jadwin Hall, Princeton University, Princeton, NJ 08544, USA
and Department of Physics, Bar-Ilan University, Ramat-Gan 52100, Israel*

The convergence time of parallel dynamics is compared analytically with that of random sequential dynamics. In parallel dynamics all elements are updated simultaneously, whereas in random sequential dynamics the elements are updated only on average at the same speed. One-dimensional and infinite-range interactions are studied. The relevance of the results to neural networks and Monte Carlo simulations is discussed.

1. Introduction

A few dynamical systems can be solved exactly under parallel or random sequential dynamics [1, 2]. In parallel dynamics all system elements are updated simultaneously according to a given set of rules. In random sequential dynamics all of the elements are updated only on average at the same speed. In random sequential dynamics, there is no global clock which synchronizes the dynamics.

The convergence time to the fixed point or to the solution plays an important role in neural networks and Monte Carlo simulations. An important question is whether the convergence time depends on the type of dynamics and which of them converges faster. It is also interesting to verify whether the efficiency of different types of dynamical rules is a function of the dimensionality of the system and if it depends on the type of the interactions.

In order to examine these questions we first solve the dynamics of a few 1D stochastic systems for any initial configuration and temperature, either under parallel dynamics or under random sequential dynamics. These solutions enable us to make an analytical comparison between the two dynamics. The solution is later extended to the mean-field limit, where each pair of elements interacts. The applications of the results to neural networks and Monte Carlo simulations are also discussed.

2. Solvable 1D systems

The 1D symmetric equal-strength Ising system provides the simplest example of 1D probabilistic cellular automata whose parallel dynamics can be solved analytically. Each Ising spin, S_i, is a discrete variable which can take the value ± 1 for $i = 1, 2, \ldots, N$. This system, which obeys the detailed balance, is characterized by two conditional probabilities $P(S_{i,t+1}|S_{i-1,t}, S_{i+1,t})$,

$$x_2 = P(-1|1,1) = P(1|-1,-1) \tag{1}$$

and

$$x_0 = \tfrac{1}{2} = P(\pm 1|1,-1) = P(\pm 1|-1,1), \tag{2}$$

where $1 - x_2 = P(1|1,1) = P(-1|-1,-1)$. It is convenient for the clarity of the following discussion to define the following quantities. The symbol $N_\pm(t)$ stands for the fraction of up/down spins at time t and the symbol $N_\pm^k(t)$ stands for the fraction of up/down spins at time t with a local field k. The local field on the ith spin, for instance, is defined as $S_{i-1} + S_{i+1}$. The local field in this particular system could be -2, 0 or 2.

The evolution equation for the fraction of up spins at time $t+1$ as a function of the state of the system at time t under parallel dynamics is given by

$$N_+(t+1) = \left[N_+(t) - N_+^0(t) - N_+^{-2}(t)\right](1-x_2)$$
$$+ \tfrac{1}{2}N_+^0(t) + \tfrac{1}{2}N_-^0(t) + N_+^{-2}(t)x_2$$
$$+ N_-^{-2}(t)(1-x_2)$$
$$+ \left[1 - N_+(t) - N_-^0(t)\right.$$
$$\left. - N_-^{-2}(t)\right]x_2, \qquad (3)$$

from which one can find

$$N_+(t+1) = x_2 + N_+(t)(1-2x_2)$$
$$+ \tfrac{1}{2}(1-2x_2)\left[N_-^0(t) - N_+^0(t)\right.$$
$$\left. + 2N_-^{-2}(t) - 2N_+^{-2}(t)\right]. \qquad (4)$$

It is easy to verify that the square brackets in eq. (4) vanish for any spin configuration (with periodic boundary conditions) and therefore

$$N_+(t+1) = x_2 + N_+(t)(1-2x_2). \qquad (5)$$

Solution of the recursion equation (5) for any initial configuration gives

$$N_+(t) = \tfrac{1}{2}\left[1 - (1-2x_2)^t\right] + N_+(0)(1-2x_2)^t \qquad (6)$$

and $N_+(\infty) = \tfrac{1}{2}$. This result proves the lack of a finite-temperature phase transition to a ferromagnetic state.

One can easily generalize this result to any uniform 1D symmetric or asymmetric system with any local rule which depends only on the two nearest-neighbor spins. By asymmetry we mean that the influence of spin i on spin j is not necessarily equal to the influence of spin j on spin i. For example, in the absence of reflection symmetry where an up spin is updated according to the right spin and a down spin is updated according to the left spin one can easily verify that

$$N_+(t) = \tfrac{1}{2}\left[1 - (1-2x_1)^t\right] + N_+(0)(1-2x_1)^t, \qquad (7)$$

where

$$x_1 = P(-1|\pm 1, 1) = P(1|-1, \pm 1) \qquad (8)$$

and

$$1 - x_1 = P(1|\pm 1, 1) = P(-1|-1, \pm 1). \qquad (9)$$

Another solvable example is the 1D Ising system with alternating interactions J_1 and J_2. The evolution equation for this system where the transition probabilities obey detailed balance is

$$N_+(t) = x_{1+2} + N_+(t)(1-2x_{1+2})$$
$$+ \left(\tfrac{1}{2} - x_{1+2}\right)\left[2N_-^{-1-2}(t) - 2N_+^{-1-2}(t)\right.$$
$$+ N_-^{1-2}(t) - N_+^{1-2}(t)$$
$$\left. + N_-^{2-1}(t) - N_+^{2-1}\right]$$
$$+ \left(\tfrac{1}{2} - x_{1-2}\right)\left[N_-^{2-1}(t) - N_+^{2-1}(t)\right.$$
$$\left. + N_+^{1-2}(t) - N_-^{1-2}(t)\right]. \qquad (10)$$

Here the transition probability x_{r+s} stands for the probability of flipping from a state where the local field is $J_r + J_s$. The symbol N_+^{-r-s}, for instance, stands for the fraction of up spins with a local field equal to $-J_r - J_s$. The terms in the first square brackets vanish for any spin configuration as for the uniform chain. The terms in the second square brackets vanish, as on the right-hand side of a spin of type N_+^{2-1} should appear a spin of type N_-^{2-1} or N_+^{1-2} among N_-^{2-1}, N_+^{1-2}, N_+^{2-1} and N_-^{1-2}. On the right-hand side of a spin of type N_+^{1-2} should appear a spin of type N_+^{2-1} or N_-^{1-2}. Hence, the master equation (10) has the form

$$N_+(t+1) = x_{1+2} + N_+(t)(1-2x_{1+2}) \qquad (11)$$

and the solution of this recursion equation is of the form (6) but with x_{1+2} instead of x_2.

The last example we would like to present is 1D equal-strength nearest-neighbor Ising system with an additional self-coupling term. In this system, each spin is updated according to its two nearest-neighbor spins and the spin itself in the previous time step. For simplicity, we assume that the conditional probabilities obey reflection symmetry, namely $P(k|lmn) = P(k|nml)$. In this case the master equation is given by

$$N_+(t+1) = x_2 + N_+(t)(1 - 2x_2)$$
$$+ \Delta I(t)(2x_0 - x_2 - x_{-2})$$
$$+ (x_2 - x_0)[N_+^0(t) - N_-^0(t)$$
$$+ 2N_+^{-2}(t) - 2N_-^{-2}(t)], \quad (12)$$

where the conditional probabilities $P(S_{i,t+1}|S_{i-1,t}, S_{i,t}, S_{i+1,t})$ are given by $x_2 = P(-1|1,1,1) = P(1|-1,-1,-1)$, $x_{-2} = P(1|1,-1,1) = P(-1|-1,1,-1)$ and $x_0 = P(-1|1,1,-1) = P(1|1,-1,-1)$. ($x_0$ is not necessarily equal to $\tfrac{1}{2}$.) The symbol k in the terms $N_\pm^k(t)$ and x_k is the induced local field from the nearest-neighbor spins, and $\Delta I(t) \equiv N_+^{-2}(t) - N_-^{-2}(t)$ is the difference in the fraction of isolated up and down spins. Here again it is clear that the square brackets vanish for any spin configuration. For the general case we do not know how to write explicitly the time evolution master equation for $\Delta I(t)$. However, for the subspace of the conditional probabilities

$$2x_0 - x_2 - x_{-2} = 0, \quad (13)$$

the time evolution can be solved exactly as for the previous cases. Condition (13) is exactly the same condition for which the Glauber dynamics (a master equation for the probability) is solvable [1]. As mentioned above, the time evolution for $\Delta I(t)$ is known only for special configurations. Nevertheless, one can conclude two interesting results from eq. (12). In general, the magnetization ($\equiv N^{-1}\Sigma S_i$) is *not a monotonic* function of time and of the noise, x_i, even in the presence of only ferromagnetic (attractive) interactions. In order to see these behaviors explicitly let us concentrate now on a simple example. The system is defined by

$$x_0 = x_2 = 1 - x_{-2} \quad (14)$$

and hence eq. (12) is reduced to

$$N_+(t+1) = x_2 + [N_+(t) - \Delta I(t)](1 - 2x_2) \quad (15)$$

and

$$dN_+(t+1)/dx_2 = -m(t) + 2\Delta I(t). \quad (16)$$

This model is known as the majority votes model, where the spin follows the majority among the spin itself and its two nearest-neighbor spins with probability $1 - x_2$. For a configuration which alternates with two up spins and one down spin, $m(0) = -\Delta I(0) = \tfrac{1}{3}$ and

$$m(1) - m(0) = \tfrac{2}{3}(1 - 3x_2). \quad (17)$$

This quantity is greater than zero for $x_2 < \tfrac{1}{3}$ although the fixed point is $m = 0$ [3]. Hence, the magnetization is not necessarily a monotonic function of time. In order to examine the monotonic behavior as a function of the noise, let us concentrate on the following configuration: one half of the system alternates with two down spins and one up spin, and the second half is constructed from one up cluster and one down cluster with a net magnetization equal to m_1. For this configuration $m(0) = -\tfrac{1}{6} + m_1$, $N_+(0) = \tfrac{1}{6} + \tfrac{1}{4}(1 + 2m_1)$ and $\Delta I = \tfrac{1}{6}$. It is now easy to verify that for $x_2 < \tfrac{1}{3}(1 - 2m_1)$, $m(1) < m(0)$ but

$$dN_+(1)/dx_2 > 0. \quad (18)$$

Hence, $m(1)$ is an *increasing* function of x_2 although $m(1) < m(0)$.

In a similar way one can find the master equation for 1D nearest-neighbor Ising system with additional self-coupling terms but in the absence of reflection symmetry.

Generalization of the abovementioned results to more complicated systems is not simple because the time evolution of the system depends on the correlations among the spins and not only on the average magnetization. Nevertheless, in the following we use these ideas to compare analytically the convergence time in parallel dynamics with random sequential dynamics.

3. Parallel dynamics versus random sequential dynamics

We first derive the average magnetization for the random sequential dynamics (Glauber dynamics) [1] from the result for parallel dynamics. Afterwards, the comparison of the convergence time between sequential and parallel dynamics is discussed.

From eq. (6) one can verify that the magnetization for a 1D uniform chain at time t is

$$m_p(t) = m_p(0)(1 - 2x_2)^t \tag{19}$$

and

$$\Delta m(t) \equiv N[m_p(t+1) - m_p(t)]$$
$$= -2Nx_2 m_p(t), \tag{20}$$

where N is the size of the system and $m_p(t)$ stands for the magnetization at time t under parallel dynamics. The physical meaning of $\Delta m(t)$ is the actual slope of the magnetization at $m(t)$ where one averages over all the spins in the system. Therefore, one can write the following equation

$$\frac{dm_s}{dt} = \frac{\Delta m(t)}{N} = -2x_2 m_s, \tag{21}$$

where m_s stands for the magnetization under random sequential dynamics and we divide $\Delta m(t)$ by N in order to rescale the time scale between parallel and sequential dynamics. The solution of eq. (21) gives

$$m_s(t) = m_s(0) e^{-2x_2 t}, \tag{22}$$

which is exactly the result for the average magnetization of uniform 1D system under Glauber dynamics [1].

From eqs. (19) and (22) one can easily verify that for any initial configuration and for any time

$$m_s(t) > m_p(t). \tag{23}$$

This result can be well understood for the reason that $dm(t)/dt$ is a decreasing function of time. Therefore, in one-step for all the spins the magnetization decreases faster compared to the change in small steps, which is the case of random sequential dynamics.

It is important to stress that the decay of the average magnetization under sequential dynamics with a fixed order of updating could be faster than for parallel dynamics. For instance, a uniform chain, eqs. (1), (2), where we assume that $m(0) = 1$. It is obvious that after one spin is updated, its neighbors have a probability $\frac{1}{2}$ ($> 1 - x_2$) to flip. Therefore, the average magnetization after one step per spin is smaller than $m_p(1)$, which is equal to $1 - x_2$ (see eq. (5)).

The same result that the decay of the average magnetization under parallel dynamics is faster than under random sequential dynamics holds for all the previous solvable 1D systems.

4. Infinite-range systems

Another limit which can be treated analytically is the mean-field limit, where the interaction between each pair of spins is $J = 1/N$, where N is the size of the system. The dynamical evolution of the system is governed by the transition probabilities (which obey detailed balance for sequential dynamics)

$$P(S_i) = \exp(\beta h_i S_i)/2\cosh(\beta h_i), \quad (24)$$

where $\beta = 1/T$, T is the temperature of the system and $h_i = N^{-1}\Sigma_{i \neq j} S_j$. One can verify from eq. (24) that the average magnetization under parallel dynamics is

$$m_p(t+1) = \tanh[\beta m_p(t)]. \quad (25)$$

It is also easy to verify that for $T > T_c = 1$

$$\Delta m(t) > \Delta m(t+1) \quad (26)$$

and for $T < T_c$

$$|\Delta m(t)| > |\Delta m(t+1)| \quad \text{for } m > m^*,$$
$$|\Delta m(t)| < |\Delta m(t+1)| \quad \text{for } m < m^*, \quad (27)$$

where

$$m^* = T\log\left[\sqrt{\beta} + \sqrt{\beta - 1}\right] \quad (28)$$

and $\Delta m(t)$ is defined in eq. (20).

Here again $|\Delta m(t)|$ is proportional to the absolute average number of spins which flip up in the $(t+1)$th step. Hence, the actual change in the magnetization under random sequential dynamics is given by

$$\frac{dm_s}{dt} = \frac{\Delta m(t)}{N} = -m_s + \tanh(\beta m_s), \quad (29)$$

which is the result of Glauber dynamics for the mean-field case.

The convergence time to a fixed point of parallel dynamics versus sequential dynamics depends in general on $m_p(t)$. If the average magnetization is a monotonic function of time in both dynamics then in the region

$$|\Delta m_p(t)| > |\Delta m_p(t+1)| \quad (30)$$

parallel dynamics converges faster than random sequential dynamics, for the same initial magnetization. For the region

$$|\Delta m_p(t)| < |\Delta m_p(t+1)| \quad (31)$$

sequential dynamics converges faster than parallel dynamics. (A similar criterion one can find for $d^2m_s(t)/d^2t$.) Hence, for any initial magnetization above the transition temperature and for $m > m^*$ below the transition temperature, parallel dynamics converges faster than random sequential dynamics. For the case $T < T_c$ and $m(0) < m^*$, $m_s(t) > m_p(t)$ at least in the region where $m_s(t)$, $m_p(t) < m^*$. For $m(t) > m^*$ it was found in the numerical solutions of eqs. (25) and (29) that in general the relation between m_s and m_p is a function of the initial magnetization and the number of steps. Nevertheless, for $m(0) \ll m^*$ it was found that m_p becomes equal to m_s for m much greater than m^*. For instance for $m(0) = 0.02$ and $T = 0.5$, $m_s \simeq m_p$ at $t = 7$ and $m = 0.93$, where $m^* = 0.44$. For $t > 7$, $m_p > m_s$ and $m(\infty) = 0.957$. However, the relative difference between m_s and m_p for $t < 7$ could be equal to $\frac{1}{2}$, where for $t > 7$ the relative difference is less than 10^{-3}. Hence, for low enough initial magnetization $m_s > m_p$, except for the long tail of the convergence time. Nevertheless, it is clear that a combination of random sequential dynamics for $m < m^*$ and parallel dynamics for $m > m^*$ is a more efficient dynamics.

The result for the mean-field case also holds for the Hopfield model with a finite number of memories [4, 5]. In this case the interactions represent the synaptic efficiencies between pairs of neurons, while the spins $s_i = \pm 1$ represent the state (active and passive) of neurons. The interaction between each pair (ij) of spins is given explicitly by

$$\frac{1}{pN}\sum_{l=1}^{p}\tau_i^l\tau_j^l, \quad (32)$$

where τ_i^l is a random number which can take the values ± 1 and p is the number of the embedded memories. This result is due to the fact that the noise coming from the patterns which have a microscopic overlap with the actual configuration of the system can be negligible in this limit and ferromagnetic mean-field equations are correct.

In the case where $p \propto N$, eq. (25) can be replaced by the relation

$$m_p(t+1) = \frac{1}{\sqrt{\pi}} \int_{-\infty}^{+\infty} e^{-x^2} \times \tanh[\beta(Ax + m_p(t))], \quad (33)$$

where Ax stands for the Gaussian noise coming from the patterns which have a microscopic overlap with the actual configuration of the system. In general eq. (33) is only an approximation because the noise as a function of time deviates from a simple Gaussian and the width of the noise could also be a function of time [6, 7]. Nevertheless, eq. (33) indicates that even in this limit, for a large enough overlap, parallel dynamics converges faster than random sequential dynamics. Furthermore, for low enough temperatures the noise decreases the range of magnetizations under which parallel dynamics converges faster (in comparison to the same system without noise). This result was confirmed in numerical solutions of eq. (33).

5. Diluted infinite-range asymmetric systems

The case of a highly diluted infinite-range asymmetric Hopfield model [8] can also be easily treated analytically. The system consists of N Ising spins where the strength of the interactions depend on the p stored patterns and is given by

$$J_{ij} = C_{ij} \sum_{\mu=1}^{p} \tau_i^\mu \tau_j^\mu.$$

The symbol J_{ij} stands for an asymmetric interaction from i to j. The p stored patterns $\{\tau_i^\mu\}$ ($\mu = 1, \ldots, p$ and $i = 1, \ldots, N$) are certain configurations of the network which were fixed by the learning process. They assume to be random with equal probability for $\tau_i^\mu = \pm 1$. The C_{ij} are random parameters which obey the following constraints:

$$\sum_{j \neq i} C_{ij} = k, \quad i = 1, 2, \ldots, N. \quad (34)$$

In this system each spin receives k fixed inputs. It is important to stress that C_{ij} is independent of C_{ji}. For simplicity, we will concentrate here only on the case $p = 1$ (a ferromagnetic system) and $k = 3$. Generalization of the results to any distribution of inputs is straightforward. The average magnetization for this system at time $t + 1$ is given by

$$m(t+1) = \tfrac{1}{4}[\tanh(3\beta) - 3\tanh(\beta)]m^3(t) + \tfrac{3}{4}[\tanh(3\beta) + \tanh(\beta)]m(t), \quad (35)$$

where $\beta = 1/T$. It is easy to verify that the transition temperature, T_c, is fixed by

$$0 = \tfrac{4}{3} - \tanh(3\beta) - \tanh(\beta) \quad (36)$$

and the fixed magnetization as a function of the temperature is given by

$$(m^*)^2 = \frac{3[\tanh(3\beta) + \tanh(\beta)] - \tfrac{4}{3}}{3\tanh(\beta) - \tanh(3\beta)}. \quad (37)$$

The convergence time, as we explained above, is related to the quantity

$$\frac{d \Delta m(t)}{dm(t)} = \tfrac{3}{4}\{\tfrac{4}{3} - \tanh(3\beta) - \tanh(\beta) + [3\tanh(\beta) - \tanh(3\beta)]m^2(t)\}. \quad (38)$$

This quantity is equal to zero at m_0 given by

$$m_0 = m^*/\sqrt{3}. \quad (39)$$

The behavior of the convergence time of this system is very similar to the fully connected one. For $T > T_c$ and for $T < T_c$ and $m > m_0$, parallel dynamics converges faster than random sequential dynamics, where for $T < T_c$ and $m < m_0$ random sequential dynamics converges faster.

6. Conclusion

In conclusion, we solved exactly the dynamics of a few 1D stochastic systems under parallel and random sequential dynamics. In all the discussed systems, parallel dynamics converges faster than random sequential dynamics for any temperature and for any initial configuration. In the mean-field limit the behavior is different. For each system and a given temperature there is a critical initial magnetization above which parallel dynamics converges faster than random sequential dynamics. Nevertheless, for small enough initial magnetization (or overlap with a pattern), random sequential dynamics converges faster to some fixed final magnetization. The presence of noise in such mean-field systems helps to enlarge the range of magnetization under which random sequential dynamics converges faster, beside the advantage of the noise to escape from a local minimum near the edge of the basin of attraction.

These results indicate that for small initial magnetization it is better to use random sequential dynamics in Monte Carlo simulations, but for large magnetization, parallel dynamics is more efficient. The critical magnetization above which parallel dynamics is more efficient depends on the details of the model.

Biological neural networks do not seem to have a clock such that all the neurons are updated at the same time. One way to understand the lack of a clock in neural networks is that a clock is a *global* constraint which is difficult to implement in these systems. Furthermore, we have shown only that parallel dynamics converges *faster* than random sequential dynamics for some initial conditions. However, it does not indicate that parallel dynamics converges faster than any possible type of updating. Indeed, as we explained in the examined uniform chain, sequential dynamics with a fixed order of updating converges faster than parallel dynamics. Hence, it is possible that neurons are updated under more efficient dynamics than parallel dynamics, even when the dynamical rules are governed by local constraints. There are not enough actual biological data to characterize precisely the dynamics of neural networks and to examine the efficiency of the dynamics. Nevertheless, one can predict dynamical rules which are more efficient than parallel dynamics. For instance, in the Hopfield model, where $p \propto N$, it is clear that, on the average, $h_i S_i$ on a spin which is parallel to the pattern is m and $h_i S_i$ on a spin which is anti-parallel to the pattern is $-m$, where m is the macroscopic overlap. Hence, it is obvious that it is better to update the spins according to their induced local fields $h_i S_i$, which is a *local* constraint. For instance, dynamics where spins with small $h_i S_i$ are updated faster than spins with larger $h_i S_i$ converges faster than parallel dynamics. One can also imagine more efficient dynamical rules which depend on the history of the system or systems with inhomogeneous temperature [9, 10]. In general, the efficiency of the dynamical rules are a function of the system (model) and initial conditions and can also be changed as a function of time. It would be interesting to find such evidence for efficient dynamical rules in neural networks.

Finally, we would like to comment that the dynamics of a finite-dimension system is much more complicated than in one dimension. The dynamical evolution of the system does not depend only on the average magnetization. Nevertheless, for low enough temperatures and for $m(0) = 1$ one can show that for some first few steps per spin $m_s > m_p$. The reason is that the dominant clusters in low enough temperatures are clusters of size 1 and 2. In random sequential dynamics there is a probability $1/N$ to update again an isolated spin or spins which belong to small clusters, before updating the rest of the system. Hence, the magnetization under random sequential dynamics is greater than under parallel dynamics, at least when the dominant clusters are small. The comparison of the convergence time between different types of dynamics for finite-dimensional systems and for any finite temperature is still an open question.

Acknowledgements

I would like to thank P.W. Anderson and D.S. Fisher for numerous stimulating discussions. The work is supported by a Weizmann Fellowship and also in part by the NSF Grant No. 8719523.

References

[1] R.G. Glauber, J. Math. Phys. 4 (1963) 294.
[2] K. Kawasaki, in: Phase Transitions and Critical Phenomena, Vol. 2, eds. C. Domb and M.S. Green (Academic Press, New York, 1972) p. 443.
[3] G. Lawrence, unpublished.
[4] J.J. Hopfield, Proc. Natl. Acad. Sci. US 79 (1982) 2554; 81 (1984) 3088.
[5] W.A. Little, Math. Biosci. 19 (1974) 101.
[6] E. Gardner, B. Derrida and P. Mottishaw, J. Phys. (Paris) 48 (1987) 741.
[7] I. Kanter, Phys. Rev. A 40 (1989) 2611.
[8] B. Derrida, E. Gardner and A. Zippelius, Europhys. Lett. 4 (1987) 167.
[9] I. Kanter, Phys. Rev. Lett. 60 (1988) 1891.
[10] M. Lewenstein and A. Nowak, Phys. Rev. Lett. 62 (1989) 225.

ON THE NATURE OF EXPLANATION: A PDP APPROACH*

Paul M. CHURCHLAND

Department of Philosophy, B-002, UCSD, La Jolla, CA 92093, USA

Neural network models of sensory processing and associative memory provide the resources for a new theory of what *explanatory understanding* consists in. That theory finds the theoretically important factors to reside not at the level of propositions and the relations between them, but at the level of the activation patterns across large populations of neurons. The theory portrays *explanatory understanding*, *perceptual recognition*, and *abductive inference* as being different instances of the same more general sort of cognitive achievement, viz. *prototype activation*. It thus effects a unification of the theories of explanation, perception, and ampliative inference. It also finds systematic unity in the wide diversity of *types* of explanation (causal, functional, mathematical, intentional, reductive, etc.), a chronic problem for theories of explanation in the logico-linguistic tradition. Finally, it is free of the many defects, both logical and psychological, that plague models in that older tradition.

1. Introduction

In contemporary philosophy of science, the notion of explanation figures centrally in most accounts of scientific knowledge and rational belief. Explanation is usually cited, along with prediction, as one of the two principal functions of our factual beliefs. And the rationality of such beliefs is commonly said to be measured, at least in part, by the relative range or quality of the explanations they make possible. If something like this is correct, then it is important for us to try to understand what explanation is, and what distinguishes a good explanation from a poor one.

Several existing accounts attempt to meet this challenge. They will be addressed below. The present paper proposes a new account of the matter – the *prototype activation model* – an account distinguished, for starters, by its being grounded in a novel and unorthodox conception of what cognition consists in. That conception derives from current research in cognitive neurobiology, and from parallel distributed processing (PDP) models of brain function (see Rumelhart et al. [1, 2]; Churchland [3, 4].) These PDP models are noteworthy for many reasons, but first among them in the present context is their almost complete dissociation from the *sentential* or *propositional* conception of what knowledge consists in, and from the *rule-governed inference* conception of human information processing. Those venerable conceptions play a central role in all of the older accounts of explanation, and in orthodox accounts of cognition generally. They will play almost no role in the account to be proposed.

The prototype activation model is focused first and foremost on what it is to have *explanatory understanding* of a problematic thing, event, or state of affairs. The linguistic expression, exchange, or production of such understanding, should there be any, is an entirely secondary matter. We shall approach the topic with the aims of an empirical scientist rather than with the aims of a logician or conceptual analyst. The goal is to

---

*A more comprehensive presentation of this material is scheduled to appear as chap. 10 of P.M. Churchland, A Neurocomputational Perspective: The Nature of Mind and the Structure of Science (MIT Press, Cambridge, MA, 1989).

outline a substantive empirical theory of what explanatory understanding really is, rather than to provide an analysis of the concept of explanation as it currently is or ideally should be used. What concerns us is the nature of the cognitive process that takes place inside the brain of the creature for whom explanatory understanding suddenly dawns, and in whom it is occasionally reactivated.

On the prototype activation (PA) model, a close approximation to this process is the process of *perceptual recognition*, as when one suddenly recognizes an indistinct outline as the face of a close friend, or as when one finally recognizes the faint motion under the hedge as a foraging mouse. On the PA model, essentially the same kind of computational achievement underlies both perceptual recognition and explanatory understanding. The latter is distinguished primarily by being a response to a wider variety of cognitive situations: it is not limited to sensory inputs.

A close connection between perception and explanation is by now a familiar theme in both psychology and philosophy. One's perceptual judgments, and perhaps even one's perceptual experiences themselves, have often been portrayed as the perceiver's best *explanatory* account of the peripheral stimuli (Gregory [6, 7]; Rock [8]). In this tradition the notion of explanation is used in hopes of explicating the phenomenon of perception. The strategy of the present paper will reverse the order of things somewhat: we shall exploit a novel PDP account of perceptual recognition in hopes of explicating the phenomenon of explanatory understanding. I remain faithful to the earlier tradition, however, since my basic aim will be to show that both phenomena are fundamentally the same.

Let me open the discussion by trying to motivate the search for a new account of explanation, and for trying to launch it in the specific directions indicated. We may begin by recalling the "covering-law" or "deductive–nomological" (D–N) model (Hempel [9]), since almost all of the current models are just artful restrictions or generalizations of that basic and very elegant idea.

According to the D–N model, to explain a singular event or a general state of affairs is just to deduce its description from a law of nature plus other premises (the initial or boundary conditions) that relate the general law to the case to be explained. This plausible model appears to capture both low-level cases of mundane singular explanation (Query: "Why is this infant crying?" Explanation: "Because all infants cry when they are overtired, and this one is definitely overtired") and also high-level cases of general intertheoretic explanation (as when we show that Kepler's three laws of planetary motion are all deductive consequences of Newtonian mechanics plus the gravitation law plus the assumption that the mass of each planet is negligible compared to the sun's). For singular cases, a typical logical structure would be:

$(x)((Gx \& Hx \& Ix) \supset Fx)$	Law
$Ga \& Ha \& Ia$	Initial Conditions
$\therefore Fa$	Event-to-be-Explained

While much attention has been paid to the *logical* virtues and vices of this model, relatively little has been paid to its shortcomings when evaluated from a *psychological* point of view. In fact, the D–N model, and the sentence-crunching conception of cognition it serves, is psychologically unrealistic in several important ways. If someone has just come to understand why a is F, the D–N model requires that we ascribe to that person knowledge of some universally quantified conditional statement having Fx as its consequent, plus knowledge of a series of initial conditions adequate to discharge the conjuncts in the antecedent of that conditional, plus the successful deduction of Fa from this assembled information, or at least the appreciation that the deductive relation exists.

However, while people have an explanatory understanding of much of what goes on around them on a minute-by-minute and even a second-by-second basis, people are decidedly and regularly inarticulate when asked to voice either the general law on which their understanding is presumably based, or the set of initial conditions that tie that law to the explanandum then at issue. Even in simple cases, the premises that people are typically able to supply, when queried, often fall dramatically short of the full requirements of the D–N model. There is no objection, of course, to the idea that people might have large amounts of inarticulable knowledge. What is suspicious here is the idea that it is both inarticulable *and* propositional in character. Furthermore, the logical acumen we must ascribe to people on the D–N account is often substantially in excess of what university students with formal training in logic can display.

Further still, the identification of relevant factual premises from a vast store of prior beliefs, and the search for relevant deductive relations, is a process that will in any system consume time, usually a good deal of time. All of this sits poorly with the great speed with which explanatory understanding is commonly achieved. It is often achieved almost instantaneously, as when one understands at a glance why one end of the kitchen is filled with smoke: the toast is burning! Such swiftness is not confined to mundane cases. If one has the relevant conceptual skills, the same speed is also displayed in more esoteric cases, as when one appreciates at a glance why Jupiter is an oblate spheroid: it is a plastic object spinning rapidly; or as when one appreciates at a glance why some red giant close-binary star has the shape of an egg pointed at its more compact blue companion: it is a very large object free-falling in a gravitational field.

At the other end of the spectrum, non-human animals provide a further illustration of these difficulties. Animals too display behavior that indicates the achievement of explanatory understanding, as when a frustrated coyote bites and paws at the leg-trap whose jaws have captured its mate. The coyote understands why its mate cannot leave. Animals too can anticipate elements of the future and understand elements of the present and past, often in some detail. But the assembly of discursive premises and the execution of formal inferences is presumably beyond their capacities, especially at the speeds that faithfulness to their insight and behavior requires.

These particular criticisms of the D–N model are unusual in being empirical and psychological, rather than logical in character. Even so, they are highly general. They will apply to all of the accounts of explanation that require, as the original D–N model requires, extensive propositional knowledge, relevant retrieval of same, and keen deductive insight. For it is precisely these features that give rise to the difficulties. Is there some alternative way of characterizing the way knowledge is represented in living creatures, and the way it is deployed or accessed in specific cases to provide local explanatory understanding? Yes, there is.

2. Conceptual organization in PDP networks

Recent years have seen an upswell of research into the functional properties of "neural" networks. These are artificial networks that simulate or model certain salient features of the neuronal organization of the brain. What is interesting is that even simple versions of these networks have shown themselves capable of some very striking computational achievements, and they perform these computations in a fashion that recalls important features of animal cognition.

The audience of this journal does not need a summary of how feedforward neural networks can be trained on a corpus of input–output pairs to discriminate the complex and subtle features represented in its input vectors. Let us therefore move straight to a prototypical example of the technique. Its features will help to illustrate the theory of explanatory understanding here to be advanced. The example is Gorman and Sejnowski's network [10], trained on many examples to discriminate between (a) sonar echoes returned from metal mines, and (b) sonar echoes returned from

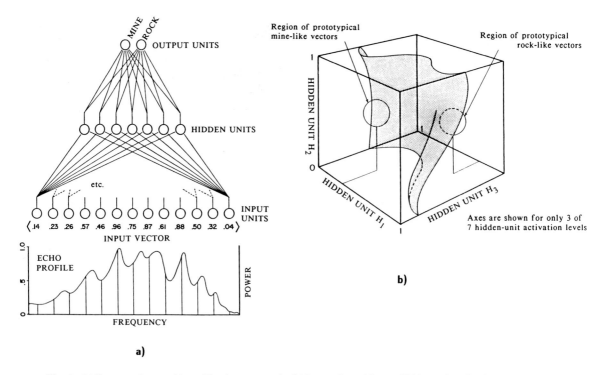

Fig. 1. (a) Perceptual recognition with a large network. (b) Learned partition on hidden unit activation vector space.

rocks. A simplified schematic of that network appears as fig. 1a.

Here we have a binary discrimination between a pair of diffuse and very hard-to-define acoustic properties. Indeed, no one ever did define them. It is the network that has generated an appropriate internal characterization of each type of sound, fueled only by examples and driven only by its learning algorithm. If we now examine the behavior of the hidden units during discriminatory acts in the trained network, we discover that the training process has partitioned the space of possible activation vectors across the hidden units (see fig. 1b). That space is split into a minelike part and a rocklike part, and within each subvolume the training process has generated a *similarity gradient* that culminates in two "hot spots" – two rough regions that represent the range of hidden-vector codings for a *prototypical* mine and a *prototypical* rock. The job of the top half of the network is then just the relatively simple one of discriminat-

ing any vector's proximity to one or other of the two prototypical hot spots.

Several features of such networks beg emphasis. First, the output verdict for any input is produced very quickly, for the computation occurs in parallel. The global computation at each layer of units is distributed among many simultaneously active processing elements: the weighted synapses and the summative cell bodies. Most strikingly, the speed of processing is entirely independent of both the number of units involved and the complexity of the function executed. Speed is determined solely by the number of distinct *layers* in the network. This makes for very swift processing indeed. In a living brain, where a typical information-processing pathway has something between five and fifty layers, and each pass through that hierarchy takes something between ten and twenty milliseconds per layer, we are looking at feedforward processing times, even for complex recognitional problems, of between one-twentieth of a second and one second. Empirically, this is the right range for living creatures.

Second, such networks are functionally persistent. They degrade gracefully under the scattered destruction of synapses or units. Since each synapse supplies such a small part of any computation, its demise leaves the network essentially unchanged.

Third, and very important for our purposes, the network will regularly render correct verdicts given only a degraded version or a smallish part of a familiar input vector. This is because the degraded or partial vector is relevantly *similar* to a prototypical input, and the internal coding strategy generated in the course of training is exquisitely sensitive to such similarities among possible inputs.

And exactly which similarities are those? They are whichever similarities meet the joint condition that (a) they unite some significant portion of the examples in the training set, and (b) the network managed to become tuned to them in the course of training. The point is that there are often many overlapping dimensions of similarity being individually monitored by the trained network: individually they may be modest in their effects, but if several are detected together their impact can be decisive. Here we may recall Ludwig Wittgenstein's 1953 description [11] of how humans can learn, by ostension, to detect "family resemblances" that unite diverse cases around some typical examples, but that defy easy definition. PDP networks recreate exactly this phenomenon.

Since that early period, various theorists have independently found motive to introduce similar notions in a number of cognitive fields. They have been called "paradigms" and "exemplars" in the philosophy of science (Kuhn [12]), "stereotypes" in semantics (Putman [13, 14]), "frames" (Minsky [15]) and "scripts" (Schank and Abelson [16]) in AI research, and finally "prototypes" in psychology (Rosch [17], Posner and Keele [18]) and linguistics (Lakoff [19]). These vague but persistent ideas appear to find microstructural explication, and systematic vindication, in the partitions and similarity gradients produced across the vector spaces of trained networks.

Finally, such networks can learn functions far more complex than the rock/mine case, and make discriminations far beyond the binary example portrayed. In the course of learning to pronounce English text, Rosenberg and Sejnowski's NETtalk [20] partitioned its hidden-unit vector space into fully 79 subspaces, one for each of the 79 letter-to-phoneme transformations that characterize the phonetic significance of English spelling. Since there are 79 distinct phonemes in English speech, but only 26 letters in the alphabet, each letter clearly admits of several different phonetic interpretations, the correct one being determined by context. Despite this ambiguity, the network also learned to detect which of several possible transforms is the appropriate one, by being sensitive to the various letters that flank the target letter inside the word. All of this is a notoriously irregular matter for English spelling, but the "rules" were learned by the network even so.

Other networks have learned to recognize the complex configuration and orientation of curved surfaces, given only gray-scale pictures of those surfaces as input. That is, they solve a version of the classic shape-from-shading problem in visual psychology. Still others learn to divine the grammatical elements of sentences fed as input, or to predict the molecular folding of proteins given amino acid sequences as input. These networks perform their surprising feats of learned categorization and perceptual recognition with only the smallest of "neuronal" resources – usually much less than 10^3 units. This is only one hundred millionth of the resources available in the human brain. With such powerful cognitive effects being displayed in such modest artificial models, it is plausible that they represent a major insight into the functional significance of our own brain's microstructure. That, in any case, is the working assumption on which the following discussion will proceed.

Two final caveats concerning the models under discussion: while the general idea of gradient descent in some sort of weight/error space is a plausible idea of how biological brains might learn, the popular "back-propagation" algorithm is not at all plausible as a model of the factors that modify the weight configurations of biological creatures. Some other account of synaptic weight change is needed. But this unsolved problem need not detain us. What concerns us here is how networks function *after* their weights have been suitably configured. And finally, but for a few parenthetical qualifications, the importance of recurrent pathways goes undiscussed. This decision is purely tactical: the basic character of the theory to be advanced can be understood and argued for without addressing the dynamical complexities that recurrence introduces.

3. Recognition and understanding

Let me now try to highlight those functional features of PDP networks that will lead us back toward the topic of explanation. The first feature to emphasize is the partitioning, in a suitably trained network, of its hidden-unit activation vector space into a system of prototype representations, one each for the general categories to which the network has been trained (see again fig. 1b for the simplest sort of case). Any prototype representation is in fact a specific vector (= pattern of activations) across the network's hidden units, but we may conceive of it more graphically as a specific point or small volume in an abstract state-space of possible activation vectors, since that portrayal highlights its geometrical relations with representations of distinct prototypes, and with activation vectors that are variously close to (= similar to) the prototype vector.

The second point to emphasize is that a single prototypical point or activation vector across the hidden units represents a wide range of quite different possible sensory activation patterns at the input layer: it represents the extended family of relevant (but individually perhaps non-necessary) features that collectively unite the relevant class of stimuli into a single kind. Any member of that diverse class of stimuli will activate the entire prototype vector at the hidden units. Also, any input-layer stimulus that is relevantly *similar* to the members of that class, in part or in whole, will activate a vector at the hidden units that is fairly close, in state-space, to the prototype vector.

Relative to scattered inputs, the prototype position is an attractor. We may think here of a wide-mouthed funnel that will draw a broad but delicately related range of cases into a single narrow path. This process is instanced in your ability to recognize a friend's face in any of a wide variety of expressions, positions, and conditions of viewing. Or in your ability to recognize a horse in almost any posture and from almost any perspective. These are exactly the sorts of capabilities displayed by suitably trained PDP networks.

A third point is to emphasize again that PDP networks are extraordinarily fast. Once trained, they achieve the "recognitions" at issue in a matter of milliseconds. And they will make distinct recognitions, one after another, as fast as you can feed them appropriately distinct stimuli.

Turn now to the units at the output layer. In the stick-figure account of cognition I am trying to outline, these are to be conceived as driving or initiating some specific motor activity: perhaps something relatively simple, as in NETtalk, where the output vector codes a phoneme and actually produces, via a speech synthesizer, an audible sound. In a living creature, however, the output will typically be more complex, as when a dog's sudden olfactory recognition of a gopher initiates a routine of rooting and digging at the favored location; or as when a bird's sudden visual recognition of a stalking cat prompts it to initiate a sequence of wing motions that launch it into the air.

This portrait of the cognitive lives of simple creatures ascribes to them an acquired "library" of internal representations of various prototypical perceptual situations, situations to which prototypical *behaviors* are the computed output of the well-trained network. The prototypical situations include feeding opportunities, grooming demands, territorial defense, predator avoidance, mating opportunities, offspring demands, and other similarly basic situations, to each of which a certain broad class of behaviors is appropriate.

To return to the basic issue: we can now see how the brain can command a large and sophisticated repertoire of prototype activation vectors, each one representing some complex prototypical situation in the external world. We have seen how such vectors can be activated by the perceptual apprehension of even a smallish portion of the relevant external situation, and how those vectors can activate in turn behaviors appropriate to the entire external situation, and not to just the smallish part that was initially coded in perception.

I wish to suggest that those prototype vectors, when activated, constitute the creature's recognition and concurrent *understanding* of its objective situation, an understanding that is reflected in the creature's subsequent behavior. Of course, a creature may *fail* to recognize/understand its current perceptual or cognitive situation. The vector activated at the relevant layer of hidden units may fall well outside any of the prototypical volumes of the relevant state-space, and the behavior subsequently produced will therefore not be drawn from its well-honed repertoire. The resulting behavior may be just confused. Or it may be a default routine of flight from the unknown. Or perhaps it will be a default routine of stumbling *exploration*, one that may finally find either a physical or a cognitive perspective from which the situation suddenly does activate one of the creature's many prototype vectors. It may find, that is, a perspective from which the situation suddenly does make sense.

By way of whatever learning algorithm governs synaptic adjustments, such failures and subsequent successes, especially in quantity, will modify the character and state-space location of the creature's internal prototype representations, so that situations of the puzzling kind just solved will successfully activate a prototype vector more readily in future.

4. Prototype activation: a unified theory of explanation

The aim of the preceding sections was to illustrate the initial plausibility of a novel conception of cognitive activity, a conception in which vector coding and vector-to-vector transformation constitute the basic forms of representation and computation, rather than sentential structures and inferences made according to structure-sensitive rules. Let us assume, for the sake of argument, that this conception is basically accurate even for human brains. If so, then we must immediately be impressed by the range of conceptual resources such systems can command, given the neuronal machinery available.

With roughly 10^{11} non-sensory neurons, the human brain commands a global state space of fully 10^{11} dimensions. Each brain subsystem will typically be operating with something like one-thousandth of that number, which gives a typical specialized state space approximately 10^8 proprietary dimensions to play with. This will allow for some very complex and finely grained representations, since a single vector with 10^8 elements can code the contents of an entire book. A state space of 10^8 dimensions will also allow for a similarly impressive *variety* of coding vectors. If we assume that each neuron admits of only ten distinct levels of activation (a serious underestimation), then that typical specialized state space must have at least 10^{10^8} or $10^{100000000}$ functionally distinct positions within it. This is the number of distinct possible *activation vectors*. To appreciate the magnitude of this number, recall that the total number of elementary particles in the entire physical universe, photons included, is only about 10^{87}. And recall that, on the above assumptions, your brain commands something like a thousand of these specialized state spaces.

We should not balk, therefore, at the premise of the following discussion, which regards it as unproblematic that the brain should command intricate prototype representations of such things as stellar collapse, cell meiosis, positron–positron

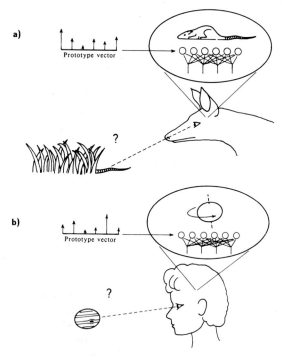

Fig. 2. Prototype activation with vector completion: two examples.

collision, redox reaction, gravitational lens, oceanic rift, harmonic oscillator, intentional action, and economic depression. Such phenomena, intricate though they are, are not beyond reach of the representational resources described.

The discussion to this point has all been a preamble to the following suggestion: explanatory understanding consists in the activation of a specific prototype vector in a well-trained network. It consists in the apprehension of the problematic case as an instance of a general type, *a type for which an experienced creature has a detailed and well-informed representation*. It is detailed in virtue of the very large number of distinct elements in the vector; it is well informed in virtue of the instructional history that produced it. Such a representation allows the creature to anticipate aspects of the case so far unperceived, and to deploy practical techniques appropriate to the case at hand (see figs. 2a and 2b). Given the preceding discussion, this idea has some plausibility already.

It is my aim in the remainder of this essay to illustrate how much illumination and unity this suggestion can bring to a wide range of cognitive phenomena.

Let me open the exposition by responding to a possible objection, which response will highlight an important feature of the vectorial representations here at issue. "What you have outlined", runs the objection, "may be a successful account of spontaneous *classification*; but explanatory understanding surely involves a great deal more than mere classification".

The objection evokes a process of mere "labeling", a process that puts the apprehended situation into a featureless pigeonhole, a process in which most of the complex information contained in the input is lost in its reduction to a canonical neural response. Yet this is precisely the wrong way to view the process of recognition and the character of the representation activated.

What we must remember is that the prototype vector embodies an enormous amount of information. Its many elements – perhaps as many as 10^8 elements was our earlier guess – each constitute one dimension of a highly intricate portrait of the prototypical situation. That vector has structure, a great deal of structure, whose function is to represent an overall *syndrome* of objective features, relations, sequences, and uniformities. Its activation by a given perceptual or other cognitive circumstance does not represent a loss of information. On the contrary, it represents a major and speculative *gain* in information, since the portrait it embodies typically goes far beyond the local and perspectivally limited information that may activate it on any given occasion. That is why the process is useful: in logicians' terms, it is quite dramatically *ampliative*. On each such occasion, the creature ends up understanding (or perhaps *mis*understanding) far more about the explanandum situation than was strictly presented in the explanandum itself. What makes this welcome talent of ampliative recognition possible is the many and various examples the creature has already encountered, and its successful generation of a unified prototype representation of them during the course of training.

This view entails that different people may have different levels or degrees of explanatory understanding, even though they "classify" a given situation in what is extensionally the "same" way. The reason is that the richness of their respective prototype representations may differ substantially. This is a welcome consequence, since explanatory understanding does indeed come in degrees. On the present view, its measure is just the richness and accuracy of the creature's prototype.

With these points in hand, let us now turn to a larger issue. One prominent fact, ill-addressed by any existing account of explanation, is the variety of different *types* of explanation. We have causal explanations, functional explanations, moral explanations, derivational explanations, and so forth. Despite some Procrustean analytical attempts, no one of these seems to be the basic type to which all of the others can be assimilated. On the prototype activation model, however, we can unify them all in the following way. Explanatory understanding is the same thing in all of these cases: what differs is the character of the prototype that is activated. Here follows a very brief summary.

4.1. Property-cluster prototypes

I begin with what is presumably the simplest, most common, and most superficial kind of explanatory understanding, and with the simplest and most superficial kind of prototype: the cluster of typically co-occurrent properties. Think of the typical cluster of features that constitutes a cat, or a cookie, or a tree, or a bird. These prototypes comprehend the vast majority of one's conceptual population, and they are activated on a regular basis in the course of one's mundane affairs. Because of their familiarity to everyone, cases involving them are seldom puzzling to anyone. But the explicit questions of inquiring children reflect the background explanatory role these prototypes continue to play for all of us. "Why is its neck so long, Daddy?" "It's a *swan* dear; swans have very long necks". "Why is he all spotted, Mummy?" "He's a *leopard* dear; leopards are always spotted, except when they're young".

4.2. Etiological prototypes

These are what lie behind *causal* explanations. An etiological prototype depicts a typical temporal *sequence* of event types, such as the cooking of food upon exposure to heat, the deformation of a fragile object during impact with a tougher one, the escape of liquid from a tilted container, and so on and so on. These sequences contain prototypical elements in a prototypical order, and they make possible our explanatory understanding of the temporally extended world. (Here, of course, the account may require recurrent networks, since the relevant prototypes may consist in a *sequence* of vectors.)

Note that the temporal inverse of an etiological prototype is generally not an etiological prototype as well. This means that causal explanations are generally asymmetric in character. The height of a flagpole and the altitude of the sun may jointly explain the length of the pole's shadow. But the length of the pole's shadow plus the height of the pole will not serve to explain the altitude of the sun, even though its altitude is easily deduced from these two facts. That asymmetry, a major problem for other accounts of explanation, is a natural consequence of the present account.

4.3. Practical prototypes

These, I suggest, are what lie behind *functional* explanations. One thing humans understand very well, being agents in the world, is complex means/ends relations between possible situations realizable by us and expectable consequences thereof that may be desirable to us. To portray any temporal configuration of actual or potential situations in this means/ends way is to make graphic or salient for us certain of the causal relations that unite them. It is a way of portraying a causal structure in the world in the guise of an actual or figurative practical problem, with the explanandum event or feature as its practical solution. Practical prototypes, like etiological prototypes, also depict sequences of event types or feature dependencies, but in the case of practical prototypes the explanandum *begins* the explanatory sequence whereas in etiological prototypes the explanandum *concludes* the explanatory sequence. Thus, a functional explanation does provide some entirely objective information, but in a constitutionally metaphorical form.

4.4. Superordinate prototypes

Some explanations, typically those in science, logic, and mathematics, concern not singular facts but *general* truths. Why do planets move on ellipses? Why are the theorems of the sentential calculus all tautologies? Why do the interior angles of any triangle always sum to exactly one straight angle? Here the objects of puzzlement are not singular situations; they are prototypical syndromes themselves. In such cases, explanatory understanding consists in the apprehension of the subordinate prototype as being an instance of some superordinate prototype. Explanations of this subordinate/superordinate kind are typically displayed in intertheoretic reductions, where one general fact or entire theory is subsumed by a more general theory. They are also displayed when our scattered understanding in some domain is successfully axiomatized, which in the present view is just another instance of the same process. The three standard axiom schemata of the propositional calculus, for example, are superordinate prototypes: each has endless different instances, each of which has endless subinstances. The same is true of its standard rule of inference: modus ponens. Thus, explanatory unification by axiomatization and deduction, as the old D-N model portrays, is entirely legitimate. But it is just one instance of the more general account here advocated.

4.5. Social-interaction prototypes

These underwrite *ethical*, *legal*, and *social-etiquette* explanations. "Why shouldn't I disperse this bunch of trouble-makers?" asks the red-neck cop approaching a so-far peaceful demonstration. "Because that would be a case of violating their constitutional right to peaceful assembly" is an explanatory reply. "Why", asks the gossip, "shouldn't I discuss Mary's marital problems with Doris?" "Because that would be violating a confidence" is an explanatory reply. "Aw, why can't I go play outside in the rain?", asks the seven-year old boy. "Because you have your new clothes on, and company will be here for Thanksgiving dinner in half an hour", is an explanatory reply. Here we appeal to various prototypical misbehaviors – denying a constitutional right, betraying a confidence, being inappropriately turned out for a family fete – of which the contemplated behavior is claimed to be an instance. Acquiring a legal, moral, or social sensibility is a matter of mastering a large system of such social-interaction prototypes, and of coming to perceive the social world from within that framework.

Though I very much doubt it is complete, I shall bring this catalogue of prominent kinds of prototypes to a close. You can see how they allow the prototype activation model to account in a unified fashion for the most familiar and widely discussed types of explanatory understanding, and for some previously undiscussed types as well.

5. Inference to the best explanation

The idea of prototype activation throws some much-needed light on the popular idea of "inference to the best explanation", a process that has often been invoked to account for the fixation of many of our beliefs, even our observational beliefs (see, for example, Harman [21, 22]). That idea is appealing, since it does seem to address what distinguishes the beliefs we do acquire from the many we might have acquired: the former have better explanatory power relative to the overall circumstance that occasioned them.

But the idea is also problematic, since it suggests a choice made from a range of considered alternatives. As a matter of psychological fact, alternatives are rarely even present. And in any case, our beliefs are typically fixed so swiftly that there is no time for the comparative evaluation of complex matters like the relative explanatory power of each of a range of alternatives.

On the PDP approach, we can begin to explicate the crude notion of "inference to the best explanation" with the more penetrating notion of "activation of the most appropriate prototype vector". Activating the most appropriate available prototype is what a well-trained network does as a matter of course, and it does it directly, in response to the input, without canvassing a single alternative vector. In the end, the process is not one of "inference" at all, nor is its outcome generally a sentence. But the process is certainly real. It just needs to be reconceived within the more penetrating framework of cognitive neurodynamics. When it is, both the "alternatives" problem and the "speed" problem disappear.

C.S. Peirce, who called the process "abduction", found the former problem especially puzzling. Peirce, one of the pioneers of modern formal logic, appreciated very early that for any set of observations there is a literal infinity of possible hypotheses that might be posed in explanation. But how can we possibly search a space of infinite size? Indeed, how can we even *begin* to search it effectively when its elements are not well ordered? Peirce marveled that human scientists are able so regularly to produce, from this infinite sea of mostly irrelevant and hopeless possible candidates, hypotheses that are both relevant and stand some non-trivial chance of being true. From the sentential perspective, Peirce was right to marvel. But from the neurocomputational perspective, the situation is not so mysterious.

We do not search an infinite space of possible explanations. In general, we do not search at all: in familiar cases a suitable prototype is activated directly. And if the novelty of the case foils our waiting categories, and thus forces us into search mode, then we search only the comparatively tiny space comprising the set of our own currently available prototype vectors. Even here the search is mostly blind, and probably stops at the first success. If one's initial encounter with the problematic case fails to activate directly a familiar and subsequently successful prototype vector, then one repeatedly re-enters the problematic input in a variety of different cognitive contexts, in hopes of finally activating some prototype vector or other, or some vector close enough to an existing prototype to give one at least some handle on the problem. (Here again we will require recurrent projections to variously modify the cognitive context encountered by the input.)

Since the range of concurrently possible "understandings" is closed under the relation "is at least within hailing distance of an existing prototype", then *of course* any element from that range will appear both "relevant" and "potentially true". Peirce, and we, are the victims of a perspectival effect. Our hypotheses will look at least passably hopeful to us, because they are drawn from a source that collectively defines what will be found plausible by us. We should thus be wary of assuming, as Peirce seems to have assumed, that we have any special *a priori* nose for truth.

6. Conclusion

Recall again the venerable D–N model of explanation. It was correct in insisting that explanatory understanding requires the deployment of some information that is general in character. Beyond this insight, almost nothing is correct. The model's commitment to a sentential or propositional mode of knowledge representation renders it unable to account for explanatory understanding outside of that narrow context, and it generates a host of problems even within that context. Slow access, inarticulation of laws, and deductive inappreciation were discussed at the beginning of this essay. To which we may add the well-known problems of explanatory asymmetries (the "flagpole" problem) and the scattered variety of distinct types of explanation, some of which do not fit the D-N pattern at all.

None of these difficulties attend the prototype activation model. Concerning the matter of access, relevant understanding is usually accessed in milliseconds. Concerning our inability to articulate laws, the PA model does not even suggest that we should be able to articulate them. For what gets accessed is not a stored universal conditional, but a complex prototype. Similarly, while our deductive incompetence is a problem for the D-N view, on the PA model deductive inference typically plays no role at all in the process of prototype activation. Moreover, as noted in section 4, etiological prototypes are in general temporally asymmetric. On the PA model, explanatory asymmetries are thus only to be expected.

Finally, the PA model brings a welcome and revealing unity into a stubborn diversity of explanation types, and the model is itself an integral part of a highly unified background conception of cognitive activity, one that encompasses with some success the general structure and activity of biological brains, and the structure and cognitive behavior of a new class of artificial computing systems. For this reason, if no other, we should be moved to explore it further.

References

[1] D.E. Rumelhart, G.E. Hinton and R.J. Williams, Learning representations by back-propagating errors, Nature 323 (1986) 533.

[2] D.E. Rumelhart and J.L. McClelland, eds., Parallel Distributed Processing: Explorations in the Microstructure of Cognition (MIT Press, Cambridge, MA, 1986).

[3] P.M. Churchland, Some reductive strategies in cognitive neurobiology, Mind 95 (379) (1986) 279.

[4] P.M. Churchland, On the nature of theories: A neurocomputational perspective, in: The Nature of Theories, Minnesota Studies in the Philosophy of Science, Vol. 14, ed. W. Savage (University of Minnesota Press, Minneapolis, 1990), reprinted in: A Neurocomputational Perspective: The Nature of Mind and the Structure of Science (MIT Press, Cambridge, MA, 1989) p. 153.

[5] P.M. Churchland, A Neurocomputational Perspective: The Nature of Mind and the Structure of Science (MIT Press, Cambridge, MA, 1989).

[6] R.L. Gregory, Eye and Brain (McGraw-Hill, New York, 1966).

[7] R.L. Gregory, The Intelligent Eye (McGraw-Hill, New York, 1970).

[8] I. Rock, The Logic of Perception (MIT Press, Cambridge, MA, 1983).

[9] K. Hempel, Studies in the logic of explanation, in: Aspects of Scientific Explanation, ed. K. Hempel (The Free Press, New York, 1965) p. 3.

[10] R.P. Gorman and T.J. Sejnowski, Learned classification of sonar targets using a massively-parallel network, IEEE Trans. Acoustics, Speech, and Signal Processing 36 (1988) 1135-1140.

[11] L. Wittgenstein, Philosophical Investigations (Basil Blackwell, Oxford, 1953) secs. 66-77.

[12] T.S. Kuhn, The Structure of Scientific Revolutions (University of Chicago Press, Chicago, 1962).

[13] H. Putnam, Is semantics possible?, in: Mind, Language and Reality (Cambridge Univ. Press, Cambridge, 1975) p. 139.

[14] H. Putnam, The meaning of 'meaning', in: Language, Mind, and Knowledge, Minnesota Studies in the Philosophy of Science, Vol. 7, ed. K. Gunderson (University of Minnesota Press, Minneapolis, 1975) p. 215.

[15] M. Minsky, A framework for representing knowledge, in: Mind Design, ed. J. Haugeland (MIT Press, Cambridge, MA, 1981).

[16] R. Schank and R. Abelson, Scripts, Plans, Goals, and Understanding (Wiley, New York, 1977).

[17] E. Rosch, Prototype classification and logical classification: The two systems, in: New Trends in Cognitive Representation: Challenges to Piaget's Theory, ed. E. Scholnick (Erlbaum, London, 1981).

[18] M. Posner and S. Keele, On the genesis of abstract ideas, J. Exp. Psychology 77 (1968) 353.

[19] G. Lakoff, Women, Fire, and Dangerous Things (University of Chicago Press, Chicago, 1987).

[20] C.R. Rosenberg and T.J. Sejnowski, Parallel networks that learn to pronounce english text, Complex Systems (1987) 145.

[21] G. Harman, Inference to the best explanation, Philos. Rev. 74 (1965) 88.

[22] G. Harman, Thought (Princeton Univ. Press, Princeton, 1973).

PARALLEL SIMULATED ANNEALING TECHNIQUES

Daniel R. GREENING

*University of California, Los Angeles
and IBM T.J. Watson Research Center, P.O. Box 704, Yorktown Heights, NY 10598-0704, USA*

Simulated annealing is a stochastic algorithm for solving discrete optimization problems, such as the traveling salesman problem and circuit placement. To reduce execution time, researchers have parallelized simulated annealing. *Serial-like* algorithms identically maintain the properties of sequential algorithms. *Altered generation* algorithms modify state generation to reduce communication, but retain accurate cost calculations. *Asynchronous* algorithms reduce communication further by calculating cost with outdated information. Experiments suggest that asynchronous simulated annealing can obtain greater speedups than other techniques. It exhibits the properties of cooperative phenomena: processors asynchronously exchange information to bring the system toward a global minimum. This paper provides a comprehensive, taxonomic survey of parallel simulated annealing techniques, highlighting their performance and applicability.

1. Introduction

Several interesting combinatorial optimization problems are *NP*-hard – that is, they require at least non-deterministic polynominal time to obtain optimal solutions. Mathematicians have exerted considerable effort trying to determine whether *P* (deterministic polynominal time) = *NP*, with no success. Thus, present-day optimal solutions for *NP*-hard problems require exponential time, rendering them intractable. That explains why sub-optimal, polynomial-time algorithms, like simulated annealing, have attracted interest.

Simulated annealing seeks to minimize a cost function for a system of interacting state variables [1]. Often the cost function presents a difficult landscape, with many local minima. Simulated annealing tries to escape local minima by randomly following cost-*increasing* paths. One cannot guarantee polynomial-time convergence to a global minimum for *NP*-complete problems (that would prove *P = NP*); however, evidence shows that simulated annealing often produces good results.

Simulated annealing is often applied to VLSI circuit placement. In VLSI design, reducing circuit area decreases fabrication price, and shortening wires increases circuit speed. Rearranging circuit elements (called "cells") on a plane will change those values. Optimizing the arrangement is VLSI circuit placement: the cost function typically includes a linear combination of total circuit area and total wire-length.

Variants of VLSI placement fall into three categories. In "gate-array placement", all cells have a uniform rectangular shape. In "row-based placement", cells have a constant height, but varying width. In "macro-cell placement", cells can vary in size and shape.

A popular row-based placement and routing program, called TimberWolfSC, uses simulated annealing [2]. In a benchmark held at the 1988 International Workshop on Placement and Routing, TimberWolfSC produced the smallest placement for the 3000-element *Primary2* circuit – 3% smaller than its nearest competitor. Moreover, it completed earlier than all other entrants. TimberWolfSC also routed *Primary2*; no other entrant completed that task.

Even approximate solutions to *NP*-hard problems require substantial time: simulated annealing is no exception. On a Sun 4/260, *Primary2* requires approximately 3 h to place. Larger circuits require more time; for 30 000-element circuits, placement runs commonly take 36 h.

Such onerous run times have driven researchers to implement simulated annealing on multiprocessors. Several techniques have been tried. Organized under the taxonomy described in this paper, their similarities should become clear.

Parallel implementations, particularly the asynchronous simulated annealing programs, exhibit properties of cooperative phenomena: individual processors make decisions based on incomplete or delayed information, yet together they approach a common goal [3]. As a result, this paper can suggest research directions and implementation techniques for areas outside simulated annealing.

1.1. Problem formulation

Problems amenable to simulated annealing typically have these features:

(1) One can construct an initial solution, or "state," $s_0 \in S$, where S is the set of all feasible states, and evaluate its cost-function $E_{s_0} = f(s_0)$.

(2) One can construct an inexpensive mapping, through a neighborhood relation g, from a single-feasible state s into a set of feasible states $g[\![\{s\}]\!]$.

(3) One can inexpensively compute the cost-difference $\Delta E_{s,s'}$ for any state $s' \in g[\![\{s\}]\!]$, so that $E_{s'} = \Delta E_{s,s'} + E_s$.

(4) A finite number, i, of recursive applications of g to any state s, $g \circ \ldots \circ g[\![\{s\}]\!]$, covers the entire state space, so $g^i[\![\{s\}]\!] = S$.

Sequential implementations follow the general form in fig. 1, algorithm SSA. It uses a pseudorandom number generator to create a random starting state (line 4), to generate a random state change for consideration (line 8), and to decide whether to accept the generated state (line 10).

The *accept* function (line 1) uses ΔE to decide whether to keep the new configuration. If ΔE is negative, the perturbed state $\hat{s}$ is better than s, the program always accepts the new configuration (lines 11–12). If ΔE is positive, state $\hat{s}$ is worse

```
1.   function accept(ΔE, T)
2.     return((ΔE ≤ 0) ∨ (e^(-ΔE/T_i) > random( )));
3.   read(P);
4.   s ← some randomly constructed initial state for P;
5.   E ← f(s);
6.   T ← ∞;
7.   loop for i ← 0 to ∞
8.     ŝ ← a randomly selected element of g[[{s}]];
9.     ΔE ← Δf(s, ŝ);
10.    if accept(ΔE, T) then
11.      s ← ŝ;
12.      E ← E + ΔE;
13.    end if;
14.    if done(T, other statistics) then
15.      write(s);
16.      stop;
17.    end if;
18.    T ← update(T, other statistics);
19.  end loop;
```

Fig. 1. Algorithm SSA: sequential simulated annealing.

than state s, and the program accepts the new configuration with probability $e^{-\Delta E/T}$.

Higher T values and lower ΔE values increase the likelihood that a cost-increasing configuration will be accepted. However, if $T > 0$, any cost-increasing configuration has some probability of being accepted.

The procedure for updating T is called the *temperature schedule*. The *equilibrium cost* is the mean value of E we would obtain from running the simulated annealing algorithm forever at some fixed temperature T. Most programs first set T at a high value, then reduce T while attempting to keep E close to the equilibrium cost. One common temperature schedule has $T \leftarrow \gamma T$, where $0 < \gamma < 1$.

Intuitively, simulated annealing first explores the entire state space and then reduces its scope. Each lowering of the temperature restrains state exploration further. While the temperature is high, the algorithm can easily jump out of local minima; at its lowest temperatures, the algorithm usually moves toward lower cost.

1.2. Parallel algorithms

Since a new state contains modifications to the previous state, simulated annealing is often con-

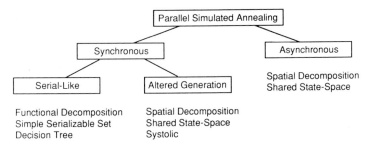

Fig. 2. Parallel simulated annealing taxonomy.

sidered an inherently sequential process. However, researchers have eliminated some sequential dependencies, and have developed several parallel annealing techniques. To categorize these algorithms, we ask several questions:

(1) How is the state space divided among the processors?

(2) Does the state generator for the parallel algorithm produce the same neighborhood as the sequential algorithm? How are states generated?

(3) Can moves made by one processor cause cost-function calculation errors in another processor? Are there mechanisms to control these errors?

(4) What is the speedup? How does the final cost vary with the number of processors? How fast is the algorithm, when compared to an optimized sequential program?

A parallel algorithm exhibits so-called "superlinear" speedup when the speed improvement over a sequential algorithm exceeds the number of processors. Simulated annealing researchers frequently see this suspicious property.

Three factors can explain most superlinear speedup observations. First, changes to state generation wrought by parallelism can improve annealing speed or quality [4]. If this happens, one can reconcile the sequential algorithm by mimicking the properties of the parallel version [5, 6]. Second, a speed increase might come with a solution quality decrease [7]. That property holds for sequential annealing, as well [8]. Third, annealing experimenters often begin with an optimal initial state, assuming that high-temperature randomization will annihilate the advantage. But if the parallel implementation degrades state-space exploration, high-temperature may not totally randomize the state: the parallel program, then, more quickly yields a better solution [9].

Knowledge of such pitfalls can help avoid problems. Superlinear speedup in cooperating systems, such as parallel simulated annealing, should raise a red flag: altered state exploration, degraded results, or inappropriate initial conditions may accompany it.

The taxonomy presented here divides parallel annealing techniques into the three major classes shown in fig. 2: serial-like, altered generation, and asynchronous. We call an algorithm *synchronous* if adequate synchronization ensures that cost-function calculations are accurate. Two major classes, *serial-like* and *altered generation*, are synchronous algorithms. *Serial-like convergence* algorithms identically maintain the convergence properties of sequential annealing. *Altered generation* algorithms modify state generation, but retain accurate cost calculations. *Asynchronous* algorithms, the third major class, eliminate some synchronization and tolerate the resulting errors to get a better speedup.

Each class makes some trade-off between cost-function accuracy, state generation, parallelism or communication overhead.

2. Serial-like algorithms

Three synchronous parallel algorithms preserve the convergence properties of sequential simulated annealing: *functional decomposition*, *simple serializable set*, and *decision tree decomposition*. We call these *serial-like* algorithms.

2.1. Functional decomposition

Functional decomposition algorithms exploit parallelism in the cost function f. In the virtual design topology problem, for example, the cost function must find the shortest paths in a graph. One program computes that expensive cost function in parallel, but leaves the sequential annealing loop intact [10]. Published reports provide no speedup information.

Another program evaluates the cost function for VLSI circuit placement in parallel [11]. Simultaneously, an additional processor selects the next state. Fig. 3, algorithm FD, shows the details.

One can obtain only a limited speedup from algorithm FD. Ideally, the parallel section from line 4 to line 13 dominates the computation, each process executes in uniform time, and communication requires zero time. One can then extract a maximum speedup of $1 + 2j + k$, where j is the average number of cells affected per move, and k is the average number of wires affected per move. Researchers estimate a speedup limitation of 10, based on experience with the VLSI placement program TimberWolfSC [2].

Since cost-function calculations often contain only fine-grain parallelism, communication and synchronization overhead can dominate a functional decomposition algorithm. Load-balancing poses another difficulty. Both factors degrade the maximum speedup, making functional decomposition inappropriate for many applications.

2.2. Simple serializable set

If a collection of moves affect *independent* state variables, distinct processors can independently compute each ΔE without communicating. We call this a "serializable set" – the moves can be concluded in any order, and the result will be the same. The simplest is a collection of rejected moves: the order is irrelevant, the outcome is always the starting state.

The *simple serializable set* algorithm exploits that property [11]. At low annealing temperatures, the acceptance rate (the ratio of accepted states to tried moves) is often very low. If processors compete to generate one accepted state, most will generate rejected moves. These can all be executed in parallel.

Fig. 4, algorithm SSS, shows such a technique [12]. P processors grab the current state in line 5. Each processor generates a new state at line 7. If the new state is accepted (line 8) *and* the old state has not been altered by another processor (line 10), the move is made. Otherwise the move is discarded.

If the acceptance rate at temperature T is $\alpha(T)$, then the maximum speedup of this algorithm, ignoring communication and synchronization costs, is $1/\alpha(T)$. At high temperatures, where the acceptance rate is close to 1, the algorithm provides little or no benefit. But since traditional

```
1.    m' ← select random state;
2.    loop for i ← 0 to ∞
3.        m ← m';
4.        parallel block begin
5.            m' ← generate(m);
6.            E_0 ← block-length-penalty(m);
7.            E_{1,0} ← overlap for affected cell c_0 before move;
8.            ... E_{1,j} ← overlap for affected cell c_j before move;
9.            E_{2,0} ← overlap for affected cell c_0 after move;
10.           ... E_{2,j} ← overlap for affected cell c_j after move;
11.           E_{3,0} ← length change for affected wire w_0;
12.           ... E_{3,k} ← length change for affected wire w_k;
13.       end parallel block;
14.       ΔE ← E_0 + (E_{1,0} + ... + E_{1,j})
              -(E_{2,0} + ... + E_{2,j}) + (E_{3,0} + ... + E_{3,k});
15.       if accept(ΔE, T) then
16.           parallel block begin
17.               update overlap values;
18.               update blocks and cells;
19.               update wire w_0;
20.               ...update wire w_k;
21.           end parallel block;
22.       end if;
23.       recompute T, evaluate stop criteria, etc.
24.   end loop;
```

Fig. 3. Algorithm FD: functional decomposition for VLSI placement.

```
1.  shared variable s, semaphore sema;
    ...
2.  parallel loop for i ← 1 to P;
3.    loop for j ← 0 to ∞
4.      wait(sema);
5.      s_old ← s;
6.      signal(sema);
7.      ⟨ŝ, ΔE⟩ ← generate(s_old);
8.      if accept(ΔE, T) then
9.        wait(sema);
10.       if s_old = s then
11.         s ← ŝ;
12.         T ← new T;
13.       end if;
14.       signal(sema);
15.     end if;
16.     change T, evaluate stop criterion, etc.
17.   end loop;
18. end parallel loop;
```

Fig. 4. Algorithm SSS. Simple serializable set algorithm.

annealing schedules spend a majority of time at low temperatures, algorithm SSS can improve overall performance.

Algorithm SSS has limitations. Some recent annealing schedules maintain $\alpha(T)$ at relatively high values, throughout the temperature range, by adjusting the generation function. Lam's schedule, for instance, keeps $\alpha(T)$ close to 0.44 [8]. With that schedule, algorithm SSS provides a maximum speedup of approximately 2.3, regardless of the number of processors.

2.3. Decision tree decomposition

A third serial-like algorithm, called decision tree decomposition, exploits parallelism in making accept-reject decisions [13]. Consider the tree shown in fig. 5a. If we assign a processor to each vertex, cost evaluation for each suggested move can proceed simultaneously. Since a sequence of moves might be interdependent (i.e. not serializable), however, we *generate* the states in sequence.

Fig. 5b shows vertex dependences. A vertex generates a move in time t_m, evaluates the cost in time t_e, and decides whether to accept in time t_d. Note that vertex 2 cannot begin generating a move until vertex 1 generates its move and sends it to vertex 2.

Research has provided hypothetical execution times, but no experimental confirmation. A simple implementation results in predicted speedups of $\log_2 P$, where P is the number of processors. By skewing tree evaluation toward the left when $\alpha(T) \geq 0.5$, and toward the right when $\alpha(T) < 0.5$, researchers predict a maximum speedup of $\frac{1}{2}(P + \log_2 P)$ [13].

In numeric simulations, however, the speedups fall flat. With 30 processors and $t_m = 16 t_e$, the estimated speedup was 4.7. Unfortunately, in VLSI placement problems $t_m \gg t_e$, and in traveling salesman problems $t_m \approx t_e$. Reconciling t_m leads to a speedup of less than 2.5 on 30 processors. As a result, this approach holds little promise for such applications.

3. Altered generation algorithms

Even if a parallel annealing algorithm computes cost functions exactly, it may not mimic the statistical properties of a sequential implementation. Often, state generation must be modified to reduce

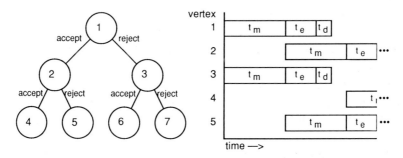

Fig. 5. Decision tree decomposition. (a) Annealing decision tree. (b) Functional dependence.

interprocessor communication. These *altered generation* methods change the pattern of state space exploration, and thus change the expected solution quality and execution time.

3.1. Spatial decomposition

In spatial decomposition techniques, we distribute state variables among the processors, and variable updates are transmitted between processors as new states are accepted. Spatial decomposition techniques are typically implemented on message-passing multiprocessors.

In *synchronous* decomposition, either processors must carefully coordinate move generation, or processors must not generate moves that affect processors' state variables. We call the resulting two techniques *cooperating processors* and *independent processors*.

3.1.1. Cooperating processors

A cooperating processor algorithm disjointly partitions state variables over the processors. A processor that generates a new state notifies other affected processors. Then, those processors synchronously evaluate and update the state. If a proposed move could interfere with another in-progress move, the proposed move is either delayed or abandoned.

One such program minimizes the number of routing channels (the slots where wires lie) for a VLSI circuit [9]. The cost is the total number of routing channels that contain at least one wire; two or more wires can share the same routing channel, if they do not overlap.

The program first partitions a set of routing channels across the processors of an iPSC/2 Hypercube; that processor assignment henceforth remains fixed. Processors proceed in a lockstep communication pattern. At each step, all processors are divided into master-slave pairs. The master processor randomly decides among four move classes:

Intra-displace: The master and slave each move a wire to another channel in the same processor.

Inter-displace: The master processor moves one of its wires to a channel in the slave processor.

Intra-exchange: Each master and slave each swap two wires in the same processor.

Inter-exchange: The master swaps a wire from one of its channels with a wire in the slave.

Experiments indicate superlinear speedups, from 2.7 on 2 processors to 17.7 on 16 processors. These apparently stem from a nearly optimal initial state and more constrained parallel moves, making the reported speedups untenable. However, the decomposition method itself is sound.

3.1.2. Independent processors

In independent processor techniques, each processor generates state changes which affect only its own variables. Under this system, a fixed variable assignment would drastically limit state-space exploration, and produce an inferior result; it requires periodic state variable redistribution.

One such technique optimizes traveling salesman problems [14]. A traveling salesman problem (TSP) consists of a collection of cities and their planar coordinates. A tour that visits each city and returns to the starting points forms a solution; the solution cost is its total length.

We construct an initial, poor-quality solution by putting the cities into a random sequence: the tour visits each in order and returns to the first city. We stretch this string of cities out like a rubber band, and evenly divide the two parallel tracks among the processors, as shown in fig. 6. The state variables consists of the endpoints of each two-city segment.

Each processor anneals the two paths in its section by swapping corresponding endpoints. After a fixed number of tries in each processor, the total path length is computed, and a new temperature and a shift count are chosen. Each processor then shifts the path attached to its top left node to the left, and the path attached to its bottom right node to the right by the shift count. This operation redistributes the state variables, ensuring that the whole state space is explored.

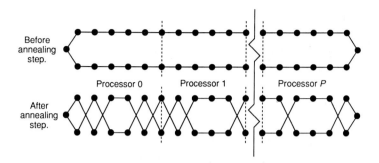

Fig. 6. Rubber band TSP algorithm.

Annealing continues until it satisfies the stopping criterion.

In one experiment, the 30-processor versus 2-processor speedup ranged from about 8 for a 243 city TSP, to 9.5 for a 1203 city TSP. Unfortunately, a single processor example was not discussed. The paper does not show final costs; final cost probably increases as the number of processors increases. Other spatial decomposition techniques exhibit similar behavior and speedups [15, 16].

3.2. Shared state-space

Shared state-space algorithms make simultaneous, independent moves on a shared-memory state-space: no cost-function errors can occur.

One such algorithm optimizes VLSI gate-array placement [7]. Changes in the state generation function, resulting from the locking of both cells and wires, caused the algorithm to generate poor convergence. Maximum speedup was 7.1 for 16 simulated RP3 processors, solving a uniform 9 × 9 grid problem. Improving the parallel algorithm's convergence would reduce its speedup below 7.1.

A similar algorithm for minimizing the equal partition cut-set (see section 4.2) obtained a dismal speedup close to 1 on 16 processors [17].

Another shared state-space algorithm constructs conflict-free course timetables [18]. Before evaluating a move, the algorithm must lock the instructors, courses and rooms for two time periods, then swap them. If the locks conflict with an in-progress move, the locks are abandoned and another move is generated. Speedup was compared against an optimized sequential algorithm. With 8 processors, a speedup of 3.2 was obtained in scheduling 100 class periods, while 6.8 was obtained in scheduling 2252 class periods.

3.3. Systolic

The systolic system relies on the property that simulated annealing brings a thermodynamic system toward the Boltzmann distribution [19, 20].

Suppose we have P processors, and we maintain the same temperature for a chain of N generated states. We would like to divide these moves into P subchains of length $L = P/N$, and execute them on different processors. Fig. 7 shows a corresponding data flow graph for this decomposition.

At any PICK node on processor p, we must decide between state $s_{(n-1, p)}$ computed by processor p at temperature T_{n-1}, and state $s_{(n, p-1)}$ computed by processor $p-1$ at temperature T_n. We make the choice according to the Boltzmann distribution. The relative probability of picking $s_{(n-1, p)}$ is

$$\rho_0 = \frac{1}{Z(T_{n-1})} \exp\left(\frac{f(s\downarrow) - f(s_{(n-1,p)})}{T_{n-1}} \right) \quad (1)$$

and the relative probability of picking $s_{(n, p-1)}$ is

$$\rho_1 = \frac{1}{Z(T_n)} \exp\left(\frac{f(s\downarrow) - f(s_{(n,p-1)})}{T_n} \right), \quad (2)$$

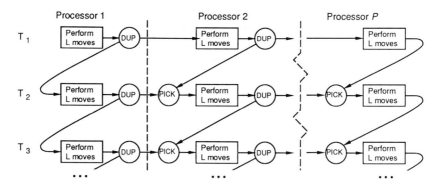

Fig. 7. Systolic algorithm.

where S is the entire state space and $s\downarrow$ is a minimum cost state. $Z(T)$ is the *partition function* over the state space, namely

$$Z(T) = \sum_{s \in S} \exp[-f(s)/T]. \qquad (3)$$

The PICK node then selects $s_{(n-1,p)}$ and $s_{(n,p-1)}$ with probabilities

$$p(s_{(n-1,p)}) = \frac{\rho_0}{\rho_0 + \rho_1},$$

$$p(s_{(n,p-1)}) = \frac{\rho_1}{\rho_0 + \rho_1}. \qquad (4)$$

If we do not know the minimum cost, we cannot evaluate $f(s\downarrow)$. A lower bound must suffice as an approximation. Choosing a lower bound far from the infimum will increase execution time or decrease solution quality [8].

The partition function, Z, requires the evaluation of every state configuration. The number of state configurations is typically exponential in the number of state variables, making exact computation of Z unreasonable.

As a result, the systolic method uses an approximate Z. In the temperature regime where the exponential function dominates, ρ_0 and ρ_1 are almost completely determined by their numerators in eqs. (1) and (2). The influence of $Z(T)$ thus becomes small, and it can be approximated by the normal distribution.

How does the algorithm perform? With eight processors operating on a 15×15 uniform grid of cities, the systolic algorithm obtained a mean path length of 230, at a speedup of about 6.2, while the sequential algorithm obtained an average of about 228.5. Accounting for the less optimal parallel result, the effective speedup is something less than 6.2.

4. Asynchronous algorithms

Without sufficient synchronization, different processors can simultaneously read and alter dependent state-variables, causing cost-function calculation errors. Such algorithms are *asynchronous*. Imprecise cost-function evaluation accelerates *sequential* simulated annealing under certain conditions [21, 22]; a similar effect accompanies asynchronous parallel simulated annealing.

These algorithms use a method related to chaotic relaxation – processors operate on out-dated information [23]. Since simulated annealing randomly selects hill-climbing moves, it can tolerate some error; under the right conditions, annealing algorithms can evaluate the cost using old state information, but still converge to a reasonable solution. This property holds for genetic algorithms, as well [24].

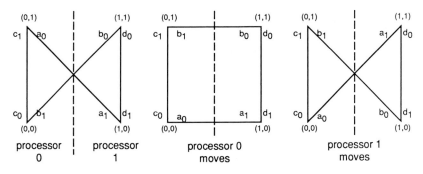

Fig. 8. Cost-function errors in spatial decomposition.

Error tolerance provides a great advantage in multiprocessing: when processors independently operate on different parts of the problem, they need not synchronously update other processors. A processor can save several changes, then send a single block to the other processors. The processor sends less control information and compresses multiple changes to a state variable into one, reducing total communication traffic. In addition, if updates can occur out-of-order, synchronization operations are reduced. Asynchronous algorithms require a minimum synchronization: two processors acting independently must not cause the state to become inconsistent with the original problem.

Fig. 8 shows how errors arise in a spatially decomposed traveling salesman problem. In the figure, variables a_0 and a_1 denote the endpoints of edge a. The simulated annealing algorithm swaps endpoints to generate a new state. The algorithm partitions the cities over two processors. A processor may only swap endpoints that point to its vertices, ensuring problem consistency. However, to reduce synchronization time, processors do not lock edges while they evaluate the cost-function.

While processor 0 considers swapping endpoint a_0 with b_1, processor 1 considers swapping endpoint a_1 with b_0. Processor 0 sees a path-length change for its move of $\Delta E = 2(1 - \sqrt{2}) \approx -0.818$. Processor 1 also sees $\Delta E \approx -0.818$, for its move.

Processor 0 makes its move, by swapping a_0 and b_1. Now, processor 1 makes its move, thinking its $\Delta E \approx -0.818$ (a good move) when the effect is $\Delta E \approx +0.818$ (a bad move). At low temperatures, the error will degrade the final result unless corrected by a later move. So, simulated annealing does not have an unlimited tolerance for errors.

Cost-function errors usually degrade convergence quality, when all other factors are fixed: note the contrast with altered state generation. For example, experiments have shown that VLSI

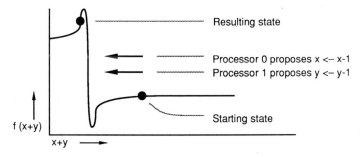

Fig. 9. Errors can cause annealing failure.

placement quality decreases as errors increase [4, 25].

Several authors have conjectured that annealing properties might be preserved when the errors are small. Experimental evidence bears this out [4, 7, 26–28]. However, we can easily construct a problem which converges well under sequential simulated annealing, but will *likely* converge to a bad local minimum in an asynchronous program.

Consider a system with two state variables x and y, so some state $s = \langle x, y \rangle \in S$. Let the cost function be $f(x + y)$, shown in fig. 9. Now put x and y on two separate processors. Each processor proposes a move: processor 0 generates $x \leftarrow x - 1$, while processor 1 generates $y \leftarrow y - 1$. In both cases, $\Delta E < 0$, so each move will be accepted.

The cost-function error causes the state to jump to a high local minimum. At low temperatures, the annealing algorithm probably will not escape this trap.

4.1. Asynchronous spatial decomposition

Asynchronous spatial decomposition methods, like the synchronous methods in section 3.1, partition state variables across different processors. However, in asynchronous algorithms each processor also maintains read-only copies of state variables from other partitions.

When a processor evaluates a new state, it uses only local copies of state variables. In some programs, when a move is accepted the new state information is immediately sent to other processors [26]. In other programs, a processor completes a fixed number of tries, called a "stream," before transmitting the modifications [4, 25]. Longer streams increase the execution-to-communication ratio, gaining a speedup, but they also increase calculation errors, reducing the solution quality.

4.1.1. Clustered decomposition
The clustered decomposition technique solves two concurrent optimization problems: the specified target problem and assigning the state variables to processors.

In one example, the target problem is VLSI macro-cell placement, and the assignment problem is cell partitioning [29]. Overlap penalties in the VLSI cost function generate the largest errors – when two cells owned by different processors are moved to the same empty location, neither processor will see an overlap, but the overlap error might be huge. This leads to a clustering problem: divide state variables (macro-cells) equally among the processors, while putting dependent variables (adjacent or connected macro-cells) on the same processor.

We compute the assignment cost function, for VLSI macro-cell placement, as follows. Let C be the set of cells, let $C = \{C_1, \ldots, C_P\}$ be the partition of C over P processors, let $\bar{c}$ be a cell's vector center and let $|c|$ be its scalar area. For each processor p, we can compute the center of gravity X_p of its partition C_p,

$$X_p = \frac{1}{\sum_{c \in C_p} |c|} \sum_{c \in C_p} \bar{c}|c|, \tag{5}$$

and its inertial moment

$$\Gamma_p = \sum_{c \in C_p} \|\bar{c} - X_p\|^2 |c|. \tag{6}$$

The assignment cost function for partition C is

$$f_c(C) = w_c \sum_{i=1}^{P} \Gamma_i, \tag{7}$$

where w_c is a weighting factor.

Experiments used the same temperature for both partitioning and placement: independent temperature schedules would probably improve the result. A 30-macro-cell problem, running on an 8-processor, shared-memory Sequent 8000, reached a speedup of 6.4 against the same algorithm running on a single processor.

Clustering improved convergence. We express a result's *excess cost* as $E_{final} - E_{min}$, where E_{final} is the result's cost, and E_{min} is the best solution known (presumably close to optimal). Clustering reduced the excess cost in a 101-cell, 265-wire problem by about 15%.

4.1.2. Rectangular decomposition

A simple approach, rectangular decomposition, tries to accomplish the same goals. It divides the grid of a VLSI placement problem into disjoint rectangles, then shifts the boundaries after each stream [4]. At low temperatures, interdependent state variables typically share a rectangle.

Different variants were tried: placing restrictions on the minimum width of a rectangle and "fuzzing" the rectangle boundaries. All rectangular decomposition schemes produced small errors and converged close to the minimum. In contrast, random landing point assignment on identical problems produced greater errors and converged to a much higher final cost [25].

One rectangular decomposition experiment fixed the number of generated states in a block of PN tries, where N is the stream length and P is the number of processors. Fig. 10 displays the resulting errors. The error value is $\varepsilon = |\Delta E - \Delta E_P|$, where ΔE is the actual cost change after completion of a stream, and ΔE_P is the sum of the apparent cost changes observed by the processors. Increasing P also increases ε, as one might expect.

4.2. Asynchronous shared state-space

Asynchronous shared state-space algorithms keep all state variables in shared memory. Processors competitively lock state variables, make moves, and unlock. Unlike synchronous algorithms, processors need not lock all affected state variables; they need only lock those variables required for the problem consistency.

One experiment compared synchronous and asynchronous algorithms for VLSI gate-array placement [7]. Under a simulated RP3 environment, three methods were tried. Method A, a synchronous shared state-space algorithm, is described in section 3.2. Each processor locked two circuits and any attached wires before attempting a swap.

In method B1, an asynchronous algorithm, processors lock only the two cells in the proposed move, and calculate the new wire length with possibly changing information. Each processor maintains a local copy of the histogram, which holds a collection of intermediate cost function variables [1]. A move updates only the local histogram; at the completion of a stream, each processor corrects its histogram with global state information. Method B2 operates like B1, except that it never corrects the local histograms. Thus, histogram information becomes progressively outdated as the temperature falls. Method B1 converged well with a maximum of 8 processors. Method B2 converged imperfectly, but surprisingly enough it converged better than a random spatial decomposition technique [25].

Using extrapolated simulation measurements for a 900-cell placement problem running on 90 processors, researchers estimated a speedup of about 45 for method A, and 72 for methods B1 and B2 [30].

Another experiment compared synchronous and asynchronous shared state-space algorithms for

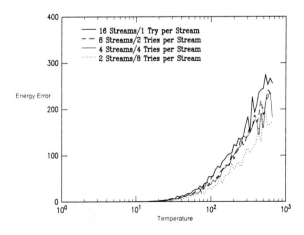

Fig. 10. Spatial decomposition, 16 tries per block.

the equal partition cut-set problem [17]. Given a graph with an even number of vertices, such algorithms partition the vertices into two equal sets, and minimize the number of edges which cross the partition boundary. The synchronous algorithm locked both vertices and edges, while the asynchronous algorithm locked only vertices.

On a 250-vertex graph, the synchronous algorithm ran more slowly than a sequential implementation, except at 16 processors where the speedup was close to 1. The asynchronous algorithm ran faster than the sequential algorithm, yielding 16-processor speedups from 5 on a graph with mean vertex degree 10, to 11 on a graph with mean vertex degree 80.

These two experiments indicate that asynchronous execution may be very beneficial in simulated annealing.

5. Hybrid algorithms

Hybrid algorithms recognize that different schemes may be more appropriate at different temperatures. We provide only a cursory review, since previous sections provide algorithmic details.

5.1. Modified systolic and simple serializable set

One hybrid combines a modified systolic algorithm and a simple serializable set algorithm [12]. In the modified systolic algorithm, independent processors copy the current state, then complete a stream of moves at the same temperature. The PICK operation chooses among the results, as per eqs. (1) and (2). Equal temperatures for PICK simplify the computations.

At high temperatures, where most moves are accepted, algorithm SSS provides little benefit – here only the systolic algorithm is used. As the lower temperatures reduce the acceptance rate, the program combines algorithm SSS with systolic. Finally, at extremely low acceptance rates, the program uses algorithm SSS exclusively.

Researchers claim this hybrid is slightly faster than the systolic algorithm alone [19].

5.2. Random spatial decomposition and functional decomposition

Another approach combines asynchronous spatial decomposition with functional decomposition [31]. This program randomly distributes the state variables across processors in an iPSC hypercube, to perform VLSI macro-cell placement.

With a 20 macro-cell problem, on a 16-processor iPSC, the algorithm obtained speedups of between 4 and 7.5. Considering the small problem size and the message-passing architecture, the speedup appears very good.

5.3. Heuristic spanning and spatial decomposition

One implementation uses heuristic spanning, a non-simulated annealing technique, and asynchronous rectangular decomposition to perform VLSI placement [28]. The heuristic spanning algorithm chooses several random starting states, and iteratively improves each. For the high-cost regime, heuristic spanning shows better convergence behavior than simulated annealing. In the low-cost regime, rectangular decomposition refines the state space to a lower final cost than heuristic spanning could achieve. The rectangular decomposition method showed a speedup of 4.1 on 5 processors, and an extrapolated speedup of 7.1 on 10 processors. Using the hybrid technique, researchers estimate speedups of 10–13 on 10 processors, when compared to a standard simulated annealing algorithm.

5.4. Functional decomposition and simple serializable set

In another hybrid algorithm, functional decomposition operates at high temperatures, and simple serializable set operates at low temperatures [11]. The poor behavior of algorithm SSS at high temperatures justifies a different algorithm.

In this early work, researchers sought to avoid convergence problems by using only serial-like algorithms – little was known of altered-generation or asynchronous algorithms. On a 100-cell gate-array placement placement, they achieved a maximum speedup of 2.25 on a 4 processor VAX 11/784.

6. Conclusion

We can neatly categorize parallel simulated annealing techniques into serial-like, altered generation, and asynchronous algorithmic classes. Experimental comparisons of these different techniques have appeared only recently, and have been limited in scope [4, 7, 11, 28].

Based on this survey, it appears that asynchronous and altered generation algorithms have provided the best overall speedup, while one serial-like technique, simple serializable set, has been incorporated advantageously at low temperatures. Several experiments indicate promising speedups in asynchronous algorithms.

Fruitful areas of parallel simulated annealing research include the following: empirical comparisons of the three algorithmic classes, using identical problems; characterization of annealing state spaces amenable to altered generation and asynchronous parallel annealing; the development of tuned temperature schedules which compensate for errors in asynchronous algorithms; and adapting work in related areas, such as computational ecologies, to parallel annealing. My colleagues and I are currently exploring these areas.

Acknowledgements

Miloš D. Ercegovac, Frederica Darema, Stephanie Forrest, Dyke Stiles, Steve R. White, Andrew Kahng, Richard M. Stein, M. Dannie Durand, Andrea Casotto, James Allwright, and Jack B. Hodges reviewed an early draft of this paper, and provided many helpful comments. Responsibility for errors rests with the author.

References

[1] S. Kirkpatrick Jr., C.D. Gelatt and M.P. Vecchi, Optimization by simulated annealing. Science 220 (1983) 671–680.

[2] C. Sechen, K.-W. Lee, B. Swartz, J. Lam and D. Chen, TimberWolfSC version 5.4: Row based placement and routing package, Technical report, Yale University, New Haven, CT (July 1989).

[3] B.A. Huberman and T. Hogg, The behavior of computational ecologies; in: The Ecology of Computation (North-Holland, Amsterdam), pp. 77–113.

[4] D.R. Greening and F. Darema, Rectangular spatial decomposition methods for parallel simulated annealing, in: Proceedings of the International Conference on Supercomputing, Crete, Greece (June 1989) pp. 295–302.

[5] M. Jones and P. Banerjee, Performance of a parallel algorithm for standard cell placement on the intel hypercube, in: Proceedings of the ACM/IEEE Design Automation Conference, (1987) pp. 807–813.

[6] V. Faber, O.M. Lubeck and A.B. White Jr., Superlinear speedup of an efficient sequential algorithm is not possible, Parallel Computing 3 (1986) 259–260.

[7] F. Darema, S. Kirkpatrick, and A.V. Norton, Parallel algorithms for chip placement by simulated annealing, IBM J. Res. Devel., 31 (1987) 391–402.

[8] J. Lam, An efficient simulated annealing schedule, Ph.D. Thesis, Yale University, New Haven, CT (December 1988).

[9] R. Brouwer and P. Banerjee, A parallel simulated annealing algorithm for channel routing on a hypercube multiprocessor, in: Proceedings of the International Conference on Computer Design (1988) pp. 4–7.

[10] J. Bannister and M. Gerla, Design of the wavelength-division optical network, Technical Report CSD-890022, UCLA Computer Science Department (May 1989).

[11] S.A. Kravitz and R.A. Rutenbar, Placement by simulated annealing on a multiprocessor, IEEE Trans. Computer-Aided Design 6 (1987) 534–549.

[12] F.M.J. de Bont, E.H.L. Aarts, P. Meehan and C.G. O'Brien, Placement of shapeable blocks, Philips J. Res. 43 (1988) 1–27.

[13] R.D. Chamberlain, M.N. Edelman, M.A. Franklin and E.E. Witte, Simulated annealing on a multiprocessor, in: Proceedings of the International Conference on Computer Design (1988) pp. 540–544.

[14] J.R.A. Allwright and D.B. Carpenter, A distributed implementation of simulated annealing for the travelling salesman problem, Parallel Computing 10 (1989) 335–338.

[15] E. Felten, S. Karlin and S.W. Otto, The traveling salesman problem on a hypercubic, mimd computer, in: Proceedings of the 1985 International Conference on Parallel Processing, St. Charles, PA (1985) pp. 6–10.

[16] S. Devadas and A.R. Newton, Topological optimization of multiple level array logic: On uni and multi-processors, in: Proceedings of the International Conference on Computer-Aided Design, Santa Clara, CA (November 1986) pp. 38–41.

[17] M.D. Durand, Cost function error in asynchronous parallel simulated annealing algorithms, unpublished manuscript (1989).

[18] D. Abramson, Constructing school timetables using simulated annealing: Sequential and parallel algorithms, Technical Report TR 112 069, Department of Communication and Electrical Engineering, Royal Melbourne Institute of Technology, Melbourne, Australia (January 1989).

[19] E.H.L. Aarts, F.M.J. de Bont, E.H.A. Habers and P.J.M. van Laarhoven, A parallel statistical cooling algorithm, in: Proceedings of the Symposium on the Theoretical Aspects of Computer Science, Vol. 210 (January 1986) pp. 87–97.

[20] N. Metropolis, A.W. Rosenbluth, M.N. Rosenbluth, A.H. Teller and E. Teller, Equations of state calculations by fast computing machines, J. Chem. Phys. 21 (1953) 1087–1091.

[21] S.B. Gelfand and S.K. Mitter, Simulated annealing with noisy or imprecise energy measurements, J. Optimization Theory Appl. (1989), to appear.

[22] L.K. Grover, Simulated annealing using approximate calculation, in: Progress in Computer Aided VLSI Design, Vol. 6, Ablex Publishing Corp., (1989); also as Bell Labs Technical Memorandum 52231-860410-01 (1986).

[23] D. Chazan and W. Miranker, Chaotic relaxation, Linear Algebra Its Applications 2 (1969) 199–222.

[24] P. Jog and D. Van Gucht, Parallelisation of probabilistic sequential search algorithms, in: Genetic Algorithms and Their Applications: Proceedings of the Second International Conference on Genetic Algorithms (1987) pp. 170–176.

[25] R. Jayaraman and F. Darema, Error tolerance in parallel simulated annealing techniques, in: Proceedings of the International Conference on Computer Design, (IEEE Computer Soc. Press, Silver Spring, MD, 1988) pp. 545–548.

[26] P. Banerjee and M. Jones, A parallel simulated annealing algorithm for standard cell placement on a hypercube computer, in: Proceedings of the International Conference on Computer-Aided Design (November 1986) pp. 34–37.

[27] L.K. Grover, A new simulated annealing algorithm for standard cell placement, in Proceedings of the International Conference on Computer-Aided Design (IEEE Computer Soc. Press, Silver Spring, MD, 1986) pp. 378–380.

[28] J.S. Rose, W.M. Snelgrove, and Z.G. Vranesic, Parallel standard cell placement algorithms with quality equivalent to simulated annealing, IEEE Trans. Computer-Aided Design 7 (1988) 387–396.

[29] A. Casotto, F. Romeo and A. Sangiovanni-Vincentelli, A parallel simulated annealing algorithm for the placement of macro-cells, IEEE Trans. Computer-Aided Design 6 (1987) 838–847.

[30] F. Darema, S. Kirkpatrick and V.A. Norton, Parallel techniques for chip placement by simulated annealing on shared memory systems, in: Proceedings of the International Conference on Computer Design (October 1987) pp. 87–90.

[31] R. Jayaraman and R.A. Rutenbar, Floorplanning by annealing on a hypercube multiprocessor in: Proceedings of the International Conference on Computer-Aided Design (November 1987) pp. 346–349.

GEOMETRIC LEARNING ALGORITHMS

Stephen M. OMOHUNDRO

International Computer Science Institute, 1947 Center Street, Suite 600, Berkeley, CA 94704, USA

Emergent computation in the form of geometric learning is central to the development of motor and perceptual systems in biological organisms and promises to have a similar impact on emerging technologies including robotics, vision, speech, and graphics. This paper examines some of the trade-offs involved in different implementation strategies, focusing on the tasks of learning discrete classifications and smooth nonlinear mappings. The trade-offs between local and global representations are discussed, a spectrum of distributed network implementations are examined, and an important source of computational inefficiency is identified. Efficient algorithms based on k-d trees and the Delaunay triangulation are presented and the relevance to biological networks is discussed. Finally, extensions of both the tasks and the implementations are given.

1. Introduction

Intelligent systems must deal with complex geometric relationships whenever they interact with their physical environment. The relationship between the motor signals sent to a set of muscles and the corresponding effect on the configuration of an organism's body in space is extremely complex and yet must be faithfully represented internally if the organism is to be able to effectively plan and carry out effective actions in the world. The relationship between the pattern of stimuli in different sensory modalities and the physical properties of the stimulus source is similarly geometrically complex. As if this intrinsic complexity was not enough, the perceptual and motor relationships change dramatically during an organism's lifetime as its size grows and its morphology matures. On an evolutionary time scale the change is even more dramatic and alternations to the organism's somatic form must be accompanied by altered internal models of its interaction with the world. A powerful strategy in the face of this variability is to base systems at least in part on learned instead of hardwired relationships. Aspects of the computations required must emerge during the interaction of the organism with the world.

As we attempt to build ever more complex and adaptive machines, we as engineers are faced with similar pressures toward systems in which at least part of the computational behavior is emergent. If successful, such systems should be far more robust and flexible than current hardwired systems. It is clear, however, that we are still at an early stage in the development of a discipline which will tell us which aspects of such systems can be made emergent and what the best approaches to implementation are. The three terms used in this paper's title suggest that such a discipline will be a symbiosis of the fields of mathematics (geometric), statistics (learning), and computer science (algorithms). While the joint interaction of these three fields together appears to just be beginning, there has been significant recent activity between them in pairs. Stochastic geometry studies statistical properties of geometric entities and much of classical statistics is being reformulated in a geometric coordinate-free form, the statistical analysis of probabilistic algorithms has become a central topic in theoretical computer science and computational issues have begun to be of critical importance to modern statistical analysis, and computational geometry and computational learning theory have flourished in the last decade. The time appears ripe for the development of a deeper understanding of emergent computation for the geometrical tasks essential to perception and motor control.

In this paper we will focus on the learning of relationships which have a geometric character to them. This is a small subclass of all possible emergent behavior, but it is one which is relevant to perceptual and motor tasks. We will examine a spectrum of implementation choices ranging from artificial neural networks to computational geometry algorithms and will identify some important trade-offs between them. We will see that in the network approaches a fundamental choice is the extent to which individual units are localized in their response. The use of localized units leads to faster and more reliable learning and is more robust in general, but global units may be more efficient if they are specifically tuned to the task at hand. We present a new biologically plausible implementation for a localized unit.

It is particularly clear in the systems with localized representations, but also true for those with global ones, that much of the computation involved in a typical retrieval is not actually used in determining the answer. We examine computational techniques for avoiding this computational inefficiency and describe some powerful data structures which have proven to be efficient, powerful, and easy to implement in practice. We describe a geometric construction known as the Delaunay triangulation and show that it is an optimal decomposition in a certain sense and that it may be efficiently implemented computationally. The algorithmic ideas presented apply to both serial and parallel implementations and the underlying concepts are generally useful for the construction of emergent systems. We describe how the basic concepts naturally lead to networks which employ strategies, such as focus of attention, which are commonly seen in animal brains.

We also discuss extensions to both the emergent tasks and their implementations.

2. Two geometric learning tasks

Our motivating goal is the construction of the components of an intelligent system which interface with the physical world. On the input side this includes components for visual, auditory, and somatosensory perception, and on the output side, graphics, sound production and robotics. These components must provide the interface between the primarily geometric nature of the world and the apparently symbolic nature of higher intelligence. As Harnad [11] discusses, it is these interface systems which provide the conceptual grounding for internal symbols.

An example task of great importance is visual object recognition. It takes you perhaps half a second to visually recognize the volume in your hands as a book, yet no current engineered systems are capable of this task. Because this kind of processing happens so quickly and is mostly below the level of consciousness, people notoriously underestimate its difficulty. Artificial intelligence researchers in the early sixties consistently underestimated the time and computing power needed for intelligent computation. Partly motivating this optimism was a belief in a kind of "holy grail" of intelligence. The idea was that if one just found the right clever inference mechanism, a simple system could exhibit human level intelligence (apparently John McCarthy believed that a PDP-1 was a sufficiently powerful computer for human level performance [7].) Unfortunately, the early AI systems were found to be quite rigid and to lack common sense [3].

During the seventies interest shifted away from fancy inference mechanisms and toward the representation of task-specific knowledge. The success of a variety of expert systems showed that with enough knowledge even systems with quite limited inferential power could perform quite well. The phrase "knowledge is power" was used to describe the new emphasis. It is now believed that large amounts of knowledge about a domain are needed to achieve human level performance. Recent estimates suggest that people know at least 70 000 "chunks" of information in each of the domains in which they are expert [30]. Most AI systems are currently constructed entirely by hand and an enormous effort would be needed to give them anywhere near this amount of knowledge. It is perhaps this realization which has stimulated the great interest in machine learning, neural networks, and other systems with emergent computation during the past few years.

We would like machines to build up their knowledge bases through experience, as people do. Simply remembering past experiences is a conceptually trivial task. The key to effective learning is the ability to *generalize* previous experience to new situations. It is difficult to precisely identify the nature of desirable generalizations because of the philosophical obstacles in inductive logic and the philosophy of science as discussed by Churchland [5]. Any criterion which a procedure uses to prefer one generalization over another, such as Occam's razor, is called the *inductive bias* of the procedure. Many different inductive biases have been studied in purely symbolic domains [22].

Unlike general domains, there is a natural inductive bias in geometric domains such as those of interest here. We might call this bias the *principle of continuity*. Unless a system explicitly knows otherwise, it should assume that geometrically nearby perceptions correspond to nearby states of the world. A system faced with a problem can apply previous experiences which were geometrically near to it. We will call problems in which this inductive bias is applicable "geometric learning" problems. Learning tasks are generally separated into "supervised learning", in which a teacher provides example inputs with corresponding desired outputs, and "unsupervised learning", in which there is no teacher present and the nature of the task is more amorphous. There are a large number of important tasks for which the basic point made here is relevant, but we will focus on the two most natural supervised geometric learning tasks: classification learning and smooth nonlinear mapping learning.

For both of our tasks, we will assume that the input to the system may be represented as a point in an n-dimensional Euclidean feature space $\mathbb{R}^n$. The geometric structure of this space is meant to capture the geometry of the input in the sense that inputs which are close in Euclidean distance in the space should have a similar structure in the world. For classification tasks, the output of the system is one of a discrete set of classes. For mapping tasks, the output lies in another continuous space and varies smoothly with changes in the input.

A topical example of classification is *optical character recognition*. There are now several products available for scanning documents as bitmap images and converting them to machine-readable text. After the images of individual characters have been isolated, the essential recognition step is to classify them as characters. Most of the commercial products work by extracting about ten or twenty real-valued features from the image of the character, such as the ratio of its width to height, the density of its darkened pixels, or Fourier components of its image density projected onto the horizontal and vertical axes. The extracted feature vector is sent to a classifier whose job it is to choose the ASCII character which produced the image. A wide variety of fonts are used in modern publications and current systems attain robustness by using *learning*. A system is trained by presenting it with a sample document along with the identity of its characters. The system must adjust its classification procedure on the basis of this example input. Any classifier defines a partition of the input space with a region corresponding to each possible output (see fig. 1). An adaptive classifier must learn this partition from a set of labelled examples.

An example of a system for the evaluation of smooth nonlinear mappings which was studied in my laboratory is provided by MURPHY, a visually guided robot arm [21] (see fig. 2). The system

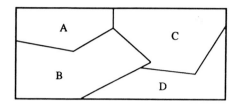

Fig. 1. The input space partition defined by a classifier.

controls three of the joints of robot arm and receives visual feedback from a video camera. Six white dots are painted on the arm and a real-time image processing system is used to identify their x and y coordinates in the image. The system thus has two descriptions of the state of the arm: a kinematic one represented by a vector in the three-dimensional space of joint angles, and a visual one represented by a vector in the twelve-dimensional space of image dot locations. Because many problems are specified in visual coordinates, but the system only has direct control over kinematic coordinates, one important task is for the system to predict the visual state corresponding to any given kinematic state. This is a smooth nonlinear mapping from $\mathbb{R}^3$ to $\mathbb{R}^{12}$. In ref. [21], this mapping was used as the basis for a variety of behaviors including reaching around visually defined obstacles to reach a visually defined goal.

The traditional approach to implementing a system of this type [27] would be to write down a set of model equations for the kinematics of the arm and for the imaging geometry of the camera, to form their composition analytically, and to evaluate the resulting mapping numerically. If the system is precisely constructed, such an approach can work well. With inexpensive or changing components, however, such an approach is likely to end up smashing the arm into the table. Biology must deal with the control of limbs which change in size and shape during an organism's lifetime. An important component of the biological solution is to build up the mapping between sensory domains by learning. A baby flails its arm about and sees the visual consequences its motor actions. Such an approach can be much more robust than the analytical one. MURPHY's camera had an auto-focus lens which caused the imaging geometry to vary dynamically as the arm moved. The system cheerfully succeeded in learning the effects of this complexity.

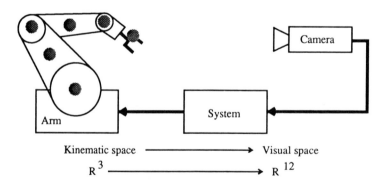

Fig. 2. The components of MURPHY, a visually guided robot arm.

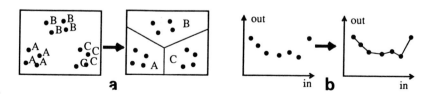

Fig. 3. (a) Classification learning. (b) Mapping learning.

We have i[...]   ition there is an intermediate layer of units
learning tasks.[...]   ally called the *hidden layer*), each of which
learner must f[...]   ives input from each input neuron and sends
based on label[...]   utput to each output neuron. The output of
tive bias shou[...]   model neuron is obtained by composing a
points similar[...]   r combination of its inputs with a nonlinear
gions with sm[...]   ashing function" such as a sigmoid. In form-
more like inte[...]   he linear combination, each input to a neuron
to smoothly [...]   ultiplied by a "weight" which is meant to
input. In ref.[...]   esent the strength of a biological synapse. For
geometric lear[...]   setting of the weights the whole network
guided robot a[...]   putes a mapping from the input space to the
                                  ut space. As we vary the weights, the mapping
                                  s in a smooth but complex way. During learn-
                                  he system is presented with inputs and the
                                  sponding desired outputs. For any setting of
3. Artificial ne[...]   veights the mapping implemented by the net-
                                  will make some errors compared to the de-
How might [...]   mapping. If we estimate this error by taking
learning tasks?[...]   nean-squared error on the training set, we
to use an artif[...]   n an error function on the space of weight
propagation le[...]   gs of the network. The backpropagation
approach the s[...]   ing procedure just performs gradient descent
neurons conne[...]   e error in this weight space. One cycles
be specific, le[...]   gh the training set and on each example
described abov[...]   es each weight proportionally to its effect on
a smooth mapp[...]   ing the error. One may compute the error
to a twelve-di[...]   ent using the chain rule and the information
the input vect[...]   propagates backwards through the network, which
three input neurons and the output vector by the   accounts for the procedure's name.
real-valued activities of twelve output neurons. In

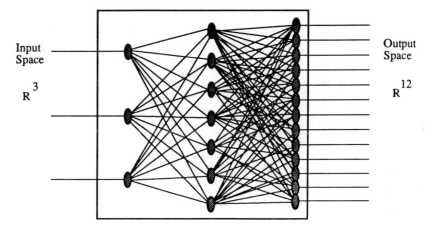

Fig. 4. An example backpropagation neural network for the robot arm learning task.

There has been much recent interest in this learning procedure partly because it is easy to implement and apply to a wide variety of problems. Unfortunately, a number of difficulties with it have become evident with experimentation. If one has an engineering task to solve, there are at present no techniques for obtaining bounds on how well a specific network will do on the task. The choice of how many hidden units to use is something of a black art. If there are too many, the network tends to just store the training examples and generalizes inappropriately. If there are too few, then it will not have sufficient representational power to approximate the desired mapping. The training procedure tends to be very slow and unpredictable, often depending strongly on the random starting conditions, and getting stuck at suboptimal local minima. Preliminary experiments indicate that the procedure does not scale well with the size of the problem [33]. If one is thinking of constructing large systems, it is a disadvantage that the activity of the units typically does not have a well defined "meaning" in terms of the input patterns. The procedure also appears to be biologically implausible.

Most of the problems with backpropagation in geometric domains stem from the fact that the units have global "receptive fields". By the receptive field of a unit we mean the region of the input space in which that unit has significant activity. The units in the class of networks we have described tend to have very large receptive fields. A single linear-threshold unit is active in the entire region of the input space which is bounded by a hyperplane. As the weights are varied, the hyperplane moves about. When the mapping is incorrect in a part of the space, it is the units whose receptive fields overlap that portion which have their weights adjusted. When receptive fields are global, most of the units must be adjusted for every kind of error. In correcting an error in one part of the space, there is a tendency to disrupt improvements made in another part. The large receptive fields are also largely responsible for the problems with local minima. Because a unit contributes to so many regions of the mapping, it is possible for it to get wedged in a position where any weight change is detrimental, even though there is a better state available. Several authors have discussed advantages of localized receptive fields [1, 35, 23].

The extreme of a network with localized receptive fields would have a separate input neuron for each small region of the input space, with no overlap. In fig. 5 we only show one output neuron. There will be as many such neurons as there are output dimensions. Learning a mapping in such a network is trivial. The partition of the input space shows the regions within which each input neuron fires. For any given input, only one input neuron is active. This makes the learning procedure extremely simple and fast. For each input/output pair, the system need only set the weight for the active input neuron at the value which produces the desired output level. This simple learning rule gives a piecewise constant approximation to the

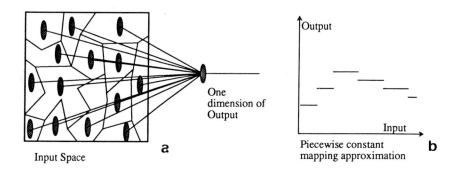

Fig. 5. (a) One of the output neurons in an extreme localist representation. (b) The resulting mapping approximation is piecewise constant.

mapping. It is easy to give explicit error bounds for a set of units representing a mapping with bounded Jacobian [24]. The receptive fields may also be adaptive to the underlying mapping, being larger in regions of small variation and smaller in regions of high variation. There are several approaches to adjusting the receptive fields automatically [15]. Such a system generalizes by assuming that the output should be constant over the disjoint receptive fields of the input units. For smooth functions, we can do much better than this. Consequently, the extreme localist representation tends to require a large number of units for a given level of accuracy.

We may take advantage of the smoothness of the desired mappings while still retaining many of the advantages of localization by allowing the input receptive fields to overlap somewhat. In fig. 6 we see the activities of six input neurons as a function of a one-dimensional input. The response is peaked at the center of each receptive field and dies down linearly away from it. Neighboring input neurons have overlapping receptive fields. We again assume a single linear output neuron for each output dimension which receives input from each input neuron. A simple learning procedure will allow the system to approach a piecewise linear approximation of the desired mapping. If each unit has the correct output when the input is at its center, then as we move from one center to the next, the output linearly interpolates between the two central values. Similar behavior may be achieved in higher dimensions by making a unit's receptive field be the star of a vertex in a triangulation and its response be linearly decreasing to zero in each bordering simplex. Again the receptive fields may be made adjustable, though now they need only be small in regions of high curvature rather than high variation. Again, precise error bounds may be obtained for mappings with bounded Hessian. The learning procedure is no longer one-shot, but must only simultaneously adjust the weights of units with overlapping receptive fields and so is fast, reliable, and analyzable. The degree of approximation and level of generalization of this kind of representation is far greater than with non-overlapping units, and yet most of the advantages of localization are retained. This kind of coding appears to be used extensively in biological nervous systems, especially near the periphery.

4. Why networks can be inefficient

We can see immediately the origin of computational inefficiency in networks with localized units. For each input presentation we must evaluate the state of each unit in the network, even though only a few of them contribute to the desired output. If you are only worried about the state of your hand, all of the computational capacity in the sensory system of your legs is wasted. This same kind of inefficiency occurs in the networks with global receptive fields, though it is less obvious. As an example, consider a perceptron classifier made up of linear threshold units. The hyperplanes corresponding to the input units partition the input space into regions. Within each region the classifier must produce the same response. The

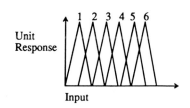
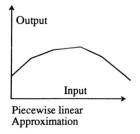

Fig. 6. The activities of six input neurons as a function of a one-dimensional input and the resulting piecewise linear approximation of the mapping.

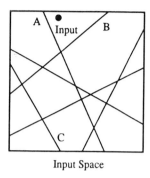

Fig. 7. A partition of the input space defined by perception hyperplanes.

perceptron learning algorithm adjusts the hyperplanes to fit the desired classification regions as well as possible. To determine the classification of an input we need only know which partition region the input lies in. A perceptron network determines this by computing the activity of each unit. This corresponds to testing the input against each hyperplane. Looking at fig. 7, however, we see that this task might be amenable to that basic trick of computer science: divide and conquer. Here the input space is two-dimensional and the hyperplanes corresponding to input units are just lines. Once we have determined that the input point is above lines A and B, there is no longer any need to compare it with line C; it must lie above it. In this manner we can use the results of earlier queries to rule out possibilities and avoid having to do computation. In section 5 we will discuss techniques for ruling out about half of the remaining possible questions for each question that is asked. In this way the number of questions whose answer must actually be computed is only the logarithm of the number of possible questions.

To see what effect such an approach might have on realistic computational problems, consider the robot arm task. Let us analyze the approach based on units with overlapping localized receptive fields. If we assign 50 units to represent each dimension, we need a total of $50 \times 50 \times 50$ or about 100 000 input units. Each input unit synapses on each of the 12 output units, so there will be over 1 000 000 weights. On each input presentation, there is a multiplication and an addition associated with each weight, so there are at least two million floating point operations involved in each query. For any particular query, only about 50 of these operations are actually relevant to the output. Variants of the algorithmic approaches discussed in section 5 need only about 100 floating point operations to complete the task. The extra operations are used in the determination of which operations are actually relevant to the output. We see that even on this relatively small problem the speedup can be a factor of 20 000. For larger tasks, speedups can be even more spectacular.

5. Algorithmic implementations

Let us now consider algorithmic approaches to the two problems under consideration. We will present specific solutions for illustration's sake, but there are many variations which may be superior in specific situations. The presentation is intended to emphasize algorithmic techniques whose basic principles are applicable to a wide variety of specific procedures.

5.1. Classification

There are many different approaches to learning a classifier from examples. In the asymptotic limit of a large number of examples, however, there is a universal classifier which does nearly as well as any other. Nearest neighbor classification simply assigns to a new input the class label of the nearest training example. A theorem proved by Cover and Hart [6] shows that if we describe the classes by probability distributions on the input space, then the probability of a classification error using nearest neighbor classification is asymptotically at most twice that of any other classifier. There are many variants on the basic scheme. For example, if there are labelling errors in the training set, it may be advantageous to let the nearest m neighbors vote on the class label.

While this basic procedure appears to depend on the notion of distance used, it is in practice quite robust. There are two length scales associated with a classification problem: the scale of significant variation in the underlying probability distributions and the typical spacing of training examples. Any of a wide variety of distance metrics which preserve the distinction between these two scales will give similar results. For a fixed number of samples, scaling one of the dimensions by a large enough number that the sample spacing becomes comparable to the scale of class variation along other dimensions will dramatically lower classification accuracy. For any fixed metric, however, such effects wash out as the number of samples increases. While there are many such practical issues to consider, we will simply consider an algorithmic technique for the basic task of finding the nearest neighbor.

Algorithms for finding nearest neighbors are studied in the field of *computational geometry*. This discipline was developed less than 15 years ago but has blossomed in recent years [10, 16]. Most of the work has been concerned with algorithms with good worst case performance. The most straightforward approach to nearest neighbor finding is to measure the distance from the test sample to each training sample and choose the closest one. This requires a time which is linear in the number of samples. Unfortunately, it does not appear that in high-dimensional spaces the worst case performance can be much better than this. If there are twice as many samples in a system, the retrieval will take twice as long. Most commercial speech recognition and optical character recognition systems on the market today use the straightforward approach and so are limited in the number of training samples they can deal with.

In practical applications one is usually concerned with good performance on average rather than with worst case performance, but theoretical analyses are then faced with the choice of distribution to average with respect to. The field of nonparametric statistics has developed techniques for analyzing the average properties of sets of samples drawn from *unknown* underlying distributions. Friedman et al. [13] used some of these techniques to develop a nearest neighbor finding algorithm which asymptotically runs in a time which on average is only logarithmic in the number of stored samples.

The algorithm relies on a data structure known as a *k-d tree* (short for *k*-dimensional tree). This is a binary tree with two pieces of information stored at each node: a dimension number d, and a value v. The nodes correspond to hyper-rectangular regions of the input space which we shall refer to as *boxes*. The root of the tree corresponds to the whole space. The two children of a node correspond to the two pieces of the parent's box that result when the dth dimension is cut at the value v. The left child corresponds to the portion of the box in which $x_d \leq v$ and the right to the portion in which $x_d > v$. As we descend the tree, the root box is whittled down to smaller and smaller boxes. The boxes of all the nodes at a given level in the tree form a partition of the input space. In particular, the leaf boxes form the *leaf partition*. Each box may be cut along any of the k dimensions and may be cut at any value which it includes. As a simple example, fig. 8 shows the partitioning of a three-dimensional input space ($k = 3$). There are four leaves in the tree whose corresponding box regions are labelled A, B, C, and D.

k-d trees are extremely simple and efficient to represent in the computer and yet they directly support a simple kind of geometric access. If we want to know in which leaf box a sample point lies, we need only descend the tree, at each node comparing the point's dth component with the value v. If it is less than or equal to v, we proceed to the left child, otherwise to the right, halting when we reach a leaf. The number of comparisons involved is equal to the depth of the tree. If the tree is balanced, it is logarithmic in the number of nodes.

For nearest neighbor finding, we need a k-d tree which is adapted to the training data. Each node is associated with the set of training samples contained in its box. The root is associated with the

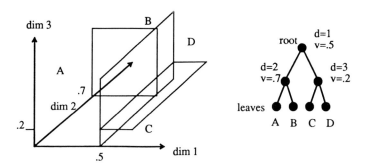

Fig. 8. The leaf partition of a three-dimensional input space and the corresponding k-d tree.

entire set of data. If we build the tree top down, for each node we need to decide which dimension to cut and where to cut it. There are several useful strategies here but the simplest is to cut the dimension in which the sample points associated with the node are most spread out and to cut it at the median value of those samples in that dimension. One might continue chopping until only one sample remains in each leaf box, but it is often useful in practice to leave a small number (e.g. 10) of samples in each leaf bucket. The expected shape of the leaf boxes under this procedure is asymptotically cubical because the long dimension is always cut. If the points are drawn from an underlying probability distribution, the expected probability contained in each leaf box is the same because cuts are made with an equal number of samples on each side. Thus the leaf partition is beautifully adapted to the underlying probability distribution. It chops the space into cubes which are big in the low-density regions and small in the high-density regions.

It is interesting to relate this strategy to Linsker's information theoretic analysis of neural adaptation strategies [18]. He suggests that a powerful way for networks to adapt is to adjust their properties in such a way that the mutual information between their inputs and outputs is maximized subject to any architectural constraints on the units. If the samples are subject to small additive noise, then the k-d cutting procedure maximizes the mutual information between an input sample and the "yes/no" question of which side of a cut it lies on. Using an argument similar to that in ref. [17] one can see that cutting at the median makes the output most informative and cutting the long dimension makes the volume of the cut as small as possible, minimizing the region of confusion caused by the noise. Of all the allowed k-d cuts at a given stage, the procedure chooses the one which is expected to give the most information about the input sample as we descend from the root. Successive "info max" cuts like this do not necessarily yield the "info max" tree but the resulting tree should be fairly close to it.

How is this adapted tree used to find the nearest neighbor of a test sample? As described above we find which leaf box the sample lies in log expected time. For many practical purposes, simply using the stored samples in this leaf box will be sufficient, yielding an extremely fast classifier. The nearest neighbor might not be in the same leaf box, however, if it is close to the edge of a neighboring box. To find it, Friedman et al. [13] present a branch and bound technique. We maintain the distance to the nearest sample point seen at any point in the algorithm. We descend the tree, pruning away any subtrees which cannot possibly contain the nearest point. If the entire box associated with a node is further from the test point than the nearest point seen so far, then the nearest neighbor cannot possibly lie within it. This branch of the tree need then be explored no further. Eventually all branches will have been pruned and the

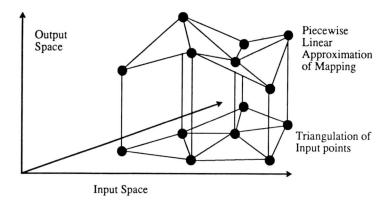

Fig. 9. An approximation of a mapping from a two-dimensional input space to a one-dimensional output space.

closest point seen so far is guaranteed to be the nearest neighbor. Both the nearest neighbor balls and the leaf boxes are big in low density regions and small in high density regions. Using a precise version of this observation, one may prove that on average only a constant number of leaves need be explored for any number of stored samples. The entire process is then asymptotically dominated by the initial logarithmic search.

5.2. Mapping learning

How might we apply similar ideas to the problem of learning nonlinear mappings? As we have discussed, we would like the system to interpolate between nearby training examples in evaluating the output for a test sample. Simple techniques, such as linearly interpolating between the values at the $k + 1$ nearest neighbors can work well in certain circumstances [8, 12, 24], but leads to discontinuous approximations. A more well behaved approximating mapping may be constructed from any *triangulation* of the sample points. If the input space is k-dimensional, $k + 1$ vertices are needed to define each primary simplex (i.e. higher-dimensional tetrahedron) in a triangulation. $k + 1$ output values are also needed to perform linear interpolation. If we choose a triangulation, then linear interpolation of the vertex values within each simplex yields a continuous approximating function. Fig. 9 shows an approximation of a mapping from a two-dimensional

input space to a one dimensional output space. There are two problems immediately suggested by this approach. The first is to find a criterion for selecting a good triangulation for approximation. In general, long skinny triangles will be bad, because the mapping may vary significantly in a nonlinear way along the long dimension. The second problem is to efficiently find the simplex in the chosen triangulation which contains the test sample, so that the linear interpolation may be performed. A nice solution to both of these problems may be had by using a special triangulation called the *Delaunay triangulation* [29].

The Delaunay triangulation is based on the fact that in a k-dimensional Euclidean space, $k + 1$ points generically determine a sphere as well as a simplex. For example (see fig. 10) in two dimensions, three points determine both a circle and a

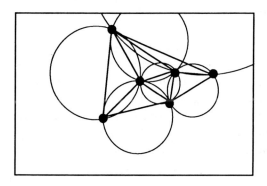

Fig. 10. The Delaunay triangulation of a set of points in two dimensions.

triangle. A set of $k+1$ sample points form the vertices of a simplex in the Delaunay triangulation if and only if the sphere which they determine does not contain any other sample points. The bounding spheres of the simplices in the triangulation are thus as small as possible and the triangles are as equilateral as possible. We have proven that among all mappings with a given bound on their second derivative, the piecewise-linear approximation based on the Delaunay triangulation has a smaller worst case error than that based on any other triangulation. The intuitive idea behind the result is quite simple. The worst error arises in approximating functions whose second derivative is everywhere equal to the maximum, i.e. with quadratic functions whose graph is a paraboloid of revolution and whose level sets are spheres. When we linearly interpolate through the values at the vertices of a simplex, the error will vanish on those vertices. The worst error function therefore has a level-set which is the sphere determined by the vertices of the simplex. The worst error occurs at the center of the sphere and is proportional to the square of the radius. At any input point the worst possible error is smaller when the spheres are smaller, so the Delaunay triangulation is best. It is straightforward to make this argument rigorously.

The Delaunay triangulation is also useful because the determination of which simplices are included may be made locally in a region of the input space. If we decompose the input space with a k-d tree built around the input samples, then the leaf boxes are small where the Delaunay spheres are small and large where they are large. A branch and bound algorithm very similar to the nearest neighbor finding algorithm may be used to determine the Delaunay simplex containing an input point in logarithmic time. Again we maintain the smallest Delaunay sphere whose simplex contains the point in question. We descend the tree, pruning away any branches whose box does not intersect the current Delaunay sphere and obtain the provably correct simplex in log expected time. It is not as easy to analytically obtain the probability

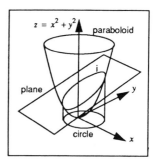

Fig. 11. The intersection of a plane and a paraboloid projects to a circle.

distribution of Delaunay spheres as it is of nearest neighbor spheres, but even coarse bounds on the distribution of the radius yield the desired algorithmic performance.

To compute the sphere determined by $k+1$ points, it is convenient to map the input space onto a paraboloid of revolution in a space of one higher dimension [29] by sending $(x_1, \ldots, x_k)$ to $(x_1, \ldots, x_k, x_1^2 + \ldots + x_k^2)$. Hyperplanes in this space intersect the paraboloid in the image of spheres under the mapping. To find the sphere determined by $k+1$ points one need only determine the hyperplane that their images determine. The Delaunay triangulation corresponds under this mapping with the convex hull of the images of the points. It is interesting to note that this construction shows that if a linear-threshold model neuron is given an extra input which is the sum of the squares of the other inputs, then its receptive field becomes a localized sphere. As the weights of the unit are varied, the center and radius of the sphere change. The exact form of the nonlinearity is not crucial, and in this way one may obtain biologically plausible neurons with localized receptive fields whose location and shape varies with simple linear weights. Fig. 11 shows a two-dimensional version of this construction.

6. Extensions

The basic ideas outlined here may be extended in a variety of ways to give algorithmic techniques for solving other important geometric learning tasks. Some of these are discussed in ref. [24]. Here we will briefly mention a few extensions that are under active investigation. The nonlinear mapping problems we have discussed have well defined input and output spaces. In many situations the system knows the values of different features at different times and from them would like to predict the values of the others. The relationship between the features may not be in the form of a mapping but instead may be a multi-sheeted surface or a probability distribution. The first case naturally leads to the task of *submanifold learning*. The system is given samples drawn from a constraint surface relating a set of variables. It must induce the dimension of the surface and approximate it. Typical queries might include predicting the values of unknown features when a subset of the features are specified (partial match queries). Geometrically this corresponds to finding the intersection of the surface with affine subspaces aligned with the axes. In situations with error, it should be able to find the closest point on a constraint surface to a given set. One approach to this problem is closely related to the Delaunay approach to mapping learning. If the surface to be learned is not self-intersecting and has bounded curvature (the analog of bounded Hessian), then we may prove that asymptotically a bounded radius variant of the Delaunay triangulation will converge on the surface with high probability. We only include spheres whose radius is less than the bound on the radius of curvature. In general, only simplices with some dimension less than $k + 1$ can be formed with this constraint and asymptotically this dimension gives the dimension of the surface. Branch and bound may be used to quickly find the closest point on the surface to a given point. This gives a nonlinear analog of linear pseudo-inverses.

k-d tree like structures are used in ref. [2] for a variety of statistical tasks. For many of these problems a generalization of k-d trees which we call *boxtrees* is superior. These structures store an entire box at each node and the boxes of siblings are allowed to intersect, yet they may be efficiently constructed and manipulated [26]. They are useful for learning and manipulating not only smooth constraint surfaces, but also smooth probability distributions using adaptive kernel estimation [9]. They support efficient random variate generation and Bayesian and maximum-likelihood inference. For distributions with fractal support they may be used to efficiently estimate the Hausdorff and information dimensions. For classification, variants which use information theoretic measure in construction are useful [14]. Boxtree structures also provide a powerful generalization of standard techniques for image decomposition [31].

These structures are also useful as components in larger systems. Graph structures which represent dependency relationships have become important in both probabilistic domains, where they are called influence diagrams or Bayesian networks [28, 32] and in deterministic domains, where they are called constraint networks [19]. Both of these domain types may be efficiently extended to continuous variables by using boxtree structures at the nodes. It appears that in this context algorithmic techniques may improve the speed and performance of the information propagation as well [4].

7. Relevance to parallel and network systems

While we have only described serial algorithms, many of the concepts are parallelizable. Omohundro [24] describes a massively parallel algorithm for k-d tree retrieval with almost optimal speedup on computers like the Connection Machine. When more exotic technologies are considered, algorithmic comparisons become more complex. While these algorithms eliminate much of the computation that is needed in straightforward network approaches, they still use as much memory to store the structures. In some technologies the cost of memory structures may be equivalent to the cost of computation structures, possibly nullifying some of the gains. In massively parallel systems, however, communication tends to be more of a bottleneck than computation. In this case, the wasted communication required by the straightforward network approaches will still cause systems which can prune away useless work to make better use of their hardware.

It is interesting to ask what relevance these considerations have for biological networks. Even though brains are large and highly interconnected, they are still extremely constrained by computational limitations. For example, a straightforward representation of the space of line segments on the retina might assign a segment unit to each pair of endpoints. Since there are about 10^6 fibers in the optic nerve, this would require 10^{12} units, more than using up our entire brain (this observation is due to Jerry Feldman). Similarly, it is easy to show that the completely connected networks favored by many backpropagation practitioners would lead to brains of enormous size if applied to any sizable chunks of cortex. Considerations such as these show that there are tremendous pressures on nervous systems for efficiency in representation, computation, and communication between areas. Neural hardware has properties which are very different from those of current computers and one expects the algorithmic trade-offs to be different. In particular, the locus of memory and computation appear to be the same in biological networks.

Some of the techniques that appear to be used in biological systems have the flavor of the algorithms described here. Each of the sensory modalities makes use of some form of focus of attention. Presumably this is a mechanism to devote higher-level hardware to only a portion of the data produced by lower-level systems. In this way a single piece of high-level hardware can be serially applied to different parts of the parallel sensory input. If this were not done, the high-level hardware would need to be replicated for each sensory focus. This kind of pruning of sensory input is very similar to the pruning of branches in the k-d tree to focus on only relevant knowledge. In general, hierarchical representations may be used to prune away possibilities at a coarse level and the strategy of proceeding from coarse to fine in the process of recognition is a common one. There appears to be psychological and neurophysiological evidence of some form of hierarchical representation in virtually every representation area of the brain.

The emergent aspect of the approaches described here arise from the construction algorithm which attempts to form a hierarchical representation which is optimally adapted to the system's experiences. One would like a general theory of such algorithms which would apply to both general algorithmic approaches and those bound by hardware constraints. The particular algorithms discussed here work in a top down manner and are off line (in that they need to be presented with all the data before they build their structures). Related structures described in ref. [26] may be constructed in a bottom up on-line fashion and are probably more similar in character to what is possible with dynamic networks. Understanding the formation of emergent hierarchical structure in realistic networks is a task of fundamental importance.

We have tried to give insights, in the context of specific examples, into some aspects of the nature of emergent computation in geometric domains. It is clear that a fundamental understanding of both the algorithmic and physical underpinnings of this kind of emergent computation will be critically important for both future science and technology.

Acknowledgements

I would like to thank Bill Baird, Kim Crutchfield, Herbert Edelsbrunner, Doyne Farmer, Jerry Feldman, Sally Floyd, Hans Guesgen, Ralph Linsker, Bartlett Mel, Nelson Morgan and Norm Packard for discussions of these ideas. This work was supported by the International Computer Science Institute.

References

[1] D.H. Ballard, Interpolation coding: a representation for numbers in neural models, Biological Cybernetics 57 (1987) 389–402.

[2] L. Breiman, J.H. Friedman, R.A. Olshen and C.J. Stone, Classification and regression trees, Wadsworth International Group, Belmont, CA (1984).

[3] E. Charniak and D. McDermott, Introduction to Artificial Intelligence (Addison–Wesley, Reading, MA, 1985).

[4] P.B.-L. Chou, The theory and practice of bayesian image labeling, University of Rochester Department of Computer Science Technical Report No. 258 (1988).

[5] P.M. Churchland, Physica D 42 (1990) 281–292, these Proceedings.

[6] T.M. Cover and P.E. Hart, Nearest neighbor pattern classification, IEEE Trans. Information Theory IT-13 (1967) 21–27.

[7] J.D. Cowan, personal communication (1989).

[8] J.P. Crutchfield and B.S. McNamara, Equations of motion from a data series, Complex Systems 1 (1987).

[9] L. Devroye and L. Gyorfi, Nonparametric Density Estimation: The L1 View (Wiley, New York, 1985).

[10] H. Edelsbrunner and J. van Leewen, Multidimensional data structures and algorithms, A bibliography, IIG, Technische Universität Graz, Austria, Report 104 (1983).

[11] S.R. Harnad, Physica D 42 (1990) 335–346, these Proceedings.

[12] J.D. Farmer and J.S. Sidorowich, Predicting chaotic time series, Los Alamos National Laboratory Technical Report No. LA-UR-87-1502 (1987).

[13] J.H. Friedman, J.L. Bentley and R.A. Finkel, An algorithm for finding best matches in logarithmic expected time, ACM Trans. Math. Software 3 (1977) 209–226.

[14] D.R. Hougen and S.M. Omohundro, Fast texture recognition using information trees, University of Illinois Department of Computer Science Technical Report No. UIUCDCS-R-88-1409 (1988).

[15] T. Kohonen, Self-Organization and Associative Memory (Springer, Berlin, 1984).

[16] D.T. Lee and F.P. Preparata, Computational geometry – A survey, IEEE Trans. Computers, C-33 (1984) 1072–1101.

[17] R. Linsker, Towards an organizing principle for a layered perceptual network, in: Neural Information Processing Systems – Natural and Synthetic, ed. D.Z. Anderson (American Institute of Physics, Denver, 1987) pp. 485–494.

[18] R. Linsker, Self-organization in a perceptual network, IEEE Computer (March 1988) 105–117.

[19] A.K. Mackworth, Consistency in networks of relations, Artificial Intelligence 8 (1977) 99–118.

[20] J.L. McClelland and D.E. Rumelhart, Parallel Distributed Processing: Explorations in the Microstructure of Cognition. Vols. 1, 2. Psychological and Biological Models (MIT Press, Cambridge, MA, 1986).

[21] B.W. Mel, MURPHY: A neurally-inspired connectionist approach to learning and performance in vision-based robot motion planning, University of Illinois Center for Complex Systems Research Technical Report No. CCSR-89-17A (1989).

[22] R.S. Michalski, J.G. Carbonell and T.M. Mitchell, eds., Machine Learning: An Artificial Intelligence Approach, Vols. I, II (Kaufmann, Los Altos, CA, 1986).

[23] J. Moody and C. Darken, Learning with localized receptive fields, Research Report No. YALEU/DCS/RR-649 (1988).

[24] S.M. Omohundro, Efficient algorithms with neural network behavior, Complex Systems 1 (1987) 273–347.

[25] S.M. Omohundro, Foundations of geometric learning, University of Illinois Department of Computer Science Technical Report No. UIUCDCS-R-88-1408 (1988).

[26] S.M. Omohundro, Five balltree construction algorithms, International Computer Science Institute Technical Report, TR-89-063.

[27] R.P. Paul, Robot Manipulators, Mathematics, Programming and Control (MIT Press, Cambridge, MA, 1981).

[28] J. Pearl, Probabilistic Reasoning in Intelligent Systems: Networks of Plausible Inference (Kaufmann, San Mateo, CA, 1988).

[29] F.P. Preparata and M.I. Shamos, Computational Geometry, An Introduction (Springer, Berlin, 1985).

[30] R. Reddy, Foundations and grand challenges of artificial intelligence, AI Magazine (Winter, 1988) 9–21.

[31] H. Samet, The quadtree and related hierarchical data structures, ACM Computing Surv. 16 (1984) 187–260.

[32] R.D. Shachter, Evaluating influence diagrams, Operations Res. 34 (1986) 871.

[33] G. Tesauro, Scaling relationships in back-propagation learning: dependence on training set size, Complex Systems 1 (1987) 367–372.

[34] L.G. Valiant, A theory of the learnable, Commun. ACM 27 (1984) 1134–1142.

[35] D. Walters, Response mapping functions: classification and analysis of connectionist representations, in: Proceedings of the First IEEE Conference on Neural Networks Vol. III, San Diego, CA (1987) pp. 79–86.

THE EMERGENCE OF UNDERSTANDING IN A COMPUTER MODEL OF CONCEPTS AND ANALOGY-MAKING

Melanie MITCHELL and Douglas R. HOFSTADTER
Center for Research on Concepts and Cognition, Indiana University, 510 North Fess, Bloomington, IN 47408, USA

This paper describes Copycat, a computer model of the mental mechanisms underlying the fluidity and adaptability of the human conceptual system in the context of analogy-making. Copycat creates analogies between idealized situations in a microworld that has been designed to capture and isolate many of the central issues of analogy-making. In Copycat, an understanding of the essence of a situation and the recognition of deep similarity between two superficially different situations emerge from the interaction of a large number of perceptual agents with an associative, overlapping, and context-sensitive network of concepts. Central features of the model are: a high degree of parallelism; competition and cooperation among a large number of small, locally acting agents that together create a global understanding of the situation at hand; and a computational temperature that measures the amount of perceptual organization as processing proceeds and that in turn controls the degree of randomness with which decisions are made in the system.

1. Introduction

How, when one is faced with a new situation, does understanding emerge in the mind? How are we guided by a multitude of initially unconnected and novel perceptions to a coherent and familiar mental representation, such as "a coffee cup", "the letter 'A'", "French Baroque style", or "another Vietnam"? And how are such representations structured so that they are flexible, fluid, and thus adaptable to many different situations, rather than brittle, rigid, and inextensible? In our research, we are investigating these questions by building a computer model of what we think are the mental mechanisms underlying the fluid and adaptable nature of human concepts. As can be seen from the above examples, the mental phenomena we are studying range over the spectrum of recognition, categorization, and analogy-making: all are instances of high-level, abstract, or "deep", perception [1] as opposed to low-level modality-specific perception. The essential philosophy behind our model is that high-level perception emerges from a system of many independent processes running in parallel. These compete with and support each other by creating and destroying temporary perceptual constructs and by changing the activation levels and degrees of overlap in an associative network of permanent concepts with blurry conceptual boundaries. Such a system has no global executive deciding which processes should run next and what each process should do; rather, all processing is done by many small independent agents that make their decisions probabilistically. The system is self-organizing, with coherent and focused behavior being a statistically emergent property of the system as a whole [1]. Our computer model, called "Copycat", is an attempt to implement and test such a system in the realm of analogy-making, a realm in which the necessity of constructing fluid and adaptable mental representations is particularly apparent.

2. A microworld for analogy-making

In order to isolate and model the mechanisms underlying perception and analogy-making, we have developed a microworld in which analogies can be made between idealized situations consisting of strings of letters. A simple analogy problem in this domain is the following: If the string **abc** changes to **abd**, what is the analogous change for **ijk**? In the microworld, knowledge about letters and strings is quite limited. The 26 letters are known, but only as abstract categories; shapes, sounds, words, and all other linguistic and graphic facts are unknown. The only relations explicitly known are predecessor and successor relations between immediate neighbors in the alphabet. Ordinal positions in the alphabet (e.g. the fact that **S** is the 19th letter) are not known. (Notational note: Boldface capitals (e.g. **S**) denote the 26 abstract *categories* of the alphabet, and never appear in strings; boldface smalls (e.g. **abc**) denote *instances* of these categories, and appear only in strings.) The strings represent idealized situations containing objects, relations, and events; these small situations serve as *metaphors* for more complex, real-world situations. In the above problem, the *initial string* **abc** and the *target string* **ijk** are two situations, each with its own objects and relationships. The change to **abd** highlights a fragment of the first situation, and the challenge is to find the "same" fragment in **ijk** and to highlight and modify it in "the same way". What has been highlighted, how it has been highlighted, and what "the same way" means in the second situation are all up to the analogy-maker to decide.

For the sample problem given above, a reasonable description of the initial change is "Replace the rightmost letter by its successor", and straightforward application of this rule to the target string **ijk** yields the commonsense answer **ijl** (other, less satisfying answers, such as **ijd**, are of course possible). However, other problems are not so simple. For instance, given the same initial change and an alternate target string, **iijjkk**, a straightforward, rigid application of the original rule would yield **iijjkl**, which ignores the strong similarity between **abc** and **iijjkk** when the latter is seen as consisting of three *letter groups* rather than as six *letters*. If one perceives the role of *letter* in **abc** as played by *letter group* in **iijjkk**, then in making a mapping between **abc** and **iijjkk** one is forced to let the concept *letter* "slip" into the similar concept *letter group*. The ability to make appropriate *conceptual slippages* of this sort – in which concepts in one situation are identified with similar concepts in a different but analogous situation – is central to analogy-making and to cognition in general [2], and our research centers on investigating how concepts must be structured and how perception must interact with concepts to allow the fluidity necessary for insightful slippages.

Consider now the same initial change but with target **srqp**. Here a rigid application of the original rule would yield **srqq**, which again ignores a more abstract similarity between **abc** and **srqp**. One of the two answers people tend to prefer is **trqp** ("Replace the *leftmost* letter by its successor"), which is based on seeing **abc** as a left-to-right string and **srqp** as a right-to-left string (where each string increases alphabetically); here there is a slippage from the concept *right* to the concept *left*, which in turn gives rise to the slippage *rightmost* ⇒ *leftmost*. The other answer given by many people is **srqo** ("Replace the rightmost letter by its *predecessor*"), for which **abc** is seen as increasing and **srqp** as decreasing (both viewed as moving rightwards), yielding a slippage from *successor* to *predecessor*.

The letter-string microworld was designed to capture the essence of the issues of concepts and high-level perception that we are investigating. Although the analogies in this microworld involve only a small number of concepts, they often require considerable insight. An example of such an analogy is the following: if **abc** changes to **abd**, what does **xyz** change to? At first glance, this problem is essentially the same as the one with target string **ijk** discussed above, but there is a snag: **Z** has no successor. Many people answer **xya**, but in our microworld the alphabet is not circular; this answer is intentionally excluded in order to force analogy-makers to restructure their original view, to make conceptual slippages that were not initially considered, and hopefully to discover a more useful and insightful way of understanding the situation. One such way is to notice that *xyz* is "wedged" against the far end of the alphabet, and *abc* is similarly wedged against the beginning of the alphabet. Thus the *Z* in *xyz* and the *A* in *abc* can be seen to correspond, and then one naturally feels that the *X* and the *C* correspond as well. Underlying these object correspondences is a set of concept slippages that are conceptually parallel: *alphabetic-first* ⇒ *alphabetic-last*; *right* ⇒ *left*, and *successor* ⇒ *predecessor*. Taken together, these concept slippages convert the original rule into a rule adapted to the target string *xyz*: "Replace the *leftmost* letter by its *predecessor*", which yields an insightful answer: *wyz*.

These examples illustrate how solving problems in the microworld can require many of the mental abilities needed for analogy-making and high-level perception in general. In the process of understanding a situation, the mind must permit competition among the large number of possible ways in which the objects in the situation can be described and related to one another, and in which similarity to other situations can be perceived. As mental processing proceeds, exploration of each of these various possibilities should be given the cognitive resources it seems to deserve as determined moment to moment, and the partial and uncertain information obtained as this exploration proceeds must be used to narrow the range of possibilities. One's already-existing concepts should direct perception in a top-down manner, and whatever is perceived should affect one's concepts in return, by activating concepts that seem relevant and by changing the perceived similarity between concepts in order to reflect the current situation. One must decide what is salient and what can be ignored, how abstractly to describe objects, relations, and events, which descriptions to take literally and which to allow to slip, etc. And if these choices lead to an impasse that seems to block progress towards understanding, then one may be required to restructure one's original perceptions, perhaps shifting one's view in unexpected ways, in order to arrive at a deeper, more essential understanding of the situation. (For a detailed discussion of the microworld and a large number of sample analogy problems, see ref. [2].)

3. The architecture of Copycat

The Copycat program solves analogy problems in the letter-string microworld, and we believe it embodies many of the mechanisms underlying analogy-making in general. The architecture of the model consists of three parts: (a) the *Slipnet*: a network of nodes and links representing the permanent concepts of the system; (b) a working area in which perceptual structures representing the system's current understanding of the problem are built and modified in statistically emergent processes arising from large numbers of various special-purpose agents and running in asynchronous parallel: this area can be thought of as corresponding to the locus of creation and modification of mental representations that occur in the mind as one formulates a coherent understanding of a situation; and (c) a pool of various perceptual and higher-level structuring agents ("codelets") waiting to run. We will explain how understanding in Copycat emerges from the interaction of these three components.

3.1. The Slipnet

Copycat's *Slipnet*, a small part of which is shown in fig. 1, contains the permanent concepts of the system. Here a node represents the "core" of a

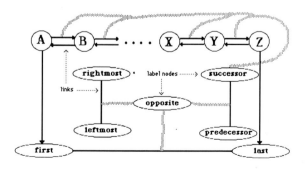

Fig. 1. A small part of Copycat's Slipnet, showing nodes, links (solid lines), and labels on links (thickly dotted lines).

concept (e.g. the node labeled "first" represents the core of the concept *first*) and a link simultaneously represents an *association* between two nodes and a potential *slippage* from one to the other. For example, *first* is the opposite of *last*, and thus in some circumstances these concepts are similar and one can be slipped to the other. Each link has a *label* that roughly classifies the type of association the link encodes. Each type of label is itself represented by a node. For instance, the link connecting nodes *first* and *last* has label *opposite*, which is itself a node; likewise, the links connecting nodes *A* and *B* have labels *successor* and *predecessor*. During a run of the program, nodes become activated when perceived to be relevant (by structure-building agents; see section 3.3), and decay when no longer perceived as relevant. Nodes also spread activation to their neighbors, and thus concepts closely associated with relevant concepts also become relevant. The amount of similarity encoded by a link can vary during a run. Since the plausibility of slippage between two concepts depends on context (e.g. *right* ⇒ *left* is plausible in "*abc* ⇒ *abd*, *srqp* ⇒ ?", but not in "*abc* ⇒ *abd*, *ijk* ⇒ ?"), the degree of similarity encoded by a link depends on the relevance of the link's label to the given problem, which is determined by the level of activation of the node representing the label. For example, the level of activation of the node *opposite* affects the degree of similarity between nodes linked by an *opposite* link.

Decisions about whether or not a slippage can be made from a given node – say, *successor* – to a neighboring node – say, *predecessor* – are made *probabilistically*, as a function of the degree of similarity between the two nodes. (Such decisions are made by codelets, as described in section 3.3 below.) Thus, in our model, a concept like *successorship* is identified not with a single node but rather with a region in the Slipnet, centered on a particular node, and having blurry rather than sharp boundaries: neighboring nodes can be seen as being included in the concept probabilistically, as a function of their degree of similarity to the core node of the concept. Since the degree of similarity between two nodes is context-dependent, concepts in the Slipnet are emergent rather than explicitly defined. They are associative and dynamically overlapping (here, overlap is modeled by links), and their time-varying behavior (through dynamic activation and degree of similarity) reflects the essential properties of the situations encountered. Thus concepts are able to adapt (in terms of relevance and similarity to one another) to different situations. Note that we are not modeling *learning* in the usual sense: the program neither retains changes in the network from run to run nor creates new permanent concepts; however, our work does involve learning if that term is taken to include the generalization from experience that humans perform in novel contexts.

3.2. Perceptual structures

At the beginning of a run, Copycat is given the three strings of letters; it initially knows only the category membership of each letter (e.g. *a* is an instance of category *A*), which letters are spatially adjacent to one another, and which letters are leftmost, rightmost, and middle in each string. In order to formulate a solution, the program must perceptually organize each situation, as well as the various relationships among situations. To accomplish this, the program gradually builds various kinds of structures that represent its high-level perception of the problem. (This process is similar to the way the Hearsay II speech-understanding system built layers of increasingly abstract perceptual structures on top of raw representations of sounds; see ref. [3].) These structures correspond to Slipnet concepts of various degrees of generality being brought to bear on the problem, and accordingly, each of these structures is built of parts copied from the Slipnet. The flexibility of the program rests on the fact that concepts from the Slipnet can be borrowed for use in perceiving situations, and on the fact that the Slipnet itself is not rigid but fluid, adjusting itself (via dynamic activation and degrees of similarity) to fit the

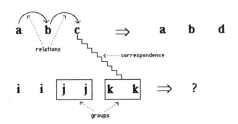

Fig. 2. Perceptual structures, including relations, groups, and a correspondence.

situation at hand. An essential part of our model is this interaction of top-down and bottom-up processing: while the program's perception of a given problem is guided by the properties of concepts in the Slipnet, those properties themselves are influenced by what the program perceives.

Fig. 2 shows examples of perceptual structures that could be built in the process of solving the problem "*abc* ⇒ *abd*, *iijjkk* ⇒ ?". The types of structures built by the program include *descriptions* of objects (e.g. the *C* in *abc* is the string's *rightmost* letter), *relations* between objects (e.g. the *B* in *abc* is the *successor* of its left neighbor, the *A*), *groups* of objects (e.g., *jj* is a group of adjacent identical letters; the entire string *abc* could be seen as a group of adjacent letters that increase alphabetically), and *correspondences* between objects (e.g. the *C* in *abc* corresponds to the group *kk* in *iijjkk*). (See section 5 for examples of these structures in a run of the program.)

3.3. Codelets, randomness, and the parallel terraced scan

The actual building (and sometimes destroying) of perceptual structures is carried out by large numbers of simple agents we call "codelets". A codelet is a small piece of code that carries out some local task that is part of the process of building a structure. For example, one codelet might estimate how important it is to describe the *A* in *abc* as "alphabetic-first" (taking into account the level of activation of that node in the Slipnet), another might notice that the *B* in *abc* is the alphabetic successor of its left neighbor in the string, and another might build the structure representing that fact. The program has several different types of codelets, each corresponding to one of the different types of structures that can be built (descriptions, relations, groups, and correspondences). Codelets are able to calculate the *strength* of a potential structure at a given time, based on a number of factors. For example, the strength of a potential successor relation between the *A* and the *B* in *abc* at a given time is a function of the degree of association encoded by the successor link from node *A* to node *B* in the Slipnet, the number of successor relations that have already been built nearby in the same string, and the level of activation (i.e. the currently perceived *relevance*) of the node *successor*. Since all these factors can change as processing proceeds, a structure's strength at a given time is only an estimate of how useful that structure will be in the long run.

For each type of structure there are two kinds of codelets: those that investigate the possibility of building that structure at a probabilistically chosen location ("musing" codelets), and those that actually build the structure ("builder" codelets). A structure is built by a series of codelets running in turn, each deciding probabilistically on the basis of its estimation of some aspect of the structure's strength whether to continue by generating one or more follow-up codelets, or to abandon the effort at that point. If the decision is made to continue, the running codelet assigns an "urgency" value (based on the strength of the structure) to each follow-up codelet. This value helps determine how long each follow-up codelet will have to wait before it can run and continue the evaluation of that particular structure. If the effort is not abandoned, the series ends with a builder codelet that tries to build the structure, possibly having to compete with already existing, incompatible structures, which may prevent it from completing its goal. The outcome of such a fight is decided probabilistically on the basis of the strengths of the competing structures.

Attempts at building many different structures are interleaved, as follows. All codelets waiting to run are placed in a single pool, and at each time step, the system probabilistically chooses one codelet to run. The choice is based on the relative urgencies of all codelets in the pool. When a codelet is chosen to run, it is removed from the pool. At a given time, the codelet pool generally contains many different types of codelets, but can also contain multiple copies of the same type of codelet – say, one that tries to make a grouping of objects already determined to be related. At the beginning of a run of the program, the pool contains a standard initial population of codelets, and as codelets run and are removed from the pool, new codelets are added both by codelets that run and by active nodes in the Slipnet. Thus the pool's population changes, as processing proceeds, in response to the needs of the system as judged by previously run codelets and by activation patterns in the Slipnet.

The fine-grained breakup of acts of structure building in our model has two purposes: (1) it allows many such acts to be carried out in parallel, by having their components interleaved; and (2) it allows the speed of and computational resources allocated to each such act to be dynamically regulated by moment-to-moment evaluations of the structure being constructed. It is important to understand that in this system, such processes, each of which consists of many codelets running in a series, are themselves emergent entities: rather than being predetermined and then broken up into small components, they are instead "post-determined", being the pathways visible, after the fact, leading to some given macroscopic act of construction or destruction of perceptual or organizational structure. In other words, only the codelets themselves are predetermined; the macroscopic *processes* of the system are emergent.

Since these processes are interleaved, attempts at building many different structures proceed simultaneously, but at different speeds. The speed of such a process emerges dynamically from the urgencies of its component codelets. Since those urgencies are determined by moment-to-moment estimates of the promise of the structure being built, the result is that structures of greater promise will tend to be built more quickly than less promising ones. There is no top-level executive directing processing here; all processing is carried out by codelets. Codelets that take part in the process of building a structure send activation to the areas in the Slipnet that represent the concepts associated with that structure. These activations in turn affect the makeup of the codelet population, since active nodes (e.g. *successor*) are able to add codelets to the pool (e.g. a codelet that tries to find a successor relation between some pair of objects). Note that though Copycat runs on a serial computer and thus only one codelet runs at a time, the system is roughly equivalent to one in which many activities are taking place in parallel at different spatial locations, since codelets work locally and to a large degree independently.

Copycat's distributed asynchronous parallelism was inspired by the similar sort of self-organizing activity that takes place in a biological cell. In a cell, all activity is carried out by large numbers of widely distributed enzymes of various sorts. These enzymes depend on random motion in the cell's cytoplasm in order to encounter substrates (molecules such as amino acids) from which to build up structures. Complex structures are built up through long chains of enzymatic actions, and separate chains proceed independently and asynchronously in different spatial locations throughout the cytoplasm. Moreover, the enzyme population in the cell is itself regulated by the products of the enzymatic activity, and is thus sensitive to the moment-to-moment needs of the cell. In Copycat, codelets roughly act the part of enzymes. All activity is carried out by large numbers of codelets, which choose objects in a probabilistic, biased way for use in building structures. As in a cell, the processes by which complex structures are built are not explicitly programmed, but are emergent outcomes of chains of codelets working in asynchronous parallel throughout Copycat's structure-building area (its "cytoplasm"). And just as in a cell, the population of codelets in the codelet pool is self-regulating and sensitive to the moment-to-moment needs of the system. And to carry this analogy further, the Slipnet could be said to play the role of DNA, with active nodes in the Slipnet corresponding to genes currently being expressed in the cell, controlling the production of enzymes. The purpose of this metaphor was to draw inspiration from the mechanisms of self-organization in a fairly well-understood natural system, and to use these ideas in thinking about the mechanisms of abstract perception. The mechanisms of enzymes and DNA in a cell are not to be taken literally as a model of perception; rather, general principles can be abstracted and carried over from the workings of cells to the workings of perception. Distributed asynchronous parallelism, emergent processes, the building-up of coherent complex structures from initially unconnected parts, self-organization, self-regulation, and sensitivity to the ongoing needs of the system are all central to our model of perception, and thinking about the workings of the cell has helped us to devise mechanisms underlying these principles in Copycat. (For further discussion of the cell metaphor, see ref. [1].)

In summary, in Copycat, processes that build up structures are interleaved, and many such processes – some mutually supporting, some competing – progress in parallel at different rates, the rate of each being set by the urgencies of its component codelets. Almost all codelets make one or more probabilistic decisions (for example, whether to continue building a given structure, whether to destroy a competing structure, etc.) and the high-level behavior of the system emerges from the combination of hundreds of these very small choices. The result is a *parallel terraced scan* [1], in which many possible courses of action are explored simultaneously, each at a speed and to a depth proportional to moment-to-moment estimates of its promise. (Note that since the program arrives nondeterministically at a solution, different answers are possible on different runs, and even runs leading to the same answer follow very different pathways on a microscopic level.) The above discussion of Copycat will be made clearer by the sample run of the program given in section 5.

4. Computational temperature

A fourth element of Copycat's architecture is a *temperature* variable, which *measures the amount of disorganization* ("entropy") in the system's understanding of the problem: the value of the temperature at a given time is a function of the amount and quality of perceptual or organizing structure that has been built so far. In particular, the temperature at a given time is an inverse function of the sum of the current strengths of all existing structures. Thus temperature starts high, falls as structures are built, and rises again if structures are destroyed or if their strengths decrease. In turn, the value of temperature *controls the degree of randomness* used both by codelets in making decisions (such as whether to add follow-up codelets to the pool, whether to break a competing structure, whether to allow a particular slippage, etc.) and by the system in choosing which codelet to run next. The idea is that when there is little perceptual organization (and thus high temperature), the information on which decisions are based (such as the urgency of a codelet or the strength of a particular structure) is not very reliable, and decisions should be more random than would seem to be indicated by this information. When a large amount of structure deemed to be good has been built (and thus temperature is low), the information is considered to be more reliable, and decisions based on this information should be more deterministic.

The solution to the well-known "two-armed bandit" problem (Given a slot machine with two arms, each with an unknown payoff rate, what strategy of dividing one's play between the two arms is optimal for profit-making?) is an elegant mathematical verification of these intuitions [4]. The solution states that the optimal strategy is to sample both arms but with probabilities that diverge increasingly fast as time progresses. In particular, as more and more information is gained through sampling, the optimal strategy is to exponentially increase the probability of sampling the "better" arm relative to the probability of sampling the "worse" arm (note that one never knows with certainty which is the better arm, since all information gained is merely statistical evidence). Copycat's parallel terraced scan can be likened to such a strategy extrapolated to a many-armed bandit – in fact, a "bandit" with a dynamically changing number of arms, where each arm represents a potential path of exploration. (This is similar to the search through schemata in a genetic algorithm; see ref. [4].) There are far too many possible paths to do an exhaustive search, so in order to guarantee that in principle every path has a non-zero chance of being explored, paths have to be chosen and explored probabilistically. Each step in exploring a path is like sampling an arm, in that information is obtained that can be used to decide the rate at which that path should be sampled in the near future[#1]. The role of temperature is to cause the exponential increase in the speed at which promising paths are explored as contrasted with unpromising ones; as temperature decreases, the degree of randomness with which decisions are made decreases exponentially, so the speed at which good paths crowd out bad ones grows exponentially as more information is obtained. This strategy, in which information is used as it is obtained in order to bias probabilistic choices and thus to speed up convergence toward some resolution, but to never *absolutely* rule out any path of exploration, is optimal in any situation in which there is a limited amount of time to explore an intractable number of paths. Such a strategy appears to be a central principle in many kinds of adaptive systems [4], which supports our belief that the temperature-controlled parallel terraced scan is a plausible description of how perception takes place in humans.

Temperature allows Copycat to close in on a good solution quickly, once parts of it have been discovered. In addition, since high temperature means more randomness, temporarily raising the temperature gives Copycat a way to get out of ruts or to deal with snags; it can allow old structures to break and restructuring to occur so that a better solution can be found. That is, when the system faces an impasse, the temperature can go up in spite of the fact that seemingly good organizing structures exist. Such a use of temperature is illustrated in the run of the program given in section 5. Note that the role of temperature in Copycat differs from that in simulated annealing, a technique used in some connectionist networks for finding optimal solutions [5–7], in which temperature is used exclusively as a top-down randomness-controlling factor, its value falling monotonically according to a predetermined, rigid "annealing schedule". By contrast, in Copycat, the value of temperature reflects the current quality of the system's understanding, so that temperature acts as a feedback mechanism that determines the degree of randomness used by the system. Ideas about such a role for temperature were first presented by Hofstadter [1].

5. A run of the program

The following screen dumps are from a run of Copycat on "*abc* ⇒ *abd*, *xyz* ⇒ ?". This run produced the desired answer *wyz*. Since the program is nondeterministic, different answers are possible on different runs. The program currently produces

---

[#1] It should be made clear that in Copycat, "paths of exploration" are defined as any of the possible ways in which the program could structure its perceptions of the problem in order to construct an analogy. Thus possible paths are not laid out in advance for the program to search, but rather are constructed by the program as its processing proceeds, just as in a game of chess, where paths through the tree of possible moves are constructed as the game is played. The evaluation of a given move in a game of chess blurs together the evaluations of many possible look-ahead paths that include that move. Similarly, any given action in building a structure by a codelet in Copycat is a step included in a large number of possible paths toward a solution, and an evaluation obtained by a codelet of a proposed structure blurs together the estimated promise of all these paths.

this answer infrequently; it more commonly produces *xyd* (as a consequence of generating the rule "Replace the rightmost letter by ***D***"), *xyz* (for which the rule is "Replace all ***C***'s by ***D***'s"), and *yyz* (in which the original rule is translated for the target string as "Replace the *leftmost* letter by its successor"). The relative merits of these and several other possible answers are discussed by Hofstadter [2]. The current version of Copycat can solve all the problems discussed in this paper; that is, for each problem, the program is able on some runs to find the answer or answers that most people agree are the best. Plans for extending the program are given in ref. [8].

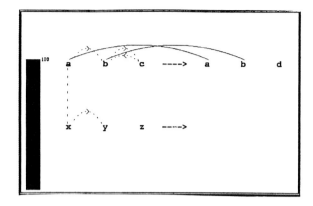

2. Codelets begin to build up structures. The two solid arcs across the top line represent correspondences from the *A* and *B* in ***abc*** to their counterparts in ***abd***. The dashed arcs inside each string represent potential *successor* and *predecessor* relations in the process of being built, and the vertical dashed line represents a potential correspondence between the *A* and the *X*.

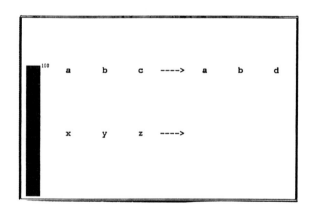

1. The program is presented with the three strings. The temperature, initially at its maximum of 100, is represented by a "thermometer" at the left. As structures are built, their representations will be drawn on the screen. Dashed lines and arcs represent structures in the process of being built, and solid lines and arcs represent fully built structures. Once fully built, a structure is able to influence the building of other structures and the temperature. A fully built structure is not necessarily permanent; it may be knocked down by competing structures.

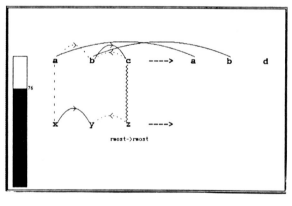

3. Some relations between letters within each string have been built and others continue to be considered. Copycat, unlike people, has no left-to-right or alphabetic-forwards biases, and in general is equally likely to perceive relations in either direction, although here, *successor* tends to be activated early when the *C*-to-*D* change is noticed, causing the system to tend to perceive the letters as having left-to-right successor relations rather than right-to-left predecessor relations. A correspondence between the *C* and the *Z* (jagged vertical line) has been built. Both letters are *rightmost* in their respective strings: this underlying concept mapping is displayed beneath the correspondence. In response to the growing amount of structure, the temperature has dropped to 76.

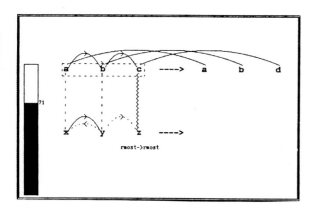

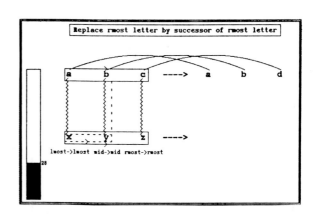

4. More relations have been built. Note that the potential predecessor relation between the *Z* and the *Y* shown in the previous screen has fizzled, and a potential successor relation has taken its place. This demonstrates the top-down pressure on the system to perceive the situation in terms of concepts it has already identified as relevant: since successor relations have been built elsewhere, the node *successor* in the Slipnet has become active, causing the system to more easily notice new successor relations. The program is also considering a left-to-right grouping of the letters in *abc* (represented by a dashed rectangle with a right arrow at the top), and other correspondences between *abc* and *xyz*. The temperature has dropped to 71.

6. *xyz* has now been described, like *abc*, as a left-to-right successor group. (The direction of a group is indicated by an arrow at the top or bottom of the rectangle representing the group.) A rival grouping of just the *X* and *Y* is also being considered (dashed rectangle), but will have little chance of defeating the stronger already-existing group. A strong set of correspondences has been made between the letters in *abc* and *xyz*, and a correspondence between the two groups *as wholes* (dashed vertical line) is being considered. The temperature has fallen to the low value of 28, reflecting the high degree of perceptual organization, and virtually ensuring that this highly coherent way of seeing the situations will win out.

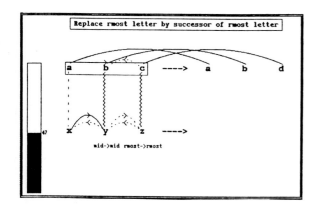

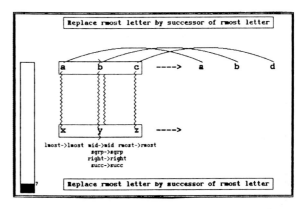

5. *abc* has been identified as a *successor group*, increasing alphabetically to the right (the relations between the letters still exist, but are not displayed). A *B–Y* correspondence has been built, and a rule (top of screen) has been constructed to describe the *abc–abd* change. Note there is no internal structuring of *abd*. Copycat currently expects the change from the *initial string* (here *abc*) to the *modified string* (here *abd*) to consist of exactly one letter being replaced. Thus no structures are built in the modified string except to identify what has changed and what has stayed the same. The program constructs the rule by filling in the template "Replace ____ by ____". As was indicated at the beginning of section 5, there are several possible rules for describing this change. Note that a right-to-left predecessor relation between the *B* and the *C* in *abc* is being considered (dashed arc), and will have to compete against the already built left-to-right successor group. The latter, being much stronger than the former, will survive, especially since the temperature is now fairly low, reflecting that a high-quality mutually consistent set of structures is taking over.

7. All the correspondences have been made. The group-level correspondence is supported by concept mappings expressing the facts that both are successor groups (displayed as "sgrp ⇒ sgrp") based on successor relations ("succ ⇒ succ") and both are increasing alphabetically toward the right. The concept mappings listed below the correspondences can be interpreted as instructions on how to translate the rule describing the initial change so it can be used on the target string. Here all the concept mappings are identities, so the translated rule (appearing at the bottom of the screen) is the same as the original rule: "Replace rightmost letter by its successor". The temperature is almost at zero, indicating the program's satisfaction in its understanding of the situation. But as an answer-building codelet attempts to carry out the translated rule, it hits a snag: *Z* has no successor.

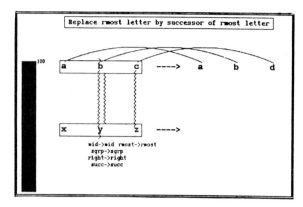
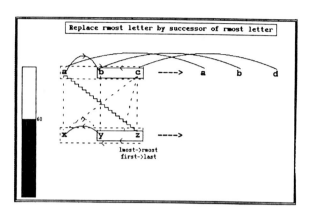

8. Being unable to take the successor of *Z*, the program hits an impasse, which causes the temperature to go up to 100. This causes competitions between structures to be decided much more randomly, and allows structures to be destroyed more easily (as can be seen, the *A–X* correspondence has been broken). In addition, since the answer-building codelet identified *Z* as the cause of the impasse, the node *Z* in the Slipnet becomes highly activated, which spreads activation to its neighbor *alphabetic-last*, making this concept relevant to the problem. In turn, *alphabetic-last* spreads activation to *alphabetic-first*.

10. Many possible ways of restructuring the situation are being considered simultaneously, but the program is beginning to develop an understanding of the situation based on the *A–Z* correspondence. Under pressure from this correspondence, the program is now beginning to perceive *xyz* as a right-to-left *predecessor* group (*yz* has already been perceived as such, and the direction of the relation between the *X* and the *Y* has reversed). The new structures have caused the temperature to fall to 60, indicating that this new way of structuring the problem seems promising to the program.

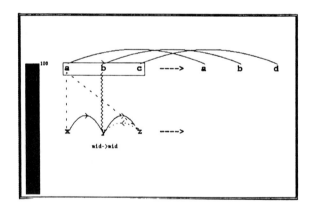

9. After breaking more structures and making other ineffectual attempts at restructuring (not shown), the program has noticed the relationship between the letters *A* and *Z*, and is trying to build a correspondence between them. Underlying it are two *slippages*: "leftmost ⇒ rightmost" and "first ⇒ last". Before the impasse was reached, the descriptions *first* and *last* were neither seen as relevant nor considered conceptually close enough to be the basis for a correspondence. But the combination of high temperature and the focus on the *Z* makes this mapping possible, though still not easy, to make. In fact, on most runs of program on this problem, this mapping is either never made, or quickly destroyed once made. But in this run, this correspondence, once made, survives.

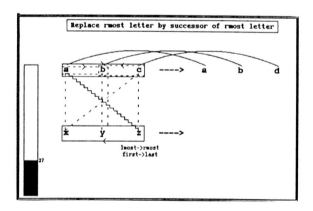

11. The "first ⇒ last" slippage has engendered a complete restructuring of the program's perception of *xyz* (which is now understood as a right-to-left predecessor group, opposite in direction from the group *abc*) and the program is closing in on a solution. Alternative ways of structuring the situation are still being considered, but the low temperature reflects the program's satisfaction with its current understanding, and will make it hard for any alternatives to complete at this point.

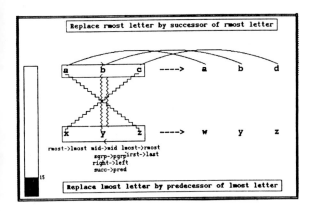

12. The mapping is complete, and all attempts at building rival structures have ceased. The concept mappings listed underneath the correspondences give the slippages needed to translate the rule. The translated rule ("Replace *leftmost* letter by *predecessor* of *leftmost* letter") appears at the bottom of the screen, and the answer *wyz* appears at the right.

6. The architecture of Copycat as a style of emergent computation

Forrest [9] has outlined a paradigm for the phenomenon of emergent computation. According to this paradigm, emergent computation occurs when explicit local instructions interact to form implicit, meaningful global patterns. Information processing is done at the level of the implicit global patterns rather than at the level of the explicit instructions. In addition, the global patterns have the possibility of looping back down and influencing the behavior of the local instructions. This account has much in common with Hofstadter's [10] discussion of emergent *symbols* in the brain, in which the top level (the *symbolic* level) reaches back down towards the lower levels (the *subsymbolic* levels) and influences them, while at the same time being itself determined by the lower levels. Since the Copycat project itself is an attempt to model the mechanisms underlying emergent symbols in the brain (i.e., concepts), it is not surprising that the features of emergent computation posited by Forrest can be identified in our model. In Copycat, individual nodes and links in the Slipnet and individual codelets can be thought of as constituting explicit local instructions. The interaction among codelets, nodes, and links gives rise to two types of implicit global patterns: (1) *concepts*, which are identified not with any one node but rather with a core node surrounded by a probabilistic "halo" of neighboring nodes, where inclusion of a node in a concept is a probabilistic function of the context-dependent conceptual distance to the core node; and (2) the *parallel terraced scan*, in which many possible ways of construing the problem are considered simultaneously, each at a rate and to a depth proportional to moment-to-moment estimates of its promise. These implicit global patterns in turn reach back and influence the lower levels (and thus each other): the emergent structure of concepts influences how codelets will evaluate possible perceptual structures (since the evaluation of a structure takes into account activations and conceptual distances in the Slipnet) and affects the population of codelets and their urgencies; and the emergent parallel terraced scan, by determining how the search through possible structurings proceeds, affects the activation of nodes and thus the conceptual distances encoded by links in the Slipnet. It is through this complex interaction of explicit and emergent levels that we are attempting to capture the fluidity and adaptability of concepts and perception.

Acknowledgements

We thank David Chalmers, Robert French, Liane Gabora, Jim Levenick, Gary McGraw, David Moser, and Peter Suber for ongoing contributions to this project and for many helpful comments on this paper. The paper was also improved by helpful suggestions from two anonymous reviewers. This research has been supported by grants from Indiana University, the University of Michigan, and Apple Computer, Inc., as well as a grant from Mitchell Kapor, Ellen Poss, and the Lotus Development Corporation, and grant DCR 8410409 from the National Science Foundation.

References

[1] D.R. Hofstadter, The architecture of Jumbo, in: Proceedings of the International Machine Learning Workshop, Monticello, Illinois (1983).

[2] D.R. Hofstadter, Analogies and roles in human and machine thinking, in: Metamagical Themas (Basic Books, New York, 1985) p. 547.

[3] L.D. Erman, F. Hayes-Roth, V.R. Lesser and D. Raj Reddy, The Hearsay-II speed-understanding system: Integrating knowledge to resolve uncertainty, Computing Surveys 12 (1980) 213.

[4] J.H. Holland, Adaptation in Natural and Artificial Systems (University of Michigan Press, Ann Arbor, MI, 1975).

[5] S. Kirkpatrick, C.D. Gelatt Jr. and M.P. Vecchi, Optimization by simulated annealing, Science 220 (1983) 671.

[6] G.E. Hinton and T.J. Sejnowski, Learning and relearning in Boltzmann machines, in: Parallel Distributed Processing, eds. J. McClelland and D. Rumelhart, (Bradford/MIT Press, Cambridge, MA, 1986) p. 282.

[7] P. Smolensky, Information processing in dynamical systems: Foundations of harmony theory, in: Parallel Distributed Processing, eds. J. McClelland and D. Rumelhart, (Bradford/MIT Press, Cambridge, MA, 1986) p. 194.

[8] D.R. Hofstadter, M. Mitchell and R.M. French, Fluid concepts and creative analogies: A theory and its computer implementation, Technical Report 10, Cognitive Science and Machine Intelligence Laboratory, University of Michigan, Ann Arbor, MI (1987).

[9] S. Forrest, Physica D 42 (1990) 1–11, these Proceedings.

[10] D.R. Hofstadter, Gödel, Escher, Bach: an Eternal Golden Braid (Basic Books, New York, 1979).

THE SYMBOL GROUNDING PROBLEM

Stevan HARNAD
Department of Psychology, Princeton University, Princeton, NJ 08544, USA

There has been much discussion recently about the scope and limits of purely symbolic models of the mind and about the proper role of connectionism in cognitive modeling. This paper describes the "symbol grounding problem": How can the semantic interpretation of a formal symbol system be made *intrinsic* to the system, rather than just parasitic on the meanings in our heads? How can the meanings of the meaningless symbol tokens, manipulated solely on the basis of their (arbitrary) shapes, be grounded in anything but other meaningless symbols? The problem is analogous to trying to learn Chinese from a Chinese/Chinese dictionary alone. A candidate solution is sketched: Symbolic representations must be grounded bottom-up in nonsymbolic representations of two kinds: (1) *iconic representations*, which are analogs of the proximal sensory projections of distal objects and events, and (2) *categorical representations*, which are learned and innate feature detectors that pick out the invariant features of object and event categories from their sensory projections. Elementary symbols are the names of these object and event categories, assigned on the basis of their (nonsymbolic) categorical representations. Higher-order (3) *symbolic representations*, grounded in these elementary symbols, consist of symbol strings describing category membership relations (e.g. "An X is a Y that is Z").

Connectionism is one natural candidate for the mechanism that learns the invariant features underlying categorical representations, thereby connecting names to the proximal projections of the distal objects they stand for. In this way connectionism can be seen as a complementary component in a hybrid nonsymbolic/symbolic model of the mind, rather than a rival to purely symbolic modeling. Such a hybrid model would not have an autonomous symbolic "module," however; the symbolic functions would emerge as an intrinsically "dedicated" symbol system as a consequence of the bottom-up grounding of categories' names in their sensory representations. Symbol manipulation would be governed not just by the arbitrary shapes of the symbol tokens, but by the nonarbitrary shapes of the icons and category invariants in which they are grounded.

1. Modeling the mind

1.1. From behaviorism to cognitivism

For many years the only empirical approach in psychology was behaviorism, its only explanatory tools input/input and input/output associations (in the case of classical conditioning [42]) and the reward/punishment history that "shaped" behavior (in the case of operant conditioning [1]). In a reaction against the subjectivity of armchair introspectionism, behaviorism had declared that it was just as illicit to theorize about what went on in the *head* of the organism to generate its behavior as to theorize about what went on in its *mind*. Only *observables* were to be the subject matter of psychology; and, apparently, these were expected to explain themselves.

Psychology became more like an empirical science when, with the gradual advent of cognitivism [17, 25, 29], it became acceptable to make inferences about the *unobservable* processes underlying behavior. Unfortunately, cognitivism let mentalism in again by the back door too, for the hypothetical internal processes came embellished with subjective interpretations. In fact, semantic interpretability (meaningfulness), as we shall see, was one of the defining features of the most prominent contender vying to become the theoretical vocabulary of cognitivism, the "language of thought" [6], which became the prevailing view in cognitive theory for several decades in the form of the

0167-2789/90/$03.50 © Elsevier Science Publishers B.V.
(North-Holland)

"symbolic" model of the mind: The mind is a symbol system and cognition is symbol manipulation. The possibility of generating complex behavior through symbol manipulation was empirically demonstrated by successes in the field of artificial intelligence (AI).

1.2. Symbol systems

What is a symbol system? From Newell [28], Pylyshyn [33], Fodor [6] and the classical work by von Neumann, Turing, Gödel, Church, etc. (see ref. [18]) on the foundations of computation, we can reconstruct the following definition:

A symbol system is:

(1) a set of arbitrary *physical tokens* (scratches on paper, holes on a tape, events in a digital computer, etc.) that are

(2) manipulated on the basis of *explicit rules* that are

(3) likewise physical tokens and *strings* of tokens. The rule-governed symbol-token manipulation is based

(4) purely on the *shape* of the symbol tokens (not their "meaning"), i.e. it is purely *syntactic*, and consists of

(5) *rulefully combining* and recombining symbol tokens. There are

(6) primitive *atomic* symbol tokens and

(7) *composite* symbol-token strings. The entire system and all its parts – the atomic tokens, the composite tokens, the syntactic manipulations (both actual and possible) and the rules – are all

(8) *semantically interpretable*: The syntax can be *systematically* assigned a meaning (e.g. as standing for objects, as describing states of affairs).

According to proponents of the symbolic model of mind such as Fodor [7] and Pylyshyn [32, 33], symbol strings of this sort capture what mental phenomena such as thoughts and beliefs are. Symbolists emphasize that the symbolic level (for them, the mental level) is a natural functional level of its own, with ruleful regularities that are independent of their specific physical realizations. For symbolists, this implementation independence is the critical difference between cognitive phenomena and ordinary physical phenomena and their respective explanations. This concept of an autonomous symbolic level also conforms to general foundational principles in the theory of computation and applies to all the work being done in symbolic AI, the field of research that has so far been the most successful in generating (hence explaining) intelligent behavior.

All eight of the properties listed above seem to be critical to this definition of symbolic[#1]. Many phenomena have some of the properties, but that does not entail that they are symbolic in this explicit, technical sense. It is not enough, for example, for a phenomenon to be *interpretable* as rule-governed, for just about anything can be interpreted as rule-governed. A thermostat may be interpreted as following the rule: Turn on the furnace if the temperature goes below 70°F and turn it off if it goes above 70°F, yet nowhere in the thermostat is that rule explicitly represented. Wittgenstein [45] emphasized the difference between *explicit* and *implicit* rules: It is not the same thing to "follow" a rule (explicitly) and merely to behave "in accordance with" a rule (implicitly)[#2]. The critical difference is in the compositness (7) and systematicity (8) criteria. The explicitly represented symbolic rule is part of a formal system, it is decomposable (unless primitive), its application and manipulation is purely formal (syntactic, shape dependent), and the entire system must be semantically interpretable,

---

[#1] Paul Kube (personal communication) has suggested that (2) and (3) may be too strong, excluding some kinds of Turing machine and perhaps even leading to an infinite regress on levels of explicitness and systematicity.

[#2] Similar considerations apply to Chomsky's [2] concept of "psychological reality" (i.e. whether Chomskian rules are really physically represented in the brain or whether they merely "fit" our performance regularities, without being what actually governs them). Another version of the distinction concerns explicitly represented rules versus hard-wired physical constraints [40]. In each case, an explicit representation consisting of elements that can be recombined in systematic ways would be symbolic whereas an implicit physical constraint would not, although *both* would be semantically "interpretable" as a "rule" if construed in isolation rather than as part of a system.

not just the chunk in question. An isolated ("modular") chunk cannot be symbolic; being symbolic is a systematic property.

So the mere fact that a behavior is "interpretable" as ruleful does not mean that it is really governed by a symbolic rule[#3]. Semantic interpretability must be coupled with explicit representation (2), syntactic manipulability (4), and systematicity (8) in order to be symbolic. None of these criteria is arbitrary, and, as far as I can tell, if you weaken them, you lose the grip on what looks like a natural category and you sever the links with the formal theory of computation, leaving a sense of "symbolic" that is merely unexplicated metaphor (and probably differs from speaker to speaker). Hence it is only this formal sense of "symbolic" and "symbol system" that will be considered in this discussion of the grounding of symbol systems.

1.3. Connectionist systems

An early rival to the symbolic model of mind appeared [36], was overcome by symbolic AI [27] and has recently re-appeared in a stronger form that is currently vying with AI to be the general theory of cognition and behavior [23, 39]. Variously described as "neural networks", "parallel distributed processing" and "connectionism", this approach has a multiple agenda, which includes providing a theory of brain function. Now, much can be said for and against studying behavioral and brain function independently, but in this paper it will be assumed that, first and foremost, a cognitive theory must stand on its own merits, which depend on how well it explains our observable behavioral capacity. Whether or not it does so in a sufficiently brainlike way is another matter, and a downstream one, in the course of theory development. Very little is known of the brain's structure and its "lower" (vegetative) functions so far; and the nature of "higher" brain function is itself a theoretical matter. To "constrain" a cognitive theory to account for behavior in a brainlike way is hence premature in two respects: (1) It is far from clear yet what "brainlike" means, and (2) we are far from having accounted for a lifesize chunk of behavior yet, even without added constraints. Moreover, the formal principles underlying connectionism seem to be based on the associative and statistical structure of the causal interactions in certain dynamical systems; a neural network is merely one possible implementation of such a dynamical system[#4].

Connectionism will accordingly only be considered here as a cognitive theory. As such, it has lately challenged the symbolic approach to modeling the mind. According to connectionism, cognition is not symbol manipulation but dynamic patterns of activity in a multilayered network of nodes or units with weighted positive and negative interconnections. The patterns change according to internal network constraints governing how the activations and connection strengths are adjusted on the basis of new inputs (e.g. the generalized "delta rule", or "backpropagation" [23]). The result is a system that learns, recognizes patterns, solves problems, and can even exhibit motor skills.

1.4. Scope and limits of symbols and nets

It is far from clear what the actual capabilities and limitations of either symbolic AI or connectionism are. The former seems better at formal and language-like tasks, the latter at sensory, motor and learning tasks, but there is considerable overlap and neither has gone much beyond the stage of "toy" tasks toward lifesize behavioral capacity. Moreover, there has been some disagree-

---

[#3]Analogously, the mere fact that a behavior is *interpretable* as purposeful or conscious or meaningful does not mean that it really is purposeful or conscious. (For arguments to the contrary, see ref. [5].)

[#4]It is not even clear yet that a "neural network" needs to be implemented as a net (i.e. a parallel system of interconnected units) in order to do what it can do; if symbolic simulations of nets have the same functional capacity as real nets, then a connectionist model is just a special kind of symbolic model, and connectionism is just a special family of symbolic algorithms.

ment as to whether or not connectionism itself is symbolic. We will adopt the position here that it is not, because connectionist networks fail to meet several of the criteria for being symbol systems, as Fodor and Pylyshyn [10] have argued recently. In particular, although, like everything else, their behavior and internal states can be given isolated semantic interpretations, nets fail to meet the compositeness (7) and systematicity (8) criteria listed earlier: The patterns of interconnections do not decompose, combine and recombine according to a formal syntax that can be given a systematic semantic interpretation[#5]. Instead, nets seem to do what they do *non*-symbolically. According to Fodor and Pylyshyn, this is a severe limitation, because many of our behavioral capacities appear to be symbolic, and hence the most natural hypothesis about the underlying cognitive processes that generate them would be that they too must be symbolic. Our linguistic capacities are the primary examples here, but many of the other skills we have – logical reasoning, mathematics, chess playing, perhaps even our higher-level perceptual and motor skills – also seem to be symbolic. In any case, when we interpret our sentences, mathematical formulas, and chess moves (and perhaps some of our perceptual judgments and motor strategies) as having a systematic *meaning* or *content*, we know at first hand that this is literally true, and not just a figure of speech. Connectionism hence seems to be at a disadvantage in attempting to model these cognitive capacities.

Yet it is not clear whether connectionism should for this reason aspire to be symbolic, for the symbolic approach turns out to suffer from a severe handicap, one that may be responsible for the limited extent of its success to date (especially in modeling human-scale capacities) as well as the uninteresting and ad hoc nature of the symbolic "knowledge" it attributes to the "mind" of the symbol system. The handicap has been noticed in various forms since the advent of computing; I have dubbed a recent manifestation of it the "symbol grounding problem" [14].

2. The symbol grounding problem

2.1. The Chinese room

Before defining the symbol grounding problem I will give two examples of it. The first comes from Searle's [37] celebrated "Chinese room argument", in which the symbol grounding problem is referred to as the problem of intrinsic meaning (or "intentionality"): Searle challenges the core assumption of symbolic AI that a symbol system capable of generating behavior indistinguishable from that of a person must have a mind. More specifically, according to the symbolic theory of mind, if a computer could pass the Turing test [43] in Chinese – i.e. if it could respond to all Chinese symbol strings it receives as input with Chinese symbol strings that are indistinguishable from the replies a real Chinese speaker would make (even if we keep testing for a lifetime) – then the computer would understand the meaning of Chinese symbols in the same sense that I understand the meaning of English symbols.

Searle's simple demonstration that this cannot be so consists of imagining himself doing everything the computer does – receiving the Chinese input symbols, manipulating them purely on the basis of their shape (in accordance with (1) to (8) above), and finally returning the Chinese output symbols. It is evident that Searle (who knows no Chinese) would not be understanding Chinese under those conditions – hence neither could the computer. The symbols and the symbol manipulation, being all based on shape rather than mean-

---

[#5] There is some misunderstanding of this point because it is often conflated with a mere implementational issue: Connectionist networks can be simulated using symbol systems, and symbol systems can be implemented using a connectionist architecture, but that is independent of the question of what each can do *qua* symbol system or connectionist network, respectively. By way of analogy, silicon can be used to build a computer, and a computer can simulate the properties of silicon, but the functional properties of silicon are not those of computation, and the functional properties of computation are not those of silicon.

斑馬　　帶有斑紋的馬

Fig. 1. Chinese dictionary entry. For translation, see footnote 17.

ing, are systematically *interpretable* as having meaning – that, after all, is what it is to be a symbol system, according to our definition. But the interpretation will not be *intrinsic* to the symbol system itself: It will be parasitic on the fact that the symbols have meaning for *us*, in exactly the same way that the meanings of the symbols in a book are not intrinsic, but derive from the meanings in our heads. Hence, if the meanings of symbols in a symbol system are extrinsic, rather than intrinsic like the meanings in our heads, then they are not a viable model for the meanings in our heads: Cognition cannot be just symbol manipulation.

2.2. The Chinese / Chinese dictionary-go-round

My own example of the symbol grounding problem has two versions, one difficult, and one, I think, impossible. The difficult version is: Suppose you had to learn Chinese as a second language and the only source of information you had was a Chinese/Chinese dictionary. The trip through the dictionary would amount to a merry-go-round, passing endlessly from one meaningless symbol or symbol-string (the definiens) to another (the definiendum), never coming to a halt on what anything meant[#6]. (See fig. 1.)

[#6]Symbolic AI abounds with symptoms of the symbol grounding problem. One well-known (though misdiagnosed) manifestation of it is the so-called "frame" problem [22, 24, 26, 34]: It is a frustrating but familiar experience in writing "knowledge-based" programs that a system apparently behaving perfectly intelligently for a while can be foiled by an unexpected case that demonstrates its utter stupidity: A "scene-understanding" program will blithely describe the goings-on in a visual scene and answer questions demonstrating its comprehension (who did what, where, why?) and then suddenly reveal that it does not "know" that hanging up the phone and leaving the room does not make the phone disappear, or something like that. (It is important to note that these are not the kinds of lapses and gaps in knowledge that people are prone to; rather, they are such howlers as to cast serious doubt on whether the system has anything like "knowledge" at all.) The "frame" problem has been optimistically defined as the problem of formally specifying ("framing") what varies and what stays constant in a particular "knowledge domain," but in reality it is the problem of second-guessing all the contingencies the programmer has not anticipated in symbolizing the knowledge he is attempting to symbolize. These contingencies are probably unbounded, for practical purposes, because purely symbolic "knowledge" is ungrounded. Merely adding on more symbolic contingencies is like taking a few more turns in the Chinese/Chinese dictionary-go-round. There is in reality no ground in sight: merely enough "intelligent" symbol manipulation to lull the programmer into losing sight of the fact that its meaningfulness is just parasitic on the meanings he is projecting onto it from the grounded meanings in his own head. (I have called this effect the "hermeneutic hall of mirrors" [16]; it is the reverse side of the symbol grounding problem.) Yet parasitism it is, as the next "frame problem" lurking around the corner is ready to confirm. (A similar form of over-interpretation has occurred in the ape "language" experiments [41]. Perhaps both apes and computers should be trained using Chinese code, to immunize their experimenters and programmers against spurious over-interpretations. But since the actual behavioral tasks in both domains are still so trivial, there is probably no way to prevent their being decrypted. In fact, there seems to be an irresistible tendency to overinterpret toy task performance itself, preemptively extrapolating and "scaling it up" conceptually to lifesize without any justification in practice.)

The only reason cryptologists of ancient languages and secret codes seem to be able to successfully accomplish something very like this is that their efforts are *grounded* in a first language and in real world experience and knowledge[#7]. The second variant of the dictionary-go-round, however, goes far beyond the conceivable resources of cryptology: Suppose you had to learn Chinese as a *first* language and the only source of information you had was a Chinese/Chinese dict-

[#7]Cryptologists also use statistical information about word frequencies, inferences about what an ancient culture or an enemy government are likely to be writing about, decryption algorithms, etc.

ionary[#8]! This is more like the actual task faced by a purely symbolic model of the mind: How can you ever get off the symbol/symbol merry-go-round? How is symbol meaning to be grounded in something other than just more meaningless symbols[#9]? This is the symbol grounding problem[#10].

2.3. Connecting to the world

The standard reply of the symbolist (e.g. Fodor [7, 8]) is that the meaning of the symbols comes from connecting the symbol system to the world "in the right way". But it seems apparent that the problem of connecting up with the world in the right way is virtually coextensive with the problem of cognition itself. If each definiens in a Chinese/Chinese dictionary were somehow connected to the world in the right way, we would hardly need the definienda! Many symbolists believe that cognition, being symbol manipulation, is an autonomous functional module that need only be hooked up to peripheral devices in order to "see" the world of objects to which its symbols refer (or, rather, to which they can be systematically interpreted as referring)[#11]. Unfortunately, this radically underestimates the difficulty of picking out the objects, events and states of affairs in the world that symbols refer to, i.e. it trivializes the symbol grounding problem.

It is one possible candidate for a solution to this problem, confronted directly, that will now be sketched: What will be proposed is a hybrid nonsymbolic/symbolic system, a "dedicated" one, in which the elementary symbols are grounded in two kinds of nonsymbolic representations that pick out, from their proximal sensory projections, the distal object categories to which the elementary symbols refer. Most of the components of which the model is made up (analog projections and transformations, discretization, invariance detection, connectionism, symbol manipulation) have also been proposed in various configurations by others, but they will be put together in a specific bottom-up way here that has not, to my knowledge, been previously suggested, and it is on this specific configuration that the potential success of the grounding scheme critically depends.

Table 1 summarizes the relative strengths and weaknesses of connectionism and symbolism, the two current rival candidates for explaining *all* of cognition single-handedly. Their respective strengths will be put to cooperative rather than competing use in our hybrid model, thereby also remedying some of their respective weaknesses. Let us now look more closely at the behavioral capacities such a cognitive model must generate.

3. Human behavioral capacity

Since the advent of cognitivism, psychologists have continued to gather behavioral data, although to a large extent the relevant evidence is already in: We already know what human beings

---

[#8] There is of course no need to restrict the symbolic resources to a dictionary; the task would be just as impossible if one had access to the entire body of Chinese-language literature, including all of its computer programs and anything else that can be codified in symbols.

[#9] Even mathematicians, whether Platonist or formalist, point out that symbol manipulation (computation) itself cannot capture the notion of the intended interpretation of the symbols [31]. The fact that formal symbol systems and their interpretations are not the same thing is hence evident independently of the Church–Turing thesis [18] or the Gödel results [3, 4], which have been zealously misapplied to the problem of mind-modeling (e.g. by Lucas [21]) – to which they are largely irrelevant, in my view.

[#10] Note that, strictly speaking, symbol grounding is a problem only for cognitive modeling, not for AI in general. If symbol systems alone succeed in generating all the intelligent machine performance pure AI is interested in – e.g. an automated dictionary – then there is no reason whatsoever to demand that their symbols have intrinsic meaning. On the other hand, the fact that our own symbols do have intrinsic meaning whereas the computer's do not, and the fact that we can do things that the computer so far cannot, may be indications that even in AI there are performance gains to be made (especially in robotics and machine vision) from endeavouring to ground symbol systems.

[#11] The homuncular viewpoint inherent in this belief is quite apparent, as is the effect of the "hermeneutic hall of mirrors" [16].

Table 1
Connectionism versus symbol systems.

Strengths of connectionism

(1) *Nonsymbolic function*: As long as it does not aspire to be a symbol system, a connectionist network has the advantage of not being subject to the symbol grounding problem.

(2) *Generality*: connectionism applies the same small family of algorithms to many problems, whereas symbolism, being a methodology rather than an algorithm, relies on endless problem-specific symbolic rules.

(3) *"Neurosimilitude"*: Connectionist architecture seems more brain-like than a Turing machine or a digital computer.

(4) *Pattern learning*: Connectionist networks are especially suited to the learning of patterns from data.

Weaknesses of connectionism

(1) *Nonsymbolic function*: Connectionist networks, because they are not symbol systems, do not have the systematic semantic properties that many cognitive phenomena appear to have.

(2) *Generality*: Not every problem amounts to pattern learning. Some cognitive tasks may call for problem-specific rules, symbol manipulation, and standard computation.

(3) *"Neurosimilitude"*: Connectionism's brain-likeness may be superficial and may (like toy models) camouflage deeper performance limitations.

Strengths of symbol systems

(1) *Symbolic function*: Symbols have the computing power of Turing machines and the systematic properties of a formal syntax that is semantically interpretable.

(2) *Generality*: All computable functions (including all cognitive functions) are equivalent to a computational state in a Turing machine.

(3) *Practical successes*: Symbol systems' ability to generate intelligent behavior is demonstrated by the successes of Artificial Intelligence.

Weaknesses of symbol systems

(1) *Symbolic function*: Symbol systems are subject to the symbol grounding problem.

(2) *Generality*: Turing power is too general. The solutions to AI's many toy problems do not give rise to common principles of cognition but to a vast variety of ad hoc symbolic strategies.

are able to do. They can (1) *discriminate*, (2) *manipulate*[#12], (3) *identify* and (4) *describe* the objects, events and states of affairs in the world they live in, and they can also (5) *produce descriptions* and (6) *respond to descriptions* of those objects, events and states of affairs. Cognitive theory's burden is now to explain *how* human beings (or any other devices) do all this[#13].

3.1. Discrimination and identification

Let us first look more closely at discrimination and identification. To be able to *discriminate* is to be able to judge whether two inputs are the same or different, and, if different, *how* different they are. Discrimination is a relative judgment, based on our capacity to tell things apart and discern their degree of similarity. To be able to *identify* is to be able to assign a unique (usually arbitrary) response – a "name" – to a class of inputs, treating them all as equivalent or invariant in some respect. Identification is an absolute judgment, based on our capacity to tell whether or not a given input is a member of a particular *category*.

Consider the symbol "horse". We are able, in viewing different horses (or the same horse in different positions, or at different times) to tell

---

[#12] Although they are no doubt as important as perceptual skills, motor skills will not be explicitly considered here. It is assumed that the relevant features of the sensory story (e.g. iconicity) will generalize to the motor story (e.g. in motor analogs [20]). In addition, large parts of the motor story may not be cognitive, drawing instead upon innate motor patterns and sensorimotor feedback. Gibson's [11] concept of "affordances" – the invariant stimulus features that are detected by the motor possibilities they "afford" – is relevant here too, though Gibson underestimates the processing problems involved in finding such invariants [44]. In any case, motor and sensory-motor grounding will no doubt be as important as the sensory grounding that is being focused on here.

[#13] If a candidate model were to exhibit all these behavioral capacities, both *linguistic* (5)-(6) and *robotic* (i.e. sensorimotor) (1)-(3), it would pass the "total Turing test" [15]. The standard Turing test [43] calls for linguistic performance capacity only: symbols in and symbols out. This makes it equivocal about the status, scope and limits of pure symbol manipulation, and hence subject to the symbol grounding problem. A model that could pass the total Turing test, however, would be grounded in the world.

them apart and to judge which of them are more alike, and even how alike they are. This is discrimination. In addition, in viewing a horse, we can reliably call it a horse, rather than, say, a mule or a donkey (or a giraffe, or a stone). This is identification. What sort of internal representation would be needed in order to generate these two kinds of performance?

3.2. Iconic and categorical representations

According to the model being proposed here, our ability to discriminate inputs depends on our forming *iconic representations* of them [14]. These are internal analog transforms of the projections of distal objects on our sensory surfaces [38]. In the case of horses (and vision), they would be analogs of the many shapes that horses cast on our retinas[#14]. Same/different judgments would be based on the sameness or difference of these iconic representations, and similarity judgments would be based on their degree of congruity. No homunculus is involved here; simply a process of superimposing icons and registering their degree of disparity. Nor are there memory problems, since the inputs are either simultaneously present or available in rapid enough succession to draw upon their persisting sensory icons.

So we need horse icons to discriminate horses, but what about identifying them? Discrimination is independent of identification. I could be discriminating things without knowing what they were. Will the icon allow me to identify horses? Although there are theorists who believe it would ([30]), I have tried to show why it could not [12, 14]. In a world where there were bold, easily detected natural discontinuities between all the categories we would ever have to (or choose to) sort and identify – a world in which the members of one category could not be confused with the members of any another category – icons might be sufficient for identification. But in our underdetermined world, with its infinity of confusable potential categories, icons are useless for identification because there are too many of them and because they blend continuously[#15] into one another, making it an independent problem to *identify* which of them are icons of members of the category and which are not! Icons of sensory projections are too unselective. For identification, icons must be selectively reduced to those *invariant features* of the sensory projection that will reliably distinguish a member of a category from any nonmembers with which it could be confused. Let us call the output of this category-specific feature detector the *categorical representation*. In some cases these representations may be innate, but since evolution could hardly anticipate all of the categories we may ever need or choose to identify, most of these features must be learned from experience. In particular, our categorical representation of a horse is probably a learned one. (I will defer till section 4 the problem of how the invariant features underlying identification might be learned.)

Note that both iconic and categorical representations are nonsymbolic. The former are analog copies of the sensory projection, preserving its "shape" faithfully; the latter are icons that have been selectively filtered to preserve only some of the features of the shape of the sensory projection: those that reliably distinguish members from nonmembers of a category. But both representations are still sensory and nonsymbolic. There is no

---

[#14]There are many problems having to do with figure/ground discrimination, smoothing, size constancy, shape constancy, stereopsis, etc., that make the problem of discrimination much more complicated than what is described here, but these do not change the basic fact that iconic representations are a natural candidate substrate for our capacity to discriminate.

[#15]Elsewhere [13, 14] I have tried to show how the phenomenon of "categorical perception" could generate internal discontinuities where there is external continuity. There is evidence that our perceptual system is able to segment a continuum, such as the color spectrum, into relatively discrete, bounded regions or categories. Physical differences of equal magnitude are more discriminable across the boundaries between these categories than within them. This boundary effect, both innate and learned, may play an important role in the representation of the elementary perceptual categories out of which the higher-order ones are built.

problem about their connection to the objects they pick out: It is a purely causal connection, based on the relation between distal objects, proximal sensory projections and the acquired internal changes that result from a history of behavioral interactions with them. Nor is there any problem of semantic interpretation, or of whether the semantic interpretation is justified. Iconic representations no more "mean" the objects of which they are the projections than the image in a camera does. Both icons and camera images can of course be *interpreted* as meaning or standing for something, but the interpretation would clearly be derivative rather than intrinsic[#16].

3.3. Symbolic representations

Nor can categorical representations yet be interpreted as "meaning" anything. It is true that they pick out the class of objects they "name", but the names do not have all the systematic properties of symbols and symbol systems described earlier. They are just an inert taxonomy. For systematicity it must be possible to combine and recombine them rulefully into propositions that can be semantically interpreted. "Horse" is so far just an arbitrary response that is reliably made in the presence of a certain category of objects. There is no justification for interpreting it holophrastically as meaning "This is a [member of the category] horse" when produced in the presence of a horse, because the other expected systematic properties of "this" and "a" and the all-important "is" of predication are not exhibited by mere passive taxonomizing. What would be required to generate these other systematic properties? Merely that the grounded names in the category taxonomy be strung together into *propositions* about further category membership relations. For example:

(1) Suppose the name "horse" is grounded by iconic and categorical representations, learned from experience, that reliably discriminate and identify horses on the basis of their sensory projections.

(2) Suppose "stripes" is similarly grounded.

Now consider that the following category can be constituted out of these elementary categories by a symbolic description of category membership alone:

(3) "Zebra" = "horse" & "stripes"[#17]

What is the representation of zebra? It is just the symbol string "horse & stripes". But because "horse" and "stripes" are grounded in their respective iconic and categorical representations, "zebra" inherits the grounding, through its grounded *symbolic* representation. In principle, someone who had never seen a zebra (but had seen and learned to identify horses and stripes) could identify a zebra on first acquaintance armed with this symbolic representation alone (plus the nonsymbolic – iconic and categorical – representations of horses and stripes that ground it).

Once one has the grounded set of elementary symbols provided by a taxonomy of names (and the iconic and categorical representations that give content to the names and allow them to pick out the objects they identify), the rest of the symbol strings of a natural language can be generated by symbol composition alone[#18], and they will all inherit the intrinsic grounding of the elementary

---

[#16]On the other hand, the resemblance on which discrimination performance is based – the degree of isomorphism between the icon and the sensory projection, and between the sensory projection and the distal object – seems to be intrinsic, rather than just a matter of interpretation. The resemblance can be objectively characterized as the degree of invertibility of the physical transformation from object to icon [14].

[#17]Fig. 1 is actually the Chinese dictionary entry for "zebra", which is "striped horse". Note that the character for "zebra" actually happens to be the character for "horse" plus the character for "striped." Although Chinese characters are iconic in structure, they function just like arbitrary alphabetic lexigrams at the level of syntax and semantics.

[#18]Some standard logical connectives and quantifiers are needed too, such as not, and, all, etc. (though even some of these may be learned as higher-order categories).

set[19]. Hence, the ability to discriminate and categorize (and its underlying nonsymbolic representations) have led naturally to the ability to describe and to produce and respond to descriptions through symbolic representations.

4. A complementary role for connectionism

The symbol grounding scheme just described has one prominent gap: No mechanism has been suggested to explain how the all-important categorical representations could be formed: How does the hybrid system find the invariant features of the sensory projection that make it possible to categorize and identify objects correctly[20]?

Connectionism, with its general pattern learning capability, seems to be one natural candidate (though there may well be others): Icons, paired with feedback indicating their names, could be processed by a connectionist network that learns to identify icons correctly from the sample of confusable alternatives it has encountered by dynamically adjusting the weights of the features and feature combinations that are reliably associated with the names in a way that (provisionally) resolves the confusion, thereby reducing the icons to the *invariant* (confusion-resolving) features of the category to which they are assigned. In effect, the "connection" between the names and the objects that give rise to their sensory projections and their icons would be provided by connectionist networks.

This circumscribed complementary role for connectionism in a hybrid system seems to remedy the weaknesses of the two current competitors in their respective attempts to model the mind single-handedly. In a pure symbolic model the crucial connection between the symbols and their referents is missing; an autonomous symbol system, though amenable to a systematic semantic interpretation, is ungrounded. In a pure connectionist model, names are connected to objects through invariant patterns in their sensory projections, learned through exposure and feedback, but the crucial compositional property is missing; a network of names, though grounded, is not yet amenable to a full systematic semantic interpreta-

---

[19] Note that it is not being claimed that "horse", "stripes", etc. are actually elementary symbols, with direct sensory grounding; the claim is only that *some* set of symbols must be directly grounded. Most sensory category representations are no doubt hybrid sensory/symbolic; and their features can change by bootstrapping: "Horse" can always be revised, both sensorily and symbolically, even if it was previously elementary. Kripke [19] gives a good example of how "gold" might be baptized on the shiny yellow metal in question, used for trade, decoration and discourse, and then we might discover "fool's gold", which would make all the sensory features we had used until then inadequate, forcing us to find new ones. He points out that it is even possible in principle for "gold" to have been inadvertently baptized on "fool's gold"! Of interest here are not the ontological aspects of this possibility, but the epistemic ones: We could bootstrap successfully to real gold even if every prior case had been fool's gold. "Gold" would still be the right word for what we had been trying to pick out (i.e. what we had "had in mind") all along, and its original provisional features would still have provided a close enough approximation to ground it, even if later information were to pull the ground out from under it, so to speak.

[20] Although it is beyond the scope of this paper to discuss it at length, it must be mentioned that this question has often been begged in the past, mainly on the grounds of "vanishing intersections". It has been claimed that one cannot find invariant features in the sensory projection because they simply do not exist: The intersection of all the projections of the members of a category such as "horse" is empty. The British empiricists have been criticized for thinking otherwise; for example, Wittgenstein's [45] discussion of "games" and "family resemblances" has been taken to have discredited their view. Current research on human categorization [35] has been interpreted as confirming that intersections vanish and that hence categories are not represented in terms of invariant features. The problem of vanishing intersections (together with Chomsky's [2] "poverty of the stimulus argument") has even been cited by thinkers such as Fodor [8, 9] as a justification for extreme nativism. The present paper is frankly empiricist. In my view, the reason intersections have not been found is that no one has yet looked for them properly. Introspection certainly is not the way to look. and general pattern learning algorithms such as connectionism are relatively new; their inductive power remains to be tested. In addition, a careful distinction has not been made between pure sensory categories (which, I claim, must have invariants, otherwise we could not successfully identify them as we do) and higher-order categories that are *grounded* in sensory categories; these abstract representations may be symbolic rather than sensory, and hence not based directly on sensory invariants. For further discussion of this problem, see ref. [14].

tion. In the hybrid system proposed here, there is no longer any autonomous symbolic level at all; instead, there is an intrinsically *dedicated* symbol system, its elementary symbols (names) connected to nonsymbolic representations that can pick out the objects to which they refer, via connectionist networks that extract the invariant features of their analog sensory projections.

5. Conclusions

The expectation has often been voiced that "top-down" (symbolic) approaches to modeling cognition will somehow meet "bottom-up" (sensory) approaches somewhere in between. If the grounding considerations in this paper are valid, then this expectation is hopelessly modular and there is really only one viable route from sense to symbols: from the ground up. A free-floating symbolic level like the software level of a computer will never be reached by this route (or vice versa) – nor is it clear why we should even try to reach such a level, since it looks as if getting there would just amount to uprooting our symbols from their intrinsic meanings (thereby merely reducing ourselves to the functional equivalent of a programmable computer).

In an intrinsically dedicated symbol system there are more constraints on the symbol tokens than merely syntactic ones. Symbols are manipulated not only on the basis of the arbitrary shape of their tokens, but also on the basis of the decidedly nonarbitrary "shape" of the iconic and categorical representations connected to the grounded elementary symbols out of which the higher-order symbols are composed. Of these two kinds of constraints, the iconic/categorical ones are primary. I am not aware of any formal analysis of such dedicated symbol systems[#21], but this may be because they are unique to cognitive and robotic modeling and their properties will depend on the specific kinds of robotic (i.e. behavioral) capacities they are designed to exhibit.

It is appropriate that the properties of dedicated symbol systems should turn out to depend on behavioral considerations. The present grounding scheme is still in the spirit of behaviorism in that the only tests proposed for whether a semantic interpretation will bear the semantic weight placed on it consist of one formal test (does it meet the eight criteria for being a symbol system?) and one behavioral test (can it discriminate, identify (manipulate) and describe all the objects and states of affairs to which its symbols refer?). If both tests are passed, then the semantic interpretation of its symbols is "fixed" by the behavioral capacity of the dedicated symbol system, as exercised on the objects and states of affairs in the world to which its symbols refer; the symbol meanings are accordingly not just parasitic on the meanings in the head of the interpreter, but intrinsic to the dedicated symbol system itself. This is still no guarantee that our model has captured subjective meaning, of course. But if the system's behavioral capacities are lifesize, it is as close as we can ever hope to get.

[#21]Although mathematicians investigate the formal properties of uninterpreted symbol systems, all of their motivations and intuitions clearly come from the intended interpretations of those systems (see ref. [31]). Perhaps these too are grounded in the iconic and categorical representations in their heads.

References

[1] A.C. Catania and S. Harnad, eds., The Selection of Behavior. The Operant Behaviorism of B.F. Skinner: Comments and Consequences (Cambridge Univ. Press, Cambridge, 1988).
[2] N. Chomsky, Rules and representations, Behav. Brain Sci. 3 (1980) 1–61.
[3] M. Davis, Computability and Unsolvability (McGraw-Hill, New York, 1958).
[4] M. Davis, The Undecidable (Raven, New York, 1965).
[5] D.C. Dennett, Intentional systems in cognitive ethology, Behav. Brain Sci. 6 (1983) 343–390.
[6] J.A. Fodor, The Language of Thought (Crowell, New York, 1975).
[7] J.A. Fodor, Methodological solipsism considered as a research strategy in cognitive psychology, Behav. Brain Sci. 3 (1980) 63–109.

[8] J.A. Fodor, Précis of the modularity of mind, Behav. Brain Sci. 8 (1985) 1-42.
[9] J.A. Fodor, Psychosemantics (MIT/Bradford, Cambridge, MA, 1987).
[10] J.A. Fodor and Z.W. Pylyshyn, Connectionism and cognitive architecture: A critical appraisal, Cognition 28 (1988) 3-71.
[11] J.J. Gibson, An ecological approach to visual perception (Houghton Mifflin, Boston, 1979).
[12] S. Harnad, Metaphor and mental duality, in: Language, Mind and Brain, eds. T. Simon and R. Scholes (Erlbaum Hillsdale, NJ, 1982).
[13] S. Harnad, Categorical perception: A critical overview, in: Categorical Perception: The Groundwork of Cognition, ed. S. Harnad (Cambridge Univ. Press, Cambridge, 1987).
[14] S. Harnad, Category induction and representation, in: Categorical Perception: The Groundwork of Cognition, ed. S. Harnad (Cambridge Univ. Press, Cambridge, 1987).
[15] S. Harnad, Minds, machines and searle, J. Theor. Exp. Artificial Intelligence 1 (1989) 5-25.
[16] S. Harnad, Computational hermeneutics, Social Epistemology, in press.
[17] J. Haugeland, The nature and plausibility of cognitivism, Behav. Brain Sci. 1 (1978) 215-260.
[18] S.C. Kleene, Formalized Recursive Functionals and Formalized Realizability (Am. Math. Soc., Providence, RI, 1969).
[19] S.A. Kripke, Naming and Necessity (Harvard Univ. Press, Cambridge, MA, 1980).
[20] A.M. Liberman, On the finding that speech is special, Am. Psychologist 37 (1982) 148-167.
[21] J. Lucas, Minds, machines and Gödel, Philosophy 36 (1961) 112-117.
[22] J. McCarthy and P. Hayes, Some philosophical problems from the standpoint of artificial intelligence, in: Machine Intelligence, Vol. 4, eds. B. Meltzer and P. Michie (Edinburgh Univ. Press, Edinburgh, 1969).
[23] J.L. McClelland, D.E. Rumelhart and the PDP Research Group, Parallel Distributed Processing: Explorations in the Microstructure of Cognition, Vol. 1 (MIT/Bradford, Cambridge, MA, 1986).
[24] D. McDermott, Artificial intelligence meets natural stupidity, SIGART Newsletter 57 (1976) 4-9.
[25] G.A. Miller, The magical number seven, plus or minus two: Some limits on our capacity for processing information, Psychological Rev. 63 (1956) 81-97.
[26] M. Minsky, A framework for representing knowledge, MIT Lab Memo No. 306 (1974).
[27] M. Minsky and S. Papert, Perceptrons: An Introduction to Computational Geometry (MIT Press, Cambridge, MA, 1969) Reissued in an expanded edition (1988).
[28] A. Newell, Physical symbol systems, Cognitive Sci. 4 (1980) 135-183.
[29] U. Neisser, Cognitive Psychology (Appleton-Century-Crofts., New York, 1967).
[30] A. Paivio, Mental Representation: A Dual Coding Approach (Oxford Univ. Press, Oxford, 1986).
[31] R. Penrose, The Emperor's New Mind (Oxford Univ. Press, Oxford, 1989).
[32] Z.W. Pylyshyn, Computation and cognition: Issues in the foundations of cognitive science, Behav. Brain Sci. 3 (1980) 111-169.
[33] Z.W. Pylyshyn, Computation and Cognition (MIT/Bradford, Cambridge, MA, 1984).
[34] Z.W. Pylyshyn, ed., The Robot's Dilemma: The Frame Problem in Artificial Intelligence (Ablex, Norwood, NJ, 1987).
[35] E. Rosch and B.B. Lloyd, Cognition and Categorization (Erlbaum, Hillsdale, NJ, 1978).
[36] F. Rosenblatt, Principles of Neurodynamics (Spartan, New York, 1962).
[37] J. Searle, Minds, brains and programs, Behav. Brain Sci. 3 (1980) 417-457.
[38] R.N. Shepard and L.A. Cooper, Mental Images and Their Transformations (MIT Press/Bradford, Cambridge, 1982).
[39] P. Smolensky, On the proper treatment of connectionism, Behav. Brain Sci. 11 (1988) 1-74.
[40] E.P. Stabler, How are grammars represented? Behav. Brain Sci. 6 (1985) 391-421.
[41] H. Terrace, Nim (Random House, New York, 1979).
[42] J. Turkkan, Classical conditioning: The new hegemony, Behav. and Brain Sci. 12 (1989) 121-179.
[43] A.M. Turing, Computing machinery and intelligence, in: Minds and Machines, ed. A. Anderson (Prentice Hall, Engelwood Cliffs, NJ, 1964).
[44] S. Ullman, Against direct perception, Behav. Brain Sci. 3 (1980) 373-415.
[45] L. Wittgenstein, Philosophical Investigations (Macmillan, New York, 1953).

SELECTIONIST MODELS OF PERCEPTUAL AND MOTOR SYSTEMS AND IMPLICATIONS FOR FUNCTIONALIST THEORIES OF BRAIN FUNCTION

George N. REEKE Jr. and Olaf SPORNS

The Neurosciences Institute and The Rockefeller University, 1230 York Avenue, New York, NY 10021, USA

Functionalism is at present widely accepted as a working basis for cognitive science and artificial intelligence. This view holds that psychological phenomena can be adequately described in terms of functional processes carried out in the brain, and that these processes can be understood independently of the detailed structure and mode of development of the brain. In the functionalist view, the brain is analogous to a computer; both can properly be described at the level of symbolic representations and algorithms.

However, an analysis of the structure, development, and evolution of the brain makes it highly unlikely that it could be a Turing machine or that brain algorithms could be either acquired by experience in the world or transmitted between generations. An alternative view is that the brain is a selective system in which two different domains of stochastic variation, the world and neural repertoires, become mapped onto each other in an *individual, historical* manner. Neural systems capable of such mapping can generalize and can deal with novelty in an open-ended environment.

Several models have been constructed to test these ideas, including automata of a new kind that can recognize and associate patterns of sensory input by selective mechanisms. In an approach called synthetic neural modelling, the environment, the phenotype, and the nervous system of such an automaton are integrated into a single computer model. One example is Darwin III, a sessile "creature" with an eye and a multi-jointed arm having a sense of touch; its environment consists of simple shapes moving on a featureless background; its nervous system consists of some 50 000 cells of 50 different kinds connected by about 620 000 synaptic junctions. Darwin III can be trained to track moving objects with its eye, to reach out and touch objects with its arm, to categorize objects according to combinations of visual and tactile cues, and to respond in a positive or negative way to such objects depending on previous experience with similar objects. Synthetic neural models give insight into how biological pattern recognizing systems might operate and may provide paradigms for the construction of improved pattern recognizing and classifying automata.

1. Introduction

Today, it is fashionable for cognitive science to be considered a field quite separate from neurobiology, with modes of analysis based on a computational metaphor for cognition that effectively divorces the question of how the mind functions from any serious consideration of brain anatomy, physiology, or development. (The so-called "PDP" and other connectionist models [1] are no exception, inasmuch as they take only general inspiration, not functional details, from the nervous system.) Undoubtedly, the enormous complexity and difficulty of experimentally studying the brain, particularly the human brain, have contributed to this divorce. Nonetheless, as biologists interested in understanding how behavior arises from processes occurring in the nervous system, we wish to make several arguments in this paper: (1) that the brain is *not* like a computer, (2) that computers, and in fact, functionalist systems in general, have severe limitations that make it extremely unlikely they can be programmed to imitate brains, and (3) to understand cognitive function, it is probably wise to pay more attention to the details of the one working example we have, namely, the brain.

Our plan will be to identify some significant issues that can be used to distinguish among competing theories of brain function; to develop a theory that confronts these issues by proposing that a form of somatic selection occurring in the nervous system plays a major role in brain function; and then to give several examples of computer simulations based on this theory that show

how selective systems can deal head-on with the problems that the real world presents to animals with adaptive nervous systems. In the course of this presentation, evidence concerning the three arguments presented above will emerge; we will conclude the paper with a detailed discussion of these arguments.

It has often been remarked that what animals do best (e.g. recognize patterns, control movements, deal with the unexpected), computers do worst, and what computers do best (e.g. numerical computation, complex logic), animals are not so good at. Clearly, we, as humans, do carry out certain activities based on symbol manipulation, such as formal problem solving, which have often been the objects of investigations in artificial intelligence. Symbol manipulation, however, is an invention made late in evolution, in societies of individuals communicating through language. We therefore suggest that we begin by asking whether we can discover how we are able to see, to reach, to walk – in short, how we do those things which arose earlier in evolution and which computers cannot yet do well. We might discover properties of the nervous system that are prerequisite to advanced reasoning capabilities. How do we recognize and categorize objects? How does memory work, apparently without the use of preestablished codes to represent objects and events? How do we decide what to do next? How do we control our muscles to carry out actions we have chosen to perform? How does the brain acquire the ability to carry out all these functions?

A particularly important ability that can shed light on these questions is perceptual categorization, the process of grouping together objects and events encountered in the world. The ability to categorize is obviously of great adaptive value to animals, as it permits them to respond correctly to items they have never encountered before by virtue of class properties of those items that relate them to previous experience. In order to carry out perceptual categorization, an animal must first notice that the jumble of sensory signals it receives is consistent with the division of the world into separate entities (objects and events); notice that certain of these entities are similar or are related by common properties (for example, "good to eat"); and then classify new entities into one of these categories or into a new category. All of these component processes are typically carried out spontaneously, without explicit instructions about the category membership of individual items. In humans, categorization starts to develop before language is acquired; at later stages language processes and the formation of physical and functional categories are intimately intertwined [2]. Once language is present, a human teacher may increase the efficiency of the process. Animals, however, carry out perceptual categorization perfectly well without language [3], suggesting that formal symbolic processes are not necessary.

The most obvious notion of how categorization is carried out is based on the classical view that categories are generally defined by lists of individually necessary and jointly sufficient conditions [4]. However, at least since the time of Wittgenstein [5] and his celebrated example of games that have no features in common except that they are games, it has been known that this picture is inadequate to describe our everyday categories (although it may fare better with the more abstract categories employed in scientific and mathematical discourse). Experimental psychologists [6] have pointed out a number of effects commonly seen in human category judgments that are inconsistent with the classical view of concepts and categories, including disjunctions, typicality effects, the use of logically unnecessary features, and failures of superordination. Furthermore, categories often depend on context and on the current goals and history of the categorizing organism. Frequently, as suggested in fig. 1, insufficient criteria are present in the stimulus for an unambiguous categorical judgment to be made. In such cases, the influence of goals and history can have a particularly strong influence and can even lead to opposite categorizations being made by the same individual at different times.

The existence of these phenomena does not ineluctably eliminate formal reasoning as a mechanism of (human) everyday categorization, but it

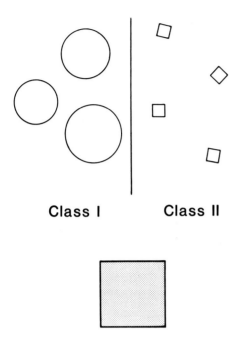

Fig. 1. Two classes [32]. The object at the bottom is ambiguous – it could be considered a member of the first class, because it is large, or a member of the second class, because it is square, or a member of neither class, because it is shaded.

raises severe difficulties which have not yet been solved after nearly 40 years of research in artificial intelligence. We will see in section 2 that by examining possible biological bases for categorization and other higher brain functions, we arrive at an alternative explanation that accounts quite naturally for the observed deviations from the classical view of categorization.

2. The theory of neuronal group selection

The search for mechanisms of perception and categorization that might operate in the nervous system must be constrained by the anatomical, physiological, and developmental properties of the brain. Some of these constraints are listed in table 1. The mechanisms of perceptual categorization and other functions of the brain must be carried out in parallel by units that are slow and incapable of carrying out highly accurate computations, but which are very numerous and densely interconnected. There is no apparent way to com-

Table 1
Some biological constraints on mechanisms of categorization.

Anatomical constraints:
1. The system is constructed of relatively simple elements (neurons) that are connected in networks and that operate in parallel. (Models based on the architecture of conventional von Neumann computers are excluded.)
2. The connections generally are unidirectional. (Models based on symmetrical connectivity are excluded.)
3. Terminal arborizations are enormously varied. (Models requiring strict prespecified point-to-point wiring are excluded.)
4. The number of connections in the nervous system is very much less than the square of the number of cells. (Models requiring n^2 connections are excluded.)

Physiological constraints:
1. The speed of neuronal responses is limited to the millisecond range. (Models requiring large numbers of iterations to produce an output are excluded.)
2. Neuronal activity is characteristically fluctuating and aperiodic. (Models in which outputs are coded as steady states of the network are excluded.)
3. There is no known mechanism to communicate desired output values to individual cells or synapses. (Models using specific learning algorithms are excluded.)

Developmental constraints:
1. The exact connectivity of the network is not prespecified. (Models that require specific connectivity for functioning are excluded.)
2. The genetic code is insufficient for transmitting algorithms between generations. (Models that assume the inheritance of effective procedures are excluded.)

municate desired outputs to this kind of system (in order to "teach" it), certainly not by clamping its processing elements to externally derived "correct" states in the presence of a particular input. The units communicate by exchanging signals which are not synchronized by an overall clock, and signals do not contain repeated patterns that could be interpreted as coded messages. Similarly, the processing elements are of very many different kinds, and the connections between them display an extraordinary degree of flexibility and variability. Examples of this variability in the nervous systems of three different kinds of animals are shown in fig. 2. It is difficult to see how this variability could be positively used by traditionally designed information processing systems, yet it is such a prominent feature of all nervous sys-

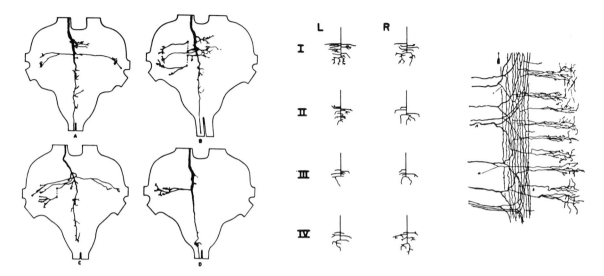

Fig. 2. Variability in the nervous system: Drawings of (a) descending lateral motion detector neurons from the common locust [7]. (b) Identified neurons from the visual system of *Daphnia magna*, from clonally identical individuals [8]. (c) Neurons from the rabbit cerebellum [9].

tems that it demands an explanation. In notable contrast with present notions of how computers need to be designed and constructed to operate reliably, it would appear that nervous systems have evolved in such a way that variance is an essential part of their adaptation, rather than a source of error requiring correction.

We do not have far to look in nature to find another situation in which variability in structure is at the basis of the formation and functioning of complex systems. One of the major contributions of Darwin's theory of natural selection was the idea that *selection*, applied to populations of entities with *preexisting diversity*, brings about functional organization [10]. In our view, this point is of fundamental importance for understanding the nervous system. To clarify how structure arises from selectional pressure, consider an example taken from the ecology of birds. In regions with relatively little selective pressure, for example, the jungles of Puerto Rico, which is an island ecosystem with relatively few bird species, the birds tend to be generalists and are found in various layers of the tree canopy, exploiting a variety of food sources. In Panama, on the other hand, where there is more competition, a greater number of species have evolved, and many of these have become specialists, confined to a narrow region of the jungle canopy and a smaller number of food sources which they are able to exploit efficiently. Thus, out of competition in a system with variance, structural and functional specialization arise.

The theory of neuronal group selection (TNGS) [11] is an attempt to apply population thinking to the nervous system. According to this theory, diverse neuronal circuitry is generated during embryonic development. The functional properties of this circuitry are not precisely prespecified for each individual component. Instead, a form of selection operates during the lifetime of the individual animal to enhance the activity of parts of the nervous system that have a competitive advantage over other parts and that produce responses having adaptive value for that individual. The neuronal networks are interconnected in such a way that selectional events in different regions are correlated, leading to the formation of a series of *mappings* between regions that ensure the consistency of their responses and permit signals originating from multiple sensory modalities to contribute to categorization events.

Fig. 3 illustrates the three main elements of the TNGS. The first is the generation of diversity in the nervous system during development. This is the result of epigenetic processes which give rise to variations in cells and networks during the developmental processes of growth, movement, cell division and cell death, all of which are under the general guidance of a genetically specified plan (fig. 3, top). Cell adhesion molecules, for example the neural cell adhesion molecule (N-CAM [12]) play a major role in these processes, not only as structural elements which allow neighboring neurons to coordinate growth processes via cell-surface interactions, but also as signalling elements which can indicate to cells when conditions for differentiation are appropriate in their immediate neighborhood. The result of all these developmental processes is the formation of various brain regions, each with its distinctive collection of cell types and local circuitry, forming what has been called the "primary repertoire".

In the second stage (fig. 3, middle), selectional processes shape neuronal function. In this stage, selection acts mainly to modify the strengths of individual connections, although activity-dependent processes continue to shape the morphology and connectivity of neurons for a certain length of time after birth [13]. The units upon which selection acts are proposed to be local collectives of neurons in the primary repertoire called *neuronal groups*. Cells within a group are connected by more numerous or stronger connections than cells in different groups; the degree to which group formation is complete in the primary repertoire differs according to the animal species and the

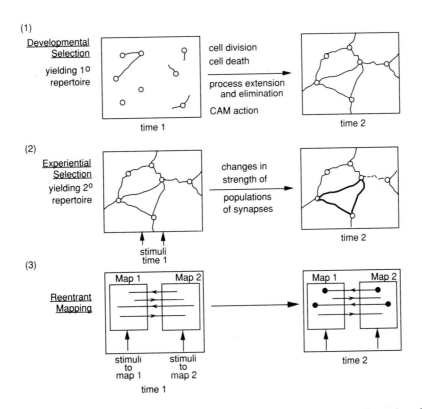

Fig. 3. Schematic diagram illustrating principles of the theory of neuronal group selection. (Top) Formation of neuronal structures and connectivity during embryonic development – epigenetic events involving selection determine the finest details of the architecture of the nervous system. (Middle) Interaction with the environment results in neuronal group selection through differential modification of synaptic strengths, enhancing the probability of behavioral responses that have adaptive value for the individual organism. (Bottom) Reentrant signalling between maps correlates responses arising from different sensory modalities, permitting categorizations based on disjunctive sampling of an unlabelled world. From The Remembered Present: A Biological Theory of Consciousness, by G.M. Edelman. Copyright © 1989 by Basic Books, Inc. Reprinted by permission of Basic Books, Inc., Publishers, New York.

brain region in question. As a result of their interconnections, cells within a group tend to have similar receptive fields and their responses tend to be correlated. Groups respond more or less well to different combinations of signals; each group has an optimal input that produces a maximum response.

Let us examine the question of how selection can shape the behavioral responses of an organism and create an adaptive order from disordered and variant circuitry. Because response specificities generally are not programmed or hard-wired during development, there is no a priori reason why the responses of any particular group should be advantageous for the animal. It is the function of selection to strengthen those responses that are advantageous, increasing their probability of reoccurring in similar situations, and to weaken those responses that are counterproductive, decreasing their probability of reoccurrence. To do so, neuronal groups not only have to encounter environmental signals, but a means of assessing the value of the resulting behavioral responses is needed, as well as a means of differentially amplifying the responses of different groups according to their value under this assessment.

The first requirement is easily met, inasmuch as groups connected in networks are constantly exposed to environmental stimulation (directly or via other groups). To assess the outcome of behavior engendered by the activity of such groups, nervous systems must have components that are able to sense the consequences of those behaviors by mechanisms that are simple and, at least in the first instance, innate. We refer to such components as "value systems". (Candidates for such systems exist, e.g. in diffuse ascending projections from the brainstem [14].) For example, a value system might contain neurons that respond to changes in blood chemistry as the animal seeks and obtains food. The responses of cells in a value system can be transmitted by either chemical or electrical means to locations in the nervous system where selection is to take place. Selection is ultimately carried out by the differential modification of the strengths of synaptic connections between cells, biased by value in such a way that responses of the groups undergoing selection are likely to be enhanced when value is positive and suppressed when value is negative. Selective synaptic change can act on connections both within a neuronal group and between groups. (In fact, competition between cells in neighboring groups determines the sizes and borders of groups. Group borders may change with time as the experience or environment of the animal vary.) Within a group, selection can lead to "tuning" of the relative responsiveness of the group to different inputs, shifting its selectivity to optimize it for particularly relevant input signals. Between groups, selection determines which groups will have access (directly or indirectly) to the motor efferents of the brain, thus changing the frequency of various behaviors in accord with value. The effect of selection is thus to modify the primary repertoires generated during development to produce so-called "secondary repertoires" that are well adapted to the needs, experiences, and environment of the individual animal. Even at this stage of adaptation, variability in neuronal repertoires is never completely depleted, and in fact, it is likely that mechanisms exist to enhance variability as a consequence of the operation of rules for selective synaptic modification [15]; this is important, since a depletion of variability would mean an end to further selection within the system.

A third important feature of the TNGS is the notion of *reentry* (fig. 3, bottom). This term refers to the exchange of signals between neuronal groups along reciprocal connections that permits consistency and coherency to be established between their responses. Cortical anatomy contains an abundance of reciprocal short-range and long-range fiber systems linking all parts of the cortical mantle (for some examples see ref. [16]). Reentry occurs in and establishes *mappings* between cortical areas. In the simplest case, neuronal mappings are based on natural separations of stimuli in the world, for example separation in spatial visual coordinates. Neural maps can also be based on higher-level abstract properties of stimuli. Perceptual categorization of sensory stimuli is carried out by the interaction of multiple neuronal maps. The TNGS proposes that it is based on the correlation of responses in different modalities connected by reentry. Pairs of neuronal networks that carry out categorization according to this principle are known as "classification couples", and they give selective systems the ability to categorize according to complex, nonclassical criteria [17].

Reentry is a process of ongoing reciprocal exchange of signals between different repertoires along parallel anatomical connections. It may occur horizontally, between repertoires at comparable levels in different sensory pathways, or vertically, between repertoires at different levels in the same pathway. Because reentrant connections in the brain form complex interlocking networks, reentrant signals often act recursively, bringing signals back to their repertoire of origin after passing through one or more intermediate mappings where they may be modified according to context. Such recursive reentry tends to suppress responses that are inconsistent with more direct inputs to any of the regions through which it passes. It forces the response patterns in each area to form a single unitary whole while still representing the individual characteristics of stimuli registered in that area. A recent model of reentrant cortical integration [18] shows how this process can operate in the visual system to achieve figural synthesis from segregated visual cues like orientation, occlusion, and motion. The model generates responses that are consistent with "illusory contour" effects seen in human psychophysics experiments.

There are a number of aspects of the TNGS which cannot be treated here in great detail. We wish to mention only a few of these details to indicate in a general way how simple selection events can be cascaded to generate complex forms of behavior. Selection is not confined to acting upon individual neuronal responses – in fact, a principle function of groups is to provide composite responses that can be subject to selection in a hierarchical fashion. Thus, as soon as simple behaviors are established by selection in young animals, these behaviors can begin to form sequences by virtue of connections between groups that can in turn be selected amongst at a higher level. Selection can also act upon value systems to generate higher-level or "secondary" values that may substitute for primary values in regulating long-range goal-seeking behavior that does not lead to immediate responses in primary value systems. Such secondary values are particularly important in humans because of the complexity of society and the indirectness of rewards that are obtained through social and linguistic behavior.

The TNGS is a scientific theory of brain function, and as such, it is subject to experimental test and possible falsification. Since the complexity and inaccessibility of the brain make such tests extremely difficult to carry out in a direct way, we have found it useful to use computer simulations to explore the implications of the abstract ideas of the theory in more detail. In so doing, we have been able to verify that the theory is internally consistent and that it in fact leads to workable arrangements for carrying out perceptual categorization and generating behavior. These arrangements provide a paradigm for the construction of artificial devices having some of the perceptual capabilities of animals. Computer simulations are particularly valuable for working out optimal arrangements for these "perceptual automata" and for comparing their performance with models based on other approaches to problems of perception.

In order to give the reader some sense of the content and results of these computer simulations, we now present two examples. The first of these investigates the consequences of reentrant signalling among visual areas and provides an explanation for recent observations of neuronal oscillations in the visual cortex of the cat. The second is a synthetic automaton incorporating both sensory and motor systems to demonstrate the process of perceptual categorization and its coupling to behavior. After presenting these two examples, we conclude by comparing the TNGS with other current approaches to perceptual problems in the light of the examples.

3. A model for oscillatory reentrant signalling in visual cortex

The visual cortex is one of the best-studied brain regions. It contains numerous functionally and anatomically defined subregions; these subregions contain local assemblies of neurons that have many of the properties of neuronal groups postulated by the TNGS. Accordingly, the visual cortex has been chosen for extensive modelling studies aimed at showing whether groups constructed according to the TNGS can display some of the observed properties of real cortical neurons [19]. The groups in this study have been modelled at the level of their constituent cells in order explicitly to observe the effect of reentrant interactions between individual cells on group behavior; the model described here contains 25 600 cells and 3.5×10^6 connections. It has been constructed with the aid of the cortical network simulator (CNS) program developed at The Neurosciences Institute and contains two areas, one with units that respond to moving oriented bars, the other with units that respond to the overall direction of motion of extended contours. We consider first the response properties of neuronal groups in the orientation area alone, then discuss the emergent effects that are obtained when the two areas interact through long-range reciprocal connections. In the model, transient dynamical links and correlations are established between neuronal groups that respond to different parts of a stimulus contour when these parts move together. With this result we are able to reach beyond the currently available experimental data to suggest a general function for reentrant interactions which may occur at multiple locations in the brain.

Recent experiments by Gray and Singer [20, 21] and Eckhorn et al. [22] are strongly indicative of group behavior and suggestive of the role of reentry in correlating the behavior of different groups. Cortical neurons in cat visual cortex undergo oscillatory activity at frequencies near 40 Hz, recorded as rhythmically varying local field potentials and multiunit activity in visual areas 17 and 18 during stimulation with optimally aligned light bars (fig. 4A). No such oscillations are found in the lateral geniculate nucleus, implying an intracortical mechanism for the generation of the oscillations. The oscillations occur in tight phase synchrony for cells in the same functional column indicating local cooperative effects [20, 21]. Phase coherency of oscillations also occurs in cells in different columns (possibly mediated by cross-columnar connections) that are separated by several millimeters of cortex and that have nonoverlapping receptive fields, if the cells are simultaneously stimulated by a single continuous light bar; correlations are weaker or absent if two separate bars are moved independently [20, 21]. Similar results were obtained by Eckhorn et al. [22]; in addition, a long-distance phase correlation between simultaneously recorded cells in areas 17 and 18 under similar conditions of stimulation was detected.

Our model, based on the properties of neuronal groups and their reentrant interactions, reproduces these experimental results. Neurons within each group are coupled through relatively sparse excitatory connections and are reciprocally connected to a layer of inhibitory neurons. Fig. 4B shows auto- (top) and cross-correlations (middle) between cells in the model recorded under optimal stimulation by a vertical bar. The oscillatory waveform of the autocorrelation is similar to that observed by Gray and Singer (fig. 4A); oscillatory intra-group cross-correlations, which are more difficult to obtain experimentally, indicate local cooperative interactions; such interactions would be expected for cells tightly coupled in a group. The bottom panel of fig. 4B shows the shifted auto-correlation, which is flat as in the actual experiment (fig. 4A). This suggests that the oscillations seen in the top two panels are a result of intrinsic mechanisms and do not depend on entrainment by other oscillatory inputs [23]. The

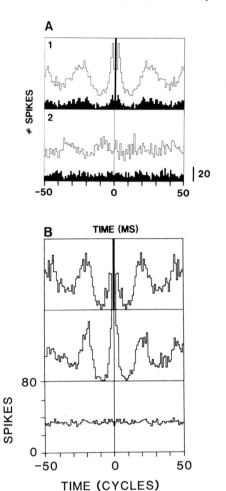

Fig. 4. (A) Temporal properties of a multi-unit activity response to a moving light bar stimulus recorded from area 17 in a five-week-old kitten. The upper panel shows the auto-correlation for both forward (filled bars) and reverse (unfilled bars) direction of stimulus movement. The lower panel shows the auto-correlation after shuffling (shifting) of the trial sequence by one stimulus period; absence of rhythmicity indicates that the oscillatory activity is not stimulus-induced. (Reproduced from ref. [20] with permission by the authors.) (B) Auto-correlation, cross-correlation and shifted auto-correlation (from top to bottom, respectively) for single spike data of simulated neurons within a single neuronal group in the orientation-selective network of our model during stimulation with an appropriately oriented light bar (responses accumulated for 10 trails). The auto-correlation shows a characteristic oscillatory waveform, very similar to the one experimentally observed. Cross-correlation is between two simultaneously recorded units within the same group, and also shows an oscillatory waveform. "Shifted" auto-correlation is averaged over all nine possible shifts (in ten trial periods); flatness indicates that correlations and oscillatory waveform are not due to stimulus-induced effects.

oscillations seen in the model are a population phenomenon that does not require regular firing of pacemaker neurons for its initiation or sustenance.

Fig. 5 presents a conceptual extension of the original experiments and suggests a functional role for reentry in correlating the responses of neuronal groups in different cortical areas. The visual stimulus used was a right-angle corner moving southeast. Neuronal groups in the orientation-sensitive area respond to parts of the moving contour within their receptive fields and detect the local direction of motion of those parts (component motion). Long-range connections from the orientation- to the motion-sensitive network lead to the combination of local component motion signals to form a response to the movement of the entire pattern (pattern motion). Pattern-motion sensitive groups can signal *back* through reentrant connections to correlate those component motion signals that contributed to the generation of the pattern motion signal (fig. 5, middle panel). As a result, positive and oscillatory cross-correlations are recorded from orientation-sensitive groups of different specificity. These groups are dynamically and transiently linked by activity in reentrant connections between the two networks. Cutting the connections running back from the motion- to the orientation-sensitive network eliminates these cross-correlations (fig. 5, bottom panel). We predict that such dynamic linkages between distant cortical sites will be found in other cortical areas as well. Coherency of responses between segregated areas could serve as a general mechanism that would, for example, permit segregated visual responses to be linked to a common object in the environment. Such coherency should be detectable in higher brain regions and facilitate the generation of a unitary percept from a number of isolated component responses.

This model provides a possible mechanism for the correlation of activity in multiple neuronal maps. The following description of the selective recognition automaton Darwin III contains another example of such a reentrant system, this

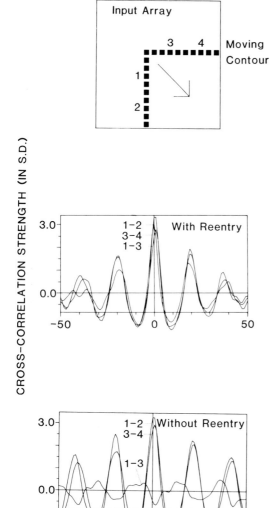

Fig. 5. Long-range reentrant connections between orientation- and motion-sensitive networks generate coherent responses to a simulated moving contour [19]. The contour is indicated in the top panel; numbers 1, 2, 3, and 4 refer to neuronal groups in the orientation-sensitive network that respond to a part of the contour. The diagrams show the cross-correlation of the average activity of neuronal groups (sum over all active cells in a given group at a given time) in the orientation-sensitive network, in the presence (middle) and absence (bottom) of reentrant connections from the motion-sensitive network. Positive cross-correlations between groups 1–2 and 3–4 are the result of local reentrant ("cross-columnar") connections linking groups of similar, but not those of orthogonal specificity. Cross-correlation strength is expressed in standard deviations (S.D.) above the background and is computed from scaled cross-correlation functions.

time modelled at a higher level of abstraction and with greater complexity of behavior.

4. Darwin III: a selective recognition automaton

Synthetic neural modelling aims at an understanding of complex neural phenomena by taking into account processes and interactions at all relevant levels, from the synaptic to the behavioral [24]. Darwin III (which consists of a simple but realistic nervous system that is embedded in a specific "phenotype" acting in a specific "environment", all of which are simulated in a computer) was constructed consistent with the principles of the TNGS. It has, in a crude form, some of the behavioral capabilities of an animal, without the use of a priori definitions of categories, codes, or information-processing algorithms. Darwin III allows us to examine, in a single system of interconnected neuronal repertoires, sensory processes involving recognition and classification, motor acts, such as visual saccades, reaching, and touch-exploration, and the combination of both in a behaving automaton in which the effects of selection on *output* can influence the perceptual categorizations of the system through the rearrangement of the environment produced by the creature's own actions. The pattern of connections, and rules for cell responses and synaptic modifications are specified by the experimenter (using the CNS program) for each network (also called repertoire). However, no instructions are given concerning the nature of particular stimuli, and no algorithms or other procedures are specified to govern the dynamics of its neuronal circuits above the single-cell level.

We will only give a brief summary of Darwin III's structure and performance; the interested reader may find a more detailed account elsewhere [24]. The present version of Darwin III receives sensory input from three modalities: vision, touch, and kinesthesia. On the motor side, there is an eye and an arm with multiple joints, each controlled by motor neurons in specified repertoires.

Every cell in Darwin III has a scalar activity state determined by a "response function". This function has terms corresponding to synaptic inputs, Gaussian noise, decay of previous activity, depression and refractory periods, and long-term potentiation (LTP). The relative magnitudes of these terms can be varied parametrically. Rules for synaptic modification contain features reflecting some of the complex properties of real neurons: in addition to depending on local activity, changes in connection strengths also may be made to depend heterosynaptically on inputs from nearby synapses. These are usually connected to cells in a repertoire, the activity of which reflects the organism's evaluation of its recent behavior according to a particular "value scheme". Synaptic changes may be applied as either increases or decreases according to whether pre- and post-synaptic activity levels are above or below certain thresholds.

Darwin III consists of four neuronal subsystems (fig. 6): (a) a saccade and fine-tracking oculomotor system, (b) a reaching system using a single multi-jointed arm, (c) a touch-exploration system using a different set of "muscles" in the same arm, and (d) a reentrant categorizing system. In this paper we will deal extensively only with the reaching system and refer to other publications for detailed descriptions of the rest of the model.

Darwin III's oculomotor system, by selection from a repertoire of eye motions, acquires the ability to move the eye toward objects in its visual field and to track moving objects. Visual signals are mapped topographically to a network containing excitatory and inhibitory neurons that loosely represents an area of the brain called the superior colliculus. Excitatory cells in this repertoire are densely connected in the model to oculomotor neurons, whose activation causes eye motion. The initial connection strengths of these connections are assigned at random, so that eye motions are initially uncorrelated to visual stimulation.

An innate value scheme imposes a global constraint on the selection of eye movements: only those motions are selected which bring visual

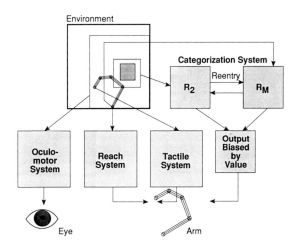

Fig. 6. Schematic diagram of the subsystems of Darwin III. The "environment" (heavy square at upper left) contains moving objects (one of which is indicated by the shaded square). A portion of the environment is viewed by the eye (large square only partially visible within border of environment: limits of peripheral vision; small square surrounding hatched object: central vision). Movements of the eye and the four-jointed arm (bottom) are controlled by the oculomotor and reach systems, respectively. The distal digit of the arm contains touch sensors used by the tactile system to trace the edges of objects. The categorization system receives sensory inputs from the central part of the visual field and from joint receptors signalling arm movements as tracing of object edges occurs. These inputs connect, via intermediate layers not shown, to a higher-order visual center, R_2, and to an area that correlates motion signals over time, R_M. R_2 and R_M are reciprocally connected to form a classification couple. Correlation of firing patterns in R_2 and R_M via reentry leads to classification of objects and eventually yields an output biased by value which can activate reflex movements of the arm.

stimuli into the foveal part of the retina. All that is needed to accomplish this is a value repertoire containing cells connected densely to the fovea and more sparsely to the periphery. The responses of these cells provide a hetero-synaptic component which modulates the modification of connections from the "colliculus" to the motor neurons. Connections which are active in a short time interval before foveation occurs are thus selected and strengthened. Thus, selection affects synaptic and cellular populations *after* they have contributed to a motor act. What is selected are populations of connections that favor the reoccurrence of motions that happen to give foveation. The selection process, however, is "blind" with respect to the individual contribution that each one of these connections makes (unlike "back-propagation" [25]). There is no a priori analysis indicating which connections may have adaptive value, and no error signal is computed and "back-propagated" into the network. The system only repeats what works for it. The oculomotor system improves considerably after only 150 presentations of an object at different locations; after training the eye quickly centers on a stimulus anywhere in the visual field. This simulation exemplifies how selection based on value can shape the development of a sensorimotor system.

Darwin III's multi-jointed arm develops smooth reaching movements by selection of such movements from a prior repertoire of spontaneous gestural motions (see fig. 7). Movements are initiated by a simple "motor cortex" (MC), and the neural activity is transmitted via an intermediate repertoire (IN), representing brainstem nuclei or an additional motor cortical layer, to the motor neurons (SG) that move arm extensors and flexors. Gestures emanating from these systems are filtered by a cerebellum-like structure, here simplified to just "granule cells" (GR) and "Purkinje cells" (PK). "Granule cell" inputs come from both kinesthesia (which senses the angular positions of the various joints in the arm) and target vision. As a result, the firing of "granule" cells corresponds to combinations of joint positions, i.e. particular conformations of the arm, and positions of the target. The system has the task of associating these arm positions with appropriate and inappropriate gestures through modification of synaptic connections, primarily those between "granule cells" and "Purkinje cells". "Purkinje cells" then act directly to inhibit inappropriate gestures at the motor cortical level (deep cerebellar nuclei are omitted here).

Nothing in these circuits prejudices the arm to move in a coordinated way, or even in a particular direction, before training. Instead, motions are selected from spontaneous movements when these are successful in getting the arm closer to the object (fig. 8). Success is evaluated by a simple value repertoire in which signals from a visual area responsive to the creature's own hand and

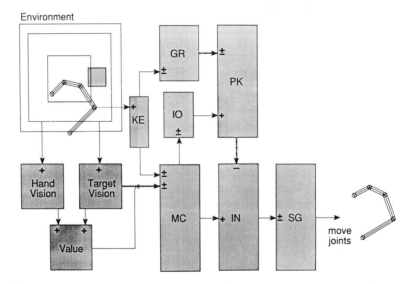

Fig. 7. Schematic diagram of one version of the arrangement mediating arm reaching in Darwin III (Reprinted from ref. [33].) The two large squares in the environment represent peripheral and foveal visual fields; the smaller shaded square is a stimulus object. The bent structure with circular links is the arm. The main motor pathways across the bottom of the diagram are responsive to both visual signals (indicating the location of the target) and kinesthetic signals (indicating the conformation of the arm). "+", "−", and "±" indicate excitatory, inhibitory, and mixed connections, respectively. Heavy dashed arrows indicate modifiable connections. The figure also indicates a biased connection between "value" and the modifiable connections efferent from the target vision repertoire. The model "cerebellum" (top) acts to inhibit inappropriate gestures and thus improves the directional specificity and smoothness of reaching.

signals from another visual area responsive to stimulus objects are combined in a common map such that its activity increases when the two visual responses map near a common location. Activity in this value repertoire is carried to the "motor cortex", where it influences the extent of synaptic modification as described earlier, as well as to the "inferior olive" (IO), where it gates activity emanating from the motor cortical network to the Purkinje cell layer. Connections whose activation leads to an increase in the value response, representing motion of the arm closer to the object, are selectively favored. The reaching problem is more complex than the visual tracking problem, however, because there are multiple degenerate solutions due to the mechanical redundancy of arm motions [26]. The nervous system needs to master such situations routinely by reducing the number of degrees of freedom to reach a controllable state. We propose that the mechanism by which the nervous system achieves this is essentially selective and leads to the amplification of gestural motions with adaptive couplings of relevant movement parameters to form motor synergies.

With training and selective amplification of synaptic connections, Darwin III's reaching system improves significantly. A systematic assessment of its behavior shows how motor synergies emerge during the training process restricting the initially very broad envelope of paths to a narrow bundle, with most of the paths intersecting the target object (fig. 8). Plotting movement variables such as joint angles shows that these variables are no longer independent of each other (fig. 9). The system has made choices and the resulting movements represent gestural "wholes", a movement gestalt; this is a strategy that is taken without external instruction and is of great adaptive value to the organism because it considerably reduces the number of independently controlled variables. The four-jointed arm model described here acts as a gestural module. Because of the lack of appropriate neuronal mappings, it is unable to perform evenly over a larger region of space. The design of such mappings was investigated in a related but non-redundant reaching model described elsewhere [24].

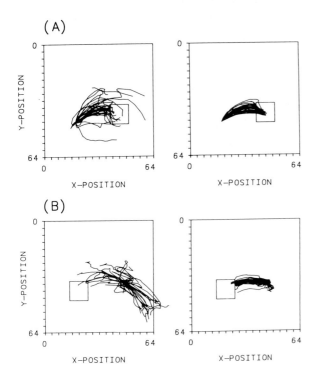

Fig. 8. Reaching trajectories. Traces of paths taken by the tip of the four-jointed arm of Darwin III before (left panels) and after (right panels) 1440 training cycles for two different combinations of initial and target positions ((A) and (B)). Training proceeded as follows: for each of 90 trials, the arm was placed in a standard position. It was then allowed to move for 16 cycles, and synaptic changes occurred depending upon the success of the movements relative to a target object whose position is shown by the square at the right in (A), and at the left in (B). After training (right panels), movements which reached the object on a direct path have been selected. Note that there is still some variability in the exact paths taken, which could serve as the basis for ongoing selective processes.

The arm of Darwin III, once it has established contact with the surface of an object, starts moving along the edges of objects guided by tactile signals. During such exploration it assumes a straightened posture by "reflex" to facilitate the tracing of objects. Tracing object contours provides Darwin III with sensory signals concerning the surface (or outline) properties of objects and these signals are used as a second sensory modality in subsequent object categorization.

In order to deal with nontrivial categorizations, a recognition system must be responsive to stimulus features of more than one kind. Subsystems

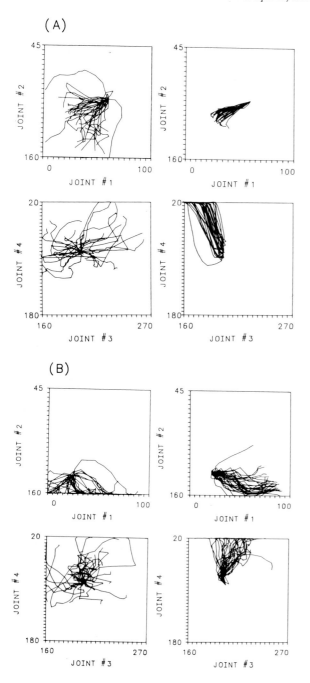

Fig. 9. Synergies. Plots of joint angles for movements of the four-jointed arm before (left panels) and after (right panels) training (same trajectories (A) and (B) as in fig. 8). Before training, changes in joint angles are uncorrelated in all cases. After training, proximal as well as distal joints are correlated and controlled as a unit. Such correlations between joints can be regarded as simple motor synergies allowing the system to master a mechanically redundant situation.

responsive to different sensory modalities (or submodalities) form "classification couples" [11], which correlate their respective neuronal signals by reentry. This arrangement reflects the need for an animal in a natural environment to combine apparently arbitrary combinations of features. Again, the issue of value is important; most likely, animal classification does not exist as an abstract faculty of "animal cognition", but subserves very definite behavioral functions. Indeed, its relevance for the organism becomes manifest only when it is coupled to appropriate behaviors having survival value for the organism. There will be, in general, neither a simple computable function dividing the stimulus space into distinct categories, nor a deterministic mapping relating perceptual categories to behavior [27]. This poses the difficult problem of how associations between sensory and motor signals can be formed.

These and other considerations, including simplicity, led to the design of the coupled categorization-response system in Darwin III. Since category formation in some form must exist prior to learning, we have made use of relatively impoverished low-level sensory networks capable of distinguishing just two categories to construct the first version of this system. The categories are "rough-striped" objects (bad for this species) and all others (good). The result of a "rough-striped" categorization is one of the simplest possible behaviors, a reflex-like arm motion (a "swat") that frequently removes the stimulus from the automaton's vicinity.

We will omit a detailed description of the connectivity and dynamic properties of the system and instead discuss its overall behavior. As a result of activity in reentrant connections between visual and tactile areas the firing thresholds of neuronal units that are activated in a correlated fashion are lowered. When no novel features have been found for a time, a pattern is triggered in the reentrant categorization area that is characteristic of the particular class to which the object belongs (although different examples of the same class will give different patterns). This pattern, if it is one that activates the ethological value system, evokes the swat reflex to remove the stimulus. Stimuli not recognized as noxious are left undisturbed. Fig. 10 shows a distribution of stimulus objects ordered by the frequency of occurrence of the rejection response. Although the design of the categorization system was kept simple, the resulting behavior of the system is fairly complex. Classification of objects seems to result in graded category boundaries and is probabilistic rather than deterministic in character. Class membership is a matter of degree (as determined by the frequency of the associated behavioral act) rather than all-or-none. These results are in agreement with many experimental studies on human and animal categorization. In this version of Darwin III the selection of rough-striped objects for rejection is built in as an evolutionary imperative; a more general solution to the problem of associating behavioral acts with categories must involve the action of "ethological" value systems, analogous to the ones used in the sensorimotor systems. Such value systems imply the ability to evaluate consequences of behavioral acts; they permit the formation of stimulus–response associations based on experience rather than innate connectivity.

Darwin III is the first example of an adaptively behaving creature synthesized from units defined at the neuronal level. Synthetic neural modelling will continue to help us bridge the gap between the neuronal and psychological levels of description of mental functions, and to test by simulation the consequences of structural and dynamical variations at all levels. Selection of neuronal responses contributing to adaptive behavior seems to be all that is required to "bootstrap" an artificial organism into states in which it can make categorical discriminations.

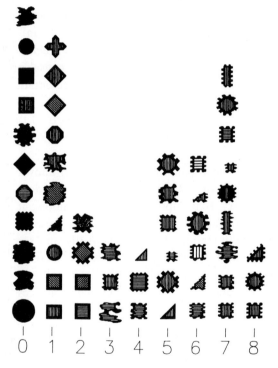

Fig. 10. Distribution of categorization frequencies [24]. Objects are grouped according to the responses of one version of Darwin III's categorization system. Each object was presented eight times and a maximal tracing time of 50 cycles per object was allowed. Tracing started at different positions on the object (usually close to the center) and the occurrence of the reflex output was recorded. In the case of a rejection response the automaton's arm hit the object and removed it from the environment. An activation of the rejection response at any time within the 50-cycle time interval was counted as a "rejection" response. If no response occurred within the 50-cycle time limit, the trial was ended and a new object was entered. Objects are arranged in nine columns depending on the frequency with which they met with a response. In this version of Darwin III, an intrinsic negative ethological value is attached to "bumpy and striped" objects.

5. Is the brain a computer?

Before we attempt to answer this question, we should make clear what we mean here by a computer. It is often assumed that if signals enter some device, are transformed in some way within the device, and the altered signals emerge from the device, then that device is a computer. If that definition is accepted, then of course the brain is a computer, but we believe that such a loose definition fails to capture some of the essential nature of the concept of computation which is critical to an understanding of the philosophical implications of the brain/computer analogy. Computation is best defined in terms of the properties of a Turing machine [28], which is a finite-state automaton with a "tape" bearing encoded information and a fixed set of rules governing its transitions between states. Those who claim that the brain "computes" are in fact claiming that the brain uses symbolic representations and algorithms.

S. Forrest, in her introductory remarks at this Conference on Emergent Computation, attempted to extend this concept of computation to include certain organized events occurring in natural as well as in artificial systems. Her definition of "emergent computation" included these elements: (1) the existence of explicit instructions; (2) the interaction of components to form implicit global patterns, and (3) a natural interpretation of these events as the processing of information. We agree with the general form of this definition, although we note that insofar as the instructions in (1) arise as an incidental result of the operation of physical laws and not according to some organizing principle, they would lack the element of direction towards a goal that is inherent in the ordinary notion of a computation. Although the brain may arguably be considered to fulfill conditions (2) and, in part, (3), it certainly does not fulfill condition (1). No one has succeeded in explicitly stating a set of instructions (i.e. a Turing machine transition table) that would serve as a plausible basis for intelligent behavior, and our discussion of the perceptual categorization problem suggests that, in fact, no such set of instructions can exist.

The conclusion that the brain is not a Turing machine is important because the assumption that the brain *is* a kind of computer is at the heart of functionalism, the philosophy of mind which underlies most current research in cognitive science and in artificial intelligence. Functionalism [29] holds that psychological phenomena can be best understood by describing them in terms of "functional processes" carried out in the brain. Of course, what functionalists have in mind when they speak of these processes is that they are a form of symbol manipulation carried out according to formal rules like those implemented in Turing machines. The functional efficacy of these processes then depends on the *interpretation* of the symbols they act upon and not upon any details of the structure or mode of development of the nervous system itself (except that it must have adequate speed and memory capacity). According to the functionalist view, there is a symbolic "language of thought" [30] that is innate and that can be described separately from the underlying neural implementation.

It is seen as an advantage of functionalism that mental functions can be analyzed independently of their physical instantiations; the results of this abstract analysis are immediately applicable to robots, extraterrestrial aliens, and any other intelligent systems one may come across. However, this ability of functionalist programs to "run" on a variety of different hardware systems is achieved by postulating the use of elementary causal processes that operate upon symbolic representations of information. Both the *processes* and the *representations* present difficulties with regard to their possible mechanisms of origin in biological systems [31].

Symbol-manipulating processes (also known as algorithms) operate by carrying out sequences of elementary steps upon strings of symbols; it is essential that the precise sequence of these steps be reliably maintained throughout life and somehow transmitted to progeny. However, there is no evidence for the existence in the brain of a means for recording and playing back such sequences of steps, nor is there any evidence that such sequences of steps could be transmitted between generations by DNA encoding. Even supposing that a mechanism for transmitting algorithms does exist, but has not been found by current experimental approaches, it seems virtually impossible that complex algorithms could emerge from the accidental shuffling of steps that would take place in evolution; mutations of the DNA encoding the steps for mental algorithms would result almost certainly in the destruction of the efficacy of the algorithms in the absence of a mechanism operating in somatic time to correct such errors.

Similarly, any mechanisms one might postulate for the origin and transmission of symbolic codes would need to incorporate means to translate any changes which occurred in those codes into the appropriate changes in the associated processing algorithms needed to maintain consistency of results; it is difficult to see how this could be done by functionalist machinery having no knowledge (other than that embedded implicitly in the system's formal rules) of the *meanings* of the symbols. Yet changes must surely occur if the representation system is to adapt to changes in the animal itself due to growth and experience, as well as the inevitable changes in the environment. According to the view of perceptual categorization we have put forward, categorization is highly dependent on individual life history as well as on the context of each moment; thus it is only through *interactions of representations with the world* that suitable response patterns can be selected. These interactions would appear to require representations based on multiple, reentrantly connected mappings of sensory signals. Their very existence is incompatible with one fundamental postulate of functionalism, viz that representations have meaning independent of their physical instantiation.

If one gives up the view that the brain is a computer, then it becomes possible to seek alternatives that are not subject to these difficulties of functionalist systems. Selectionism, as we have described it, is one such alternative that is based firmly on biological principles. It provides a rather different set of conclusions about the nature of mental states and processes than that provided by functionalism; a few of these contrasts are outlined in table 2. Whether or not the selectionist approach turns out upon experimental test to be right in detail, we believe that a serious consideration of the difficulties afflicting mainstream functionalist philosophy provides a strong incentive for studying the brain as it is rather than to continue searching for answers amidst the ruins of the failed computer analogy.

Acknowledgements

This research has been supported in part by the Neurosciences Research Foundation, the Office of Naval Research, the John D. and Catherine T. MacArthur Foundation, the Lucille P. Markey Charitable Trust, the Pew Charitable Trusts, the van Ameringen Foundation, and the Charles and Mildred Schnurmacher Foundation. O.S. is a Fellow of the Lucille P. Markey Charitable Trust, Miami, FL. Some of this research was carried out using facilities of the Cornell National Supercomputer Facility, a resource of the Center for Theory and Simulation in Science and Engineering at Cornell University, which is funded in part by the National Science Foundation, New York State, and the IBM Corporation and members of the Corporate Research Institute.

Table 2
Differences between functionalist and selectionist approaches.

Functionalism	Selectionism
Mental states are determined by functional or causal properties of physical processes	Causality, but macro-indeterminacy due to degeneracy and underlying variance
Functional processes are indifferent to physical instantiation and history	Brain function is critically dependent upon development and historical events
Functionally isomorphic systems have identical cognitive states	No 1:1 mapping of states of nervous system to cognitive states
Concepts and beliefs are computational states	Concepts and beliefs acquire meaning by reference to a physical and social environment
Learning is a form of programming	Learning emerges from selection upon value

References

[1] G.E. Hinton and J.A. Anderson, Parallel Models of Associative Memory (Erlbaum, Hillsdale, 1981);
J.L. McClelland and D.E. Rumelhart, Parallel Distributed Processing, Vols. 1, 2 (MIT Press, Cambridge, 1986).
[2] E.M. Markman, Categorization and Naming in Children (MIT Press, Cambridge, 1989).
[3] S.E.G. Lea and S.N. Harrison, Quart. J. Exp. Psychol. 30 (1978) 521;
J. Cerella, J. Exp. Psychol. Human Percept. Perf. 5 (1979) 68;
R.J. Herrnstein, Phil. Trans. R. Soc. London Ser. B 308 (1985) 129.

[4] E.E. Smith and D.L. Medin, Categories and Concepts (Harvard Univ. Press, Cambridge, 1981);
D.L. Medin and E.E. Smith, Ann. Rev. Psychol. 35 (1984) 113.
[5] L. Wittgenstein, Philosphical Investigations (Macmillan, New York, 1973).
[6] E. Rosch and C.B. Mervis, Cogn. Psychol. 7 (1975) 573;
C.B. Mervis and E. Rosch, Ann. Rev. Psychol. 32 (1981) 89.
[7] K.G. Pearson and C.S. Goodman, J. Comp. Neurol. 184 (1979) 141.
[8] E.R. Macagno, V. Lopresti and C. Levinthal, Proc. Natl. Acad. Sci. US 70 (1973) 57.
[9] S. Ramón y Cajal, Histologie du système nerveux de l'homme et des vertébrés, transl. by L. Azoulay (Maloine, Paris, 1904), reprinted (Instituto Ramón y Cajal, Madrid, 1952).
[10] C. Darwin, The Origin of Species, 6th Ed. (John Murray, London, 1872);
E. Mayr, Animal Species and Evolution (Harvard Univ. Press, Cambridge, 1963).
[11] G.M. Edelman, in: The Mindful Brain, eds. G.M. Edelman and V.B. Mountcastle (MIT Press, Cambridge, 1978) p. 51;
G.M. Edelman, Neural Darwinism (Basic Books, New York, 1987).
[12] G.M. Edelman, Ann. Rev. Cell Biol. 2 (1986) 81;
G.M. Edelman, Topobiology (Basic Books, New York, 1988).
[13] W. Singer, in: The Neural and Molecular Bases of Learning, eds. J.-P. Changeux and M. Konishi (Wiley, New York, 1987) p. 301.
[14] S.L. Foote and J.H. Morrison, Annu. Rev. Neurosci. 10 (1987) 67.
[15] L.H. Finkel and G.M. Edelman, Proc. Natl. Acad. Sci. US 82 (1985) 1291.
[16] E.G. Jones and T.P.S. Powell, Brain 93 (1970) 793;
L.L. Symonds and A.C. Rosenquist, J. Comp. Neurol. 229 (1984) 39;
S. Zeki and S. Shipp, Nature 335 (1988) 311.
[17] G.M. Edelman and G.N. Reeke Jr., Proc. Natl. Acad. Sci. US 79 (1982) 2091;
G.N. Reeke Jr. and G.M. Edelman, Ann. NY Acad. Sci. 426 (1984) 181.
[18] L.H. Finkel and G.M. Edelman, J. Neurosci. 9 (1989) 3188.
[19] O. Sporns, J.A. Gally, G.N. Reeke Jr. and G.M. Edelman, Proc. Natl. Acad. Sci. US 86 (1989) 7265.
[20] C.M. Gray and W. Singer, Proc. Natl. Acad. Sci. US 86 (1989) 1698.
[21] C.M. Gray, P. König, A.K. Engel and W. Singer, Nature 338 (1989) 224.
[22] R. Eckhorn, R. Bauer, W. Jordan, M. Brosch, W. Kruse, M. Munk and H.J. Reitboeck, Biol. Cybern. 60 (1988) 121.
[23] D.H. Perkel, G.L. Gerstein and G.P. Moore, Biophys. J. 7 (1967) 391, 419.
[24] G.N. Reeke, L.H. Finkel, O. Sporns and G.M. Edelman, in: Signal and Sense: Local and Global Order in Perceptual Maps, eds. G.M. Edelman, W.E. Gall and W.M. Cowan (Wiley, New York, 1990), in press.
[25] D.E. Rumelhart, G.E. Hinton and R.J. Williams, Nature 323 (1986) 533.
[26] N.A. Bernstein, The Coordination and Regulation of Movements (Pergamon, Oxford, 1967);
H.T.A. Whiting, ed., Human Motor Actions. Bernstein Reassessed (North-Holland, Amsterdam, 1984).
[27] M. Bongard, Pattern Recognition (Spartan Books, New York, 1970).
[28] A.M. Turing, Proc. Lond. Math. Soc. 42 (1937) 230.
[29] H. Putnam, in: Minds and Machines, ed. S. Hook (Collier, New York, 1960) p. 138;
H. Putnam, Representation and Reality (MIT Press, Cambridge, 1988).
[30] J.A. Fodor, The Language of Thought (Thomas Y. Crowell, New York, 1975).
[31] G.N. Reeke Jr. and G.M. Edelman, Daedalus 117 (1988) 143.
[32] G.N. Reeke and G.M. Edelman, in: Structure and Dynamics of Nucleic Acids, Proteins and Membranes, eds. E. Clementi and S. Chin (Plenum, New York, 1986) pp. 392–353.
[33] G.N. Reeke, L.H. Finkel and G.M. Edelman, Selective recognition automata, in: An Introduction to Neural and Electronic Networks (Academic Press, New York, 1990), pp. 203–226.

BIFURCATION AND CATEGORY LEARNING IN NETWORK MODELS OF OSCILLATING CORTEX

Bill BAIRD

Department of Biophysics, University of California, Berkeley, CA 94720, USA

A generic model of oscillating cortex, which assumes "minimal" coupling justified by known anatomy, is shown to function as an associative memory, using previously developed theory. The network has explicit excitatory neurons with local inhibitory interneuron feedback that forms a set of nonlinear oscillators coupled only by long-range excitatory connections. Using a local Hebb-like learning rule for primary and higher-order synapses at the ends of the long-range connections, the system learns to store the kinds of oscillation amplitude patterns observed in olfactory and visual cortex. In olfaction, these patterns "emerge" during respiration by a pattern forming phase transition which we characterize in the model as a multiple Hopf bifurcation. We argue that these bifurcations play an important role in the operation of real digital computers and neural networks, and we use bifurcation theory to derive learning rules which analytically guarantee CAM storage of continuous periodic sequences – capacity: $N/2$ Fourier components for an N-node network – no "spurious" attractors.

1. Introduction

There is a large gap still between physiologically detailed network models of real vertibrate cortical memory systems [31, 26, 21], and analytically understood "artificial" neural networks for associative or content addressable memory (CAM) (see ref. [3] for anthology). Prominent features observed in real cortical networks are recurrent anatomical connectivity, distinct excitatory and inhibitory neural populations that exhibit sigmoidal pulse frequency input/output relations [21, 23, 16], oscillatory (possibly chaotic) dynamics [27, 15, 24, 34], and sharp transitions of dynamical behavior called "bifurcations" [21, 28, 54, 17, 4, 41]. These features are not found together in current artificial memory networks for which analytic results may be obtained. In this paper we construct networks with these characteristics and exhibit learning algorithms that allow us to obtain analytic guarantees of associative memory performance. The goal is to obtain a clear mathematical account of the principles of operation of the biological systems, so that they may be extracted and observed in other scientific contexts, or applied for engineering purposes in artificial systems. The learning rules developed here to explain biological phenomena also provide precise methods for synthesis of artificial networks with analytically guaranteed CAM storage of continuous analog spatio-temporal sequences. The plan of the paper is to introduce biological models and give verbal accounts of the mathematical results that have been proved in earlier work. This is intended to allow nontechnical access to the basic ideas of the approach. The next paragraphs provide an overview of the biological models and the results of the mathematical theory published elsewhere [5, 7, 8].

Two "minimal" models that are justifiable caricatures of the anatomy of the primary and secondary sensory cortex of the olfactory system – the olfactory bulb and prepyriform cortex – are introduced and analyzed. The prepyriform cortex is thought to be one of the clearest cases of a real biological network with associative memory function [31]. The models are intended to assume the least anatomically justified coupling sufficient to allow function as an oscillatory associative memory. Explicit excitatory neural populations connect with local inhibitory interneuron populations in separate feedback loops that form a set of unconnected oscillators. These are then cross-coupled only by long-range excitatory connections W_{ij} with additional "higher-order" synaptic weights W_{ijkl} [13, 42]. This may be considered a simple generalization of the analog "Hopfield" [35] network to store periodic instead of fixed point attractors. The long-range cross-coupling is from excitatory to excitatory neural populations in the case of the prepyriform cortex, and from excitatory to inhibitory populations in the model of the olfactory bulb.

Analysis shows that a local Hebb-like outer product rule for these long-range connections alone will establish, as eigenvectors of this coupling matrix W, the kinds of oscillation amplitude patterns observed experimentally in the olfactory system. If further higher-order synaptic weights from the projection algorithm (described below) are added, these amplitude patterns are analytically guaranteed by the projection theorem [5, 7] to be the only attractors in the network vector field. The frequency of each stored oscillation is a specified function of the eigenvalue corresponding to its amplitude pattern and the strength of the fixed feedback connections within the local oscillators. It is argued that some dimensionality of the higher-order synapses in the mathematical model may be realized in a biological system by synaptic interactions in the dense axo-dendritic interconnection plexus (the "neuropil") of local neural populations – given the information immediately available on the primary long-range connections W_{ij}. Latest theoretical work shows that only N^2 of the N^4 possible higher-order weights are required in principle to approximate the performance of the projection algorithm, and this has already been incorporated into a new learning rule. The model neuron pools here can be viewed as operating in the linear region of the usual sigmoidal *axonal* nonlinearities, and multiple memories are stored instead by the highly programmed multiplicative *synaptic* nonlinearities.

The "normal form projection algorithm" introduced in previous theoretical work [5, 7] is applied to these models. The general projection algorithm supplies an analytic formula for determining the weights in recurrent networks with these higher order correlations or "sigma-pi" units [18] in order to store precise analog patterns in memory. The approach allows programming of aspects of the network vector field independent of the patterns to be stored. Stability, basin geometry, and rates of convergence may be determined. For a network of N nodes, N static or $N/2$ single-frequency periodic attractors of different frequencies may be stored – without unintented or "spurious" attractors. If the oscillation amplitude patterns to be stored can be assumed to be orthogonal, as may be reasonable for a sparsely activated biological system, the learning operation reduces to a kind of periodic or phase-dependent outer product rule that permits local, incremental learning. The system can be truly self-organizing because a synapse of the net can modify itself according to its own activity during external driving by an input pattern to be learned. Between units of equal phase, or for static patterns, learning reduces further to the usual Hebb rule.

Subsets of the $N/2$ possible single-frequency periodic patterns may be chosen as the principle Fourier components of a set of desired multifrequency trajectories. This permits the storage and retrieval of complex pattern sequences as attracting periodic trajectories [9]. In the proper coordinate system, a strict Lyapunov or energy function may be explicitly constructed to describe and guarantee the approach of the network to such a trajectory. Elsewhere we argue that sequence storage in a minimal model may provide a portrait of how action sequences might be stored in motor cortex [8]. $N/4$ chaotic attractors may also be stored to mimic the more complex patterns often observed in the olfactory bulb [24].

1.1. Justification of the approach

In real vertebrate neural networks, there may be no such things as two state neurons which go into hard saturation on either side of a threshold and stay there. An action potential is a brief pulse, not a step change of state. If repetitive firing is an "on" state, then the observed variation of firing rate with different levels of suprathreshold stimulation is unaccounted for [38, 37]. When pulse frequency is considered to be the neural state variable, there is often a sizable linear region of operation about some baseline rate. When the units of the network are taken to be neural populations, as we do here, then continuous local pulse densities and cell voltage averages are the natural state variables to use. Smooth sigmoidal population input–output functions, whose slope increases with arousal of the animal, have been measured in the olfactory system [23, 16]. The approach here describes how associative memory can be programmed and work precisely with continuous dynamics and smooth sigmoids that are weakly nonlinear, when it is operated in the vicinity of a "critical point" (discussed below). This may be thought of as operation in the low gain limit near the origin, as opposed to the high gain limit on a hypercube – where analytic results are usually obtained [35, 48, 51, 39, 30, 32, 33]. The effective decision making nonlinearity in this system is in the higher-order synapses, not in the axons.

The learning rules described here permit the construction of biological models and the exploration of engineering or cognitive networks – with analytically guaranteed associative memory function – that employ the recurrent architectures and the type of dynamics found in the brain. There is recurrence in cortical networks on all levels: molecular, intercellular, intercolumnar, cortico-thalamic, and between cortical areas. Patterns of 40 to 80 Hz oscillation have been observed in the large-scale activity of the olfactory bulb and cortex [21] and visual neocortex [27, 15], and shown to predict the olfactory [25] and visual pattern recognition responses [20] of a trained animal. A discussion of the anatomy and physiology of the olfactory system and the experimental work that led to this approach is given in ref. [4].

It appears that cortical computation in general may occur by dynamical interaction of resonant modes, as has been thought to be the case in the olfactory system. The olfactory bulb converts static input into oscillatory output to the prepyriform cortex. Given the sensitivity of neurons to the location and arrival times of dendritic input [36], the successive volleys of pulses (Freeman's "wave packet" [21]) that are generated by the collective oscillation of a neural net may be ideal for reliable long-range transmission of the collective activity of one cortical area to another. The oscillation can serve a macroscopic clocking function and entrain the relevant microscopic activity of disparate cortical regions into phase-coherent macroscopic collective states which override irrelevant microscopic activity. If this view is correct, then oscillatory network modules, possibly of the style discussed in this paper, form the actual cortical substrate of the diverse sensory, motor, and cognitive operations now studied in static networks, and it must ultimately be shown how those functions can be accomplished with these networks.

In the olfactory system, the oscillatory modes emerge during respiratory inspiration and subside at exhalation through a kind of phase transition or pattern formation process [21, 54, 43, 45, 28] which we characterized mathematically as a multiple Hopf bifurcation [29, 4]. A "bifurcation", for example, is an important qualitative change in the dynamical behavior of a system as a parameter is varied – the creation or destruction of attractors, or a change of stability of an attractor to become a repeller. A phase transition may be viewed mathematically as a bifurcation in a stochastic dynamical system. Where "dynamics" describes the continuous temporal evolution of a system, "bifurcations" describe discontinuous changes in the dynamical possibilities. A bifurcation parameter in the networks to be described here could be a respiratory input bias, the sigmoid slope (known to vary with attention in the olfactory system), a general scaling of the coupling matrix, or the membrane decay constant τ. Variation of any of these parameters in the network models to be discussed here can cause all stored attractors to arise from a single equilibrium at the origin.

This mechanism for the "emergence" of this kind of collective computation – the bifurcation of an excitable network into oscillation – is formally similar to that for the bifurcation of the excitable nerve axon membrane to form an impulse [36, 54, 17]. Aspects of the functional behavior of the individual neuron, of value for reliable information transmission at the neuron level, are thus recovered at the level of the neural mass. The bifurcation can act as a critical decision point for the selection of stored memories by small input patterns, in the same way that small fluctuations are amplified to change liquid/gas phase near the critical "triple point" for water [46]. Near the bifurcation point, in the presence of noise input alone, the networks to be described here will make random jumps between stored attractors, or "daydream", as an analog physical network like the optical system of Anderson [1] is observed to do. This behavior could be employed in a search process such as "babbling", or finding an attractor to match, in one area of cortex, the input from another. At high values of a bifurcation parameter, the system may be locked into the neighborhood of a particular attractor regardless of further changes in inputs. At lower, nearly critical values of the bifurcation parameter, new inputs may easily switch the network to a new attractor, but the previous state of the system can determine which transition is possible [34]. This sensitivity to the previous internal state may then allow the system to function as an automaton, and give distinct outputs to distinct but overlapping input sequences. At values below critical, the previous state is erased altogether, and new input can have unbiased access to all attractors, or force the system into a new state to be learned, without interference from previously learned attractors.

Mathematical analysis is simplified, and the power of learning algorithms increased, when this additional dimension of organization about a critical point is considered in the design of the network. We can determine how attractors of the net appear or disappear during learning or with changes in the slope of sigmoids. The network can be organized so that its dynamics may be turned on and off gracefully to readdress the full set of memory attractors with a new input. This is hypothesized to happen in the olfactory system when an animal resamples its environment by the next sniff. Here the pattern recognition computation itself emerges de novo on each sniff, and new pattern recognition capabilities emerge on a personal evolutionary time scale as learning proceeds throughout the life of the individual. In this view, evolution is population learning, and learning is individual evolution.

In a digital computer, the binary flip–flop elements are continuous dynamical systems at the level of their transistors, which are made to bifurcate, say from the 1 equilibrium state being stable to the 0 state being stable, by changes in the signal or clock inputs. The associative networks of the neocortex might (very crudely) be viewed as N-dimensional flip–flops, "clocked" to change oscillatory attractors by the alpha rhythm at a 10 Hz "framing rate" under thalamic attention control – much as the olfactory system appears to be clocked to change attractors between 40 Hz "bursts" by the 3–8 Hz respiratory rhythm [4]. This is a digital on/off kind of clocking by bifurcation, on a longer time scale than the continuous macroscopic "entrainment" clocking of microscopic activity within a 40 Hz burst. During learning, similar 3–8 Hz driving rhythms from a structure in the brain called the hippocampus, known to be essential for long-term memory storage, may present key patterns and coordinate activation of separate cortical areas. This may serve to establish a common phase relationship among the collective activities required for storage of the entire "gestalt" by the kind of phase-dependent Hebb rules to be discussed in this paper.

We are interested here in modeling category learning and object recognition, *after* feature preprocessing. Here equivalence classes of ratios of feature outputs in feature space must be established as prototype "objects" or "concepts" that are invariant under endless particular sensory instances. This is the kind of function generally hypothesized for prepyriform cortex in the olfactory system, or inferotemporal cortex in the visual system. Somewhere in the loop from sensation to action, an animal filters out many details of a perception to make a response choice that is invariant to those details. The hungry cat, seeing a mouse, engages in stalking behavior that is unaffected by most details of the mouse's appearance. Without categories, the world never repeats.

This is a different oscillatory network function from the feature "binding", or clustering role that is hypothesized for "phase labels" in primary visual cortex [27, 15, 11], or from the "decision states" hypothesized for the olfactory bulb by Li and Hopfield [41]. In the visual system, the segmenting of the visual scene into different figures versus background is thought to be accomplished through the grouping of locally related features by independent local phase locking of cortical activity in separate regions. There is no modification of connections, and no learning of particular perceptual events. Intuitively, this kind of behavior might be realized in the minimal network presented below, if only fixed short-range coupling were allowed, so that the cross-coupling matrix of excitatory connections was banded. Then, for sufficiently weak coupling, sparse and disparate input could establish areas of independently phase locked activity in the network.

More cross-coupling appears to be required for the system to reach a unitary global decision (fully phase locked) about the nature of an input object. For category learning, complete adaptive cross-coupling allows all possible input vectors to be potential attractors. This is the kind of anatomical structure that characterizes prepyriform and inferotemporal cortex. The columns there are less structured, and the associational fiber system is more prominent than in primary cortex. Man shares this same high-level "association" cortex structure with cats and rats. Phylogenetically, it is the preprocessing structures of primary and secondary cortex that have grown and evolved to give us our expanded capabilities. While the bulk of our pattern recognition power may be contributed by the clever feature preprocessing that has developed, the object classification system seems the most likely locus of the learning changes that underlie our daily conceptual evolution. That is the phenomenon of ultimate interest in this work. The kind of preprocessing hypothesized above for primary cortex would be ideal for input to the associative memory discussed here, which is especially sensitive to periodic driving by phase-locked collective activity, and will filter incoming signals by ignoring uncorrelated activity.

2. Minimal model of oscillating cortex

Analog state variables, feedback, oscillation, and bifurcation have already been discussed as important features of biological networks which we explore in this approach. Explicit modeling of known excitatory and inhibitory neurons and use of only known long-range connections is also a basic requirement to have a biologically feasible network architecture. We analyze a biological model that is intended to assume the least anatomically justified coupling required to capture essential features of real cortical structure, and use simulations and analytic results proved in refs. [5, 7, 8], to argue that an oscillatory associative memory function can be realized in such a system. This network is meant only as a cartoon of the real biology, which is designed to reveal the general mathematical principles and mechanisms by which the actual system might function. Such principles can then be observed or applied in other contexts as well.

While there are many uncertainties about local processing in the nervous system, it is fairly well established that transmission of neural activity over distances greater than 1 mm is largely by action potentials on excitatory axons. We take the position here that this is for the advantages that a digital encoding (pulse frequency modulation) affords in the reliable transmission of analog signals in the presence of noise. Long-range excitatory to excitatory connections are well known as "associational" connections in prepyriform cortex [31] and cortico–cortico connections in neocortex. Since our units are neural populations, we know that some density of full cross-coupling exists in the system [31], and our weights are the average synaptic strengths of these connections. There is little problem at the population level with coupling symmetry in these average connection strengths emerging from the operation of an outer product learning rule on initially random connections. The existence of the local inhibitory "interneurons" (shown in this network of fig. 1) is a ubiquitous feature of the anatomy of cortex

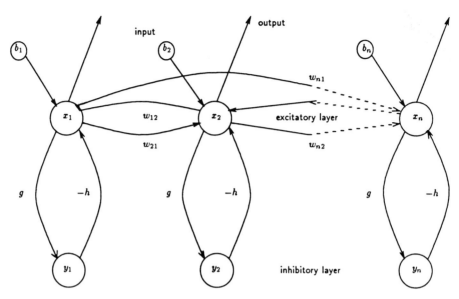

Fig. 1. Minimal network of excitatory cell populations x_i, inhibitory cell populations y_i, inputs b_i, adaptive excitatory to excitatory connections W_{ij}, and constant local inhibitory feedback connections g and $-h$.

throughout the brain [27]. It is unlikely that they make long-range connections (> 1 mm) by themselves. These connections, and even the debated interconnections between them, are therefore left out of a minimal model. The resulting network is actually a fair caricature of the well studied anatomical circuitry of prepyriform cortex [31]. It may provide analytic insight into aspects of the simulation results of the "bottom up" models of these systems which include considerable physiological detail [53, 26, 19]. Although neocortex is far more complicated, it may roughly be viewed as two prepyriform cortices stacked on top of each other. We expect that analysis of this system will lend insight into mechanisms of associative memory there as well. In refs. [9, 8] we show that this model is capable of storing complicated multifrequency spatio-temporal sequences, and argue that it may serve as a model of memory for sequences of actions in motor cortex.

For an N-dimensional system, this primary coupling structure is described mathematically by the matrix

$$T = \begin{pmatrix} W & -hI \\ gI & 0 \end{pmatrix},$$

where W is the $N/2 \times N/2$ matrix of excitatory interconnections, and gI and hI are $N/2 \times N/2$ identity matrices multiplied by the positive scalars g and h. These give the magnitudes of coupling around the local inhibitory feedback loops. A state vector is composed of local average cell voltages for $N/2$ excitatory neuron populations x and $N/2$ inhibitory neuron populations y (hereafter notated simply as $x, y \in \mathbb{R}^{N/2}$). Standard network equations with this coupling might be, in component form,

$$\dot{x}_i = -\tau x_i - h\sigma(y_i) + \sum_{j=1}^{N/2} W_{ij}\sigma(x_j) + b_i, \tag{1}$$

$$\dot{y}_i = -\tau y_i + g\sigma(x_i), \tag{2}$$

where $\sigma(x) = \tanh(x)$ or some other sigmoidal function symmetric about zero. Intuitively, since the inhibitory units receive no direct input and give no direct output, they act as hidden units that create oscillation for the amplitude patterns stored in the excitatory cross-connections W. This may be viewed as a simple generalization of the analog "Hopfield" network architecture [35] designed to store periodic instead of static attractors. Rewriting the equations above more generally, in terms of T, and suppressing the distinction between excitatory and inhibitory neurons, x, y, we recover the equations of this well known network:

$$\dot{x}_i = -\tau x_i + \sum_{j=1}^{N} T_{ij}\sigma(x_j) + b_i. \tag{3}$$

The inhibitory components y are now implicitly defined as the components of the state vector x from $N/2 + 1$ to N.

We will show later that learning by an outer product (Hebb) rule for excitatory connections W alone is sufficient to establish desired orthogonal patterns as modes of the linearization of this network. The Jacobian matrix of the above equations is just $T - \tau I$, since $\sigma'(0) = 1$. We show that in this architecture, the real eigenvectors of the matrix W of excitatory cross coupling produced by a Hebb rule become "complexified" into complex eigenvectors and eigenvalues for the coupling matrix T of the full network that includes inhibitory feedback. This formalizes the intuition stated above that the negative feedback units produce oscillation for the static patterns stored in the W matrix.

For networks that store static patterns, the attractors of the nonlinear system can be intimately related to the eigenvectors of the linearization [48]. For oscillatory networks with full cross-coupling, however, the relation is more tenuous and multiple attractors are hard to get. This is when the vector field is probed with a grid of initial conditions $b_i\delta(t)$ that are the usual form of input for a content addressable memory of this type. When different *clamped* inputs b_i are used, a new vector field is produced by each input, and a single attractor will be shifted around. A large enough fixed input will produce an amplitude pattern that looks much like the input. There is no *exact* reconstruction of a stored prototype – every different ("corrupted") input gives a different (possibly less "corrupted") output. This appears to be the mode of operation of the olfactory network models investigated by most physiologists [19, 53, 26, 41]. These systems learn continuous maps from inputs to outputs – not categories as defined here. A "category" here is a map from a continuous range of inputs (the "instances") to a set of discrete, unvarying outputs – the "prototypes" or "category labels". Such maps are implemented naturally by networks with multiple attractors. The basins of attraction define the categories, since all initial condition inputs within a basin go to the same unvarying attractor (category label) as an output. We are interested here in learning rules that accomplish the storage of many periodic attractors in the same vector field.

2.1. Higher-order model

If additional higher-order synaptic weights for competitive (negative) cubic terms given by the projection algorithm described in refs. [5, 7, 8] are used, the oscillation amplitude patterns of the attractors of the full nonlinear system are analytically guaranteed by the projection theorem to be the absolute values of the eigenvectors of the matrix W of excitatory connections. General network equations with these terms added are

$$\dot{x}_i = -\tau x_i + \sum_{j=1}^{N} T_{ij}\sigma(x_j) - \sum_{jkl=1}^{N} T_{ijkl}\sigma(x_j)\sigma(x_k)\sigma(x_l) + b_i\delta(t). \tag{4}$$

Adding these higher-order weights corresponds in network terms to increasing the complexity of our

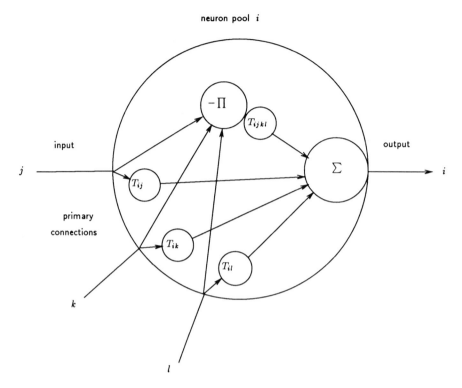

Fig. 2. Neural population subnetwork acting as a sigma-pi unit. It uses secondary higher-order synaptic weights T_{ijkl} on products of the activities of incoming primary connections T_{ij}.

neural population nodes to become "sigma-pi" units [18]. Fig. 2 shows schematically how subnetworks within a population unit must compute products of the activities on incoming primary connections T_{ij}, during higher-order Hebbian learning, to establish a weight T_{ijkl}. These secondary higher-order synapses are then used in addition to the synapses T_{ij}, during operation of the overall network, to weight the effect of triple products of inputs in the output summation of the subnet.

In the real olfactory system, as the physiological models emphasize, and as we discuss in refs. [4, 10], inputs are not impulses $b_i \delta(t)$ which establish initial conditions. In the olfactory bulb they are closer to slow sinusoids (3–8 Hz) given by respiratory inspiration. This drives the bulb into oscillation through a Hopf bifurcation to convert the slowly varying input pattern $b_i(t)$ into a 40 Hz oscillatory output pattern $d_i(t)\cos \omega t$ as input to the pyriform cortex. This input establishes both transient initial conditions and slowly varying sinusoidal steady state conditions. Analysis and simulation of the multiple attractor networks studied here, using more realistic oscillatory forcing inputs instead of initial conditions [8], show robust recall behavior in any case. The change in attractor locations produced by an additive input term $b_i(t)$ can be made arbitrarily small when the net is operated near its critical point, where small inputs may be used, and by proper use of the capabilities of the projection algorithm, which determines the higher order weights. The balance between the relative impact on final output of external input or internal dynamics can be adjusted at will. During learning, the network may be biased near critical, so that it is forced into a state largely determined by the input, and synapses may self-organize by a Hebb rule according to the activations they experience. The input may thus be encoded in memory without interference from the tendency of the system to converge to a previously learned state. Then, during recognition, the network may be biased so that it is well beyond critical when the input ramps up to its maximum, and the strong internal dynamics now "reconstruct" the proper category output for even heavily deformed inputs. It is this kind of decomposition of the mechanisms at work in complicated simulations that we seek in this approach.

In the general network (4) above, however, the higher-order terms require far from "minimal" primary coupling, since they require the physiologically unlikely long-range inhibitory connections. Fortunately, as we discuss later, analysis and simulation show that the system will perform as specified by the projection theorem and only those higher-order synapses W_{ijkl} at the ends of the known long-range excitatory–excitatory connections W_{ij}. In addition, most recent analysis shows that in principle only $(N)^2$ of *these* $(N/2)^4$ higher-order weights are required to approximate the theoretical performance. Returning to explicit x and y neural populations, the equations for this network with only biologically feasible primary connections are

$$\dot{x}_i = -\tau x_i - h\sigma(y_i) + \sum_{j=1}^{N/2} W_{ij}\sigma(x_j) - \sum_{jkl=1}^{N/2} W_{ijkl}\sigma(x_j)\sigma(x_k)\sigma(x_l) + b_i\delta(t), \tag{5}$$

$$\dot{y}_i = -\tau x_i + g\sigma(x_i). \tag{6}$$

This net has a summation of higher-order couplings only over the $N/2$ excitatory components.

If we assume that the local interconnections between the inhibitory and excitatory cell populations are operating in a linear regime, and expand this network to third order in a Taylor series about the origin, we get

$$\dot{x}_i = -\tau x_i - hy_i + \sum_{j=1}^{N/2} W_{ij}x_j - \sum_{jkl=1}^{N/2} W_{ijkl}x_jx_kx_l + b_i\delta(t), \tag{7}$$

$$\dot{y}_i = -\tau y_i + gx_i, \tag{8}$$

where $\sigma'(0) = 1$, and $\sigma'''(0)/3!$ (<0) is absorbed into W_{ijkl}. A sigmoid symmetric about zero has odd symmetry, and the even order terms of the expansion vanish, leaving the cubic terms as the only nonlinearity. The competitive (negative) cubic terms of (7) therefore constitute a directly programmable nonlinearity that is independent of the linear terms. When they are given by the projection rule to be discussed later, they serve to create multiple periodic attractors by causing the oscillatory modes of the linear term to compete, much as the sigmoidal nonlinearity does for static modes in a Hopfield network. Intuitively, these terms may be thought of as sculpting the maxima of a "saturation" landscape into which the stored linear modes whose eigenvalues have positive real parts expand. The maxima are positioned to lie in the directions specified by the eigenvectors of these modes to make them stable. In refs. [7, 8], we construct a strict Lyapunov function in a special coordinate system that precisely defines this saturation landscape.

From a physiological point of view, (7) may be considered a model of a biological network which is operating in the linear region of the known *axonal* sigmoid nonlinearities [21], and contains instead sigma-pi units or higher-order *synaptic* nonlinearities. The theoretical performance guaranteed by the projection theorem [5] is specified for a purely cubic network. The higher-order network above with sigmoids (eq. (4)) is guaranteed to give the same behavior if it has a Taylor expansion that is equal to the cubic net up to third order. This is easy to arrange, as is shown in ref. [7]. We will show later that for orthonormal static patterns x^s, the projection operation for the W matrix reduces to an outer product, or

"Hebb" rule, and the projection for the higher-order weights becomes a *multiple* outer product rule:

$$W_{ij} = \sum_{s}^{N/2} \alpha^s x_i^s x_j^s, \quad W_{ijkl} = c\delta_{ij}\delta_{kl} - d \sum_{s=1}^{N/2} x_i^s x_j^s x_k^s x_l^s. \tag{9}$$

The first rule is guaranteed to establish desired patterns x^s as eigenvectors of the W matrix with corresponding eigenvalues α^s. The second rule, with $c > d$, gives higher-order weights W_{ijkl} for the cubic terms in (7) which ensure that the oscillatory amplitude patterns defined by magnitudes of these eigenvectors appear as attractors in the network vector field.

The outer product is a *local* synapse rule for synapse ij that allows additive and incremental learning. The system can be truly self-organizing because a synapse of the net can modify itself based on its own activity, as envisioned above. The rank of the coupling matrix W and T grows as more memories are learned by the Hebb rule, and the unused capacity appears as a degenerate subspace with all zero eigenvalues. The flow is thus directed toward regions of the state space where patterns are stored. In simulations, we never found evidence that the usual Hebb rule for W in the first-order sigmoid network (1) will store *desired* periodic attractors. We are thus driven to consider the very effective higher-order network (7, 8) as a biological model.

2.2. Biological justification of higher-order synapses

Using the long-range excitatory connections available, some number of the higher-order synaptic weights W_{ijkl} can conceivably be realized locally in the axo-dendritic interconnection plexus known as "neuropil". This is a feltwork of tiny fibers so dense that its exact circuitry is impossible to investigate with present experimental techniques. Single axons are known to bifurcate into multiple branches that contribute separate synapses to the dendrites of target cells [37]. It is also well known that neighboring synapses on a dendrite can interact in a nonlinear fashion that has been modeled as higher-order synaptic terms by some researchers [14]. It has been suggested that the neuropil may be dense enough to allow the crossing of every possible combination of jkl axons in the vicinity of some dendritic branch of at least one neuron in neuron pool i [44]. Trophic factors stimulated by the coactivation of the axons and the dendrite could cause these axons to form a "cluster" of nearby synapses on the dendrite to realize a jkl product synapse. The required higher-order terms could thus be created by a Hebb-like process.

The resulting sigma-pi unit is a feedforward model of dendritic processing, but the activity within a real dendritic tree is itself so complex that it should be modeled as a recurrent subnetwork that substitutes for the simple nodes of a conventional neural net. The three-dimensional geometry of the dendritic arborization, as well as the pattern of synaptic clustering of incoming axons on the branches, must be considered as contributing to the weights of this subnetwork. The dendritic growth process is also subject to a trophic guidance by neural activity that can be considered learning [40]. The evolution of the dendritic tree is being investigated as a process of self-organization wherein the neuron seeks "features" in the input patterns that give it an optimal firing rate [12]. In the model presented here, we need the effect of the triple products of excitatory inputs to be subtracted from the effect of the primary excitatory inputs in the final cell voltage summation that determines the output firing frequency of the neurons. It is known that the location of synaptic input on the dendritic tree strongly affects its action, and excitatory synapses can have shunting inhibitory effects on other inputs. We must assume that the constants c and d in the learning rule of eq. (9) are implicit in the structure of the dendritic tree.

The dendritic processing story is yet to be fully unfolded, and may in the end support the subfunctions demanded by this theoretical system for the realization of oscillatory associative memory. The use of the competitive cubic cross terms as the system nonlinearity may thus be viewed physiologically as the use of this complicated *synaptic/dendritic* nonlinearity, as opposed to the simple sigmoidal *axonal* nonlinearity. There are more weights in the cubic synaptic terms, and there is more power to program the resulting nonlinearity.

3. Analysis

In conjunction with the values g and h of the fixed local coupling, the real eigenvalues of W determine the complex eigenvalues of T, and the real eigenvectors of W determine the magnitudes of the complex eigenvectors of T.

Theorem 3.1. If α is a real eigenvalue of the $N/2 \times N/2$ matrix W, with corresponding eigenvector x, then the $N \times N$ matrix

$$T = \begin{pmatrix} W & -hI \\ gI & 0 \end{pmatrix}$$

has a pair of complex conjugate eigenvalues

$$\lambda_{1,2} = \tfrac{1}{2}\left(\alpha \pm \sqrt{\alpha^2 - 4hg}\right),$$

for $\alpha^2 < 4hg$. The frequency ω of this complex mode is $\sqrt{4hg - \alpha^2}$, and the corresponding complex conjugate pair of eigenvectors are

$$\begin{pmatrix} x \\ (\alpha+\omega)x/2h \end{pmatrix} \pm i \begin{pmatrix} x \\ (\alpha-\omega)x/2h \end{pmatrix}.$$

The proof of this theorem is given in the appendix. To more clearly see the amplitude and phase patterns, we can convert to a magnitude and phase representation, $z = |z|\,e^{i\theta}$, where

$$|z_i| = \sqrt{\operatorname{Re}^2(z_i) + \operatorname{Im}^2(z_i)}$$

and

$$\theta_i = \arctan[\operatorname{Im}(z_i)/\operatorname{Re}(z_i)],$$

we get,

$$|z_{x_i}| = \sqrt{x_i^2 + x_i^2} = \sqrt{2}\,|x_i|,$$

$$|z_{y_i}| = \sqrt{\frac{2(\alpha^2 + \omega^2)}{4h^2}x_i^2} = \sqrt{\frac{4hg}{2h^2}}\,|x_i| = \sqrt{\frac{2g}{h}}\,|x_i|,$$

$$\theta_x = \arctan 1 = \pi/4, \qquad \theta_y = \arctan\left(\frac{\alpha - \omega}{\alpha + \omega}\right).$$

Dividing out the common $\sqrt{2}$ factor in the magnitudes, we get eigenvectors that clearly display the amplitude patterns of interest

$$\begin{pmatrix} |x|\,e^{i\theta_x} \\ \sqrt{g/h}\,|x|\,e^{i\theta_y} \end{pmatrix}, \quad \text{or} \quad \begin{pmatrix} |x|\cos\theta_x \\ \sqrt{g/h}\,|x|\cos\theta_y \end{pmatrix} \pm i \begin{pmatrix} |x|\sin\theta_x \\ \sqrt{g/h}\,|x|\sin\theta_y \end{pmatrix}.$$

An expression for the periodic attractor $X^s(t)$ established for patterns when the higher-order synapses are given by the projection algorithm is

$$X^s(t) = \begin{pmatrix} |x^s|\,e^{i\theta_x^s + i\omega^s t} \\ \sqrt{g/h}\,|x^s|\,e^{i\theta_y^s + i\omega^s t} \end{pmatrix}.$$

Because of the restricted coupling, the oscillations possible in this network are standing waves, since the phase θ_x, θ_y here is constant for each kind of neuron x and y and differs only between them. This is basically what is observed in the olfactory bulb and prepyriform cortex. There are actually small phase gradients in x and y which imply traveling waves that are so fast relative to the period of the global oscillation that they are neglected here [31, 54, 5]. The phase of inhibitory components θ_y in the bulb generally lags the phase of the excitatory components θ_x by approximately 90°. It is easy to choose α and ω in this model to get phase lags of nearly 90°. But for *exactly* this value, we need

$$\theta_x - \theta_y = \pi/4 - \theta_y = \pi/2,$$

which requires

$$\theta_y = \arctan\left(\frac{\alpha - \omega}{\alpha + \omega}\right) = -\pi/4,$$

or

$$\frac{\alpha - \omega}{\alpha + \omega} = \tan(-\pi/4) = -1,$$

which can occur here only for $\alpha = 0$. This is a feature of the simple model chosen here. The restricted coupling implies these restrictions on the type of patterns which can be stored. In a more detailed model where some range of lateral inhibitory feedback is included, for example, we expect this singularity to disappear.

3.1. Learning by the projection algorithm

For each real eigenvalue α_s and eigenvector x_s of W, a complex pair of eigenvalues and eigenvectors is generated for T. From the theory detailed in refs. [5, 7], we can program any linearly independent set of eigenvalues and eigenvectors into W by the "projection" operation $W = BDB^{-1}$, where B has the desired eigenvectors as columns, and D is a diagonal matrix of the desired eigenvalues. Because the complex eigenvectors of T follow from these learned for W, we can form a projection matrix P with those eigenvectors of T as columns, and a matrix J of the complex eigenvalues of T in blocks along the diagonal, and project directly to get $T = PJP^{-1}$. The conjugate eigenvectors shown above have the same magnitude or amplitude pattern, but are globally phase shifted from one another $\bar{\theta} = -\theta$ in the complex plane. Since there is no absolute global phase reference in this system, these shifts are irrelevant, and either member of the pair will suffice as the eigenvector that is entered in the P matrix for the projection operation. We get real and imaginary component columns for P:

$$P = \begin{pmatrix} |x^s| \sin \theta_x^s & |x^s| \cos \theta_x^s & \cdots \\ \sqrt{g/h}\, |x^s| \sin \theta_y^s & \sqrt{g/h}\, |x^s| \cos \theta_y^s & \cdots \end{pmatrix}.$$

If general cubic weights T_{ijkl}, also given by a projection operation, are added to network equations with these weights T_{ij}, as we did earlier for (3) to get (4), the complex modes (eigenvectors) of the linearization are analytically guaranteed by the projection theorem to characterize the periodic attractors of the network vector field [7]. Chosen "normal form" coefficients A_{mn} [5, 7] are projected to get the higher-order synaptic weights T_{ijkl} for these general cubic terms. Together, these operations constitute the "normal form projection algorithm":

$$T = PJP^{-1}, \quad T_{ijkl} = \sum_{m,n=1}^{n} P_{im} A_{mn} P_{mj}^{-1} P_{nk}^{-1} P_{nl}^{-1}. \tag{10}$$

The proof of the "projection theorem", which states the capabilities of this learning rule, is given in refs. [5, 7, 8]. It says that the learning rule is essentially a linear transformation which projects an ideal network with a chosen vector field in a special coordinate system (the "normal form") into network coordinates in a way that makes the simple attractors of the normal form into the attractors desired for the network. The construction begins with the specification of oscillations by coupled ordinary differential equations in this special polar coordinate system, where the equations for the amplitudes of the oscillations are independent of those for the phase rotations. These equations are called the normal form for the "Hopf bifurcation".

Amplitude coupling coefficients A_{ij} are chosen to give stable fixed points on the coordinate axes of the vector field of amplitudes, frequencies are chosen for the phase equations, and the polar system is then transformed to Cartesian complex conjugate coordinates. The axes of this system of nonlinear ordinary differential equations are then linearly transformed into desired spatial or spatio-temporal patterns by projecting the system into network coordinates (the standard basis), using the desired vectors as columns of the transformation matrix P. Any linearly independent (nonidentical) set of patterns may be projected. This method of network synthesis is roughly the inverse of the usual procedure in bifurcation theory for analysis of a given physical system. These operations are all compactly expressed in network coordinates as the formula or "learning rule" shown above for the coupling weights, given the desired patterns and frequencies.

All attractors stored in these models will arise from the origin in a multiple Hopf bifurcation as the real parts of the eigenvalues (if all are chosen to be the same) go positive under the influence of a chosen bifurcation parameter. Such a parameter might be a scaling of the linear coupling matrix T (equivalent to a sigmoid slope change in the net (4) with sigmoids) or the decay term τ, either of which directly affects the eigenvalues of the Jacobian $T - \tau I$. A spatially uniform input bias term b_i will shift the equilibrium at the origin to some nonzero value, and thus affect the eigenvalues indirectly through the new Jacobian of the Taylor expansion about this point, as discussed in ref. [6]. Such a bias is supplied in olfaction by the average input level that rises and falls with respiration.

3.2. Learning with only excitatory connections

The general terms T_{ijkl}, as we have said, require use of physiologically unlikely long-range primary inhibitory connections. Simulations of two and four oscillator networks thus far ($N = 4$ and $N = 8$), reveal that use of the higher-order terms for only the anatomically justified long-range excitatory connections W_{ijkl}, as in the cubic net (7), is effective in storing randomly chosen sets of desired patterns. The behavior of this network is close to the theoretical ideal for the network of eq. (4). There is no alteration of the stored oscillatory patterns when this reduced coupling is used. We prove in ref. [8] that this is to be expected when the higher order Hebb rule (eq. (9)) is used for orthogonal patterns. The desired normal form coefficients for the oscillation amplitudes are undisturbed, but extra coefficients are introduced which produce a dispersion of the frequencies of stored oscillations like that observed experimentally [25].

This same behavior is seen in simulations when the projection rule (eq. (10)) is used for nonorthogonal patterns. At the moment, we have at least general analytic justification for this. "Normal form" theory [29, 7, 8] guarantees that many other choices of weights will give a vector field topologically equivalent to that given by the projection operation, but does not in general say how to find them. Latest work shows that a perturbation theory calculation of the normal form coefficients for general high-dimensional cubic nets is tractable and in principle permits the removal of all but N^2 of the N^4 higher-order weights normally produced by the projection algorithm. We have already incorporated this in an improved learning rule (non-Hebbian thus far) which requires even fewer of the excitatory higher-order weights ($(N)^2$ versus $(N/2)^4$), and are exploring the size of the neighborhood of state space about the origin in which the rule is effective. This should lead as well to a rigorous proof of the capacity of the "reduced" networks, with only excitatory connections, for the storage of arbitrary linearly independent patterns.

3.3. Learning by local Hebb rules

For biological modeling, where the patterns to be stored might be assumed to be orthogonal, and for the storage of static patterns, the projection operation described above reduces to the usual Hebb-like outer product rule. To see how an outer product rule arises, we fill the columns of the W matrix with the real orthogonal vectors x^s to be learned, and for each one choose real eigenvalues u^s to be placed in the corresponding column on the diagonal of a matrix D. When the columns of B are orthogonal, then they are also orthonormal, since we are assuming that the chosen relative amplitude patterns x^s are normalized to begin with. It follows then that $B^{-1} = B^{T}$, and the formula above for the linear network coupling becomes $B = BDB^{T}$.

First performing the matrix multiplication BD gives us a matrix $\tilde{B}$ where each column (eigenvector) is now multiplied by its eigenvalue from D. If we write the remaining matrix multiplication $W = \tilde{B}B^{T}$ in component form, we get $W_{ij} = \sum_s u^s x_i^s x_j^s$ (W is the identity of course if the eigenvalues u^s are identical, but full cross-couplings arise otherwise). This is a summation over the columns of B and the rows of B^{T}, which we know by our construction to be the patterns we want to learn. Thus the W matrix can be written in the familiar form of the outer product rule

$$W_{ij} = \sum_{s=1}^{N/2} W_{ij}^s = \sum_{s=1}^{N/2} u^s x_i^s x_j^s,$$

where the eigenvalue u^s is revealed to be the "learning rate" or relative frequency of pattern s in the summation of patterns. Similarly, in refs. [7, 8] we show that the higher-order weights for storage of static patterns are given by the multiple outer product rule,

$$W_{ijkl} = c\delta_{ij}\delta_{kl} - d\sum_{s=1}^{N/2} x_i^s x_j^s x_k^s x_l^s, \quad d < c,$$

where the condition $d < c$ ensures stability of desired memories.

Thus, in the minimal net, the real eigenvectors learned for W are converted by the network structure into standing wave oscillations (constant phase) with the absolute value of those eigenvectors as amplitudes. From the mathematical perspective, there are $N!$ eigenvectors, with different permutations of the signs of the same components, which lead to the same positive amplitude vector. This means that nonorthogonal amplitude patterns may be stored by the Hebb rule on the excitatory connections, since there may be many ways to find a perfectly orthonormal set of eigenvectors for W that stores a given set of nonorthogonal amplitude vectors. Given the complexity of dendritic processing discussed previously, it is not impossible that there is some distribution of the signs of the final effect of synapses from excitatory neurons that would allow a biological system to make use of this mathematical degree of freedom.

If we restrict ourselves to common physiological dogma, however, we must follow Dale's law and require that the synapses of the excitatory neurons have only positive weights. Then only positive patterns may be used in the outer product (Hebb) rule to give positive weights in W and W_{ijkl}. The only full capacity set of orthonormal memories that allow the condition $D^{-1} = D^T$, for which the Hebb rule is theoretically valid, are the uninteresting set of vectors with only one different positive component equal to one – the standard basis. This Hebb rule only makes biological sense when less than the full capacity of amplitude patterns is stored. Individual patterns must be sparse, but may then have many nonzero values and still be orthogonal.

The assumption of orthogonality of memory patterns is often justified for biological systems by the high dimensionality of the cortical state space and the sparse nature of the activity patterns observed. For different input objects, feature preprocessing in primary and secondary sensory cortex may also be expected to orthogonalize outputs to the object recognition systems modeled here. When the rules above are used for nonorthogonal patterns, the eigenvectors of W and T are no longer given directly by the Hebb rule, and we expect that the kind of performance found in Anderson's [2] and Hopfield's [35] networks for nonorthogonal memories will obtain, with reduced capacity and automatic clustering of similar exemplars [2]. Investigation of this induction of categories from training examples will be the subject of future work [8].

3.4. Architectural variations – olfactory bulb model

Another biologically interesting architecture which can store these kinds of patterns is one with associational excitatory to inhibitory cross-coupling. This may be a more plausible model of the olfactory bulb than the one above. The olfactory bulb is the first sensory cortex in the olfactory system. It receives input directly from receptors in the nose, and gives output to the prepyriform cortex. Experimental work of Freeman suggests an associative memory function for this cortex as well [21, 22, 25, 4]. For analysis of an alternative hypothesis that the bulb is a preprocessor with only fixed short-range inhibitory interconnections, see ref. [41]. The evidence for long-range excitatory coupling in the olfactory bulb is much weaker than that for the prepyriform cortex. Long-range excitatory tracts connecting even the two halves of the bulb are known [49, 50], but anatomical data thus far show these axons entering only the inhibitory granuel cell layers [47]. Electron microscope pictures show synapses on excitatory cells of a morphological type that are argued to be inhibitory [52].

A simple network model with excitatory to inhibitory associational coupling has the previous matrix T, with W swapping places with gI,

$$T' = \begin{pmatrix} gI & -hI \\ W & 0 \end{pmatrix}.$$

Now,

$$\lambda_{1,2} = \tfrac{1}{2}\left(g \pm \sqrt{g^2 - 4\alpha g}\right),$$

and, for $g^2 < 4hg$, the frequency ω of this complex mode is $\omega = \sqrt{4\alpha g - g^2}$. Thus

$$\lambda_{1,2} = \tfrac{1}{2}(g \pm i\omega).$$

The eigenvectors are, as before,

$$\begin{pmatrix} x \\ (g+\omega)x/2h \end{pmatrix} \pm i \begin{pmatrix} x \\ (g-\omega)x/2h \end{pmatrix}.$$

In polar form,

$$P = \begin{pmatrix} |x^s|\sin\theta_x^s & |x^s|\cos\theta_x^s & \dots \\ \sqrt{\alpha/h}\,|x^s|\sin\theta_y^s & \sqrt{\alpha/h}\,|x^s|\cos\theta_y^s & \dots \end{pmatrix},$$

where $\theta_x^s = \pi/4$, and $\theta_y^s = \arctan[(g-\omega)/(g+\omega)]$.

Here the eigenvalue α of W appears only in the $4\alpha g$ term of ω, and cannot affect the real parts of the eigenvalues of T. The local excitatory self-coupling g now plays the role of the real part that α played in the previous system. Instead, α now scales the inhibitory eigenvector magnitudes by $\sqrt{\alpha/h}$, which g did before. The relative amplitude patterns of the complex eigenvectors, however, are again just the magnitudes of the eigenvectors learned for the coupling matrix W. A distinction of the present system is that the eigenvalues of W given by the learning can only affect the frequency of patterns stored in T, and not real parts of the eigenvalues of these modes. Positive real parts, here given by g, are essential for the Hopf bifurcation instability that gives rise to the limit cycles. Alternatively, positive real parts may obtain without excitatory self coupling if W has complex eigenvalues [41].

Other simple biologically plausible variations of these architectures are easy to analyze in this fashion. If we add inhibitory population self-feedback $-f$ to the first model, this additional term is added to the eigenvalue α in all expressions for the full coupling matrix. The coupling matrix is now

$$T = \begin{pmatrix} W & -hI \\ gI & -fI \end{pmatrix}.$$

Then

$$\lambda_{1,2} = \tfrac{1}{2}\left(\alpha - f \pm \sqrt{\alpha^2 + f^2 + 2\alpha f - 4hg}\right),$$

and, for $\alpha^2 + f^2 + 2\alpha f < 4hg$, the frequency ω of this complex mode is $\sqrt{4hg - \alpha^2 + f^2 + 2\alpha f}$. Thus

$$\lambda_{1,2} = \tfrac{1}{2}(\alpha - f \pm i\omega).$$

If we let $\alpha + f = \gamma$, we get the same eigenvectors as above, with γ replacing α. Now $\omega = \sqrt{4hg - \gamma^2}$, and the eigenvectors are as before,

$$\begin{pmatrix} x \\ (\gamma + \omega)x/2h \end{pmatrix} \pm i \begin{pmatrix} x \\ (\gamma - \omega)x/2h \end{pmatrix}.$$

In polar form, in columns of the P matrix,

$$P = \begin{pmatrix} |x^s| \sin \theta_x^s & |x^s| \cos \theta_x^s & \cdots \\ \sqrt{g/h}\, |x^s| \sin \theta_y^s & \sqrt{g/h}\, |x^s| \cos \theta_y^s & \cdots \end{pmatrix},$$

where $\theta_x^s = \pi/4$, and $\theta_y^s = \arctan[(\gamma - \omega)/(\gamma + \omega)]$.

Similarly, when we add these coupling terms to the olfactory bulb model,

$$T = \begin{pmatrix} gI & -hI \\ W & -fI \end{pmatrix}.$$

Now,

$$\lambda_{1,2} = \tfrac{1}{2}\left(g - f \pm \sqrt{g^2 + f^2 + 2gf - 4\alpha g}\right),$$

and, for $g^2 + f^2 + 2gf < 4hg$, the frequency ω of this complex mode is $\sqrt{4\alpha g - (g^2 + f^2 + 2gf)}$. Thus

$$\lambda_{1,2} = \tfrac{1}{2}(g - f \pm i\omega).$$

Again, if we let $g + f = \gamma$, we get the same result as above, with γ replacing α. Thus $\omega = \sqrt{4\alpha g - \gamma^2}$, and the eigenvectors are exactly as before.

Further extensions of this line of analysis will consider lateral inhibitory fan out of the inhibitory/excitatory feedback connections. The $-hI$ block of the coupling matrix T becomes a banded matrix. Similarly, the gI and $-fI$ may be banded, or both full excitatory to excitatory W and full excitatory to inhibitory V coupling blocks may be considered. We conjecture that the phase restrictions of the minimal model will be relaxed with these further degrees of freedom available, so that traveling waves may exist, and the phase leg singularity at 90° will vanish.

4. Conclusions

We have extended previous theory to analyze a simple model of cortical neural circuitry that uses only anatomically justified long-range excitatory connections between nonlinear oscillators. These are formed of local feedback loops between explicit excitatory and inhibitory neural populations that are characterized by analog activation densities. We show that a Hebbian learning rule for primary W_{ij} and higher-order

W_{ijkl} synapses at the ends of the long-range connections is sufficient for the system to function as a content addressable memory. It can store the kind of amplitude patterns of oscillation thought to mediate sensory pattern recognition in the olfactory system.

Acknowledgements

Supported by AFOSR-87-0317. It is a pleasure to acknowledge the support of Walter Freeman and invaluable assistance of Morris Hirsch.

Appendix A. Proof of theorem 3.1

The characteristic equation for this matrix is,

$$\begin{pmatrix} W & -hI \\ gI & 0 \end{pmatrix} \begin{pmatrix} x \\ y \end{pmatrix} = \begin{pmatrix} \lambda x \\ \lambda y \end{pmatrix} \quad \text{or} \quad \begin{aligned} Wx - hy &= \lambda x \\ gx &= \lambda y \end{aligned}.$$

Now $y = gx/\lambda$, and substituting this into the top equation gives $Wx - hgx/\lambda = \lambda x$ or $Wx = (\lambda + hg/\lambda)x$. This is now the characteristic equation for the matrix W – showing $(\lambda + hg/\lambda)$ to be an eigenvalue of W, and x to be the corresponding eigenvector. We can write $(\lambda + hg/\lambda) = \alpha$, where α is an eigenvalue of W, and solve for the eigenvalues λ_i of T in terms of the eigenvalues α_i of W. Thus, $\lambda^2 - \alpha\lambda + hg = 0$, which gives a conjugate pair of complex eigenvalues for T,

$$\lambda_{1,2} = \tfrac{1}{2}\left(\alpha \pm \sqrt{\alpha^2 - 4hg}\right),$$

for $\alpha^2 < 4hg$. The frequency ω of this complex mode is $\sqrt{4hg - \alpha^2}$. Thus

$$\lambda_{1,2} = \tfrac{1}{2}(\alpha \pm i\omega).$$

Now T will have a conjugate pair of complex eigenvectors $z, \bar{z}$ corresponding to these eigenvalues. The excitatory component z_x of the eigenvector z is composed of x, assumed at the outset to be the excitatory component of an eigenvector of T (it is also a real eigenvector of W), which can be multiplied by any complex scalar $a + ib$ in the complex space of the T matrix eigenvectors, and so becomes the complex eigenvector $z_x = ax + ibx$ for T. Here we will assume $a = b = 1$ for simplicity.

Now, from above, the complex eigenvector z_y of T for the inhibitory states is

$$z_y = \frac{g}{\lambda} z_x = \frac{2g}{\alpha \pm i\omega}(1 + i)x.$$

Thus the eigenvectors corresponding to $\lambda_{1,2}$ in T are composed of the eigenvector $x + ix$ of W for both excitatory states, and the two different multiples of this, $[2g/(\alpha + i\omega)](1 + i)x$, for the inhibitory states. Choosing $\lambda_1 = \tfrac{1}{2}(\alpha + i\omega)$, and multiplying the expression for y above by $\bar{\lambda}/\bar{\lambda} = (\alpha - i\omega)/(\alpha - i\omega)$ gives the eigenvector in its real and imaginary parts,

$$z_y = \frac{2g}{\alpha^2 + \omega^2}[(\alpha + \omega)x + i(\alpha - \omega)x],$$

and since $\alpha^2 + \omega^2 = \alpha^2 + 4gh - \alpha^2 = 4gh$,

$$z_y = \frac{1}{2h}(\alpha + \omega)x + \frac{i}{2h}(\alpha - \omega)x.$$

For $\lambda_2 = \bar{\lambda} = \frac{1}{2}(\alpha - i\omega)$, we get

$$\bar{z}_y = \frac{g}{\lambda}\bar{z}_x = \frac{1}{2h}(\alpha + \omega)x - \frac{i}{2h}(\alpha - \omega)x.$$

Thus our complex conjugate eigenvectors are

$$\begin{pmatrix} x \\ (\alpha + \omega)x/2h \end{pmatrix} \pm i \begin{pmatrix} x \\ (\alpha - \omega)x/2h \end{pmatrix}. \qquad \square$$

References

[1] D.Z. Anderson, Coherant optical eigenstate memory, Opt. Lett. (1986) 56.
[2] J. Anderson and G. Hinton, eds., Parallel Models of Associative Memory (LEA, NJ, 1981).
[3] J.A. Anderson and E. Rosenfeld, eds., Neurocomputing (MIT Press, Cambridge, MA, 1988).
[4] B. Baird, Nonlinear dynamics of pattern formation and pattern recognition in the rabbit olfactory bulb, Physica D 22 (1986) 150–176.
[5] B. Baird, A bifurcation theory approach to the programming of periodic attractors in network models of olfactory system, in: Advances in Neural Information Processing Systems I, ed. David S. Touretsky (Kaufman, Los Altos, 1988) p. 459.
[6] B. Baird, Bifurcation theory methods for programming static or periodic attractors and their bifurcations in dynamic neural networks, in: Proc. IEEE Int. Conf. on Neural Networks, San Diego, CA (July 1988) p. I9.
[7] B. Baird, A bifurcation theory approach to vector field programming for periodic attractors, in: Proc. Int. Joint Conf. on Neural Networks, Washington, DC (June 18, 1989) p. I381.
[8] B. Baird, Bifurcation Theory Approach to the Analysis and Synthesis of Neural Networks for Engineering and Biological Modelling, Research Notes in Neural Computing (Springer, Berlin, 1990).
[9] B. Baird, A learning rule for cam storage of continuous periodic sequences, in: Proc. Int. Joint Conf. on Neural Networks, San Diego (June 1990), to appear.
[10] B. Baird, Bifurcation analysis of oscillating neural network model of pattern recognition in the rabbit olfactory bulb, in: Neural Networks for Computing, ed. J. Denker, AIP Conf. Proc. 151 (AIP, New York, 1986) pp. 29–34.
[11] P. Baldi and A. Atiya, Oscillations and synchronizations in neural networks, an exploration of the labelling hypothesis, Technical Report JPL198-330, Jet Propulsion Laboratory, CA (1989).
[12] T. Brown, personal communication.
[13] H.H. Chen, Y.C. Lee, G.Z. Sun, H.Y. Lee, T. Maxwell and C.L. Giles, High order correlation model for associative memory, in: Neural Networks for Computing, ed. J. Denker, AIP Conf. Proc. 151 (AIP, New York, 1986) p. 86.
[14] R. Durbin and D.E. Rumelhart, Product units: A computationally powerful and biologically feasible extension to backpropogation networks, Neural Computation, 1 (1989) 133.
[15] R. Eckhorn, R. Bauer, W. Jordan, M. Brosch, W. Kruse, M. Munk and H.J. Reitboeck, Coherent oscillations: A mechanism of feature linking in the visual cortex?, Biol. Cybern. 60 (1988) 121.
[16] F.H. Eeckman, The sigmoid nonlinearity in prepyriform cortex, in: Neural Information Processing Systems, ed. D.Z. Anderson (AIP, New York, 1988) p. 135.
[17] G.B. Ermentrout and J.D. Cowan, Temporal oscillations in neuronal nets, J. Math. Biol. 7 (1979) 256.
[18] J.A. Feldman and D.H. Ballard, Connectionist models and their properties, Cognitive Sci. 6 (1982) 205.
[19] W.J. Freeman, Y. Yao and B. Burke, Central pattern generating and recognizing in olfactory bulb: A correlation learning rule, Neural Networks 1 (1988) 277.
[20] W.J. Freeman and B.W. van Dijk, Spatial patterns of visual cortical EEG during conditioned reflex in a rhesus monkey, Brain Res. 422 (1987) 267.
[21] W.J. Freeman, Mass Action in the Nervous System (Academic Press, New York, 1975).

[22] W.J. Freeman, EEG analysis gives model of neuronal template matching mechanism for sensory search with olfactory bulb, Biol. Cybern. 35 (1979) 221.
[23] W.J. Freeman, Nonlinear gain mediating cortical stimulus-response relations, Biol. Cybern. 33 (1979) 237.
[24] W.J. Freeman, Simulation of chaotic EEG patterns with a dynamic model of the olfactory system, Biol. Cybern. 56 (1987) 139.
[25] W.J. Freeman and B. Baird, Relation of olfactory EEG to behavior: Spatial analysis, Behav. Neurosci. 101 (1987) 393.
[26] R. Granger, H. Ambrose-Ingerson, J. Henry, and G. Lynch, Partitioning sensory data by a cortical network, in: Neural Information Processing Systems, ed. D.Z. Anderson (AIP, New York, 1988) p. 317.
[27] C.M. Grey and W. Singer, Stimulus dependent neuronal oscillations in the cat visual cortex area 17, Neurosci. Suppl. 22 (1987) 1301P.
[28] S. Grossberg and S. Elias, Pattern formation, contrast control, and oscillations in the short term memory of shunting on-center off surround networks, Biol. Cybern. 20 (1975) 69.
[29] J. Guckenheimer and D. Holmes, Nonlinear Oscillations, Dynamical Systems, and Bifurcations of Vector Fields (Springer, New York, 1983).
[30] I. Guyon, L. Personnaz, J.P. Nadal and G. Dryfus, Higher order neural networks for efficient associative memory design, in: Neural Information Processing Systems, ed. D.Z. Anderson (AIP, New York, 1988) p. 233.
[31] L.B. Haberly and J.M. Bower, Olfactory cortex: model circuit for study of associative memory?, Trends Neurosc. 12 (1989) 258.
[32] J. van Hemmen, Nonlinear neural networks near saturation, Phys. Rev. A 36 (1987) 1959.
[33] A. Herz, B. Sulzer, R. Kuhn and J.L. van Hemmen, Hebbian learning reconsidered: Representation of static and dynamic objects in associative neural nets, Biol. Cybern. 60 (1989) 457.
[34] M.W. Hirsch, Convergent activation dynamics in continuous time neural networks, Neural Networks 2 (1989) 331.
[35] J.J. Hopfield, Neurons with graded response have collective computational properties like those of two state neurons, Proc. Natl. Acad. Sci. US 81 (1984) 3088.
[36] J.J.B. Jack, D. Noble and R.W. Tsien, Electric Current Flow in Excitable Cells (Clarendon Press, Oxford, 1983).
[37] D. Junge, Nerve and Muscle Excitation (Sinauer, Sunderland, MD, 1981).
[38] B. Katz, Nerve, Muscle, and Synapse (McGraw-Hill, New York, 1966).
[39] D. Kleinfeld, Sequential state generation by model neural networks, Proc. Natl. Acad. Sci. 83 (1986) 9469.
[40] W.B. Levy and N.L. Desmond, The rules of elemental synaptic plasticity, in: Synaptic Modification, Neuron Selectivity, and Nervous System Organization, eds. W.B. Levy, J.A. Anderson and S. Lehmkuhle (Erlbaum, Hillsdale, NJ, 1985) p. 105.
[41] Z. Li and J.J. Hopfield, Modeling the olfactory bulb – coupled nonlinear oscillators, in: Advances in Neural Information Processing Systems I, ed. D. Touretzky (Kaufmann, Los Altos, 1989) p. 402.
[42] T. Maxwell, C.L. Giles, Y.C. Lee and H.H. Chen, Nonlinear dynamics of artificial neural systems, in: Neural Networks for Computing, ed. J. Denker, AIP Conf. Proc. 151 (AIP, New York, 1986) p. 299.
[43] H. Meinhardt, Models of Biological Pattern Formation (Academic Press, New York, 1982).
[44] B. Mel, personal communication.
[45] J.D. Murray, Nonlinear Diffusion Models in Biology (Oxford Univ. Press, Oxford, 1976).
[46] G. Nicolis and I. Prigogine, Self-Organization in Nonequilibrium Systems (Wiley, New York, 1977).
[47] J.L. Price and P.P.F. Powell, The short axon cells of the olfactory bulb, J. Comparative Neurol. 7 (1970) 631.
[48] D. Psaltis and S.S. Venkatesh, Neural associative memory, in: Evolution, Learning, and Cognition, ed. Y.C. Lee (World Scientific, Singapore, 1989).
[49] T.A. Schoenfeld and F. Macrides, Topographic organization of connections between the main olfactory bulb and the pars externa of the anterior olfactory nucleus in the hampster, J. Comparative Neurol. 227 (1984) 121.
[50] T.A. Schoenfeld, F. Marchand and J.E. Macrides, Topographic organization of tufted cell axonal projection in the hamster main olfactory bulb, J. Comparative Neurol. 235 (1985) 503.
[51] H. Sompolinsky and L. Kantner, Temporal association in asymmetric neural networks, Phys. Rev. Lett. 57 (1986) 2861.
[52] T.J. Wiley, The ultrastructure of the cat olfactory bulb, J. Comparative Neurol. 152 (1973) 211.
[53] M.A. Wilson and H. Bower, A computer simulation of olfactory cortex with functional implications for olfactory processing, in: Neural Information Processing Systems, ed. D.Z. Anderson (AIP, New York, 1988) p. 114.
[54] H.R. Wilson and J.D. Cowan, A mathematical theory of the functional dynamics of cortical and thalamic nervous tissue, Biol. Cybern. 34 (1979) 137.

NON-LINEAR DYNAMICAL SYSTEM THEORY AND PRIMARY VISUAL CORTICAL PROCESSING

R.M. SIEGEL[1]

Laboratory of Neurobiology, The Rockefeller University, New York, NY 10021, USA

The vertebrate brain consists of a large number of neurons, each with highly complex non-linear dynamics. These neurons communicate with each other with a complex nexus of connecting nerve axons. The current work examines the dynamical properties of neurons located in cat primary visual cortex from the perspective of non-linear dynamical theory. The temporal patterns of activation of such neurons achieved with periodic stimuli suggest that the dynamics are relatively simple and may be modeled using a small set of coupled non-linear equations. Predictions are made based as to the patterns of activation to be found in populations of neurons.

1. Introduction

Visual perception is the activation of a selected set of visual neurons distributed among multiple visual cortical areas. Optical illusions also initially activate part of such a set of neurons. The illusion occurs when the neurons interact to complete the perception [1].

How does one discuss the properties of a set of neurons? How does one describe the dynamics of such large numbers of cells with many different patterns of connectivities and highly non-linear interactions? The thesis advanced here is that these interactions in a *real* neural network (1) result in simple dynamics, (2) are restricted to a limited repertoire and (3) are reflected in the temporal activation of single cells.

The origin of these ideas can be traced from the work of Hebb [2] who stated:

"...timing has its effect in the functioning of the cell assembly and the interrelation of assemblies: diffuse, anatomically irregular structures that function briefly as closed systems, and do so only by virtue of the time relations in the firing of constituent cells..."

It has been known for many years that there are spatial-temporal patterns in the electroencephalogram (EEG). As early as 1954 the role of the alpha rhythm, for example, was thought to regulate the speed of transmission of information [3]. Hebb was unique for his time in that he attempted to describe how temporal patterns might be formed and what their role might be at *the single neuron level*. Experimental evidence for the temporal patterns in single cells has been accumulating over the past twenty years [4–14] although it is only recently that a theoretical description of the source of such patterns in the neural substrate has emerged.

In particular, Freeman and co-workers [15] have found evidence at the multi-neuronal level for oscillatory activity in the olfactory bulb. Freeman's major contribution is that he began to describe these temporal and spatial patterns in terms of non-linear distributed circuits. In this work they utilize the idea of a chaotic system as one from which many different possibilities can emerge.

However, a criticism of their work is that the basis of these oscillations at the single neuron level is still hidden. This viewpoint for studying the brain is important and has yielded many concrete and widely accepted results about processing in various regions of the brain during the past thirty years. In the brain the activity of the single unit encodes all the "messages" or "information" that can be passed onto another neuron due to the "all or none" nature of the conducted action

---

[1] Current address: Thomas J. Watson Research Center, IBM P.O. Box 218, Yorktown Heights, NY 10598, USA

potential in nerve axon. The information in population responses, such as those recorded with the local field potential method [15] or the EEG [16], are transmitted along many separate axons. Therefore the temporal code inherent in the single neuron's activity must be understood.

More recently two groups have demonstrated oscillatory activity in single neuron recordings [17, 18]. Both these groups see a strong 40 Hz frequency component in primary visual cortex in the cat. The appearance of these temporal patterns correlate nicely with the presentation of the visual stimulus. These two groups speak of these oscillations as reflecting "coherence" of the population of neurons, in much the same manner as the early EEG clinicians. Eckhorn et al. [18] have shown that two different cortical regions (V1 and V2) will oscillate together. These results are evidence for large numbers of cells working together.

Theoretical examination of the properties of net-like neuronal structures using excitatory connections showed that cortex was capable of sustaining waves of activity across its surface [19]. Later work added inhibition to this model showing that varied stable and unstable patterns could be found in a neural tissue [20]. The spatial distribution of neurons has been incorporated into these models and chaotic behavior has been formally described [21, 22].

Recently there has been a surge in interest in the dynamics of systems that are composed of large numbers of non-linear elements [23–25]. Many of these non-linear systems exhibit generic behaviors. One such well known behavior is period-doubling bifurcations as a function of some control parameter. As the parameter is varied, periodic oscillations occur which can turn chaotic. Such behavior is a universal property and particular universal constants can be derived [26].

The present work will explore the possibility that the cortical tissue, with its non-linearities and large number of cells will also exhibit such simple behavior. The strength of this non-linear approach, if successful, is that further quantitative predictions as to the temporal-spatial behavior of real neural networks can be made without detailed examination of the underlying circuitry. And it is these same spatial-temporal patterns many neurophysiologists would agree underlie sensation, perception, motor control and other cognitive functions.

Therefore it is suggested here that the dynamics seen in the theoretical models [19, 20] and in the experimental measurements [15, 17, 18] all can be explained as particular instances of a non-linear dynamical system which may be modelled independent of the detailed underlying circuitry. These universal dynamics may be seen as the "lingua franca" of cortical tissue.

In order to test this possibility, techniques from non-linear dynamical systems are used to study single unit activity of single cells in cat primary visual cortex. The phase plots and parameter sensitivities obtained suggest that the cortical mantle has interesting temporal patterns with features in common with known non-linear dynamical systems and has orderly transitions in temporal activity.

2. Methods

2.1. Data collection

Recordings were made from primary visual cortex (area 17) of the cat using standard methods [27]. In brief, the cat was anesthetized with sodium pentothal (20 mg/kg i.v. (intravenous), supplement as needed i.p. (intraperitoneal)) and paralyzed with succinylcholine (10 mg/kg/h i.v.). The EEG, heart rate, and expired CO_2 were monitored. The eyes of the cat were dilated with atropine, then refracted with contact lenses to focus on a tangent screen at 100 inch. A craniotomy was performed at Horsey–Clarke coordinates −5 mm AP. The dura was resected and standard tungsten electrodes were used to record from single neurons along the medial bank of area 17. The signal was ac amplified and bandpass filtered at 300–3000 kHz. Single action potentials were discriminated on their amplitude and time course. The amplified signal was also sent to an audio monitor which permitted the experimenter to listen to the neuron.

The receptive fields of the cells were first plotted using a hand held projector. Then a bar of light was projected onto the screen and a high-speed shutter was used to turn the light on and off (fig. 1). Timing pulses for the shutter (Uniblitz Co.) were obtained from Wavetek 191 waveform generator.

The time of the onset of the shutter and the time of the occurrences of single spikes were collected using Heurokon M68020 VME board with a precision of 0.1 ms. This system was attached to a Sun 3/160, which transferred the data from the slave CPU system to a disk. Data were analyzed after the experiments using either the Sun 3/160 or an IBM PC with custom written software.

2.2. Analysis

Two standard neurophysiological and one novel analysis were performed on the interspike interval data. Interspike interval histograms, which are the distribution of times between spikes, and post-stimulus time histograms, which show the correlation between the stimulus and spike, were computed. "Return maps" of the ith and the $(i + 1)$th interspike interval were plotted as described in further detail below.

3. Results

3.1. Single cell response

The effect of the flashed stimulus on one neuron is described using both the standard techniques of neurophysiology and techniques derived from non-linear dynamical system theory. The first cell to be described (fig. 2) was a complex cell; it was responsive to the onset of the stimulus.

The raw data from the first 40 s of the 20 min of recording is first shown. The plus sign indicates

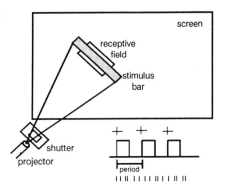

Fig. 1. Schematic representation of the recording situation. Stimuli were projected onto a tangent screen which was at a distance of 100 inch from the cat's eye. The receptive field was first mapped using a hand held projector and is marked by the open rectangle. Then an illuminated stimulus bar (shaded rectangle) was projected at the optimal angle for the cell. A shutter attached to a square wave generator was used to turn the stimulus on and off. The stimulus timing, which always had a 50% duty cycle, is illustrated at the lower right. The stimulus period was defined as the time between successive stimulus onsets. A few sample spikes are schematically displayed below the stimulus trace. The plus signs indicate the time of stimulus onset and are again seen in fig. 2.

the stimulus onset; the vertical bar indicates the time of spike occurrence. Stimuli were presented at 271.5 ms intervals. A burst of action potentials can be seen following each stimulus. By plotting the distribution of interspike intervals, it can be seen that there is a major periodicity at approximately 270 ms. There are also peaks at 175, 540, and 1080 ms. The latter two values are roughly integer multiples of the driving period. These slow frequencies can be attributed to the cell not being activated by one or more stimuli in a row.

The 175 ms peak is not directly attributable to any linear or non-linear processing thought to occur in single cells (e.g. a potassium conductance). Furthermore as will be seen below, these periodicities were dependent on the stimulus.

Post-stimulus time histograms were also constructed averaging the cell's response to each stimulus. Such histograms are often used to classify single visual neurons. For example, the present cell is called an "on-cell" because of the response to the onset of the stimulus. Other cells are called

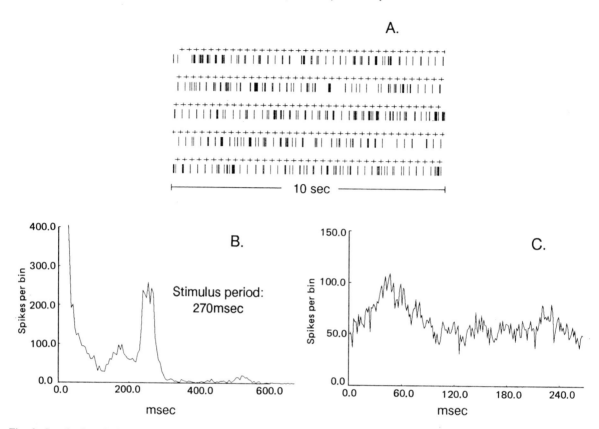

Fig. 2. Standard analysis of the interspike interval data. (A) The first 40 s of a 20 min recording session are shown. The plus sign indicates the stimulus onset; the vertical bar indicates the actual time of spike occurrence. Occasional bursts can be seen in this presentation of the data which tend to occur after the stimulus onset. (B) The interspike interval histogram is formed by counting the number of spikes that occur with different interspike intervals. (C) The post-stimulus time histogram is computed by computing a histogram of the time of spike occurrence relative to the stimulus onset. More formally it is a cross-correlogram of the point processes that constitute the stimulus and the spike train. The 270 ms peak shown correspond to the period of the stimulus.

"off cells", which respond to the off-set of the stimulus or "tonic" cells which have sustained responses [27].

The third and final analysis to be performed was the generation of "return maps". The motivation for these plots in the present work comes from non-linear dynamical theory where similar graphs from continuous time systems reveal temporal patterns which are not seen in the raw data. Such reconstructions in continuous time systems have been shown to embody the complete dynamics of the system under study [28]. Poincaré sections can be taken through these phase trajectories which can show fractal structure [26]. If the repeated stimulation of the cell was to lead to a temporal pattern of activation, then the return maps of the interspike interval should show interesting structure.

It can be seen that the return map of the interspike interval has a number of features (fig. 3). No such features were seen when the cell was spontaneous active (c.f. fig. 5a, 5b). Furthermore these patterns are reminiscent of the spatial return maps from coupled map lattices [23]. The origin of some of these features is trivial; some require deeper explanations. First there is the density close to the origin. It is caused by the rapid firing of the neuron. In this particular cell, there is no apparent structure at the shorter intervals, however with other cells (fig. 4) there are some relationships

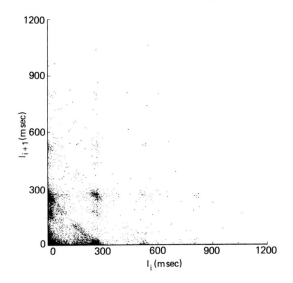

Fig. 3. "Return map" computed by plotting the ith interspike interval against the $(i+1)^{\text{th}}$ interval. A number of regions with concentrations of points can be seen. These correspond to attractors. See text for details.

between the interspike intervals. This pattern indicates that there are mechanisms which regulate the relationships between intervals at the onset of a burst as action potentials. There are two possible choices for such mechanisms – either an intrinsic property of the neuron (e.g. refractory period) or a network property.

There is also a clear repeating structure both vertically, horizontally and parallel to the lines:

$$I_i + I_{i+1} = kT,$$

where T is the period of stimulation, I_i is the ith interspike interval and k is 1, 2 or 3.

The vertically and horizontally oriented clusters of points are arranged at intervals equal to the base period. They arise from the neuron failing to fire from a given presentation of the stimulus. Some intervals approximately equal to the driving period are followed predominately by intervals at the same period. This cluster represents an 1:1 relationship between the stimulus and response respectively. In some cells, there are phase locking patterns of 2:1, 3:1 where one burst of action potentials occurs for every 2 or 3 stimuli.

A projection of this two-dimensional distribution onto one dimension results in the major peaks seen in the interspike interval histogram. The source of the peak in the interspike interval histogram at 175 ms can now be seen to be the result of a density of points clustering about a portion of the diagonal line given by eq. (1). Part of this diagonal line is stimulus driven and part is related to the temporal dynamics of the system.

The stimulus driven portion of the signal can be understood by example. Suppose every time the stimulus is turned on, a short burst of spikes occurs. Now imagine that a spike occurs at random in the interval between the two stimuli. This would result in points evenly distributed on the line given by eq. (1) with $k = 1$.

The basic diagonal structure is a result of the driving stimulus. What is interesting is that the spike data are not evenly distributed among the diagonals given by eq. (1). In this example, the points are clustered at part of the diagonal. Some process is permitting the neuron to fire with a particular clustering of interspike intervals. In the parlance of non-linear theory, the clusters seen in figs. 3, 4 and 5 are attractors.

Other neurons have other clusters of interspike interval sequences (figs. 4, 5). (These clusters can also be constructed in three-dimensional return maps.) These attractors are stable over long periods; some neurons show the same patterns for the full 1.5 h tested. These attractors do not appear to be linked to the slow electrical waves in EEG that arise when the animal is anesthetized. Furthermore, these attractors change very systematically with the stimulus period.

3.2. Sensitivity to period of stimulation

It was noted during the course of these experiments that these patterns could easily be heard on the audio monitor. Small changes in the stimulus period resulted in a change in the response of the cell from sounds like a "trotting" to a "galloping" horse. Did these patterns change in some orderly

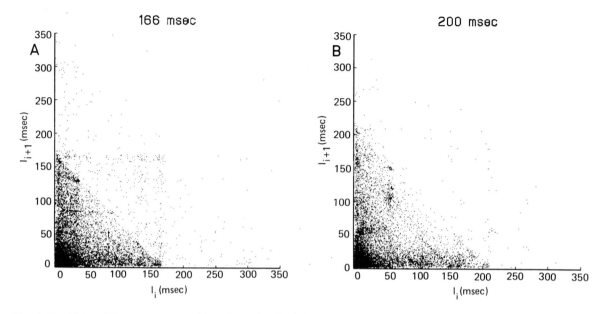

Fig. 4. Sensitivity of the return map to a 36 ms change in stimulation period. (A) The return map for a stimulation period of 166 ms. Note the multiple densities which represent attractors. Close to the origin (< 50 ms), finer patterns can be seen. The main diagonal intercepts the two axis at the driving period. There is a second weak diagonal band that intercept the axis at about twice the driving period. Other complex patterns of point densities can be seen. (B) Increasing the stimulus period by 36 ms leads to a large change in the return map patterns even though there only is a small change in the overall firing rate. Note the asymmetries in the return map.

way as function of stimulus period? The dependence of the neural response on the stimulus period was therefore examined. Fig. 4 shows the phase plots for a neuron different from that in figs. 2 and 3 with the stimulus periods of 166 and 200 ms. This neuron was of the complex type; it had almost no spontaneous activity. These particular stimulus periods were chosen due to the complexity of the auditory patterns. The "return map" also showed numerous complex periodicities. (This cell had the most complex patterns observed in the 20 cells in 5 cats studied.)

This sensitivity to stimulus period was then studied in greater detail by varying the stimulus period over a wide range. An example of such an experiment for a third cell is depicted in fig. 5 Spontaneous firing, and three stimulus periods are shown. No patterns are seen when the cell is spontaneously firing. Decreasing the stimulus period lead to additional periodicities and changes in the return map pattern.

To further quantify these effects, a finer exploration of the dependence of the stimulus period was performed. The stimulus period was set at some value for 2–3 min duration and then changed to a new value. The data is displayed by plotting the interspike interval against the stimulus interval (fig. 6A). In order to get a better representation of the interspike interval density, a small random value was added to the stimulus period prior to plotting. Such a plot reveals that at critical values, new multiples of the stimulus period are added to the interspike intervals. These data were also plotted using a normalized ordinate (fig. 6B) by dividing the interspike interval by the stimulus interval. In this representation, a value of "1" on the ordinate means that the cell was firing at an exactly the same interval as the stimulus period, "2" is twice the stimulus period, etc. The interspike interval is now seen to be phase locked to the stimulus with patterns of $2:1, 3:1, 4:1,\ldots, 8:1$. (Some cells also show $1:2, 1:3, 1:4$, etc.) It appears from the qualitative picture of fig. 6 that there is a systematic increase in interspike intervals at multiples of the stimulus period. Similar effects were seen in five other cells suggesting that

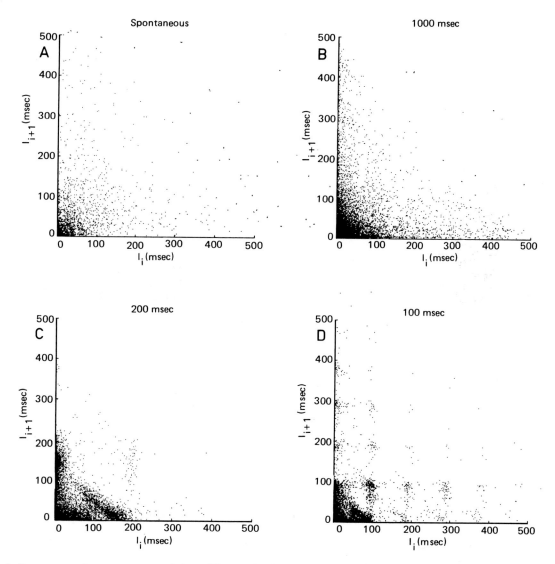

Fig. 5. Return maps for spontaneous and three different stimulus periods of a cat visual neuron. (A) Spontaneous activity. The spikes occur in bursts, but there is no pattern to the sequence of bursts that occur. (B) 1000 ms stimulus period. At this stimulus period, there begins to be some structure in the return map. Most of the points fall below the line given by eq. (1) with $k = 1/2$. (C) 200 ms stimulus period. The temporal pattern becomes more complex; an attracting cluster at around (175, 25 ms) emerges. As well, the cell begins to have longer interspike intervals at multiples of the stimulus period. (D) 100 ms stimulus period. More complex patterns are seen. The attracting cluster stays at the same relative position in the graph relative to the driving period. Longer multiples of the driving period are seen.

the change in temporal dynamics for these neurons is restricted to a particular repertoire.

In order to quantify these effects, normalized interspike interval histograms were created for each of the stimulus periods. In these plots the spike count was divided by the total stimulus duration since different durations were used for each stimulus period. This spike count per stimulus duration, defined here as the "density", was then plotted as a function of stimulus period (fig. 7A). Multiple peaks can be seen that correspond to the cluster of points seen in fig. 6B.

The relationship of the peaks to the normalized interspike intervals was next investigated. A geo-

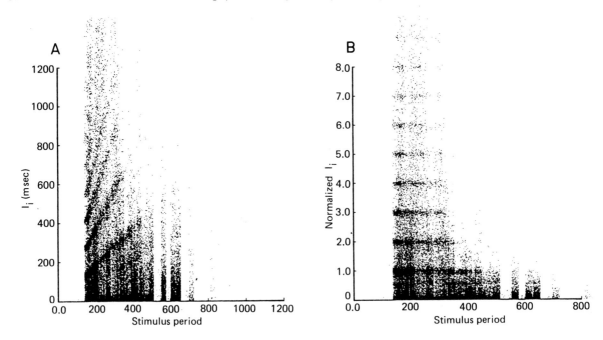

Fig. 6. The effect of the stimulus period on the interspike intervals. The cell presented here is the same as in figs. 2 and 3. The ordinate is the stimulus period; the abscissa is the interspike interval (A) or the interspike interval divided by the stimulus period (B). In order to better present the number of spikes for each stimulus condition, a random number was added to each of the stimulus periods. In both presentations, the decreases in stimulus period lead to longer interspike intervals at integer values of the stimulus period.

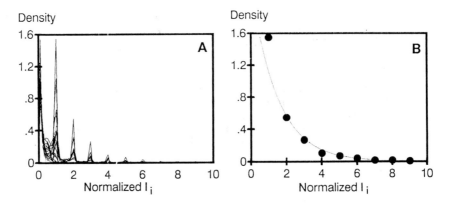

Fig. 7. Quantitative measures of the spike density as a function of the stimulus period and integer multiple of the stimulus period. (A) A series of overlaid interspike interval density plots for the 26 different stimulus periods used are shown. The abscissa is the interspike interval divided by the stimulus period; the ordinate is the spike density (see text). Successive peaks at up to seven times the base stimulus period can be seen. It appears that the peaks of the interspike interval densities decays roughly geometrically with the stimulus period. (B) This can be demonstrated by examining the data collected for one of the stimulus periods (135 ms). Again the abscissa is the interstimulus interval divided by the stimulus period; the ordinate is the spike density. Only the values of the density at the integer multiples of the interspike interval are plotted. A geometric relationship given by: $Y = 2.2 \times 0.51^X$ can be obtained (r^2 of 0.98). Similar plots are found for other stimulation periods.

metric function $Y = AB^X$ was a good fit for all the different stimulus periods used. Similar relationships were seen in other cells. The significance of these relationships with respect to underlying dynamics however remains an open question as there is little dynamical theory at present that addresses the dynamics of systems of this degree of complexity (see section 4).

4. Summary and discussions

The present work seeks to apply the theory and methodology from the study of non-linear dynamical systems to single unit activity in the brain. The underlying principle of this approach is that the dynamics of a large number of neurons (presumably determined by as many as 10^4 non-linear coupled differential equations) is not highly complex but can be simple. (The value of 10^4 is a conservative estimate of the number of cells in a local column that are interconnected and excludes all the equations that arise from the cable and membrane properties of excitable tissue.)

The present data are supportive of these claims. Interesting patterns are obtained by plotting adjacent interspike intervals. These patterns are sensitive to the period of the stimulus. These patterns exhibit multiple steady states which are not explicable based on standard membrane phenomena. Single neurons, when driven by periodic stimuli, do not show patterns of this type [29]. Preliminary work also indicates a sensitivity to the orientation of the bar, which is thought to be a result of interactions amongst neurons. These patterns are robust, repeatable and can be maintained over long periods of recording. It is suggested that the patterns seen in the return maps result from the network properties of the visual cortex.

It is remarkable that these patterns are quite similar to spatial return maps in coupled map lattices [23]. In these models, 100–1000 quadratic maps, given by $X_{i+1} = aX_i(1 - X_i)$, are coupled together. When the dynamics are explored by altering the coupling and the map parameter a, orderly changes in the dynamics are seen. These models have multiple attractors where the attractors are distributed across the lattice. It will be interesting to determine if periodic forcing of these coupled lattices can lead to patterns similar to those presented here from cerebral cortex. Also some of these analytical techniques developed for these lattices can be applied to neural data.

The sensitivity of these neural patterns to the stimulus period are qualitatively similar to period-adding bifurcations seen in electronic systems [30, 31]. Period-adding bifurcations are seen when a change in a parameter leads to an adding of a new periodicity rather than the splitting of periodicity into two periods (a pitchfork bifurcation). Again, it will be necessary to further explore these qualitative similarities with both models and experiments.

The implication of this work is that the temporal activity of a single neuron that is embedded in the matrix of approximately 10^4 neurons is relatively simple. Is it possible to reflect this temporal simplicity in a small set of equations? If so, it may not be necessary to model the visual cortex in complete detail in order to describe its activity. Rather, an emergent property of temporal dynamics can be used to describe this highly complex system. This is not to state that detailed physiological and anatomical analyses are not needed. Even with a high-level description such as that delimited here, it is still necessary to understand how the cortical circuit is formed from the constituent elements.

It is of course possible to numerically determine the relative simplicity or complexity of the neural system using dimensional analysis [32] or by computing the Lyapunov exponents [33]. Technically there are some problems since the reconstruction of the attractor requires data evenly spaced in time and the spike data are inter-event data. The one published example of a computation of the dimension using interspike interval data for somatosensory neurons [34] does not show good convergence of the computed dimension. (It is possible to perform temporal averaging to obtain a continuous time series and more properly fulfill the requirements of dimensional or Lyapunov exponent analyses [35].)

However, there is a more important conceptual problem to be surmounted even if these sorts of analyses can be performed and low dimensionality and positive Lyapunov exponents can be determined. Neither of these two quantities are sufficient to establish that non-linear dynamical theory is applicable to brain tissue [36], nor that the tissue is truly chaotic. Rather it will take a concerted effort to first demonstrate that the dynamics of brain tissue exhibit *features* of non-linear dynamical systems (e.g. universality). Following that, it is then necessary to physiologically demonstrate that these dynamics are not epiphenomena but are actually used by the brain in its processing of the world about us.

It was assumed at the onset of this work that this temporal pattern of activity is indicative of the more distributed nature of the cortical mantle. It is predicted here that similar patterns should be observable in the spatial dimension (i.e. across the cortical surface and/or through the cortical layers) by plotting the interspike intervals of one cell against the interspike intervals of a second cell. Such data have been collected by a number of researchers [37–43] and can be easily examined to test this hypothesis. One supportive result for this prediction can be found in the work of Aertsen and Gerstein [14], who have demonstrated temporally varying relationships for two cells that have significant cross-correlation.

If these results are correct, then questions as to the underlying nature of these non-linear processes arise. Are these responses caused by a completely deterministic process or by a stochastic process? The visual system is a biological system and as such will have "real" noise. Two biological systems where non-linear dynamical approaches have been applied have demonstrated that the presence of noise results in a removal of more complex phase locking patterns [44,45]. Theoretical approaches to this problem have been initiated [25, 46, 47]; also ch. IV in ref. [24].

Finally, regardless of the underlying theoretical correctness of the above approach, it is also important to note that these patterns may prove to be useful indicators for the classification (e.g. morphology, laminar distribution, simple, complex, hyper-complex and end-stopped properties) of neurons with different connectivities in the nervous system. In the past it has been quite difficult to classify a neuron's morphology based purely upon its response to light [48]. The current analysis may simplify this classification based upon the additional information that can be obtained from the "return map" representation of neural activity.

Acknowledgements

This work could not have been accomplished without the use of Dr. Torsten Wiesel's laboratory facilities. Essential discussion of this material with Drs. Mitchell J. Feigenbaum, Michael C. Mackey, Torsten N. Wiesel, John Lowenstein and George Carman are acknowledged. Programming of the powerful data collection system was performed by Kaare Christian. Bozenka Glatt provided excellent secretarial support and Peter Pierce excellent photographic support. This paper was made possible in part by funds provided by the Charles H. Revson Foundation. The statements made and the views expressed are solely the responsibility of the author. Thanks is also due to Dr. Charles Gilbert for use of his SUN computers. Computational support through the San Diego Supercomputing Center is acknowledged. Additional support was through a National Institute of Health grant 7R01 EY05253 to Dr. Torsten N. Wiesel.

References

[1] E.A. DeYoe and D.C. Van Essen, Trends Neurosci. 11 (1988) 219–226.
[2] D.O. Hebb, The Organization of Behavior (Wiley, New York, 1949)

[3] H. Gastaut, in: Brain Mechanisms and Consciousness, eds. E.D. Adrian, F. Bremer and H.H. Jasper (Blackwell, Oxford, 1954) p. 247.
[4] D.H. Perkel, G.L. Gerstein and G.P. Moore, Biophys. J. 7 (1967) 391–418.
[5] B.W. Knight, J. Gen. Physiol. 59 (1972) 767–778.
[6] S.H. Chung, J.Y. Lettvin and S.A. Raymond, J. Physiol. (London) 236 (1974) 63P–66P.
[7] G.J. Carman, B. Rasnow and J.M. Bower, Soc. Neurosci, Abstr. 12 (1986) 1417.
[8] B.L. Strehler and R. Lestienne, Proc. Natl. Acad. Sci. 83 (1986) 9812–9816.
[9] B.J. Richmond and L.M. Optican, J. Neurophysiol. 57 (1987) 132–146.
[10] B.J. Richmond and L.M. Optican, J. Neurophysiol. 57 (1987) 147–161.
[11] L.M. Optican and B.J. Richmond, J. Neurophysiol. 57 (1987) 162–178.
[12] G.L. Shaw, D.J. Silverman and J.C. Pearson, in: Systems with Learning and Memory Ability, eds. J. Delacour and J.C.S. Levy (Elsevier, Amsterdam, 1988).
[13] S.H. Chung, S.A. Raymond and J.Y. Lettvin, Brain. Behav. Evol. 3 (1970) 72–101.
[14] A. Aertsen and G.L. Gerstein, Soc. Neurosci. Abstr. 14 (1988) 651.
[15] C.A. Skarda and W.J. Freeman, Behav. Brain Sci. 10 (1987) 161–173.
[16] A. Babloyantz and A. Destexhe, Proc. Natl. Acad. Sci. 83 (1986) 3513–3517.
[17] C.M. Gray and W. Singer, Proc. Natl. Acad. Sci. 86 (1989) 1698–1702.
[18] R. Eckhorn, R. Bauer, W. Jordan, M. Brosch, W. Kruse, M. Munk and H.J. Reitboeck, Biol. Cybern. 60 (1988) 121–130.
[19] R.L. Beurle, Phil. Trans. R. Soc. London, Ser. B. 240 (1956) 55–94.
[20] H.R. Wilson and J.D. Cowan, Biophys. J. (1972) 1–23.
[21] W. von Seelen, H.A. Mallot and F. Giannakopoulos, Biol. Cybern. 56 (1987) 37–49.
[22] J. Cowan, personal communication.
[23] K. Kuneko, Physica D 34 (1989) 1–41.
[24] Cvitanović, Universality in Chaos (Adam Hilger, Bristol, 1984)
[25] L. Glass and M.C. Mackey, From Clocks to Chaos (Princeton Univ. Press, Princeton, 1988).
[26] M.J. Feigenbaum, Los Alamos Sci. 1 (1980) 4–27.
[27] D.H. Hubel and T.N. Wiesel, Proc. R. Soc. London Ser. B. 198 (1977) 1–59.
[28] F. Takens, in: Geometry Symposium, Utrecht 1980, eds. D. Rand and L.-S. Young (Springer, Berlin, 1981).
[29] K. Aihara, T. Numajiri, G. Matsumoto and M. Kotani, Phys. Lett. A 116 (1986) 313–317.
[30] S.D. Brorson, D. Dewey and P.S. Linsay, Phys. Rev. A 28 (1983) 1201–1203.
[31] T.-H. Yoon, J.-W. Song, S.-Y. Shin and J.W. Ra, Phys. Rev. A 30 (1984) 3347–3350.
[32] P. Grassberger and I. Proccaccia, Physica D 9 (1983) 189–208.
[33] A. Wolfe, J.B. Swinney, H.L. Swift and J.A. Vastana, Physica D 16 (1985) 285–317.
[34] P.E. Rapp, I.D. Zimmerman, A.M. Albano, G.C. Deguzman and N.N. Greenbaum, Phys. Lett. A 110 (1985) 335–338.
[35] R.M. Siegel, work in progress.
[36] R. Thom, Beh. Br. Sci. 10 (1987) 182.
[37] K. Toyama and K. Tanaka, in: Dynamic Aspects of Neo-cortical Function, eds. G.M. Edelman, W.E. Gall and W.M. Cowan (Wiley, New York, 1984).
[38] G.L. Gerstein and D.H. Perkel, Biophys. J. 12 (1972) 453–473.
[39] G.L. Gerstein, D.H. Perkel and K.N. Subramanian, Brain. Res. 140 (1978) 43–62.
[40] G.L. Gerstein, D.H. Perkel and J.E. Dayhoff, J. Neurosci. 5 (1985) 881–890.
[41] D.Y. Ts'O, C.D. Gilbert and T.N. Wiesel, J. Neurosci. 6 (1986) 1160–1170.
[42] D.Y. Ts'O and C.D. Gilbert, J. Neurosci. 8 (1988) 1712–1727.
[43] J. Krüger and F. Aiple, Brain Res. 477 (1989) 57–65.
[44] L. Glass, C. Graves, G.A. Petrillo and M.C. Mackey, J. Theor. Biol. 86 (1980) 455–475.
[45] M.R. Guevara, L. Glass and A. Shrier, Science 214 (1981) 1350–1353.
[46] A. Lasota and M.C. Mackey, Probabilistic Properties of Deterministic Systems (Cambridge Univ. Press, Cambridge, 1985).
[47] M.C. Mackey and A. Lasota, Physica D 28 (1987) 143–154.
[48] C.D. Gilbert and T.N. Wiesel, Vis. Res. 25 (1985) 365–374.

A DYNAMICAL SYSTEM VIEW OF CEREBELLAR FUNCTION

James D. KEELER

MCC, 3500 West Balcones Center Drive, Austin, TX 78759, USA

First some previous theories of cerebellar function are reviewed, and deficiencies in how they map onto the neurophysiological structure are pointed out. I hypothesize that the cerebellar cortex builds an internal model, or *prediction*, of the dynamics of the animal. A class of algorithms for doing prediction based on local reconstruction of attractors are described, and it is shown how this class maps very well onto the structure of the cerebellar cortex. I hypothesize that the climbing fibers multiplex between different trajectories corresponding to different modes of operation. Then the vestibulo–ocular reflex is examined, and experiments to test the proposed model are suggested.

The purpose of the presentation here is twofold: (1) To enlighten physiologists to the mathematics of a class of prediction algorithms that map well onto cerebellar architecture. (2) To enlighten dynamical system theorists to the physiological and anatomical details of the cerebellum.

1. Introduction

The cerebellar cortex contains nearly half of the total number of neurons in the entire brain and consists of five main neuronal types that are organized in a highly regular (almost crystalline), compact structure. This structure implies a very specialized and optimized function, and it has attracted much attention from neurobiologists and theoreticians. Nevertheless, after thousands of experimental investigations, the function of the cerebellum is still poorly understood.

Lesion studies of the cerebellum in adult mammals show that removal of the cerebellar cortex impairs coordination and the ability to learn certain tasks. However, lesion studies in young mammals show relatively few sustained deficiencies in motor-control tasks. This fact has prompted the jocular statement "the cerebellum compensates for its own absence" [1].

Although this statement was made in jest, it is indicative of the level of frustration with our understanding of what the function of the cerebellum is. Clearly the above statement cannot be true. The cerebellum is one of the phylogenetically oldest structures of the brain; its structure is nearly the same in fish as it is in man. Certainly evolution would have not devoted half of all of the neurons in the brain to a structure that compensates for its own absence!

What, then, is the function of the cerebellum? In the following I briefly review the neuroanatomy and physiology of the cerebellum, then I mention previous models and how they overlook some (perhaps important) facts of the neurophysiology. I will then propose a novel view of the functional mechanisms of the cerebellum based on a dynamical systems view. In particular, I claim that the role of the cerebellar cortex is to build an internal model of the dynamics of the external world and the dynamics of the sensory–motor system of the animal. This role is tantamount to being able to *predict* the dynamics of the system. Furthermore, I claim that the function of the cerebellum is determined by the interaction of the deep cerebellar nuclei with the rest of connecting subsystems, and that the exact function will be different for different subsystems.

0167-2789/90/$03.50 © Elsevier Science Publishers B.V.
(North-Holland)

I will describe how this prediction function can be achieved through local reconstruction of trajectories on an attractor, and will show how the algorithm for doing this maps onto a neural network model of cerebellar architecture (Marr [2], Albus [3], and Kanerva [4]). Finally, I will discuss the vestibulo-ocular reflex in some detail to demonstrate the utility of this view and suggest further neurophysiological experiments.

2. Structure of the cerebellum

This section contains a very brief overview of the anatomy and physiology of the cerebellum. Of course I will not be able to give details in such a brief report. Ito [5] has given an extensive compilation of many of the experimental facts of the cerebellum, and where no reference is given here, the details can be found in his book. (See also Palkovitz et al. [6–9]. The cell dimensions and counts are approximate for the cat, except where otherwise stated.)

The cerebellum sits near the back of the brain at the top of the brainstem and consists of two main parts: the cerebellar cortex (containing the vast majority of the neurons) and the deep cerebellar nuclei (DCN). Its basic structure is shown in fig. 1. The cerebellum can be divided into three main functional sections: the vestibulo cerebellum (containing the flocculus), the spinocerebellum (containing the vermis and intermediate zone) and the cerebrocerebellum (containing the hemispheres). Each of these sections has their own group of DCNs.

If we look at the input information to the cerebellum, we find that the cerebellum as a whole receives input information from almost every sensory system; it receives somatosensory information from diverse parts of the body, visual information, auditory information, vestibular information, and perhaps even olfactory information. Furthermore, this information is not distributed in a simple topological map the way it is, say, in the somatosensory cortex. Indeed, if we

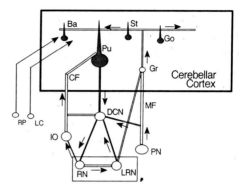

Fig. 1. The basic structure of the cerebellar circuit. The cerebellum is divided into two parts: the deep cerebellar nuclei (DCN) and the cerebellar cortex (enclosed in the bold box), which contains the granule cells (Gr), Golgi cells (Go), stellate cells (St), basket cells (Ba), Purkinje cells (Pu). The cerebellar cortex receives mossy-fiber (MF) input from the pontine nuclei (PN), climbing-fiber (CF) input from the inferior olive (IO), noradrenergic inputs from the locus coeruleus (LC), and seretonergic inputs from the Raphe nucleus (RP). A more detailed view of the cerebellar cortex is shown in fig. 2. Note the positive feedback loop formed by the DCN, the red nucleus (RN), and the lateral rectricular nucleus (LRN) (in the small box). These connections are present as part of the intermediate cerebellum, but not in the hemispheres or flocculus.

look at the somatosensory information input to the cerebellum of a rat, we see that the information is "jumbled"; inputs from the whiskers and mouth are mapped adjacent to input from the front paw, and input from the lower lip may be mapped adjacent to a patch of skin on the back. Compare this jumbled (not random) map to that of the somatosensory or motor cortex, where there is a simple, regular topological mapping.

Whereas the inputs are jumbled, the cellular structure of the cerebellar cortex is very regular. There are five main cells of the cerebellar cortex: granule, Purkinje, Golgi, basket, and stellate cells (there are also Lugaro cells, but these are sparse, and there are not many data concerning them). Of these five, only the granule cells are excitatory. The *Purkinje cells* are the largest cells and are the main processing cells of the cerebellum. These cells have a huge dendritic tree that is nearly planar in structure: the dendritic tree is about 9 µm thick by 250 µm wide and 300 µm tall, and has nearly 100 000 dendritic spines. The Purkinje cells are stacked in regular planes every 10–20 μm or so. Piercing the plane of the Purkinje cells are approximately 250 000 *parallel fibers*, which are the axons of the *granule cells*; tiny cells with an average of 4–5 dendritic "claws" each. They receive the major sensory input to the cerebellum through the *mossy fibers*. Granule cells send their axons up into the planes of the Purkinje-cell dendritic tree, splitting into two different directions forming the parallel fibers. The *Golgi cells* also receive mossy-fiber input as well as parallel-fiber input, and they send inhibitory axons back to the synaptic inputs of the granule cells. *Basket cells* receive parallel-fiber input from the lower portion of the Purkinje cell dendritic plane and send inhibitory projections to the soma of about 9 or so neighboring Purkinje cells in a direction perpendicular to the parallel fibers. The *stellate cells* also receive parallel-fiber input and send inhibitory projections to the upper part of the dendritic tree of the Purkinje cells (fig. 2).

Palkovitz [6–9] gives the following cell counts for the cat: There is a fan-out of information from about 1 million mossy-fiber cells to about 2.2 billion granule cells, then a fan-in to about 1.3 million Purkinje cells, which are the sole output of the cerebellar cortex. There are about 0.4 million Golgi cells, 8 million baskets, and 22 million stellate cells. The Purkinje cells send their axons to the deep cerebellar nuclei (DCN) and inhibit those cells. There is a further fan-in of about 26 to 1 from the Purkinje cells to the DCNs. Hence, we see that there is a tremendous fan-out to the granule-cell representation and a tremendous fan-in to the DCN level.

Another major pathway comes from the inferior olive and gives rise to climbing fibers, so-called because they climb around the dendrites of the Purkinje cell (fig. 2). Any single Purkinje cell receives input from a single climbing fiber, and the climbing fibers branch to contact up to about ten different Purkinje cells. The climbing fibers also give rise to collaterals to the DCN, and it appears that these collaterals may contact the same group of DCNs as is touched by the Purkinje cell that is synapsed on by the very same climbing fiber.

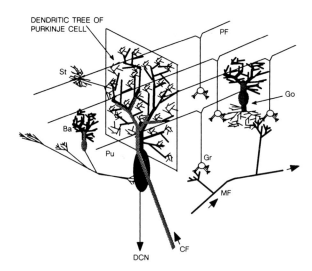

Fig. 2. Cells of the cerebellar cortex. The main input to the cerebellar cortex is from the mossy fibers (MF), which synapse on the granule cells (Gr). The Gr are small cells with an average of 4–5 small dendrites called "claws", and the Gr axons give rise to the parallel fibers (PF). These fibers contact the dendritic tree of the Purkinje cells (Pu), which are the sole output of the cerebellum. The inferior olive gives rise to climbing fibers (CF), which also synapse on the Purkinje cells. The cortex also has three types of inhibitory interneurons, including Golgi cells (Go), basket cells (Ba), and stellate cells (St), all of which receive input from parallel fibers. The Golgi cells also receive input from the mossy fibers. The Ba and St cells synapse on the Purkinje cell, and the Golgi cell synapses on the claws of the granule cell.

Although the model presented here is mostly computational, the theory attempts to account for the following details of cerebellar neuroanatomy and physiology.

(1) Fractured somatotopic input to the granule cells.

(2) Why the cerebellar cortex has so many cells.

(3) The largest cerebellar-cortex to body-mass ratio is in a species of electric fish that have incredibly simple motor activities.

In addition, the role of the Golgi, basket and stellate cells will be discussed, as well as the role of the climbing fiber temporal activity and the deep cerebellar nuclei.

3. Brief review of previous theories of cerebellar function

If the cerebellum does not just "compensate for its own absence", what does it do? This section contains a very brief overview of previous theories of cerebellar function.

Based on the work of Eccles et al. [10], several authors in the late 1960's proposed a model of the function of the cerebellar cortex as an associative memory (Marr [2], Albus [3]). The Marr and Albus models of the cerebellum are very similar in structure to the Kanerva (1988) [4] model of associative memory, and I will discuss the Marr–Albus–Kanerva (MAK) models in detail in a later section. The presumed function of the cerebellar cortex in these models is that it is an associative memory mapping of the motor commands that are required for coordinated muscle movement. The associative memory was proposed as being useful for fast readout of reflex actions to predict the appropriate trajectory of a desired movement. These models are a good first approximation to the cerebellar function, but they do not account for the temporal information processing aspects of the cerebellum, they do not discuss the processing at the deep cerebellar nuclei, and they do not give a good explanation of the function of the inhibitory cells of the cortex. Fujita [11] extended the Marr–Albus models of the cerebellum to incorporate temporal processing, and the cortex acted as a *phase-lead* (predictive) filter, but the resulting model was linear and did not account for the inherently nonlinear aspects of the cells of the cerebellum or the nonlinear nature of the task.

Several other models of cerebellar function have been proposed. Pellionisz and Llinas [12] have proposed a model of the cerebellum as a system that is performing a tensor-coordinate transformation from one sensory system to another. They point out from a computational standpoint that this nontrivial transformation must take place somewhere in the sensory–motor system, and that the system is performing a predictive mapping to achieve motor coordination. However, the coordinate transformation that they suggest is formulated as a smooth (linear) topological mapping, and this fact does not agree with the experimental evidence of a fractured somatotopy input into the cerebellar cortex.

Again from a computational point of view, Paulin [13] suggested that the cerebellum plays the role of a Kalman filter for optimal signal processing of sensory information. However, Paulin did not connect the functional and mathematical formalism of a Kalman filter with the architecture of the cerebellum. In particular, the standard formulation of the Kalman filter requires a matrix inversion, and Paulin did not shown how this operation maps onto the cerebellar architecture.

Ito [5] suggested a motor control model of the cerebellum based on previous work by Albus, Marr and Fujita. In his model, the cerebellar cortex serves as a predictive, adaptive filter in a feedback controller, Houk [14] expands on this idea and views the cerebellar cortex as an array of adjustable pattern generators that read out a prelearned motor trajectory. Both of these models are more complete than the Marr–Albus models in that they include the roles of the deep cerebellar nuclei and other connected nuclei in the cerebellar subsystem. However, these motor control models do not account for the fact that the largest cerebellum per body size is found in a species of fish with an incredibly simple motor repertoire. (Why should an animal with such a small repertoire of motor activities require a large cerebellum if the cerebellum's main function is motor control?) Nor do they fully account for how the trajectories are actually computed and learned.

A common theme throughout all of these models is the ability of the cerebellum to predict dynamics or trajectories. However, the standard known mathematical methods of obtaining a predictive device (matrix inversion, tensor transform, phase-lead Laplace transform) do not map well onto the physiology and anatomy of the cerebellar cortex. I present a different mathematical framework for performing prediction tasks and show how this mathematical framework maps naturally onto the architecture of the cerebellum, but first, we must review some recent results from dynamical systems theory.

4. Prediction through local reconstruction on an attractor

In many of the above theories the cerebellar cortex builds an internal model of the dynamics of the world. *Building an internal model of the dynamics of the world is equivalent to being able to predict the near-future dynamics of a system given the present state.* Thus, let us focus on the prediction problem as as fundamental question.

The problem of prediction or *forecasting* can be stated in the following way: suppose that we are given a certain set of variables $x(t) = \{x_1(t), x_2(t), x_3(t), \ldots, x_n(t)\}$, from discrete time steps $t = 0$ to $t = T$, and we wish to predict what will be the value of $x(t)$ at $t = T + k$. We can view this problem as a mapping from a real n-dimensional space, $\mathbb{R}^n$, into an m-dimensional space i.e., $f: \mathbb{R}^n \Rightarrow \mathbb{R}^m$, or we could also consider mappings from $\mathbb{R}^n \Rightarrow B^m$ or $B^n \Rightarrow B^m$ (where B^m is a binary m-dimensional space). The assumption is that there is such an underlying mapping that governs the process (if the process is totally random, there is no such mapping). However, the process may appear to be random to a casual observer but may actually be the result of a deterministic nonlinear (chaotic) process. If the time series comes from a random process, then there will be no correlation between the plotted points. However, if the dynamics is deterministic, then the points will generically lie on an *attractor* (for a good discussion on attractors, see e.g. ref. [15]). (Note: if one does not obtain an attractor in n dimensions, one can increase the dimension of the state space by creating other independent variables through time lags or derivatives and plot the dynamics in a higher-dimensional space, see e.g. ref. [16].)

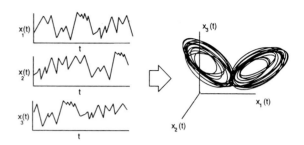

Fig. 3. Reconstruction of the dynamics of a system from a time series. An input set of data in a three-dimensional embedding space, and a hypothetical attractor that the dynamics lies on.

Suppose that such an attractor exists. How do we predict future behavior from such a system? A very simple method for doing prediction is based on local approximations on the attractor. Farmer and Sidorowich [17] suggested the following method for predicting the dynamics of a dynamical system given the time series $x(t)$: plot each of the coordinates against each other, as in fig. 3. Suppose that we have plotted all previous points up to a time $t = T$, and we wish to find where a new point goes at k time steps later. If we plot this new point in the phase space, it will land somewhere on (or near) the attractor. To predict where this new point goes, we can simply look at where the nearest neighbors of this point went. That is, we pick the p nearest neighbors to the given point $x_0(t_0)$, and, to predict where $x_0(t_0 + k)$ goes, we simply interpolate the trajectories of the p nearest neighbors of this point as in fig. 4.

Fig. 4. Local reconstruction of trajectories on an attractor. Prediction of the dynamics on the attractor from the p nearest neighbors (●) of the given point (○). The trajectories follow similar paths, so we can interpolate the future trajectory of the given point from the past information.

5. Prediction in Marr, Albus, and Kanerva cerebellar neural networks

What does this local approximation method for predicting dynamics have to do with cerebellar function? To see the correspondence, let us examine the Marr, Albus, and Kanerva (MAK) models. These are all three-layer neural-network models in which the number of hidden units is much larger than the number of input or output units. The input units correspond to the mossy-fiber cells, hidden units correspond to the granule cells, and the output units correspond to the Purkinje cells (fig. 5).

The MAK models can be viewed as a three-layer, feed-forward perceptron with an architecture like that shown in fig. 5. Given the inputs, x_k, the activations of the hidden units, y_j are determined by the connection strengths, T_{jk}, from the input to the hiddens, and the bias, a_j. That is, the hidden-unit activation is given by

$$y_j = g\left(\sum_k T_{jk} x_k - a_j\right), \quad (1)$$

where $g(x)$ is a nonlinear (threshold) gain function ($g(x) = 0$ for $x < 0$, $g(x) = 1$ for $x > 0$). Similarly, the outputs, o_i, are determined by the connections from the hiddens to the outputs, W_{ij}, and the corresponding bias, b_i,

$$o_i = g\left(\sum_j W_{ij} y_j - b_i\right). \quad (2)$$

In the MAK models, the weights T_{jk} are fixed and the weights W_{ij} are learned according to a Hebbian (Marr, Kanerva) or a least-mean-squares error propagation (Albus) learning rule.

To formulate the Kanerva model as a perceptron, we choose the inputs as plus or minus one, the weights T_{jk} from the input to the hidden layer are also plus or minus one and chosen at random, and the biases have all the same value $a_j = n - 2D$. The hidden and output values can then be chosen as binary (0, 1). This means that given an input pattern x, a hidden unit will be activated only if its weights to the input layer lie within a ball of radius D from the input (see ref. [18]). The activated hidden units then turn out to be the nearest neighbors to the input pattern, as shown in fig. 6. Thus, if we view each of the input vectors as an encoding of the embedding space of a dynamical system (i.e. the continuous values are encoded into binary patterns, e.g., by a sliding-bar encoding), the Kanerva model can be used for building a local approximation to the dynamics just as with the attractor-reconstruction method above. That is, the activated hidden units are the nearest neighbors of the input trajectory in the embedding space, and the local approximation that allows prediction is contained in the weights between the hidden and output units (i.e. W_{ij} interpolates the local dynamics through the nearest neighbors). The Kanerva model can also be used as a statistical predictor, as pointed out by Rogers [19].

There are some obvious differences between the Kanerva model and the Farmer–Sidorowich (FS) approach, however. First of all, whereas the FS

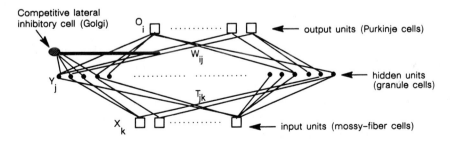

Fig. 5. The architecture of the MAK models of the cerebellum: a 3-layer feed-forward perceptron.

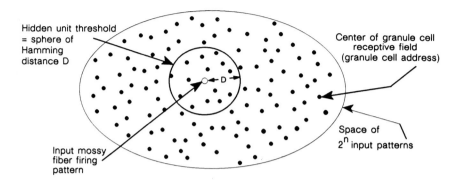

Fig. 6. Representation of input space and hidden unit activation. A view of the space of hidden unit weights in Kanerva's sparse distributed memory model. If the SDM has n inputs, the space of all possible input patterns is of size 2^n, and the n-dimensional vector of weights into each hidden unit is represented by a point in this space (the weights are binary vectors chosen at random). For a given input, all hidden units within a Hamming distance D to that input will be active, and these active hidden units correspond to the nearest neighbors of the input pattern vector. These nearest neighbors can then be used to build a local approximation of the dynamics.

algorithm relies on continuous values, Kanerva uses binary inputs and outputs and a Hamming distance D. Second, Kanerva's model chooses the hidden weights between the input and output layer at random, whereas they are chosen to lie on the trajectory of the attractor in the FS case. Third, the number of selected hidden units (nearest neighbors) in the Kanerva model is variable, whereas it is a constant in the Farmer-Sidorowich algorithm. All of these differences turn out to be minor. The binary units of the Kanerva model can be generalized to continuous units, the hidden unit weights can be chosen as a function of the input patterns by a simple competitive learning rule, and the number of activated hidden units can be kept as a constant by a simple feedback circuit [18]. Prediction is then possible through Hebbian learning as a smooth interpolation by the hidden-to-output weights in the SDM model, just as in the FS work. This generalized Kanerva model is much closer to the algorithm used by Farmer and Sidorowich, but not quite the same because the continuous inputs still have to be encoded into multiple binary units (i.e. by a thermometer for sliding-bar encoding). However, we can build a neural network algorithm with the exact same inputs as in the FS algorithm (i.e. linear inputs of the embedding space). For example, this was done by Lapedes and Farber [20], who investigated a four-layer backpropagation network to do prediction, and they showed that the receptive fields in the hidden units in the second layer were "bump functions" (the difference of two offset sigmoids). Taking this idea a step further, Moody and Darken [21] investigated three-layer neural networks based on the Albus model of the cerebellum with radial basis function (Gaussian) activation units in the single hidden layer. (This radial basis function network has the same architecture as that of the cerebellar neural network shown in fig. 5, but uses Gaussian activations of the hidden units.) The work of Lapedes and of Moody show that these networks can be used to learn to predict chaotic time series.

The Kanerva model and the Moody networks are fully connected from the input layer to the hidden layer, whereas the Marr model is very sparsely connected. Indeed, in the cerebellum, each granule cell is connected to only about 4–5 mossy fibers on average. Thus the hidden units have a very small fan-in, and this is a quite different structure than the Kanerva model. Marr postulated that if each granule cell has only 5 total inputs, it may take 3–4 active inputs to make the cell fire. He called the number of active inputs to make a granule cell fire a "codon" and pointed out that a codon representation of the input data is very useful for separating highly correlated inputs. That is, if the codons are of size R, and there are L active mossy fibers, then the number of active codons is $L!/R!(R-L)!$, and if there is another pattern that overlaps this first pattern by W units, then the number of active codons that overlap in the hidden layer will be

$$\begin{aligned}x &= \binom{W}{R} \Big/ \binom{L}{R} \\ &= \frac{W(W-1)(W-2)\ldots(W-R)}{L(L-1)(L-2)\ldots(L-R)} \\ &\sim (W/L)^R.\end{aligned} \quad (3)$$

For example, if the inputs at the mossy-fiber level are correlated by 0.8, the codons at the granule cell level will be correlated by $(0.8)^R$, which is 0.41 for $R = 4$. This implies that the codon representation is very useful for distinguishing between highly correlated patterns and expanding them into an uncorrelated, sparsely coded representation in the granule-cell layer.

If the input vectors are randomly chosen and uncorrelated, the representation in the hidden unit layer is statistically the same for the fully connected case (Kanerva) as the codon representation (Marr), if each has the same number of active hidden units (the first- and second-order statistics of these two systems are the same as n gets large). Thus, the codon representation can cover the input space of patterns (fig. 6) just as the fully connected Kanerva model does, but the codon representation covers this space with low-dimensional *slices*. That is, the granule cells ignore most of the input coordinates, and they respond to an R-dimensional slice through the input space rather than to a ball of radius D. If enough of the codon "slices" are combined, then the full space can be covered just as before, and this representation can be used for prediction of trajectories. The advantage that the Marr encoding has over Kanerva's encoding is apparent when the embedding space is large, but the dynamics in that space in on a small (3–4) dimensional attractor. In this case, the attractor dynamics can be reconstructed by taking low-dimensional slices and projections through the embedding space. This encoding may help alleviate the requirement for a tremendous number of data points to fill a large embedding space, as in the Kanerva encoding.

Now we have the following basic idea: there are very natural and simple algorithms for doing prediction problems that map easily onto three-layer neural network models similar to that of the cerebellar cortex. This view is promising because it ties together the postulated predictive, adaptive function of the cerebellum to a three-layer neural network architecture. However, we must now go a step further and examine the function of the cerebellum from a more detailed view, and tie in physiological facts into a detailed model.

6. Details of the physiology and anatomy of the cerebellum

Section 5 outlined several models of the cerebellar cortex and showed how one class of models, MAK, could be viewed from a dynamical systems perspective as building local reconstruction of trajectories on an attractor. This section goes into more detail regarding the physiology, and postulates roles for some of the cells in the cerebellar system.

6.1. Input mapping

Whereas the MAK networks and the one examined by Moody have an architecture similar to that of the cerebellar cortex, the detailed structure is quite different. First of all, the radial basis function activation is, at first glance, very different from the response of the granule cells. However, consider the following: suppose that we look at a single, continuous variable such as the frequency of sound sensed by the ear. This frequency is mapped out topologically onto the auditory cortex so that a given frequency excites a group of cells that are centered at a given position. If we look at the response of this cell as a function of frequency, the "receptive field" is localized at f_0 in frequency space, and the response falls off in an approximately Gaussian fashion from f_0 on either side. Now, if a granule cell is connected to this unit, then the response of the granule cell will also be Gaussian around f_0. Thus, the granule cell response can be viewed as a radial basis function along that coordinate. However, the response along several different coordinates is presumably additive in their responses, rather than multiplicative as used by Moody. This cell's response can presumably be modeled by a radial basis function in the input variable f, centered at f_0, and the information is presumably *distributed* in an approximately sliding-bar encoding as inputs to the granule cells.

However, if we examine the actual neurophysiological data, we see that the input encodings seem to be closer to linear than binary or Gaussian. For example, the response of vestibular cells to head rotation is approximately linear [5]. This fact presents only a minor problem for the MAK models. We can generalize the Kanerva model to linear inputs over a range of $(0, 1)$, say. The mathematics goes directly over for this case: change the hidden unit activation function to be linear rather than nonlinear. In that case the hidden unit activations define a simple surface in the input space (one can visualize it as a cone rather than a pill box, or a weighted sphere rather than a constant sphere). (We are still dealing with "nearest neighbors" but in a weighted sense.) In the Marr model, linear input also corresponds to conical sections in the input space, and the "codons" correspond to low-dimensional "slices" through the high-dimensional embedding space.

6.2. Golgi cells

The Golgi cells receive inputs from mossy fibers on their lower dendrites, inputs on their upper dendrites from granule cells via about 6000 parallel fibers, and they send their outputs to about 5000 granule cells directly underneath them. Thus, the Golgi cells are in a feedback arrangement with the granule cells. They receive input from parallel fibers, and they send output back to the granule cells. Marr and others have postulated the role of the Golgi cells as a mechanism for regulating the number of active parallel fibers: If too many parallel fibers are active, the Golgi cells become active and inhibit the granule cells. This is an important function from an information-processing point of view, because, as pointed out above, the number of active granule cells is the key parameter in the function of the three-layer neural network model. This is particularly important if the inputs are correlated, as pointed out by Keeler [18].

This view of the Golgi cell as a simple feedback regulator neglects the role of the mossy-fiber inputs. The mossy-fiber inputs are local connections and give the Golgi cell information about the average activity of the local granule cells (the 5000 cells right underneath them). The parallel-fiber inputs are most likely long-range connections (probability 5000/100 000 of being local). I view the Golgi cell in a slightly different light: that of an arbitrator in a bidding scheme (similar to the bidding in a classifier system or genetic algorithm, see e.g. ref. [22]). In the binary input models, the mossy fibers set the bidding level based on local information, and the parallel fibers set the bidding level based on long-range information. The Golgi cell could have the following effect on a granule cell: if the granule cell has five claws, it may need only two active inputs to be activated depending on the activity of the Golgi cell; if the local activity is low, and the long-range activity is low, then the granule cell will not be inhibited by the Golgi. However, if the local activity is moderate, and long-range activity is very high, then the granule cell will be inhibited so that it will only fire if 4 of the 5 inputs are active. In this sense the Golgi cell has effectively raised the "bid" (that is the number of active inputs to a granule cell) from two to four for the granule cells in its area, based on both local and long-range information. See also the simulation by Pellionisz [23].

In the linear input view, much the same thing happens: If the Golgi cell receives a lot of long-range input, it will be active and will inhibit the granule cells. This corresponds to shifting the conical section (changing the ordinate intercept), and again can be viewed as a bidding scheme, but the effects of the bidding is not winner-take-all; it is an overall penalization, or taxation of the activity. This actually is an efficient way of covering the space: Visualize the input space being covered by several conical sections, and those sections can be moved up or down depending on the input conditions to achieve the best representation.

6.3. Stellate cells

The stellate cells are small inhibitory neurons that synapse on the Purkinje cells. In almost every model of the function of the cerebellar cortex, the synaptic weights between the hidden layer and the output layer are of both signs (excitatory and inhibitory). Since the parallel-fiber–Purkinje-cell synapse is excitatory, there must be some way of getting inhibitory connections. These connections are effectively provided by the stellates: If a stellate is activated by a parallel fiber, then the stellate inhibits the Purkinje cell. This is not quite what the models require because the mapping is not one-to-one, but since the number of stellates is much larger than the number of Purkinje cells, and their size is small, they can effectively make small local averages of the inhibitory input of the parallel fibers and inhibit the Purkinje cells in a local but not one-to-one fashion.

6.4. Basket cells

Whereas Marr grouped the basket and stellate cells together in terms of function, their role and morphology are rather different. The basket cells receive their main connections from the parallel fibers (and perhaps climbing fibers and collaterals of the Purkinje cells). They send their inhibitory outputs to the cell bodies of about nine Purkinje cells in a direction perpendicular to the parallel fibers (fig. 2). Since the basket cells inhibit Purkinje cells, and since Purkinje cells inhibit DCNs, the firing of basket cells allow increased DCN activity. Suppose that there is some activity in a parallel fiber bundle ("on beam"). The basket cells in this region will be activated and will inhibit Purkinje cells perpendicular to the parallel fibers, but not their parent cell in the beam. These inhibited Purkinje cells allow more activity in the DCNs. Thus, the Purkinje cells that are "on beam" actually inhibit trajectories, and the Purkinje cells in neighboring regions allow other trajectories to be activated via the basket cell inhibition. We can view this as *multiplexing* of the trajectories, and this multiplexing occurs in part through the basket-cell interactions.

6.5. Climbing fibers

The other major input to the cerebellar cortex is the climbing fiber input. The climbing fibers arise in the inferior olive, and each fiber makes contact with about 10 different Purkinje cells, but each Purkinje cell is contacted by only one climbing fiber. This selective synaptic input prompted Marr, Albus and Kanerva to postulate that the climbing fibers provide the data necessary for learning at the parallel-fiber–Purkinje-cell interface. Many investigators have investigated the plasticity of these synapses, yet the results are still controversial (see ref. [24] for a recent discussion). I will not re-hash that issue here, but instead concentrate on the important timing properties of the climbing fibers. However, it should be pointed out that the original Hebb hypothesis of Marr and Kanerva is computationally insufficient: the synapses would quickly saturate. One needs not only activity in climbing fibers and in granule cells, but presumably a "learn signal" as well – telling the system when to be in the learn mode. It turns out that such a signal could be provided by the locus coeruleus, which secretes norepinepherine, and enhances the signal-to-noise properties of the cerebellar cells. This has important computational properties as well as important implications for learning (see ref. [25]).

Suppose that the cerebellum is actually involved in coordinating motor activities, and that the cerebellar cortex provides the predictions for the proper trajectories to coordinate. Prediction is only half of the problem – the signals for the correct behavior must somehow be *synchronized* as well. I postulate that this is part of the role of the climbing fibers. They fire at 1–4 Hz, and cells in the inferior olive tend to oscillate in synchronous, localized groups. When the climbing fibers synapse on the Purkinje cells, they cause complex spiking. After the complex spike on the Purkinje cells, there is about a 250 ms "window" where the spontaneous activity of the Purkinje cell is modulated (both enhanced and depressed, see ref. [26]). The modulation depends on the parallel-fiber, basket, stellate and Golgi cell activity in the region near the Purkinje cell. I view the modulation in this window as a "read-out" of the state of the Purkinje cell. Thus, I view the role of the climbing fibers as *multiplexing* the information contained in the Purkinje cells of the cerebellar cortex: the fibers turn on selective groups of Purkinje cells, and allow their state to be read out in a synchronized manner, and the synchronization signal arises in the local oscillations of the inferior olive. The climbing fibers presumably work in conjunction with the basket and Golgi cells to facilitate this multiplexing. The climbing fibers may also be involved in learning (in conjunction with the locus coeruleus, perhaps), but we note that even with previously learned skills, destroying the climbing fiber inputs yields almost the same defects as destroying the entire cerebellar cortex in the short term. Indeed, if climbing-fiber input is suppressed, the Purkinje cells soon begin to fire at a regular, unmodulated frequency [27]. Thus climbing fiber input is essential for proper function of the cerebellar cortex in the short term, not just changing synaptic strengths in the long term.

6.6. Purkinje cells

The Purkinje cells are some of the largest and most complicated cells of the brain, and the known details about these cells could easily fill a book. However, for our purposes here, we deal with only a few details. First of all, the Purkinje cells have active processes on the upper part of the dendritic tree. This means that the dendrites themselves can give rise to action-potential-like input to the cell body. Certainly it is a very gross approximation to view such a cell as a simple binary threshold unit as in the MAK models. Houk [14] has pointed out that such cellular dynamics may give rise to hysteresis in the firing rate of the Purkinje cell. Thus he views the Purkinje cells as hysteretic devices that can "latch and hold" the information for short intervals. This view fits in well with our multiplexing hypothesis: The climbing fibers provide contextual information and the prediction is held ready at the Purkinje cell level until a "read" command comes. Houk postulates that the read trigger comes from the DCN and red nucleus. I propose that the trigger at the DCN level is synchronized with the climbing fiber firing. In that sense, the Purkinje cells are acting very much like their description in the MAK models: The information that allows interpolation of trajectories and prediction of future events is stored in the granule-cell–Purkinje-cell synapses, and these synapses are presumably modifiable.

6.7. Putting it all together: the deep cerebellar nuclei

Lastly, I discuss the deep cerebellar nuclei, DCN, the sole targets of the Purkinje cells. Any information that was processed in the cerebellar cortex has to be funneled through this small group of cells before it is allowed to influence any other neuronal activity. The inputs to the DCN seem to be mapped topologically in piecewise smooth coordinate patches. The DCN also receive collaterals from the climbing fibers, and there is some evidence to suggest that the collaterals synapse on the same DCN that are contacted by the Purkinje cell that the very same climbing fiber synapses on. There is a large fan-in from the Purkinje cells to the DCN: there are 26 times more Purkinje cells than DCN cells. Thus, the DCN are where the processed information is concentrated, and, as such, the DCN function is central to the function of the cerebellum.

Different deep cerebellar nuclei correspond to the three different subdivisions of the cerebellum: the vestibular nuclei, the fastigial nuclei, and the lateral nuclei. Each of these nuclei can again be divided into further sub-parts. Presumably the function of each of these sub-parts may be different, so it is not possible to give a global functional explanation of the cerebellar function. However, one can give some general comments that hold for all of the nuclei. The Purkinje cell outputs are strongly inhibitory. In the fastigial and the lateral nuclei, the DCN sit in the middle of a positive feedback loop with the red nucleus and the lateral recticular nucleus (fig. 1). Thus, without the inhibitory connections from the Purkinje cells, the DCN would fire at saturation [14]. This fact must be taken into account when interpreting lesion studies: the cerebellar cortex sits in a modulatory position in the loop, and if its modulation is taken away, the cells can fire at saturation, causing many behavior defects that are not necessarily related to the actual function of the cerebellum. Given these facts, a plausible role of the cerebellar cortex is to inhibit the unwanted signals so that the proper signals can be processed through the positive feedback loop at the DCN in a very fast manner. The cerebellar cortex integrates sensory information from many different modalities and modulates the DCN signals according to the ambient sensory context. I view the function of the cortex as providing a prediction, and this prediction is used for different purposes in the different nuclei. For example, in the fastigial nucleus, the signals may be used to sculpt trajectories that allow coordinated, smooth motion, and the function of the cerebellum is motor control. In contrast, in the electric fish, the cerebellar cortex could be used as a predictive, adaptive filter to cancel noise in the sensory information processing. In both of these cases the cortex does prediction, but the prediction is used in different ways to obtain the actual functionality.

7. A computational view of the vestibulo–ocular reflex

As a concrete example of a function that is done by the cerebellum consider the vestibulo-ocular reflex (VOR). The vestibulo–ocular reflex (VOR) is an eye-muscle reflex that compensates for head movement by moving the eyes in the opposite direction of the head so that the focal point remains stationary. It is governed by a rather simple circuit: head velocity is sensed in the semi-circular canals in the ear, and this information is passed through the vestibular nuclei, the floccular target neurons, and then to eye-muscle motor neurons. The floccular target neurons are innervated by the Purkinje cell axons from the flocculus (part of the vestibulo cerebellum [24]). When one thinks about this from a computational point of view, one sees that this is a very trivial coordinate transformation from head-velocity coordinates to eye-velocity coordinates. There is a simple, direct pathway from the vestibular nucleus (the head-velocity sensors) to the eye muscles. Why then would one need the tremendous computational power that is contained in the flocculus? An obvious answer is to allow self-calibration of this system. However, since the climbing fibers give retinal slip information (error information), why could not the calibration be done without the flocculus involved? Is the flocculus compensating for its own absence? Most likely not. If one looks at the computation that is required for the *exact* VOR, one sees that the computation is much more complicated than just a simple transformation of the head coordinates to eye coordinates.

Clearly, the angle that the eyes have to move through to keep them perfectly fixated on a target depends on where they are focused; if they are focused at infinity, then the amount that they have to move is much less than if they are focused five inches in front of the nose. In particular, the angle that the eyes must move through for horizontal head motion is given by

$$\tan(\phi) = \frac{\sin(\theta)}{\cos(\theta) + r/l}, \qquad (4)$$

where θ is the head angle, ϕ is the eye angle, r is the radius of the center of the head to the center of the eye, and l is the distance from the center of the head to the point of focus (note that r changes as the animal grows). Note also that as $l \to \infty$, the transformation approaches a very simple linear transformation, $\phi = \theta$, but to achieve perfect VOR, for all l, the focal point must be taken into account. Moreover, the full transformation is much more complicated than this. If we write down the correct VOR transformation from head velocity to eye velocity, we see that the perfect transformation is

$$\phi'_x = f(\theta_x, \theta_y, \theta_z, \theta'_x, \theta'_y, \theta'_z, \phi_x, \phi_y, \phi_z, \phi'_y, \phi'_z, l, r), \quad (5)$$

where $\theta_{x,y,z}$ and $\phi_{x,y,z}$ denote head (θ) and eye (ϕ) coordinates in the x, y, and z directions respectively; f is a nonlinear, transcendental function derived from the generalization of eq. (4) to three dimensions, and the prime denotes differentiation with respect to time. If we start to include the nonlinearities of the eye muscles, then the transformation is even more complicated.

Thus, to perform the VOR perfectly, this information must be used. There are now at least four possibilities:

(1) The inaccuracies are too small to be noticed, so this information is not necessary.

(2) This information is used in the computation to obtain an accurate VOR.

(3) There is a fast feedback mechanism to correct for inaccurate VOR.

(4) The flocculus actually computes when to stop the eye motion.

Let us examine these possibilities in turn. First of all, if we focus on an object seven inches in front of our eye and turn our head at an angular velocity of 50°/s the retinal slip from the exact calculation (eq. (4)) and the calculation at $l = \infty$ is about 20°/s. This inaccuracy is not negligible. Thus, either we have significant retinal slip focusing this close, or there is some method for compensating for the close focal length. Assuming that we are able to compensate, let us examine the second possibility, i.e. that there is some mechanism for calculating an accurate VOR response. If this is the case, then somehow the eye position and eye-muscle information must be available for the flocculus to calculate the proper response. Recent experiments show that some of this information is available (see ref. [24]), but I have not seen any reports of cells that respond to different focal lengths (this may never have been investigated; one likely candidate for such projections would be from the pre-tectal area or the superior colliculus through the nucleus recticulus tegmenti pontis giving rise to mossy fibers). If this view is correct, then the computation must be performed at the flocculus, not in the vestibular nuclei. We can now see how the computation would be performed: the relevant angles, muscle feedback signals and focal length information would be mapped in piecewise smooth patches into the flocculus, and then the variables in the patches would be selected through climbing-fiber activity. The flocculus would then compute the proper eye trajectory via the model described above. Note that since there is no way to map this nine-or-more-dimensional coordinate space onto two dimensions, the best that we can do is map the variables in coordinate patches in a "jumbled" fashion that allows each of the variables to be used in the computation in the relevant way. This would be one explanation for the fractured somatotopy mapping of the cerebellar cortex.

This view can be tested experimentally by lesion studies to the flocculus (or inputs to the flocculus): the animal would show no deficits at a very long focal length viewing straight ahead, but would show deficits for short focal length and off-center vision. (Note that most of the experiments are done for straight-ahead and on-center vision, so these deficits would not have been detected.) (Note: even if this is the case, it may be very hard to detect this experimentally.)

If we are able to perform the VOR accurately for these different variables, but the motor-control hypothesis (2) is not correct, then another possibility exists: (3) there is a fast feedback loop that allows correction of the inaccurate response. The climbing fibers seem too slow to allow such a computation. They respond at 4 Hz maximum, so the time lag could be up to 250 ms. The motion could be over before the climbing fibers even responded. There is, however, enough bandwidth in the climbing fibers so that if their phase was sensitive to the error, then they could convey the correction signal. However, in this view, there is no reason to have the tremendous computational machinery available in the cerebellar cortex – indeed, it gets in the way.

A fourth possibility is that the flocculus calculates when to stop moving the eyes. This view is actually not much different than the view that the flocculus calculates the transformation (4). Somehow the flocculus would have to know when to stop the motion, and this is formally equivalent to knowing the transformation. However, from an approximation standpoint, it may be easier to learn when to stop than learning the entire transformation.

There is too little experimental evidence at this point to distinguish between each of these views, but the experiments suggested above should help us discriminate. We expect that the final answer will be a combination of all four: some of the inaccuracies will be tolerable, there will be some method for forward computation and predicting when to stop, and there will be a mechanism for correcting errors (the optokinetic reflex can be viewed as performing this role over a longer time scale); however, it is important to point out that if the transformation is done exactly, it is done by (4) and (5).

8. Summary and discussion

The view of the cerebellar function presented here is the following: *The cerebellar cortex combines its input sensory information to build an internal model of the world that allows prediction of the dynamics of the sensory-motor system.* The prediction of this internal model is then combined at the DCNs and used for modulating motor commands to achieve smooth, coordinated motion (as is the presumed function in much of the literature), or it can be used to enhance sensory information discrimination (as is the presumed function in electric fish). The cerebellar cortex is viewed as a generic sensory processing device, and the resulting processed information can be used in a variety of particular ways. The *function* of the cerebellum cannot be deduced by looking only at the cortex; the function is mainly determined by the DCNs and their interaction with the rest of the system. The main theory presented here is based on the Marr, Albus and Kanerva models of the cerebellum and is a simple extension of those models to incorporate more of the neurophysiological details, and to show the direct correspondence with predicting trajectories. The Marr, Albus and Kanerva models do not include the role of the DCN, and, as such, inadequately describe the function of the cerebellum, whereas the DCN interactions were discussed in some detail in this report.

The view presented here ties together many of the previous views of cerebellar function and in that respect is not entirely new. However, these previous theories are viewed in a new mathematical framework that allows us to connect the presumed predictive nature of the cerebellar cortex to a family of prediction algorithms that map well onto the neuroanatomical structure of the cerebellum. I also postulated that the climbing fibers synchronize and multiplex signals into the cortex, and that they allow context-dependent predictions to be read out of the Purkinje cells. I went on to discuss details of some of the cells in the cortex to explain how our theory corresponds with the experimental facts. I discussed the vestibulo–ocular reflex in some detail from a computational point of view and suggested an experiment to test this view.

Acknowledgements

The view presented here emerged out of a long study of the cerebellar architecture including an intense 10 month literature search by Egon Loebner, Coe Miles, David Rogers and myself. Although the views presented here are mainly the author's, they were influenced through many hours of discussion among this group (the multiplexing idea was originally Loebner's). I thank the Research Institute of Advanced Computer Science at NASA–Ames for their support during 1987–1988 to pursue this investigation, and the encouragement of Pentti Kanerva and his proofing of the text. I also benefitted from discussions with D.S. Tang, Jim Houk, and Jim Bower.

References

[1] J. Bower, Oral presentation given at the Emergent Computation Conference, Los Alamos, 1989.
[2] D. Marr, J. Physiol. 202 (1969) 437–470.
[3] J.A. Albus, Math. Biosci. 10 (1971) 25–61.
[4] P. Kanerva, Sparse Distributed Memory (Bradford Books, MIT Press, Cambridge, MA, 1988).
[5] M. Ito, The Cerebellum and Neural Control (Raven Press, New York, 1984).
[6] M. Palkovits. P. Magyar and J. Szentagothai, Brain Research 32 (1971) 1–13.
[7] M. Palkovits. P. Magyar and J. Szentagothai, Brain Research 32 (1971) 15–30.
[8] M. Palkovits. P. Magyar and J. Szentagothai, Brain Research 34 (1971) 1–18.
[9] M. Palkovits. P. Magyar and J. Szentagothai, Brain Research 45 (1972) 15–29.
[10] J.C. Eccles, M. Ito and J. Szentagothai, The Cerebellum as a Neuronal Machine (Springer, Berlin, 1967).
[11] M. Fujita, Biol. Cybern. 45 (1982) 195–206.
[12] A. Pellionisz and R. Llinas, Neuroscience 4 (1979) 323–348.
[13] M. Paulin, in: Competition and Cooperation in Neural Nets II, eds. M. Arbib and S. Amari (Springer, Berlin, 1988).
[14] J.C. Houk, in: Cerebellum and Neuronal Plasticity, eds. M. Glickstein, C. Yeo and J. Stein (Plenum, New York, 1987).
[15] A.J. Lichetenburg and M.A. Lieberman, Regular and Stochastic Motion (Springer, Berlin, 1983).
[16] F. Takens, in: Dynamical Systems and Turbulence, eds. D. Rand and L. Young (Springer, Berlin, 1981).
[17] D. Farmer and J. Sidorowich, Phys. Rev. Lett. 59 (1987) 845;
D. Farmer and J. Sidorowich, Exploiting chaos to predict the future and reduce noise, Los Alamos preprint 88-901 (1988).
[18] J.D. Keeler, Cognitive Sci. 12 (1988) 299–329.
[19] D. Rogers, in: Neural Information Processing Systems, eds D. Touretzky (Kauffman, Los Altos, CA, 1989).
[20] A. Lapedes and R. Farber, Nonlinear signal processing using neural networks: Prediction and system modeling, Los Alamos Technical Report LA-UR-87 (1987).
[21] J. Moody and C. Darken, in: Proceedings of the 1988 Connectionist Models Summer School, eds D. Touretzky, G. Hinton and T. Sejnowski (Kaufmann, Los Altos, 1988).
[22] J. Holland, K.J. Holyoak, R.E. Nisbett and P.R. Thagard, Induction: Process of Inference, Learning and Discovery (MIT Press, Cambridge, MA, 1986).
[23] A. Pellionisz and J. Szentagothai, Brain Res. 49 (1973) 83–99.
[24] S. Lisberger, Science 242 (1988) 728–734, and references therein.
[25] J. Keeler, E. Pichler and J. Ross, Proc. Natl. Acad. Sci. US 86 (1989) 1712–1716.
[26] N. Mano and K. Yamamoto, J. Neurophys. 3 (1980) 713–728.
[27] P.G. Montaloro, M. Palestini and P. Strata, J. Physiol. 332 (1982) 187–202.

NEUROMAGNETIC STUDIES OF HUMAN VISION: NONINVASIVE CHARACTERIZATION OF FUNCTIONAL NEURAL ARCHITECTURE

John S. GEORGE, Cheryl J. AINE and Edward R. FLYNN

Neuromagnetism Laboratory, Life Sciences and Physics Divisions, Los Alamos National Laboratory, Los Alamos, NM 87544, USA

> Biological solutions to the computational problems of vision appear to emphasize parallel processing within a hierarchy of visual areas specialized for different features of the visual array. This paper summarizes recent studies concerning the organization of the human visual system using noninvasive magnetic recording techniques. The results illustrate principles of visual system organization previously demonstrated in animal single unit studies – sequential activation of multiple visual areas – as well as new observations not previously evident in animal studies – different (apparent) cortical sources of evoked responses to different spatial frequencies.

1. Introduction

Vision is an excellent example of the rich interplay between computational and biological approaches to the understanding of complex information processing problems. Biological solutions to the computational problems of vision appear to emphasize parallel processing within multiple, anatomically segregated areas of visual cortex, which are specialized for different dimensions or features of the visual array [1, 2]. Biological systems may represent optimal or near-optimal solutions to such problems, having developed over billions of years of evolution. Organisms ranging from insects to humans exhibit a variety of visual capabilities that substantially outperform existing computational systems. The convergent evolution of functionally similar visual systems from separate genetic stem-lines suggests that the requirement to efficiently encode and process visual information can serve as a powerful selective influence [3, 4]. In model systems in which optimal information encoding was used as a criterion to guide network evolution, resulting network models exhibited functional organization similar to that seen in biological systems [5–7]. For these reasons, studies of biological solutions to the computational problems of vision may offer important clues and constraints for purely computational approaches.

Neurophysiological and anatomical studies of experimental animals have provided a wealth of data concerning the structure and function of the visual system. Several observations appear as recurring themes across species and techniques and are considered "organizational principles": The visual system employs multiple-step, hierarchical processing within and between a number of anatomically distinct areas [1, 8–11]. However, as in other biological networks, the functional architecture within the visual system is massively parallel and the system employs multiple, parallel information streams [12–16]. A major function of the neural circuitry is to produce specialized representations of information which subserve specific processing needs. Presumably, many or all of these patterns should be found in the human visual system. Thus far, however, the limitations of experimental techniques applicable for human studies have prevented the direct characterization of microscopic functional organization.

Evoked response techniques have proven useful tools for the study of sensory information processing by human subjects [17, 18]. Neuroelectric and neuromagnetic measurements provide complementary functional views of neural activity with ms temporal resolution. Scalp-recorded electrical potential measurements (event related potentials or ERPs) display a complex temporal waveform consisting of a series of positive and negative peaks referred to as "components". While a variety of stimulus and behavioural task manipulations have been shown to affect the kinetics or amplitudes of specific response components, limitations of ERP methodology have prevented unequivocal determination of neural sources for many components of interest.

Neuromagnetic field distributions appear to be produced by intracellular current and unlike electrical potentials, appear to be relatively unaffected by the skull and scalp. Neuromagnetic sources often can be modeled adequately as one or more equivalent current dipole elements in a uniformly conducting sphere. It is also possible to reconstruct a two- or three-dimensional current distribution consistent with an observed field distribution which satisfies some numerical criteria such as the minimum number of dipoles, maximum entropy, or minimum norm. To the extent that a model's assumptions are valid, sources can be localized with considerable precision from magnetic measurements. In this paper we describe how neuromagnetic mapping techniques have been used to characterize evoked responses to sinusoidal gratings presented at various locations in the visual field. In this effort, we attempt to discover salient features of the functional architecture of the human visual system.

2. Background

The neural process of vision begins in the retina, which transduces light energy into a cellular electrical response. The input state of the retina is a cellular mosaic consisting of two classes of photoreceptors, rods and cones, which are distinguished by their morphology and by their response characteristics [9, 20]. Rods are sensitive at much lower light levels than cones, producing statistically reliable responses to single photons [21, 22]; however, both rods and cones adapt their sensitivity over several orders of magnitude [23, 24]. Coupled with adaptation, the differential sensitivity of these subsystems provides the visual system with an extended dynamic range difficult to match with an electronic sensor.

There are three types of cones, each characterized by its sensitivity to a particular band of wavelengths. The broad, overlapping absorbance spectra of cones, with peaks roughly corresponding to red, green and blue, are the basis for color encoding by the visual system [25–27]. The packing density of photoreceptors, the ratio of cones to rods, and visual acuity vary across the retina, with highest values in the central, "foveal" region [28–30]. This design coupled with mechanisms to orient the eye provides information processing flexibility and efficiency while minimizing hardware requirements in the retina and subsequent processing stages. Considerable information processing occurs in the retinal neural network, which may serve to maximize the information content of the signal. One important transformation performed by the retina is encoding the analog signal of the primary receptors into the frequency-encoded format characteristic of much of the central nervous system.

Axons of the retinal ganglion cells form the optic nerve which transmits information from the retina through the optic chiasm to the Lateral Geniculate Nuclei (LGN). Relay neurons arising in this structure project to the primary visual area in occipital cortex, also known as cytoarchitectural area 17, or V1. Cortical neurons in V1 project to a chain of secondary visual areas including V2, V3, V4, MT, MST and others, spanning occipital, parietal and temporal cortex [2, 12, 14]. Although information processing and transfer between cortical areas is sequential and hierarchical, there are extensive feedback pathways, and functionally defined subsystems within each visual area which give rise to multiple parallel information streams.

A hierarchy of specialized representations: In classical neurophysiology, the function of visual system neurons is described in terms of properties of the neuron's receptive field. The receptive field is the limited region of visual space which can directly influence the firing of the cell. Retinal ganglion cells and the so-called "simple" cells of LGN and cortex are characterized by center/surround receptive fields consisting of a central region which may be excitatory or inhibitory surrounded by an annular region which is antagonistic [31]. Cells with such receptive fields typically encode information about luminance and chromatic contrast based on relative differences in light intensity or wavelength, rather than absolute levels. One consequence of this encoding at an early stage is that subsequent information processing operations are relatively insensitive to variables such as the intensity or wavelength composition of ambient illumination.

Simple cells exist at the bottom of a hierarchy of cells which respond to progressively more complex visual features. For example, some cortical neurons have receptive fields specific for oriented bars and edges, probably driven by collections of simple cells. Higher-order feature detectors such as neurons sensitive to corners or to motion in a particular direction within the receptive field are also observed in primary visual cortex. These are local operations performed in parallel by relatively specialized circuits. In secondary visual areas receptive field properties may be highly specific, and there are many varieties.

One of the most striking organizational features of the early cortical visual areas is the maintenance of retinotopic order: local spatial relationships between objects in the visual field are preserved in the cortical projection even though global order is (necessarily) distorted [32]. Columnar structures collecting groups of related neurons are superimposed onto the retinotopic map. These include ocular dominance columns, consisting of neurons driven by one eye; hypercolumns, which contain the full complement of feature detector types, driven from a particular region of retina; blobs and interblobs and thick stripes and other histologically or functionally defined microsystems within the visual cortex [31].

3. Neuromagnetic studies of vision

In an attempt to discover organizational principles (or organizational details) we have applied neuromagnetic measurement techniques to the study of human visual processing. The effort has employed traditional psychophysical paradigms as well as analytical strategies driven by computational models of multiple-source configurations. The process necessarily has been iterative; by reference to existing knowledge of neurophysiology, we have been able to develop more sophisticated analytical tools. By applying these tools we are able to make inferences concerning the neurophysiology underlying our experimental observations.

As in ERP measurements, the initial data produced by neuromagnetic measurements is a time-varying voltage representing the physical quantity being measured, in this case the second-order spatial difference or "derivative" of the radial component of the magnetic field around the head. All of the information available in the measurement is embedded in this time/space matrix. For some purposes the data are represented as functions of time at each spatial location. By taking a temporal snapshot of waveform amplitudes over the ensemble of measurement locations it is possible to construct field maps which may disclose the position and orientation of evoked neural activity at that instant. In such maps, a simple current dipole produces balanced extrema of positive and negative magnetic flux, flanking, and orthogonal to the projection of the source into the measurement surface. In our initial studies we attempted to localize neural sources associated with empirically defined temporal components identified in previous ERP studies, assuming a single equivalent dipole source. One complication which became apparent early in our analyses was the existence of multiple simultaneous sources of neuromagnetic activity. In some cases this situation was obvious in a single field map; in other instances it was inferred from a temporal sequence of maps. Because multiple source modeling appears crucial for interpretation of neuromagnetic visual evoked response data, our approach is discussed in some detail within this paper.

We expected that retinotopic order would be a primary determinant of the pattern of evoked neuromagnetic fields. Based on the retinotopic organization of primary visual cortex observed in monkeys as well as the "cruciform model" of cortical organization in humans [33–36], we expected to find significant differences in the location and/or orientation of neuromagnetic sources as a function of field of stimulation. We were also interested to see whether neuromagnetic techniques could detect the effects of more subtle stimulus manipulations such as spatial frequency (SF). Given the progressive increase in receptive field size as a function of increasing eccentricity from the center of the retina [37, 38], we expected to find "preferences" for different spatial frequencies at central and peripheral stimulus locations. Anatomical and physiological studies of animals (for reviews, see refs. [39–41]) as well as psychophysical and ERP studies in humans [42–48] suggest the existence of spatial frequency selective channels within the visual system. We hoped to distinguish such channels on the basis of latency and/or amplitude differences in temporal response components.

4. Experimental and analytical procedures

Neuromagnetic responses to visual stimuli were monitored with a 7-channel hexagonal sensor array. Each sensor consisted of a second-order gradiometer coupled to a superconducting quantum interference device (SQUID). Responses were averaged from 25 or more presentations of each stimulus. Three or more blocks of trials were conducted at each sensor array location and six or more array positions were collected during a single experiment, which typically spanned two recording sessions. Amplitude data from all sensors (42–112 in these studies) were used to compute isofield contour or pseudocolor maps of magnetic field distribution at 10 ms intervals from 80–360 ms following stimulus presentation. If two or more field extrema of opposite polarity were observed in a map, a least-squares procedure was applied to fit a single current dipole model to the empirical distribution, and the percentage of variance accounted for by the model was assessed. Maps of theoretical and residual field distributions were prepared for visual inspection. In some cases residual maps were featureless and appeared to reflect measurement noise. In other cases residual maps contained dipole-like features with amplitudes above estimated noise levels, suggesting the existence of additional sources. In such cases, or where three or more peaks were present in the empirical field maps, multiple dipoles were fit to a single map [49].

Magnetic resonance brain images (MRIs) were obtained for some subjects. Images were acquired using full volumetric scanning techniques, producing a series of 32 slices at 4.5 mm thickness in each of three views. During MRI, oil-containing capsules were attached at reference locations to a bathing cap also worn during neuromagnetic data acquisition. By identifying the inion and left and right periauricular reference points in each slice series, it was possible to reconcile coordinate systems between volumetric series and with the neuromagnetic coordinate system. Slices were selected and pixel coordinates were calculated by computer program. Neuromagnetic and MRI coordinate systems were reconciled by comparison of common anatomical landmarks, and the calculated source of neuromagnetic activity was located on the MRI images [50].

Stimuli consisted of 100% contrast, intensity-modulated sinusoidal gratings randomly presented at various locations in the visual field. Gratings ranged from 1 to 8 cycles per degree and appeared in a rectangular window 2.0° H × 1.5° V. Gratings were displayed for 100 ms at approximately 1 s intervals. Background intensity was chosen so that nominal average luminance was the same in the presence or absence of the grating. Raster displays were generated by microcomputer, and projected onto a translucent screen. A system of mirrors relayed the images into a magnetically shielded chamber where experiments were conducted. In some experiments subjects were required to respond by pressing a fiber optic coupled switch for each occurrence of a grating of specified spatial frequency and location, in order to manipulate the attentional state of the subject.

5. Results

Electrical and magnetic evoked responses were collected simultaneously. Magnetic waveforms evoked by a small 1 cpd sinusoidal grating presented at an eccentricity of 2° in the lower right quadrant of the visual field are illustrated in fig. 1. Selected electrical and magnetic waveforms are illustrated in the upper panel; for these measurement locations, approximately corresponding temporal "components" can be identified in electrical and magnetic responses. We have adopted nomenclature consistent with ERP terminology; the "P1" or "P100" ERP component is a positive peak (in occipital electrode recordings) occurring at ≈ 100 ms post-stimulus. The following negative peak is the N1; the next positive peak (at ≈ 200 ms) is the P2, which is followed by the N2. Observed event related field (ERF) distributions were typically much more focal than corresponding ERP distributions. Note the polarity inversion in the 100 ms component of the ERF waveform over 2–4 cm on the head surface.

Fig. 2 illustrates source modeling of observed neuromagnetic field distributions using one and two dipole models. Field maps for time points corresponding to peaks in ERP waveforms are illustrated in column 1. Each distribution was fit by a single current dipole model using least squares

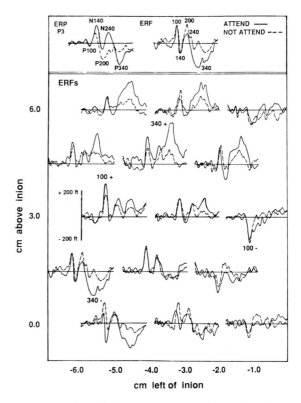

Fig. 1. Electrical and magnetic waveforms evoked by a stimulus at 2° in the lower right quadrant of the visual field. Upper panel: Selected event related potential (ERP) and event related neuromagnetic field (ERF) waveforms. Temporal components are identified by labels and the traces. Lower panel: An ensemble of magnetic response waveforms. Complete maps were constructed from at least six array placements, extending the equilateral triangular sampling grid over an area of approximately 10 × 12 cm on the occipital and left parietal surfaces of the head. Labels indicate positive and negative extrema for 100 and 340 ms components. Amplitude scales are in units of femtotesla (10^{-15} T).

procedures, and the predicted field distribution for the empirical sampling matrix was calculated. Calculated maps are shown in column 2. Fields observed at 280 ms were fit by a two-dipole model.

In most of these examples, a single equivalent current dipole accounts for a reasonable percentage of the variance (average = 85%; range = 76–93%) in the field maps. A distinct apparent source is associated with each of the time slices chosen to correspond to "component" peaks in the ERP waveform. Early sources (through ap-

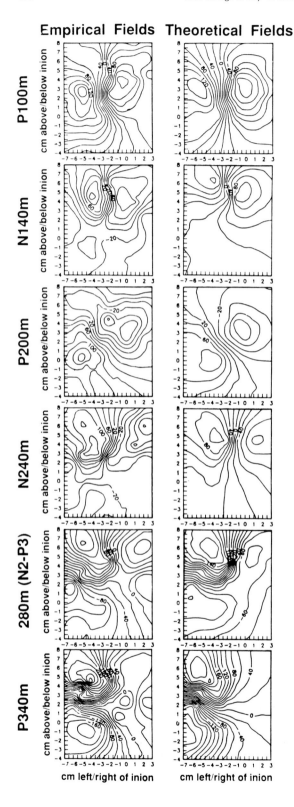

proximately 260 ms) appear to be in or very near occipital cortex. However, in field maps spanning larger regions of the head surface we found evidence for other, non-occipital sources with activity overlapping the early occipital sources (unpublished observations). The calculated source of the P1 is near the source of the P2; N1 and N2 sources are also similar. The 340 ms source (which becomes apparent at 280 ms) appears to be parietal in origin; this source was much stronger in attended than in non-attended stimuli.

5.1. Visual field effects

We examined neuromagnetic field maps associated with the P1 ERP component in all subjects. These maps disclosed magnetic field distributions consistent with a current dipole source in the left hemisphere when the right visual field was stimulated and a right hemisphere source when the left visual field was stimulated. The estimated neuromagnetic sources were located on MRI scans for each subject. In all cases this procedure suggested a contralateral source in occipital cortex, although precise locations varied between individuals. In other experiments we compared responses to central versus peripheral (right visual field, RVF) stimulation. Peripheral stimuli consistently produced deeper source fits for the earliest response

◂ Fig. 2. Left column: Neuromagnetic (isocontour) field maps for subject CA, evoked by an attended stimulus at 2° in the right visual field. Maps are Mercator (surface) projections reconstructed from field amplitude measures at a number of sensor locations over the head surface. The inion, a small boney projection at the posterior base of the scull is located at $x = 0$, $y = 0$ in these plots. Maps are at times corresponding to component peaks in the ERP temporal waveform. Right column: Calculated field distributions based on fitted parameters for a single equivalent current dipole (ECD) model. The observed field distribution at 280 ms was fit with two simultaneous dipole sources.

components, consistent with one of the qualitative predictions of the cruciform model, as well as with previous neuromagnetic data [35, 51]. At later times we often observed a more lateral source; this might reflect the cortical topology of one or more secondary visual areas.

In many sequences of field maps we were able to resolve multiple neuromagnetic sources with distinct temporal envelopes. For example, fig. 3 illustrates selected time slices for a stimulus presented in right visual field, near but not on the central fixation point. From the initial map for the stimulus we inferred a source in the hemisphere contralateral to the visual field of stimulation. In later maps we detected evidence for ipsilateral activation: a second dipole-like source which became apparent within 20 ms of initial (observed) cortical activity. With 40 ms of the initial response, the strength of these sources was well balanced so that field distributions associated with right or left visual field stimuli appeared similar. The contralateral projection of the visual field was expected based on human neuroanatomy and animal single unit electrophysiology. The subsequent ipsilateral activation may reflect interhemispheric transfer of information by the fibers of the corpus callosum, which connect regions near the cortical projection of the vertical meridian of the visual field.

In the case above, the intermediate field map clearly reflects the existence of multiple simultaneous sources. While it is possible to fit such a map in isolation using a two-dipole model, the distribution may be ambiguous. The fitting procedure will

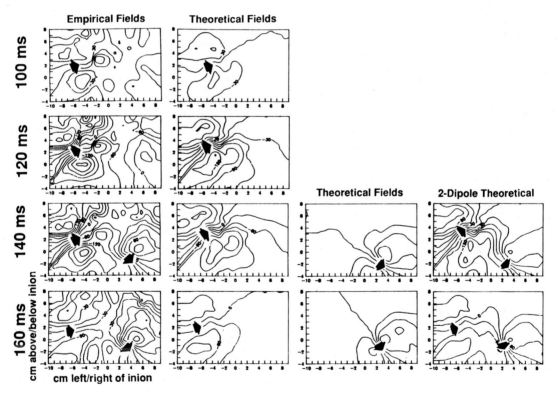

Fig. 3. A temporal sequence of field maps for subject MO evoked by a stimulus at 2° in the right visual field. Note that these maps span a larger region of right hemisphere than the previous example. Top two maps were fit with a single ECD source. Lower maps were fit with two simultaneous dipoles, and appear to reflect the delayed activation of a secondary source in ipsilateral occipital cortex.

converge on one of several possible dipole pairs, depending on the initial parameter estimates. In this case we can make a parsimonious selection between these possibilities, based on simpler field distributions observed at earlier or later times.

In other cases a single dipole-like field distribution was observed over a number of time steps, but the location and orientation of the apparent source evolved in time. Fig. 4 illustrates a sequence of ERF maps obtained in one subject in response to a 1 cpd grating presented at 2° in the right visual field. Poststimulus latency is indicated below each map. From 90–140 ms the location and strength of field extrema appeared to systematically shift. A similar sequence was observed from 160 to 240 ms. At around 280 ms another dipole-like source became apparent. This source was much stronger in attended that non-attended stimuli.

The top panel in fig. 5 illustrates the location and orientation of calculated sources for the interval 90–140 ms poststimulus. Evoked field maps for each 10 ms interval were fit with the single-dipole model. Equivalent dipole source locations and orientations are illustrated in orthographic projections (upper panels). Note the apparent path of activation suggested by successive apparent sources. In the head centered coordinate system used for these studies, the $-x$ axis is centered on the inion (a boney protrusion near the rear of the skull), the $+y$ axis passes through a reference point on the left hemisphere ≈ 2 cm above and in front of the ear canal, and the $+z$ axis exits the top of the head.

Because confidence intervals for source calculations are difficult to determine analytically, Monte Carlo techniques were used to assess the potential scatter in calculated sources due to measurement noise [52]. Noise with a distribution estimated from the prestimulus baseline was added to model amplitude measures for each sensor, and the resulting field distribution was fitted. Fifty source calculations for each timepoint are illustrated in the lower panel of fig. 5. Note that even with measurement noise, modeling procedures were able to resolve at least three distinct (non-overlapping) sources. Recent studies by Kaufman and coworkers [53] suggest that brain noise is diminished during evoked activity so that our estimate is probably an upper limit. This procedure does not account for errors in absolute localization due to uncertainty of sensor position with respect to the head; however, the experimental design insures that relative differences in source locations can be reliably determined.

Evolving field patterns observed from 90 to 140 ms might reflect a single migrating source; however, they might also be explained by the combination of field distributions associated with two or more discrete, temporally overlapping sources. Fig. 6 summarizes model studies designed to test whether observed field distributions might be explained by a combination of two distinct sources. Neuromagnetic field maps in the upper panels

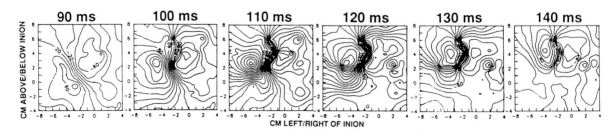

Fig. 4. A temporal sequence of field maps for subject CA evoked by a stimulus at 2° in the right visual field, from the same data set illustrated in fig. 2. Note the shift of field extrema apparent in the sequence.

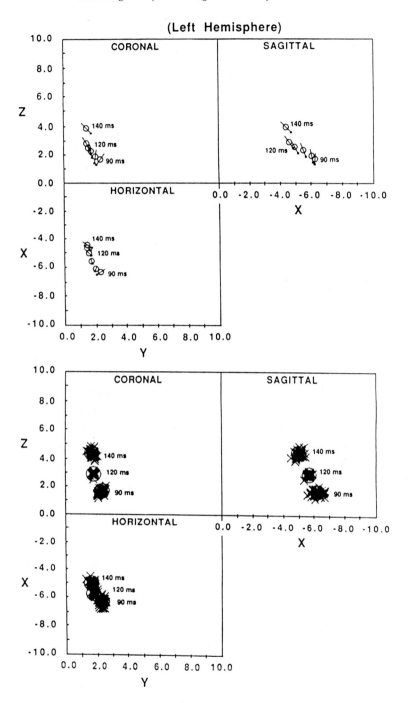

Fig. 5. Top panels: Orthographic projections of the location and orientation of ECD sources calculated from a single dipole model fit to field maps at 10 ms intervals (the data set illustrated in fig. 4). Note the apparent path of source migration suggested by this analysis. Lower panels: Scatter plots in orthographic projection of source locations calculated in Monte Carlo simulations of the effect of measurement noise. Each cluster is an ensemble of 50 calculations on field distributions derived by adding noise with a distribution estimated from the prestimulus baseline to the observed field maps at 90, 120 and 140 ms poststimulus. The stimulus is at 2° in the right visual field.

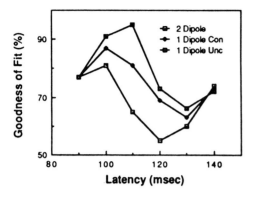

Fig. 6. Computer modeling to account for the sequence of field maps observed from 90 to 140 ms. The graph compares the percent of field variance accounted for by each of three models: the unconstrained single-source model; a single-source model, in which model ECD parameters were constrained to vary linearly between 90 and 140 ms values; and a two-dipole model using an optimal combination of 90 and 140 ms sources.

Table 1
Three source constrained fits (percent of empirical variance accounted for by model). A comparison of three classes of source model: an unconstrained single dipole fit to each field map; a single dipole constrained to vary linearly between specified locations and orientations; and a linear combination of two specified dipoles. In the upper half of the table, specified dipole pairs were best fitting single dipoles for 90/120 and 120/140 ms field maps. In the lower half, dipole pairs were 90/110 and 110/140 ms.

	Latency (ms)					
	90	100	110	120	130	140
original fit	–	92%	85%	–	66%	–
constrained 1-dipole	–	89%	85%	–	65%	–
constrained 2-dipole	–	89%	83%	–	65%	–
original fit	–	92%	–	73%	66%	–
constrained 1-dipole	–	89%	–	72%	65%	–
constrained 2-dipole	–	89%	–	70%	65%	–

illustrate the modeling process for a representative timepoint: 110 ms. The first panel is the empirical field map. The second panel is a single-dipole model where dipole location and orientation is a weighted average of fitted parameters for 90 and 140 ms field distributions. The third panel is two-dipole model where the resulting distribution is a weighted linear combination of 90 to 140 ms theoretical distributions. The weighting ratio and field amplitude multiplier were adjusted to produce optimal fits in each case. The lower panel compares goodness of fit measures between constrained single and double dipole sources over the range of times. The single migrating source model was the better explanation for observed intermediate field distributions.

The superiority of the single-dipole model lead us to ask whether three distinct, temporally overlapping sources could better explain the 90–140 ms evoked field distribution. Our modeling strategy was similar to that described above. 90 and 140 ms calculated sources were accepted as endpoints. The intermediate source was taken as the optimal single-dipole fit for 110 or 120 ms timepoints. Table 1 lists the percent of variance accounted for by each of three models applied to selected latencies: the original unconstrained single-source model, the single source with linearly constrained parameters, or the optimal linear combination of two (specified) dipole sources. Note that there is little difference in goodness of fit between constrained single-source and dual-source models, particularly when 120 ms was accepted as the intermediate source. The unconstrained single-source model performed slightly better; however, this may reflect a less than optimal selection of intermediate source parameters used in constrained fits. Thus while the observed field distributions may reflect a continuously migrating focus of neural activation, the data are reasonably explained by three discrete, temporally overlapping sources.

Neuromagnetic sources suggested by these modeling procedures were located on magnetic resonance images. Calculated sources for 90, 120 and

140 ms evoked field distributions are illustrated in fig. 7. Locations of slices are indicated by crosshairs. The silhouette was produced from the summed thresholded images of each of the slices in a series and are provided as a visual coordinate reference. We estimate the accuracy of sensor location measurements used in this study to be ±0.5 to 1.0 cm, better near the inion. Improved procedures will allow millimeter resolution. Even given the present degree of uncertainty, suggested sources are anatomically reasonable. The 90 ms source appears to lie along the calcarine fissure (primary visual cortex). The 120 ms equivalent source is deeper, located near the anterior extent of striate cortex, possibly in area 18. The 140 ms source is higher and may lie along the parietal occipital sulcus.

The general patterns of response observed in the case described above were consistently seen across subjects and under a variety of experimental conditions. We have begun to extend these spatial/temporal analyses to other subjects. During the initial cycle of activation ($\approx$ 90–150 ms), the apparent path of activation is typically curved, suggesting at least three sources must be involved. In most subjects the second cycle of activation (180–240 ms) follows a similar path, suggesting repetitive activation of the same anatomical structures.

5.2. Effects of visual field and spatial frequency

Experiments involving the manipulation of spatial frequency were intended to serve (at least) two purposes. First, we sought evidence for the existence of spatial frequency selective channels in the human visual pathway. Second, we considered the study to be a test of the capability of the neuromagnetic technique to distinguish functional differences beyond those of gross anatomical projection.

We examined amplitudes and source moment of neuromagnetic responses as a function of visual field and spatial frequency of stimulation. In each

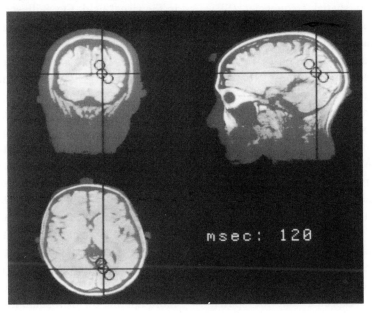

Fig. 7. Neuromagnetic sources at 90, 120 and 140 ms located on MRI. Calculated source locations are indicated by circles; the 90 ms source is the lowest in all three views. Slices were selected and pixel coordinates were calculated by software after reference points were identified in each series of images. Locations of selected slices are indicated by crosshairs. Silhouttes are provided as a visual coordinate reference. Note that by radiological convention, computed sections are displayed as though viewed from in front of (coronal), below (axial or horizontal), or to the right of (sagittal) the head.

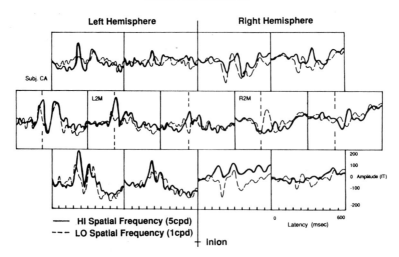

Fig. 8. Differences in evoked field waveforms as a function of spatial frequency of the sinusoidal grating stimulus. Waveforms are from two contiguous sensor array positions spanning the midline of the head, over occipital cortex.

field map we were able to identify a pair of adjacent sensor locations which contained extrema for an initial response component for both 5 cpd (HI) and 1 cpd (LO) spatial frequencies (SF). Waveform amplitudes were sampled at selected timepoints and averaged across these two locations. HI SF elicited larger responses in CVF while LO SF produced larger responses for RVF presentations. This trend was also observed for source moments (a measure of dipole strength) calculated on the basis of dipole model fits for latencies of 90–130 ms. HI SF elicited a stronger early response in CVF while LO SF elicited a stronger and earlier response in RVF. These observations were consistent with our predictions based on the variation of receptive field size across the retina.

When examining ensembles of waveforms, we expected to see differences in amplitudes and peak latencies of specific components as a function of spatial frequency. We expected that the spatial distribution of peaks would be a function of the anatomical locations of sources, and would therefore reflect only the retinotopic location of the visual stimulus, at least within primary visual areas. We were surprised to observe differential effects of spatial frequency on the shapes and spatial distributions of early ERF response components elicited by stimuli presented at the same location in the visual field. Fig. 8 illustrates averaged waveforms for one subject collected at two adjacent array locations over occipital cortex, when the central visual field was stimulated. Solid lines are responses to HI SF. Dashed lines are responses to LO SF. Such differences as a function of spatial frequency were reproducible and were consistently observed across subjects.

We also observed corresponding differences in ERF maps as a function of spatial frequency. Plate I presents field maps for subject JMG at indicated response latencies, when the central field was stimulated. Left-hand maps are for LO SF CVF stimuli; right-hand maps are HI SF CVF. All maps are displayed on the same coordinate grid. Upon initial examination, field maps for different spatial frequencies in one location were more notable for their similarities than differences. However, in this and other subjects, significant differences were consistently observed in the distributions of earliest apparent components (90–100 ms), and in response components appearing at around 190 ms. For this and other subjects,

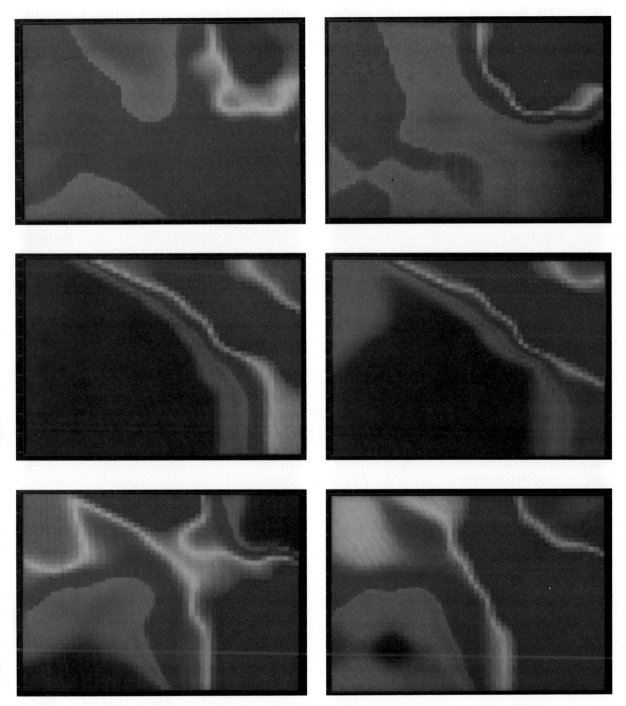

Plate I. Field maps for subject JMG at selected time points illustrating effects of spatial frequency on evoked field distributions. Left-hand maps are for LO SF CVF stimuli and right-hand maps for HI SF CVF stimuli. The first row is at 90 ms, the second row at 160 ms and the third at 200 ms. The maps are surface projections spanning an area 8 cm left and 4 cm right of, and 8 cm above and 2 cm below the inion, a boney projection near the base of the pasterior skull. Red signifies magnetic flux emerging from the head while blue signifies flux re-entering.

HI and LO SF stimuli produced very similar response maps for latencies from ≈ 130–180 ms.

By analyzing time sequences of response maps for several subjects showing differences as a function of spatial frequency, we identified three apparent reasons for discrepant maps. In the earliest maps (typically 80–100 ms) there appeared to be a difference in the relative strength or number of dipole sources. In other subjects (not illustrated here) HI and LO spatial frequencies produced qualitatively similar field maps, but Monte Carlo analyses disclosed significant differences in the location and/or orientation of a single apparent source. Based on analyses previously described, such a finding might reflect differential activation of two or more simultaneous sources with overlapping field distributions. By 120 ms these differences had largely disappeared.

In some subjects significant differences were again observed around 130 ms; however, subsequent time slices suggest that these simply reflected response latency differences as a function of spatial frequency. Observed distributions at ≈ 150 ms were similar for HI and LO spatial frequency. By 190 ms differences in the number, extent, or locations of negative peaks suggest a significant difference in the location and orientation or relative strengths of equivalent dipole sources as a function of spatial frequency.

6. Discussion

Neuromagnetic techniques permit high-resolution spatial and temporal characterization of patterns of neural activity evoked by sensory stimulation and allow discrimination of differences in evoked responses due to major (visual field) or more subtle (spatial frequency) manipulations of visual stimuli. As expected from retinotopic organization of striate cortex, we observed significant differences in ERF distribution as a function of field of stimulation. Initial evoked sources consistently appeared in the hemisphere contralateral to the field of stimulation. As predicted from classical models of primary visual cortex organization, early responses to peripheral stimuli appeared to originate deeper in cortex than responses to central stimuli. For stimuli near (but not spanning) the midline of the visual field we were able to detect subsequent ipsilateral activation, possibly mediated by the corpus collosum. Such interconnection would be expected on the basis of animal studies, and presumably prevents perceptual discontinuities which might otherwise arise. For example, motion across the midline of the visual field is perceived as continuous; an object in the central visual field can be recognized on the basis of a gestalt of features in the left and right hemispheres. While appropriate correlations might be extracted at a higher processing level, such a strategy could introduce perceptual delays for central vision, and would establish boundaries which could distort local processing.

Although we were able to estimate equivalent current dipole sources for neuromagnetic field distributions associated with each of the identified components of the ERP waveform, we have discovered that analyses which focus on peaks in the temporal waveform may be misleading. For example, source locations calculated for the P1 and N1 ERP components suggest two discrete sources; however, field maps spanning the P1–N1 components display a pattern of temporal evolution consistent with a migrating focus of activation. Alternatively, such sequences may reflect a complex of spatially discrete, temporally overlapping sources. In the example discussed here, at least three sources must be postulated to adequately explain the data. Simpson et al. [54], using a multiple-dipole model for ERP data, have postulated eight dipole sources with sequential, overlapping activation to account for a time sequence spanning 60–280 ms poststimulus. Based on our initial spatial/temporal source analyses of other subjects we conclude that at least three sources must be postulated to explain the initial cycle of observed activity (≈ 90–150 ms). In most subjects a second cycle of activation (180–240 ms) follows a similar anatomical path.

In these experiments, we estimate our error in the measurement of sensor location to be ±0.5 cm. Although this degree of uncertainty prevents unequivocable identification of the anatomical substrates of the evoked responses, the most parsimonious explanation for our observations is the sequential activation of several areas within occipital cortex. The initial calculated source was near the calcarine fissure in striate cortex, consistent with activation of primary visual cortex. Subsequent sources were in regions suspected to be secondary visual areas, on the basis of anatomical analogy with experimental primates and studies of perceptual effects of human lesions. In any case, our observations support the idea that human visual information processing is sequential to some extent, and distributed across a number of anatomically discrete loci.

If this sequence of apparent sources reflects the hierarchy of visual areas, it might be possible to dissect the pathway further through manipulation of visual stimulus parameters. For example, although retinotopic order is present in both V1 and V2, the precise pattern of projections and location and topology of the areas is significantly different. Also, a tendency toward progressive functional specialization is apparent in the characteristic information processing activities associated with secondary visual areas. Higher-order systems may discard some types of information, in order to effectively perform specialized processing tasks. Single unit recordings in monkeys have suggested that processing of color information is a primary function of area V4, and that area MT is particularly involved in processing motion in the visual field [8, 11, 12]. By manipulating the information content of stimuli we may be able to differentially activate such specialized areas, and trace the flow of information through the distributed system.

In this report, we described some consequences of manipulating the spatial frequency of stimuli at varying locations in the visual field. We chose to examine spatial frequency for a number of reasons. The size of ganglion cell receptive fields varies across the retina with larger fields in the periphery; this difference alone should establish differential sensitivity to spatial frequencies. There is also a systematic difference in the distribution of cell types across the retina. Finally, psychophysical and evoked potential studies have suggested but do not prove the existence of specialized spatial frequency channels within the visual system. In our experiments, response amplitudes, latencies and calculated current dipole moments varied as a function of eccentricity of the grating within the field and demonstrated an expected statistical interaction between spatial frequency and visual field. CVF "preferred" higher spatial frequency gratings, while RVF "preferred" LO, although responses to the LO SF grating were strong for both fields of stimulation.

We also observed differences in neuromagnetic field *distributions* as a function of spatial frequency. Based on hypotheses about the consequences of retinal organization on cortical responses, we had expected to find differences in the latencies of components of their relative strengths as a function of spatial frequency and visual field location of the experimental stimuli. We did not expect to find differences in the shapes or characteristic features of evoked field distributions as a function of spatial frequency, since we expected that source location and orientation would be dominated by retinotopic order of the cortex. However, such differences were consistently seen. These appeared to reflect differences in the number and/or relative strength of dipole sources, differences in component latencies, and differences in the location and orientation of sources.

Observed differences in evoked field maps might reflect the preferential activation of distinct subsystems within the primary visual pathway as a function of spatial frequency. Recent studies of non-human primates have provided evidence for two distinct streams of visual processing which exist in parallel through areas V1 and V2: the "parvocellular" and "magnocellular" systems. Differences between the systems begin at the retina with anatomically and functionally distinct types of ganglion cells (types A and B); type A cells are larger, have larger receptive fields and respond more transiently than type B cells. The cellular pathways are anatomically segregated into interleaved laminae at the LGN (type A cells project to magnocellular layers, type B to parvocellular) which project to anatomically segregated structures within the cortex. The magnocellular system operates at lower spatial resolution and is insensitive to color; however, it is substantially more sensitive to luminance contrast and to temporal aspects of a stimulus. Parvocellular neurons in the LGN are typically sensitive to color contrast, and this sensitivity is maintained in a subsystem of the cortical parvocellular pathway. Consistent with a proposed role in high-resolution form perception, the parvocellular system incorporates around 90% of cortical neurons. So far, our studies involving manipulation of stimulus luminance contrast is consistent with this interpretation; however, at this stage of analysis it is not clear whether observed differences arise within or between visual areas.

In lower animals, sensory information processing is often highly distributed and highly specialized. For example, the retina of the fly performs motion processing operations which are handled by much later stages of the human visual pathway. Because humans perform so many separate analyses of the visual data stream, we might assume that early processing would be generic. However, our studies of spatial frequency processing clearly suggest that specialized and anatomically segregated representations of the visual world exist in parallel at relatively low levels within the human visual system. These specialized subsystems must reflect a (local) optimal solution to the problem of encoding and processing visual information.

Acknowledgements

The authors wish to thank Ivan Bodis-Wollner for his collaboration in the design and execution of some of the spatial frequency experiments referred to here, and for his insightful discussions of their interpretation; Chris Wood for useful editorial oversight; and a cast of volunteer subjects who contributed their time and patience in extended mapping procedures. The work was supported by the US Army Research Institute, Department of Energy Contract W-7405-ENG-36, and the VA/LANL/UNM magnetoencephalography program.

References

[1] D. Van Essen, Ann. Rev. Neurosci 2 (1979) 227–263.
[2] D.C. Van Essen and J.H.R. Maunsell, Trends in Neurosci. (1983) 370–375.
[3] A. Kaneko, Ann. Rev. Neurosci. 2 (1979) 169–191.
[4] H.B. Barlow, Proc. R. Soc. London B 212 (1981) 1–34.
[5] R. Linsker, Proc. Natl. Acad. Sci. US 83 (1986) 7508–7512.
[6] R. Linsker, Proc. Natl. Acad. Sci. US 83 (1986) 8390–8394.
[7] R. Linsker, Proc. Natl. Acad. Sci. US 83 (1986) 8779–8783.
[8] P. Lennie, Vis. Res. 20 (1980) 561–594.
[9] S.M. Zeki, Nature. 274 (1978) 423–428.
[10] D.H. Hubel and T.N. Wiesel, J. Physiol. 160 (1962) 106–154.
[11] D.H. and T.N. Wiesel, J. Neurophysiol. 28 (1965) 229–289.
[12] M.S. Livingston and D.H. Hubel, Science 240 (1988) 740–749.
[13] D.C. Van Essen, in: Cerebral Cortex Vol. 3, eds. A. Peters and E.F. Jones (Plenum, New York, 1985) pp. 259–329.
[14] R.W. Rodieck, Ann. Rev. Neurosci. 2 (1979) 193–225.
[15] K.A. Macko, C.D. Jarvis, C. Kennedy, M. Miyaoka, M. Shinohara, L. Sokoloff and M. Miskin, Science 218 (1982) 394–397.
[16] M. Wilson, in: Handbook of Behavioral Neurobiology, Vol. 1, ed. R.B. Masterton (Plenum, New York, 1978) pp. 209–247.
[17] Bodis-Wollner, ed., Ann. NY Acad. Sci. 388 (1982).
[18] J.E. Desmedt, Visual Evoked Potentials in Man: New Developments (Clarendon Press, Oxford, 1977).
[19] B.B. Boycott and J.E. Dowling, Philos. Trans. R. Soc. London 255 (1969) 109–184.
[20] G. Wald, Science 162 (1968) 230–239.
[21] S. Hecht, S. Shlaer and M.H. Pirenne, J. Gen. Physiol. 25 (1942) 819–840.
[22] D.A. Baylor, T.D. Lamb and K.-W. Yau, J. Physiol. (London) 288 (1979) 589–611.
[23] G.L. Fain, J. Physiol. (London) 261 (1976) 71–101.
[24] J.A. Coles and S. Yamane, J. Physiol. (London) 247 (1975) 189–207.
[25] J.K. Bowmaker and H.J.A. Dartnall, J. Physiol. (London) 298 (1980) 501–511.
[26] P.K. Brown and G. Wald, Science 144 (1964) 45–52.
[27] W.B. Marks, W.H. Dobelle and E.F. MacNichol Jr., Science 143 (1964) 1181–1183.

[28] E.T. Rolls and A. Cowey, Exp. Brain Res. 10 (1970) 298–310.
[29] V.H. Perry and A. Cowey, Vis. Res. 25 (1985) 1975–1810.
[30] O.D. Creutzfeldt, in: Biophysics, eds. W. Hoppe, W. Loman, H. Markl and H. Ziegler (Springer, Berlin, 1983).
[31] S.W. Kuffler, J.G. Nicholls and A.R. Martin, From Neuron to Brain (Sinauer, Sunderland, MA, 1984).
[32] E.L. Schwartz, Science 227 (1985) 1066.
[33] P.M. Daniel and D. Whitteridge, J. Physiol. (London) 159 (1961) 203–221.
[34] A. Cowey, J. Neurophysiol. 27 (1961) 366–393.
[35] D.A. Jeffreys and J.G. Axford, Exp. Brain Res. 16 (1972) 1–21.
[36] T.M. Darcey, J.P. Ary and D.H. Fender, Prog. Brain Res. 54 (1980) 128–134.
[37] Y. Fukada and J. Stone, J. Neurophysiol. 37 (1974) 749–772.
[38] F.M. De Monasterio and P. Gouras, J. Physiol. (London) 251 (1975) 167–195.
[39] O. Braddick, F.W. Campbell and J. Atkinson, in: Handbook of Sensory Physiology, Vol. 8, eds. R. Held, H.W. Leibowitz and H. Teuber (Springer, Berlin, 1978).
[40] B.G. Breitmeyer and L. Ganz, Psychol. Rev. 83 (1976) 1–36
[41] L. Maffei, in: Handbook of Sensory Physiology, Vol. 8, eds. R. Held, H.W. Leibowitz and H. Teuber New York: (Springer, Berlin, 1978).
[42] C. Blakemore and F.W. Campbell, J. Physiol. (London) 203 (1969) 237–260.
[43] F.W. Campbell and L. Maffei, J. Physiol. (London) 207 (1970) 635–652.
[44] C. Blakemore, J.P.J. Muncey and R.M. Ridley, Vis. Res. 13 (1973) 1915–1931.
[45] R.L. DeValois, H. Morgan and D.M. Snodderly, Vis. Res. 14 (1974) 75–81.
[46] D.J. Tolhurst, Vis. Res. 15 (1975) 1143–1149.
[47] M.R. Harter, V.L. Towle and M.F. Musso, Vis. Res. 16 (1976) 1111–1117.
[48] R. Blake and E. Levinson, Exp. Brain Res. (1977) 221–222.
[49] C. Aine, J. George, P. Medvick, S. Supek, E. Flynn and I. Bodis-Wollner, in: Advances in Biomagnetism, eds. S.J. Williamson, M. Hoke, M. Kotani and G. Stroink (Plenum, New York), in press.
[50] J.S. George, P.S. Jackson, D.M. Ranken and E.R. Flynn, in: Advances in Biomagnetism, eds. S.J. Williamson, M. Hoke, M. Kotani and G. Stroink (Plenum, New York), in press.
[51] E. Maclin, Y.O. Okada, L. Kaufman and S.J. Williamson, Nuovo Cimento 2 (1983) 410–419.
[52] P.A. Medvick, P.S. Lewis, C. Aine and E.R. Flynn, in: Advances in Biomagnetism, eds. S.J. Williamson, M. Hoke, M. Kotani and G. Stroink (Plenum, New York), in press.
[53] B.J. Schwartz, C. Salustri, L. Kaufman and S.J. Williamson, in: Advances in Biomagnetism, eds. S.J. Williamson, M. Hoke, M. Kotani and G. Stroink (Plenum, New York), in press.
[54] G.V. Simpson, M. Scherg, W. Ritter and H.G. Vaughan Jr., Paper presented at EPIC IX: International Conference on Event-Related Potentials of the Brain, May 28, 1989, Noordwijk, The Netherlands.

COMPUTATIONAL CONNECTIONISM WITHIN NEURONS: A MODEL OF CYTOSKELETAL AUTOMATA SUBSERVING NEURAL NETWORKS

Steen RASMUSSEN[a,1], Hasnain KARAMPURWALA[b], Rajesh VAIDYANATH[b], Klaus S. JENSEN[c] and Stuart HAMEROFF[d,1]

[a]*Center for Nonlinear Studies and Theoretical Division (T-13), MS-B258, Los Alamos National Laboratory, Los Alamos, NM 87545, USA*
[b]*Department of Electrical and Computer Engineering, University of Arizona, USA*
[c]*Department of Anesthesiology, University of Colorado, USA*
[d]*Advanced Biotechnology Laboratory, Department of Anesthesiology, University of Arizona College of Medicine, Tucson, AZ 85724, USA*

"Neural network" models of brain function assume neurons and their synaptic connections to be the fundamental units of information processing, somewhat like switches within computers. However, neurons and synapses are extremely complex and resemble entire computers rather than switches. The interiors of the neurons (and other eucaryotic cells) are now known to contain highly ordered parallel networks of filamentous protein polymers collectively termed the cytoskeleton. Originally assumed to provide merely structural "bone-like" support, cytoskeletal structures such as microtubules are now recognized to organize cell interiors dynamically. The cytoskeleton is the internal communication network for the eucaryotic cell, both by means of simple transport and by means of coordinating extremely complicated events like cell division, growth and differentiation. The cytoskeleton may therefore be viewed as the cell's "nervous system". Consequently the neuronal cytoskeleton may be involved in molecular level information processing which subserves higher, collective neuronal functions ultimately relating to cognition. Numerous models of information processing within the cytoskeleton (in particular, microtubules) have been proposed. We have utilized cellular automata as a means to model and demonstrate the potential for information processing in cytoskeletal microtubules. In this paper, we extend previous work and simulate associative learning in a cytoskeletal network as well as assembly and disassembly of microtubules. We also discuss possible relevance and implications of cytoskeletal information processing to cognition.

1. Cognition: emergent computation in the brain

1.1. Connectionism

The use of connectionist models of functional brain organization has advanced understanding of cognition and linked neuroscience to computer science. "Neural networks", which approximate cortical neurons as parallel processors with variable lateral interconnections, may be simulated on computers and can provide a working model of some aspects of brain function [32, 41]. A key concept relating neural network dynamics to neuroscience was introduced by Hebb [40]. He hypothesized that spatially organized neural networks called "cell assemblies" function as reverberatory circuits which constituted elements of thought. Gazzaniga's [30] "modules", Minsky's [55] "agents", Freeman's [26] "cartels" and other conceptualized functional entities are examples of more recent, comparable proposals for anatomic neural networks. In Hebb's view, an individual neuron could participate in many cell assemblies just as an individual member of society may participate in many social groups. By strengthening or reinforcing repeatedly used connections ("synaptic plasticity"), Hebb suggested that recognition, learning and problem solving occurred

---

[1] To whom all correspondence should be sent.

through lowered thresholds of specific loops. By assigning energy levels to threshold loop patterns ("landscapes"), mathematical solutions could be applied to neural net configurations [41].

There are, however, at least two inconsistencies in the analogy between neural nets and brain function. The first is that neurons and their synapses are extremely complex and by themselves resemble computers more than fundamental switches. Accordingly, further attempts to approach mechanisms of brain cognition need to consider dynamic regulatory activities within each neuron. The second is that a number of important artificial neural networks require "back propagation": tuning of internal parameters from output conditions. Both of these inconsistencies may be resolved by consideration of the cytoskeleton, an intracellular parallel network of protein polymers which regulate synapses and performs other important functions.

We contend that rudimentary information processing occurs within and among intraneuronal cytoskeletal elements such as microtubules (MTs) (fig. 1). By subserving synaptic connectionism, such information processing can provide a lower level in a hierarchy of cognitive processes from whose highest level emerges consciousness. Further, retrograde patterns travelling through the cytoskeleton could serve a role analogous to back-propagation in some artificial neural networks. In sections 4 and 5, we describe simulation of such retrograde patterns travelling through microtubules.

In the cytoskeleton, filamentous bridges among parallel MTs and/or neurofilaments could serve functions comparable to the recognition automata of Reeke and Edelman [61]. To take maximal advantage of parallelism, these authors have modelled two parallel automata which communicate laterally and have distinct and complementary personalities. One model automaton ("Darwin") is highly analytical, keyed to recognizing edges, dimensions, orientation, color, intensity, etc. The other ("Wallace") is more "gestalt" and attempts to merely categorize objects into preconceived

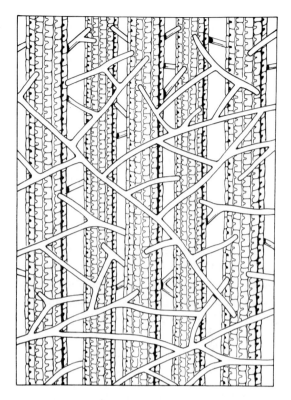

Fig. 1. Connectionist network of five parallel arrayed microtubules (MTs) interconnected by lateral crosslink filaments. Drawn by Fred Anderson from electron micrograph of neuronal dendrite cytoplasm (Aoki and Siekevitz [5]).

classifications. Lateral communications between the two parallel automata resolve conflicting output and form an associative memory. In the cytoskeleton, filamentous bridges among parallel MTs and/or neurofilaments could serve comparable functions.

Multi-level nets are consistent with connectionist brain models in which neural net "assemblies", "modules", etc. fit in an overall cognitive hierarchy of parallel layers of brain/mind organization [13, 20]. The highest cognitive level in such models is global brain function which correlates with awareness, thought, or "consciousness" (although some would argue for even higher-level social consciousness among people, political parties, societies, etc. [42, 59]). The second level appears to be comprised of anatomically and functionally recognizable brain systems and centers (i.e. respiratory center, satiety center, etc. [66]). At intermediate levels, maps such as the motor and sensory homunculi represent the body and outside world. Finally, the lowest level is thought to be the neural synaptic network, module or cartel which may operate cooperatively by utilizing dense interconnectedness, parallelism, associative memory and learning due to synaptic plasticity.

The lowest level of brain/mind organizational hierarchy, and potential correlate of the binary switch in computers, is generally considered to be the neuronal synapse. However, we suggest that information processing in the cytoskeleton could provide a basement level to the cognitive hierarchy. The cytoskeleton is a parallel, interconnected network of microtubules, actin, and intermediate filaments which also connects to membrane proteins, cell nuclei, centrioles and other structures. Capabilities for dynamic organization coupled with the cytoskeleton's grid-like lattice structure have prompted at least a dozen author groups to model microtubules and other cytoskeletal structures as information processing devices [35]. In neurons, synaptic connectionism and regulation (plasticity) depend on cytoskeletal functions as do other cognitive processes. Information representation, transduction, translocation (including retrograde signaling suitable for "back-propagation") and computation in the cytoskeleton may function just below synaptic neural networks in a cognitive hierarchy.

The next sections will describe: (1.2) evidence linking the cytoskeleton to computation, (2) a biochemical overview of microtubules and other cytoskeletal elements, and (3) models of cytoskeletal information processing. In section 3.3, our previously published model of molecular automata in microtubules [39] is reviewed. Section 4 describes a network model of connected microtubule automata capable of learning and association and section 5 covers an assembly/disassembly microtubule automata model relevant to cell division, growth and plasticity. Section 6 concludes with a discussion of the potential implications of cytoskeletal computation.

1.2. Cytoskeleton and computation

The concept of cytoskeletal computation is based on the observation that cytoplasmic interiors of living cells, particularly nerve cells, are not watery soups. The current picture of cell interiors is one of a highly organized network in which cell water is bound and governed by cytoskeletal structures rather than a freely diffusing solution of macromolecules. Two lines of evidence support this concept. The first is the cytoskeleton which interpenetrates the cytoplasm and defines its architecture and function. Some cytoskeletal components are in polymerization equilibrium, interconverting cytoplasm between "sol" (liquid, solution) and "gel" (gelatinous, viscous) states. This "sol/gel" equilibrium (mediated by Ca^{2+} and Mg^{2+} effects on cytoskeletal actin and other proteins) indicate that the interiors of living cells operate very close to the solid–liquid phase transition. Theoretical findings predict optimal computational capabilities for distributed dynamical systems near this phase transition [47, 58].

Another line of evidence supporting the possibility of a computational cytoplasm includes data from nuclear magnetic resonance (NMR), neutron diffraction and other techniques which show a high degree of bound water in the cytoplasm. Water molecules abutting cytoskeletal (and other) cytoplasmic surfaces are attracted to these surfaces and are restricted in their motions. As a result, a greater proportion has four (rather than three or less) hydrogen bonds with their neighbors. Such water is called "vicinal" water, and has unusual properties compared to "bulk" water: lower density, greater heat capacity and greater viscosity. Further, such vicinal water may cooperatively oscillate with coherent excitations in cytoskeletal proteins (section 3.2).

Further circumstantial evidence for cytoskeletal computation and communication stems from the spatial distribution of discrete sites/states in the cytoskeleton. For example, subunits in microtubules have a "density" of about 10^{17} per cm^3, very close to the theoretical limit for charge separation [34]. Thus, cytoskeletal arrays have maximal density for information storage via charge, and the capacity for dynamically coupling that information to mechanical and chemical events via conformational states of proteins (i.e. Fröhlich's mechanism of dipole excitations coupled to conformational state).

The dynamic complexity of neurons and their synaptic connections suggest that they utilize "lower" layers which may include dendritic processing, branch point conductance and molecular dynamics [14, 63]. Lynch and Baudry [51] have studied long-term potentiation ("LTP") of synaptic efficacy in glutamate–NMDA hippocampal neurons, a correlative model of learning. They find that LTP depends on rearrangement of subsynaptic cytoskeleton – a dynamic network of parallel arrayed, interconnected proteins including microtubules, actin, and intermediate filaments. Desmond and Levy [21] have studied dendritic spine structural changes during a synaptic learning paradigm. They find mechanical shape changes in spines mediated by cytoskeletal actin connected to microtubules in dendrites. Aoki and Siekevitz [5] have studied dendritic spine synaptic plasticity in development and find that synaptic down-regulation occurs in conjunction with depolymerization of microtubules in the main dendrite. They find also that signaling and regulation for dendritic spine synapse function depends on phosphorylation of a cytoskeletal protein: MAP2. Microtubule associated proteins (MAPs) attach to and connect microtubules to other microtubules, actin, intermediate filaments, organelles, membrane proteins and cell nucleus. MAP2 is found only in neuronal dendrites, and consumes a large proportion of phosphorylation energy. MAP2 phosphorylation is regulated by cyclic AMP-dependent protein kinase and calcium–calmodulin protein kinase. Theurkauf and Vallee [68] have found MAP2 to be "the major substrate for endogenous cyclic AMP-dependent phosphorylation in cytosolic brain tissue", and concluded that MAP2 phosphorylation "may be an important reaction in response to neurotransmitter stimula-

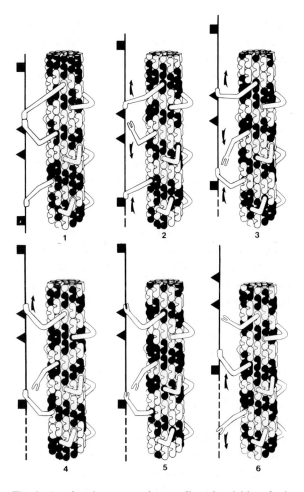

Fig. 2. Axoplasmic transport by coordinated activities of microtubule attached contratile (MAP) proteins ("dynein"). The orchestration mechanism is unknown, but shown here as the consequence of traveling conformational patterns on the MTs. By Fred Anderson.

tion". Thus, numerous mechanisms exist for coupling membrane synaptic events and cytoskeletal activities.

Maintenance and modulation of synaptic function also depend on axoplasmic transport, an MT based conveyor system which supplies structural components to synapses [48, 57]. Contractile "sidearm" MAPs attached at specific sites on microtubules pass materials in a cooperative bucket brigade which achieves a velocity of 400 mm/day (fig. 2). The contractile sidearms hydrolyze ATP for energy, but the mechanism of orchestration and signaling is unknown. Material can travel in opposite directions along a single microtubule. Other organized activities in cytoskeletal structures help control the molecular machinery of cell division, growth, differentiation, formation of synapses and dendritic spines, as well as relatively complex functions within single cell organisms like amoeba and paramecium which occur without benefit of brain, neuron, or synapse [22, 65].

Evidence for direct cytoskeletal involvement in cognitive processes comes from Mileusnic et al. [54], who have correlated production of MT subunit protein ("tubulin") and microtubule activities with peak learning, memory and experience in baby chick brain. Cronly-Dillon and co-workers [16] have shown that when baby rats first open their eyes, neurons in visual cortex begin producing vast quantities of tubulin. When the rats are 35 days old and the critical learning phase is over, tubulin production is drastically reduced. Other studies have shown that cytoskeletal proteins directly link to nerve membrane ion channels and receptors [46, 70] and that the intra-neuronal cytoskeleton is linked to nerve membrane excitability and synaptic transmission [2, 53]. Direct evidence for signal propagation in MT has been generated by Vassilev et al. [71], who suspended parallel excitable membranes in ionic solution. Only when two membranes were connected by MTs, excitation in one membrane provoked excitation in the other. The authors suggested that similar communication signals occurred routinely within the cytoskeleton. Also, Becker et al. [10] studied energy resonance transfer among fluorescent groups separately attached to different MT subunits or to membranes. They showed energy resonance transfer occurs both among MT subunits and among MT subunits and membrane proteins.

The intra-neuronal, sub-synaptic cytoskeleton is important to neural function including synaptic plasticity. In assessing cognition, low-level processing may be important. Because the cytoskeleton comprises a connectionist network within each branching neuron, it may be described as a cytoskeletal forest within each dendritic tree.

2. Microtubules and the cytoskeleton

2.1. Cytoplasm

Neurons and other cells are comprised of protoplasm, which in turn consists of membranes, organelles, nuclei, and the bulk interior medium of living cells: cytoplasm. Nineteenth century light microscopists described cytoplasm as containing or consisting of "reticular threads", "alveolar foam", or "watery soup". Development of the electron microscope through the 1960's did not initially illuminate the substructure of cytoplasm because the commonly used fixative (osmium tetroxide) was dissolving fine structure and the cell was still often perceived to be a "bag of watery enzymes".

In the early 1970's, with the event of glutaraldehyde fixation, delicate tubular filamentous structures were found in virtually all cell types and they came to be called microtubules. Originally thought to provide merely structural, or bone-like support, MTs and other filamentous structures such as actin, intermediate filaments and centrioles were collectively termed the "cytoskeleton". Recognition that complex dynamic activities of microtubules and other cytoskeletal elements were essential for organization, movement and growth of cellular cytoplasm finally brought the cytoskeleton out of the closet.

2.2. Cytoskeletal networks

Bulk cytoplasm contains networks of individual microtubules, arrayed in parallel and interconnected by filamentous strands. Other interconnecting networks of smaller filamentous proteins (actin, intermediate filaments, microtrabecular lattice, etc.) intersperse with MTs to form a dynamic gel whose activities in all types of cells (i.e. mitosis, growth and differentiation, locomotion, food ingestion or phagocytosis, synapse modulation, dendritic spine formation, cytoplasmic movement, neurotransmitter release, etc.) are essential to the living state [18]. Orientation and directional guidance of these cytoskeletal functions depend on centrioles: assemblies of nine pairs or triplets of microtubules arranged in replicating super-cylinders always found in pairs oriented perpendicular to each other. To initiate cell mitosis, centriole pairs migrate to opposite poles of the cell from where MT "mitotic spindles" separate chromosomes and establish orientation and architecture for the next generation cells [17, 19].

2.3. Microtubules

Of the various filamentous structures which comprise the cytoskeleton, microtubules are the most prominent, best characterized and appear best suited for dynamic information processing [22, 65].

MT are hollow cylinders 25 nm in diameter whose lengths may span meters in some mammalian neurons. MT cylinder walls are assemblies of 13 longitudinal protofilaments which are each a series of subunit proteins known as tubulin (fig. 3). Each tubulin subunit is a polar, 8 nm dimer which consists of two slightly different classes of 4 nm, 55 kilodalton monomers known as α and β tubulin. Genes for α and β tubulin are complex, multi-gene families which give rise to varying tubulin isozymes. During evolution some tubulins have been highly conserved for basic cell functions. However, extensive microtubule heterogeneity exists in complex cells due to genetic diversity, expression, post-translational modifications, MAPs, and assembly patterns [24, 31]. For example, two-dimensional gel electrophoresis has shown 17 different β tubulin isozymes exist in mammalian brain MTs, whereas fewer exist in other tissues [49].

The tubulin dimer subunits within MTs are arranged in a hexagonal lattice which is slightly twisted, resulting in differing neighbor relationships among each subunit and its six nearest neighbors (fig. 4).

MT self-assembly and disassembly are dynamic, complex processes which depend on various factors including temperature and calcium ion con-

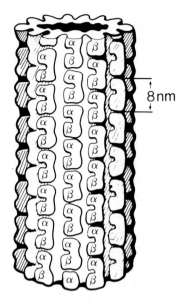

Fig. 3. Microtubules (MTs) are cylinders whose walls are 13 protofilaments, each a string of 8 nm tubulin dimers. Alpha and beta tubulin monomers form the dimers; each dimer has 6 neighbors. Drawn by Fred Anderson from X-ray crystallographic data of Amos and Klug [4].

centration. Oriented by centrioles, MT polymerization determines the architecture and form of cells which can quickly change by MT depolymerization and reassembly in another direction [45]. GTP, an energy-providing analog of ATP, binds to polymerizing tubulin; GTP hydrolysis energy is subsequently delivered to assembled MTs. MTs whose ends are GTP tubulin are stable and will continue to grow. After the GTP is hydrolyzed to GDP, GDP tubulin is exposed at MT ends and unless stabilized by MAPs, centrioles or other structures, MTs rapidly disassemble with their component subunits being utilized by assembling MTs. This dynamic instability results in a "selectionist" activity in which MT networks can probe or retreat in cellular appendages including dendritic synapses. The precise mechanism of consumption of GTP hydrolysis energy in MTs is not understood, although one possible utilization is the production of solitons or coherent lattice excitations as proposed by Fröhlich [27–29]. It is known that GTP tubulin and GDP tubulin have different conformational states [8].

When viewed in cross section by electron microscopy, MT outer surfaces are surrounded by a "clear zone" of several nm, which apparently represents an electronegative field due to excess electrons in tubulin and may also serve to organize cytoplasmic water and enzymes [67]. MTs, as well as their individual dimers, have dipoles with negative charges localized toward α monomers [17]. Thus MTs are "electrets": oriented assemblies of dipoles which are predicted to have piezoelectric properties [7, 52]. Contractile or enzymatic MAP proteins may be attached to MTs at specific dimer sites and MAP attachments can result in various helical patterns on MT surface lattices [12]. Contractile MAPs participate in axoplasmic transport while other MAPs form bridges which laterally connect specific subunits on parallel arrayed MTs.

3. Models of microtubular information processing

3.1. Cytoskeletal signal processing

The lattice polymer structure of MTs have suggested capabilities for information processing, and numerous author groups (some of which are mentioned below) have published theoretical models of MT/cytoskeletal information processing.

Atema [6] proposed that cellular sensory transduction occurred by propagated conformational changes in MT subunits. His view was an "all or none" propagation in which entire MTs were essentially trigger switches.

Barnett [9] proposed that filamentous cytoskeletal structures operated like information strings analogous to word processors. He proposed that MTs are processing channels along which strings of information can move, and that neurofilaments (intermediate filaments within neurons) are parallel arrayed memory channels. While not specifying the mechanism of pattern representation within MTs or neurofilaments, Barnett's model emphasizes the potential utility of parallel arrays of laterally interconnected molecular structures.

A model which does address patterns within MTs was proposed by Roth and Pihlaja [62]. They suggested conformational states of tubulin subunits within MTs were regulated by binding of inter-MT linkages of MAPs. They assumed five possible conformational states for each tubulin subunit (based on drug binding studies) could be induced by proximity of MAP binding and cooperative allosteric effects.

Hameroff and Watt [36] proposed that conformational states of tubulin within MTs were coupled to charge or dipole states, and interacted cooperatively with neighbor tubulin states. Such interactions were thought to be based on "molecular automata" in which the conformational states of individual tubulin dimer subunits represent "bits" of information subject to dynamic influence by neighboring subunits [36–39, 64]. The control of protein conformation is, by itself, an interesting enigma. Before considering molecular automata as a specific model of MT information processing, some aspects of protein conformational regulation theory will be mentioned.

3.2. Coherent protein dipole excitations

Proteins are vibrant, dynamic structures in physiological conditions. A variety of recent techniques have shown that proteins and their component parts undergo conformational motions over a range of time scales from 10^{-15} s to many minutes with the functionally most significant conformational vibrations appearing in the "nanosecond" range of 10^{-9}–10^{-11} s. Biologically relevant motions of globular proteins in this time scale are thought to include collective elastic modes, e.g. solitons and other nonlinear motions, and coherent excitations [43, 44, 60]. These motions are global changes in protein conformation rather than more rapid thermal fluctuations of side chains or local regions. Such global protein conformations appear suitable for computation: finite states which can be influenced by dynamic neighbor interactions.

Collective conformational states near the nanosecond time domain have been woven into a theory of coherent protein excitations by Fröhlich [27–29]. Changes in protein conformation, according to Fröhlich [28], are triggered by charge redistributions such as dipole oscillations within specific hydrophobic regions of proteins. Another concept of Fröhlich [27] is that a set of proteins connected in a common physical structure and electromagnetic field such as within a polarized membrane (or polymer electret like a microtubule) may be excited coherently if chemical energy such as ATP or GTP were supplied. Coherent excitation frequencies of the order of 10^9–10^{11} Hz are deduced by Fröhlich, who cites as evidence sharp windows of sensitivity to electromagnetic energy in this region by a variety of biological systems [33]. Circumstantial evidence for Fröhlich's activities in microtubules include the apparent electronegative field surrounding MTs, and the unexplained consumption of GTP hydrolysis energy by MT lattices. Other aspects of Fröhlich's model include metastable states (longer-lived conformational state patterns stabilized by local factors) and polarization waves, i.e. travelling regions of conformational states slightly out of phase with the majority of coherently excited states.

3.3. Molecular automata in microtubules

Conrad and coworkers [15] introduced the idea of "molecular" automata within neurons as a basis for intracellular information processing related to cognition. We believe molecular automata activities in cytoskeletal structures such as microtubules could explain much of their organizational activities.

Automaton behavior including dynamic patterns and capabilities for computation require a lattice whose subunits can exist in two or more states at discrete time steps, and transition rules which determine those states among lattice neighbor subunits. Fröhlich's [27] model of coherent excitations and cooperative coupling among proteins arrayed in an electromagnetic field may be applied to tubulin subunit dimers within MTs to serve as a "clocking" mechanism and define discrete generations for automaton behavior. We can obtain an estimate for the clocking frequency assuming one coherent "sound" wave across the MT diameter ($\Phi \simeq 25$ nm). A crude calculation, assuming that $V_{\text{sound}} \simeq 10^3$ m/s, yields a clocking frequency of approximately 4×10^{11} Hz for one wave across the MT diameter. This is in the predicted range and correlates with a clocking generation time of 2.5×10^{-11} s. However, faster and slower automata are possible, depending on the type of MT excitation being considered.

Orientation or phase of any tubulin subunit dimer relative to the coherently oscillating majority at any excitation period would depend on factors such as initial conformational states, binding of water, ions, or MAPs, bridges to other MTs, tyrosinated/glutamated state, energy-providing phosphate nucleotides (i.e. GTP) and associated proteins, genetically determined subunit factors and electrostatic dipole interactions among neighboring subunits. Many of these factors may be considered programming modes (either hereditary or environmental) whose net effects would serve to alter phases of particular subunits in coherently excited arrays. To simulate MT molecular automata, we will focus solely on electrostatic dipole interactions among MT subunits.

Our derivation of automaton neighborhoods and neighbor dynamics for computer simulation are illustrated in figs. 4 and 5. The MT cylinder (25 nm in diameter) has a circumference of 13 tubulin subunit dimers and the pitch pattern (angle from horizontal) of the leftward helix is one and one-half dimers. For automaton simulation we consider seven-member neighborhoods of tubulin dimers: a central dimer surrounded by a tilted hexagon of six neighbor dimers. Taking the cylinder axis of the MTs to define the y-axis, the dimer neighborhood is defined as follows (fig. 4): "C" is the center dimer and "N" (north) is the nearest-neighbor dimer in the positive direction of the y-axis. Similarly, dimers labeled "NE" (northeast), "SE" (southeast), "S" (south), "SW" (southwest),

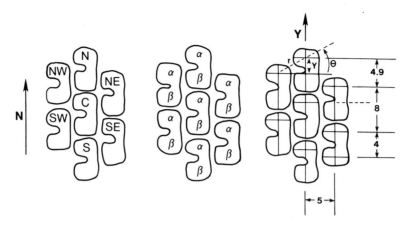

Fig. 4. Microtubule automata (MTA) neighborhood. Left: Definition of neighborhood dimers. Center: α and β monomers within each dimer are labelled. Right: Distances (in nm) and orientation among lattice neighbors. Interaction forces are calculated using $y_i = r_i \sin \theta$.

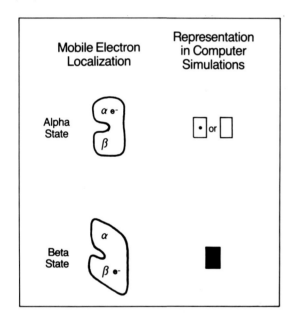

Fig. 5. Switching between conformational states in microtubule (MT) dimers. Top row: α states, bottom row: β states. Left: dimer conformation coupled to mobile electron localization, Right: representation of states in computer simulation.

and "NW" (northwest) are appropriately oriented around the "C" dimer. Within each dimer the α monomer is north of its β partner. The helical twist yielding an offset along the y-axis for NW, SW, NE, and SE neighbors leads to neighbor dimer distances and dipole interaction forces which differ among the six neighbors surrounding the C dimer.

To calculate the neighbor dipole interaction forces as a basis for automaton transition rules, each dimer may be viewed as having a mobile electron shared by the two monomers (fig. 5). At each time step the electron's average position is considered to be oriented either more toward the α monomer ("alpha state") or more toward the β monomer ("beta state") with associated changes in dimer conformation. Experimental evidence [69] suggests that a conformational shift of 29° from vertical may occur. Because MT net dipoles are negative toward the α ends, oscillations need not necessarily change direction totally to effect a conformational change. For example, a relative dipole reorientation toward the α monomer may be sufficient to induce a tubulin dimer conformational state. For our calculations, however, we have assumed that electrons localize in the centers of the two monomers.

Considering the MT lattice geometry, the y-component of the resulting electrostatic force acting on an electron in a central dimer can then be calculated as:

$$f_{\text{net}} = \frac{e^2}{4\pi\epsilon} \sum_{i=1}^{6} \frac{y_i}{r_i^3}, \tag{1}$$

Table 1
Relative forces ($= -1000 y/r^3$) for neighbor configurations. Net forces are summation of six neighbors.

Neighbor position	Central dimer α		Central dimer β	
	neighbor α	neighbor β	neighbor α	neighbor β
north	+15.625	+62.500	+6.944	+15.625
northeast	+15.205	−7.022	+9.635	+15.205
southeast	−14.250	−8.338	−7.022	−14.250
south	−15.625	−6.944	−62.500	−15.625
southwest	−15.205	−9.635	+7.022	−15.205
northwest	+14.250	+7.022	+8.338	+14.250

where y_i and r_i are defined as illustrated in fig. 4, e is the electron charge, and ϵ is the average permittivity for MT proteins, typically 10 times the vacuum permittivity [50]. We have assumed that only the y-component of the interaction forces are effective and have neglected any net force around the MT circumference.

Relative neighbor dipole coupling forces calculated from eq. (1) and MT geometry are shown in table 1. These relative forces, which we use for transition rules, may be multiplied by 2.3×10^{-14} N to obtain calculated absolute values of intersubunit forces.

To simulate MT automata, we represent MT structure as a two-dimensional grid in which the cylindrical MT has been fileted open and flattened (figs. 6 and 7). The grid consists of MT subunit dimer loci which can exist in either an α state (blank or blank with dot) or β state (solid black). At each time step (generation), resulting neighbor forces are calculated for each dimer in the grid. The extent to which each dimer is influenced by neighbor forces acting upon its mobile electron is represented by a "threshold" parameter. The higher the threshold, the greater are the summated neighbor forces necessary to induce a transition. For example, a threshold of ±9.000 means that net neighbor forces greater than $+9.0 \times 2.3 \times 10^{-14}$ N will induce an α state, and negative forces of less than $-9.0 \times 2.3 \times 10^{-14}$ N will induce a β state.

Biological factors which might affect the threshold include temperature, pH, voltage gradients, ionic concentration, genetically determined vari-

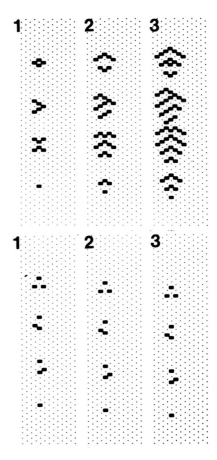

Fig. 6. Some virtual microtubule automata existing for two different threshold values. Top: objects in left panel are labelled (top to bottom) as diamond blinker, triangle glider, spider glider and dot glider. The middle and right sided panels in the top show the next two generations at threshold ±1.000. The objects either blink or move and leave "wakes" which lead to travelling wave structures. Bottom: the first three generations of some other gliders with threshold ±9.000. At this higher threshold, the gliders travel without creating a wake or travelling waves.

ability in individual dimers, and binding of molecules including MAPs and/or drugs to dimer subunits. In nerve cells, travelling membrane depolarizations could induce transient waves of lowered threshold along parallel arrayed MTs. Consequently, the frequency of depolarization for a particular nerve cell would directly influence the elaboration of patterns within that nerve cell's MT automata.

In the MT automata of Smith, Watt, and Hameroff [64] and Hameroff, Smith and Watt [37, 38] boundary conditions were chosen in which reflection of automata patterns occurred at each end of simulated MTs. In more recent studies using toroidal boundary conditions [39], we have investigated effects of varying thresholds on MT automata behavior as well as the effects of asymmetrical thresholds in which α and β transition thresholds differ. The simulations have demonstrated behaviors of both a general nature (e.g. virtual automata, gliders and blinkers) similar to behavior in other automata systems, and of a specific nature (e.g. perturbation stable gliders, wedge patterns, NE–SW orientations) derived from unique MT properties. Threshold-dependent behaviors include gliders (dot, spider, bus and giant gliders), travelling and standing wave patterns, blinkers, linearly growing patterns ("bean sprouts"), bidirectional gliders, and frozen patterns. The bifurcations found in the automata dynamics are summarized in fig. 8.

Having demonstrated the potential capability of MT automata for information processing in single MTs, in the present study we have begun to model simple MT networks as well as MT assembly and disassembly.

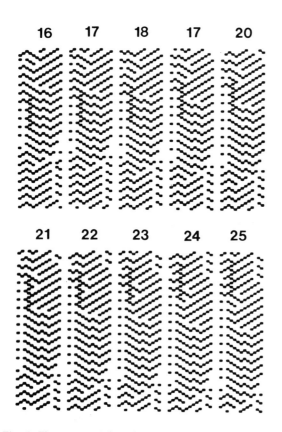

Fig. 7. Time steps 16 through 25 of a microtubule automata simulation with asymmetric threshold, demonstrating localized bilateral information transformation. The initial condition was few randomly scattered α seeds on a β background. The threshold from α to β is -20.000. From β to α the threshold is 2.000. Note that as the β bus (black line of dimers) moves north a new α glider (white line of dimers) is developed in the western region of the neighborhood. This new glider continues to expand until generation 22 when it is "released" from the β bus and starts to travel south together with the other α glider. This dynamical phenomenon may account for the observed bilateral transport on MTs as well as processes related to "back-propagation" in neural networks.

4. A cytoskeletal connectionist network

The largest portion of free energy used in neuronal dendrite cytoplasm is consumed by MAP2, which actively interconnects dendrite MTs and other structures [68]. A number of empirical studies have also shown that cytoskeletal structures within dendrites change during learning. The relation between cognitive structural changes within neurons and among neurons is enigmatic: which is cause and which is effect? In "true" neural networks (i.e. the brain), the formation of new connections, as well as the breakdown of old ones, is presumably a function of both internal neuronal states (configuration and states of the cytoskeleton, membranes, membrane proteins, second messengers, etc.), as well as signals received from other neutrons. Thus, a multi-level hierarchy of information processing is suggested in biological

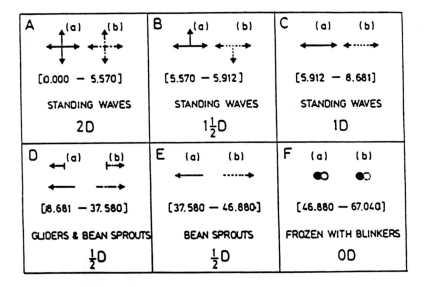

Fig. 8. Bifurcation table for toroidal microtubule automata dynamics with symmetrical dipole coupling thresholds. The upper items in each sector (A)–(F) indicate the dominant global behavior for: (a, solid arrows) β seeds on an α background, and (b, dashed arrows) α seeds induced on a β background. Below the dominant behaviors are the threshold intervals, how we describe each type of dynamic phenomena, and the dimensionality of the growth patterns.

neural networks. To explore this possibility, we have devised an adaptive network based on cytoskeletal structure and function.

Present technologies cannot yet address nanoscale actions of individual protein subunits (although scanning tunnelling microscopy and related techniques may soon be capable). Consequently, there are currently insufficient data to directly test or accurately model such actions and their collective properties.

To merely explore some possibilities, we have used what is known about cytoskeletal structure to serve as the basic architecture for a new connectionist computer. The result is an adaptive dynamical system having certain similarities with the cytoskeleton.

Parallel arrays of MT automata (MTA) are interconnected via MAP-like connections enabling different MTA to communicate with one another. The MAP connections are themselves modeled to be simple automata (MAPA) capable of transmission of signals (sequences of α and β states) from one MTA to another. In the simplest case, this transmission is unidirectional, only transferring signals from a predefined input area on one MTA to a predefined output area on another. Consequently, any tubulin dimer state (α or β) from the MAPA origin on the "input" MTA1 is transmitted to the MAPA target area on the "output" MTA2 (fig. 9).

Conditions which can contribute to dynamic behavior of the MTA network (MTAN) system include MTA thresholds as well as number, attachment locations, and directionality of MAPA. By defining inputs and outputs for each MTA (fig. 10), together with a recursive procedure for updating subunit states, we are able to construct a simple adaptive network.

To model a computationally efficient connectionist network via a feedback between connection topology, internal automata dynamics and automata rules, we use an optimization process often used by Nature: *variation* combined with *selection*. Several systems compete in parallel with respect to a given set of tasks. At each step in the optimization process the most efficient system is selected as the "mother system". At the next step in the optimization process a population of other systems is created as "daughters" or random offspring from the mother system. When one of the offspring begins to perform better than the original system, this offspring is turned into the mother

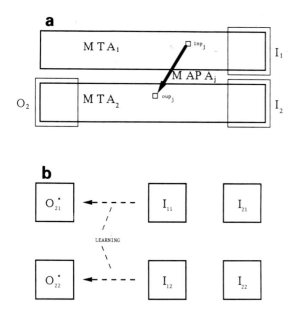

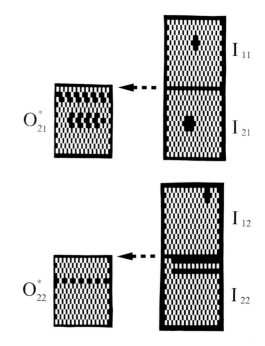

Fig. 9. (a) The simple microtubule automata network, the MTA net, consists of two toroidal MTA of length l dimers, coupled via microtubule associated protein automata, MAPA. The output for the MTA net is defined at the left end of the lower MTA indicated by O_2, and it is taken when the system is updated $l - (13 + 1)$ times after the input patterns are set at locations I_1 and I_2 at time 0. Input and output areas occupy a matrix area of $13 \times 13 = 169$ dimers. The jth MAPA transmits the dimer states from the input area inp_j at the upper MTA to the output area oup_j at the lower MTA. For each pair of input patterns, I_1 and I_2 at the upper and lower MTA, there exists a desired output pattern O_2^* at the lower MTA. (b) By means of the learning algorithm, the mappings $(I_{11}, I_{21}) \to O_{21}^*$ and $(I_{12}, I_{22}) \to O_{22}^*$ are generated.

Fig. 10. A MTA network with start configurations (I_{11}, I_{21}) and (I_{12}, I_{22}) and corresponding desired output patterns O_{21}^* and O_{22}^*. These two maps have been chosen from the learnable set for the MTA network. The structure of this set is as yet unknown. The present realization of the MTA network is not able to satisfy every input-output pattern combination.

system as the next step in the optimization process. In this way it is possible to evolve efficient networks with respect to different computational tasks.

In our initial efforts to model a cytoskeletal network, two parallel MTA are capable of being interconnected by MAPA at any subunit loci. Each MTA has an input area at the right end, and output area at the left. Output area states are compared to desired output states using their Hamming distance. For the system to recognize inputs to produce desired outputs, we have random MAPA connections between the two MTA. The learning algorithm for our network is represented both by the threshold parameter for the individual MTA and by the interconnections between the different MTA. A detailed description of the learning algorithm is given in the appendix.

The MTAN obtains truly distributed computation since it is only governed by low-level, neighborhood interactions mediated by intermolecular forces. Such a spatial locality is usually not found in other connectionist models. Another important difference is that the learning algorithm for the MTAN does not use a local learning rule as most artificial neural networks do; rather, MTAN uses an *evolutionary search* to accomplish learning. However, our current MTAN system needs input/output learning in the same sense as most artificial neural net systems do.

In the following example we have fixed the threshold parameters for the upper and the lower MTA at ± 5.900 and ± 9.000 respectively. The corresponding input-output maps used for learning are shown in fig. 10. The connection topology

$C_j = C_2$ for the first accepted input–output map is shown in fig. 11 together with the dynamics for the MTA net at time 0, 43, and 66. The Hamming distance between the desired and the actual map is $H_d = 0$. In fig. 12 the connection topology $C_k = C_4$ for the second input–output map is shown together with the dynamics at time 0, 23, and 66 ($H_d = 0$). It turns out that C_4 also satisfies the first input–output map. We had to redo pass (5) and (6) in the learning algorithm a number of times before the system found a correct C_k satisfying both the maps $(I_{11}, I_{21}) \rightarrow O_{21}^*$ and $(I_{12}, I_{22}) \rightarrow O_{22}^*$.

In the situations where the two pattern combinations can be memorized, there always seem to be several C_k's satisfying the two maps. However, the present realization of the network is not able to memorize every pattern combination at the same time. Some maps cannot coexist.

The MTA net is also able to "associate" or generalize patterns. In figs. 13a and 13b the system tries to associate or recognize the first input pattern, and in 13c and 13d the system tries to associate the second pattern. Note that H_d may be quite high due to a simple horizontal offset, even though the patterns seem very similar. The basis for this associative ability is the existence of relatively perturbation-stable virtual automata in this system. The dynamics of the system is more

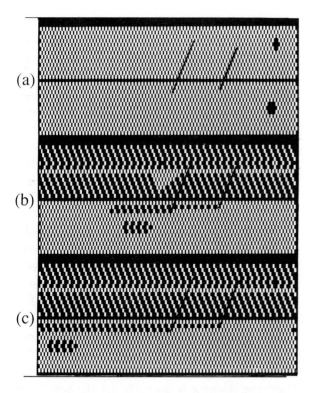

Fig. 11. MTA net at stage (3) in the learning process. The parameters for this net are $l = 80$ and $inp_j = oup_j = 1$. The connection topology $C_j = C_2$, with MAPA at the dimer locations $(60, 6) \rightarrow (55, 2)$ and $(47, 4) \rightarrow (41, 3)$, turns out to satisfy the first input–output map. The Hamming distance $H_d = |O_{21}^* - O_{21}| = 0$. The dynamics is shown at time: (a) 0, (b) 43, and (c) 66. The thresholds for the upper and lower MTA are ± 5.900 and ± 9.000 respectively.

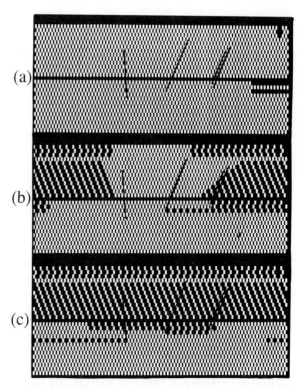

Fig. 12. The same MTA net as shown in fig. 11 at stage (5) in the learning process. $C_k = C_4$ is here given by: $(60, 6) \rightarrow (55, 2)$, $(47, 4) \rightarrow (41, 3)$, $(59, 7) \rightarrow (55, 1)$, and $(28, 7) \rightarrow (29, 5)$. The dynamics is shown at time: (a) 0, (b) 26, and (c) 66. It can be shown that C_4 both satisfies the first and the second input–output map ($H_d = 0$). However, pass (5) and (6) had to be repeated a number of times, with $D = 10$, before a correct connection topology was found. The thresholds for the upper and lower MTA are still ± 5.900 and ± 9.000 respectively.

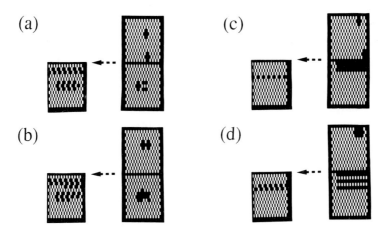

Fig. 13. Association of patterns in the MTA net with connection topology C_4. (a) and (b) are related to the first map, and (c) and (d) are related to the second map. In (a) $H_d = 26$ and in (b) $H_d = 6$ although (a) looks more right than (b). (c) is perfectly associated. In (d) we see a less correctly associated pattern, $H_d = 26$. Compare with fig. 10.

resistant to perturbations in the longitudinal (north–south) MTA direction than in the east–west direction. This property appears due to chosen threshold values which mainly support communication along the north–south axis on the MTA (see the bifurcation table in fig. 8). The reason for the preferred north–south communication for higher thresholds is due to the geometry of the MT lattice. The east–west interactions between most of the dipole configurations are simply cut off for higher thresholds. For example, table 1 shows relative forces for the 24 configuration pairs occurring. None of the pairwise east–west interactions exceed an absolute value of the relative force larger than $15.205 \times 2.3 \times 10^{-14}$ N. The pairwise north–south interactions have absolute values up to $62.500 \times 2.3 \times 10^{-14}$ N.

Selection of specific asymmetric threshold values which differ between the two connected MTA as well as selection of different background states in the two connected MTA can lead to significant behavior. For example, a larger absolute positive threshold in MTA1 and a larger absolute negative threshold in MTA2 cause north → south direction gliders in MTA1 which induce, by MAPA connections, south → north gliders in MTA2. Another example of retrograde information transport is a system with MAPA connections between MTA in different ground states. This is possible since β-gliders always travel south on an α background and α-gliders always travel north on a β background (for details on the dynamics in a single MTA, see ref. [39]). Such retrograde signaling could implement "back-propagation" within neural net neurons. MTs from separate dendritic branches merge and interconnect in the dendrite; thus dendritic MTA nets can merge information from multiple dendritic branch synapses. "Upregulation" due to increased activity in one synaptic area could thereby "down-regulate" relatively inactive synapses on separate dendritic branches.

Although our system presently is able to adapt to certain dynamics, the learning process is not optimal for several reasons. These include the basis on which the daughter systems are both constructed and chosen, and because of an inefficient use of the MTA patterns. Also, the Hamming distance is probably not the best index for selection of a connection topology because two identical patterns shifted just for one dimer often result in a huge Hamming distance. This behavior also limits the associative properties.

Several potential improvements could be implemented in the MTA net system. The resolution of different maps on the same MTA net would increase significantly if the system were able to use "coded" MAPA. A coded MAPA could be an automaton which only transmits if a certain α/β sequence ("an initiator pattern") passes the input area for the MAPA. Similarly, the MAPA should stop the transmission whenever a certain stop pattern passes the input area; thus the system would have properties very similar to the DNA transcription system. Alternatively, the MAPA could be coded to transmit a signal averaged over a certain area in the input zone. Another possible improvement which could reflect true biological activity involves MAPA connection patterns. Rather than random connection topologies, MTA patterns could determine where the MAPA connections should be located. For example, connection sites could be determined by the location of a persistent diamond blinker or some other pattern occupying roughly $2 \times 2 = 4$ dimer subunits. These possibilities have yet not been tested, but indicate a potential for adaptation and learning in structures based on cytoskeletal networks. Future work will explore these areas.

Assembly and disassembly of MTs, MAPs, centrioles and other cytoskeletal structures determine cell division, differentiation, movement, transport, and (in neurons) synaptic connectivity. Coupling between MTA dynamics and MT assembly/disassembly could explain the orchestration of these elegant and delicate functions. In section 5, we investigate this area.

5. Assembly and disassembly of microtubules

Experimental results of studies on assembly and disassembly of MTs both in vivo and in vitro are summarized in ref. [11]. In order to grow, the microtubule must have dimers containing exchangable GTP at the end. Without this GTP "cap", depolymerization occurs unless the MT is stabilized by other factors such as binding to other cytoskeletal structures. Thus, free MTs tend to

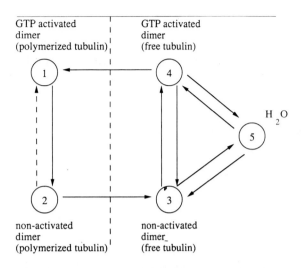

Fig. 14. State diagram for a simple model of assembly/disassembly of microtubules. State 1 is GTP activated tubulin dimer in polymerized form (microtubule geometry), state 2 is microtubule tubulin is non-activated form (GDP), state 3 is non-activated free dimer (in solution), state 4 is GTP activated free dimer, and state 5 is water. The transitions are discussed in the text. Note that this model is formulated partly as a Markov chain, and partly as a cellular automaton.

either grow or shrink. These results support the dynamic instability model of MT assembly and disassembly [45]. In a simple cellular automaton model for such a process, each boundary site can be considered to be in one of five different states: as polymerized tubulin dimers with either GTP or GDP, as free tubulin with either GTP or GDP, or as water. The possible transitions are shown in fig. 14.

In the present study, we are concerned with the elongation of already existing tubules, since formation of initial fragments is governed by complex nucleation processes and interactions with centrioles.

We assume that the GTP hydrolysis to form GDP tubulin occurs through a process $(1 \rightarrow 2)$ described by some probability P_{12} at each update of the system (fig. 14). The disassembly $(2 \rightarrow 3)$ of the MTs occurs through dissolving of the non-activated dimers. This process is assumed to depend on the concentration of activated free tubulin [11]: a polymerized boundary dimer will get free if a major part (three or more) of its neighbors is of the non-GTP free form (including water). Thus, the lower the GTP-activated free dimer concentration, the faster the disassembly. In this model both the concentration of the activated GTP tubulin and the non-activated GDP tubulin are determined by a stochastic process involving internal transitions between the three states: 3, 4, and 5. The higher the overall probability for the presence of an activated dimer among the free states, the more likely is a polymerization process in the system. Polymerization (4 → 1) takes place if an activated free dimer is aligned with at least two activated polymerized dimers. A necessary condition for polymerization is that the fractional occupation time for activated dimer among the free states is significantly higher than the decay probability P_{12} for the activated microtubule dimer. The transition 2 → 1 is not modeled at this point, but will be discussed later.

Simulations of microtubule assembly and disassembly with the simple model are shown in fig. 15. The clocking frequency in this model is significantly lower than the assumed clocking frequency for signal propagation on the microtubule (10^9–10^{11} Hz). Elongation of existing dentrite MTs in vivo may result in a MT growth of 1 mm/day corresponding to 1 nm/s (or one dimer row per 8 s), although in vitro assembly under physiological conditions may elongate at least two orders of magnitude faster. The simulation in figs. 15d–15f shows an elongation rate of approximately 6 times 8 nm for each 50 generations. This corresponds to approximately 1 nm/s. Therefore the updating of the assembly/disassembly model does not demand nearly the same frequency as the signal propagation model does (10^9–10^{11} Hz). In the present simulation we update with a real time of only 1 Hz.

If specific dynamics on the MTA are able to activate dimers at certain locations, for instance at the free ends (the 2 → 1 transition), there exists an internal mechanism for determining whether a microtubule in the network should start growing. Introducing to the simple cytoskeletal model from

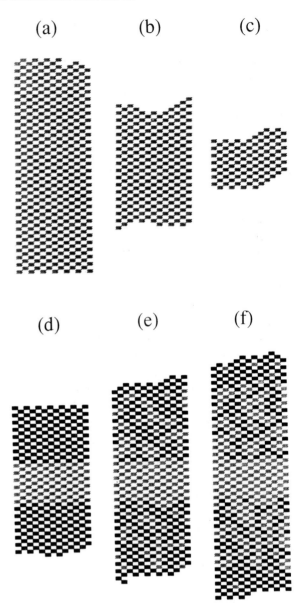

Fig. 15. (a)–(c) disassembly of a microtubule shown at time 0, 21, and 49, with hydrolysis probability $P_{12} = 0.05$ per generation, and an overall probability of 0.091 that the free states are visited by an activated dimer during the automaton generation time. Non-activated microtubule tubulin dimers (state 2) are represented by shaded squares, whereas the free tubulin and water are blank (states 3, 4, and 5). (d)–(f) Assembly of a microtubule with GTP activated caps (state 1 shown as dark squares), shown at time 0, 20, and 50, with hydrolysis $P_{12} = 0.05$ per generation, and an overall probability of 0.987 that the free states are visited by an activated dimer during the automaton generation time.

section 4, such a mechanism adds another feedback between local dynamics and meta-dynamics of our network. Relevant to section 4, selecting the sites of MAPA connections by means of MTA dynamics would also be consistent with this type of feedback.

In vivo assembly/disassembly of MTs may actually be controlled by signals transmitted on the cytoskeleton.

6. Discussion: connectionism in the cytoskeleton

Deciphering of the genetic code in DNA and RNA has led to many technological applications. The existence of a "real-time" processing code in the cytoskeleton could be comparably useful because of the important functions which could be controlled or regulated (mitosis, differentiation, aging, growth, immune function, etc.).

In the neuronal cytoskeleton, MTA can be relevant to brain/mind cognitive function. For example, MT automata may be utilized for guidance and movement of cells and cell processes (axon growth cones, dendritic spines) during morphogenesis of the brain's network architecture. In learning, axoplasmic transport and other mechanisms orchestrated by MT automata could regulate synaptic plasticity. Threshold-dependent bidirectional MT automata patterns indicate that a "back-propagation" type of synaptic regulation could occur because rapid information transport in directions opposite to the nerve impulse is possible. Combined with possible retrograde information flow across synapses ("trophism", pre-synaptic modulation) the cytoskeleton could support "neural-net" type phenomena in the brain. Beyond these morphological roles, MT automata could themselves directly participate an imprinting, pattern recognition, memory and other cognitive functions. MT automaton gliders (8 to 800 m/s) may exist as traveling waves, coupled to nerve membrane depolarizations. Coupling between membrane events and MT automaton behaviors may occur via ionic fluxes or direct connection through specialized sub-membrane cytoskeletal proteins such as ankyrin, fodrin and spectrin to the cytoskeleton. This coupling can lead to a view of the brain as a hierarchy of networks with a cytoskeletal basement level. MT automata would then comprise the "fine grain" of cognition, the "roots of consciousness".

Relegating cognition function to the level of protein conformational dynamics is supported by efforts to understand the mechanism of general anesthesia. Anesthetic gas molecules erase consciousness (while sparing other brain functions like breathing) by inhibiting conformational switching in classes of proteins such as receptors, ion channels, enzymes and cytoskeletal subunits [25]. The anesthetic molecules are thought to bind in hydrophobic regions within these proteins: the same sites proposed by Fröhlich to house dipole oscillations. Anesthetics such as halothane have also been shown (at relatively high concentrations) to cause depolymerization of microtubules [1]. Anesthesia (the reversible cessation of consciousness) may be due to interruption of communicative interactions within and among cytoskeletal networks and their membrane connections.

We have demonstrated that a population of simple networks of paired MTA is able to learn two different patterns by using an evolutionary mechanism as well as associate nearby patterns. However, many improvements may be implemented in the model which can (a) more closely resemble biological activities, and (b) increase MTA network performance. Some of these are discussed in section 4.

We have described the cytoskeleton as a connectionist information network. As such, it may be compared with other connectionist systems like artificial neural networks, classifier systems, etc. Farmer (these Proceedings [23]) has described a group of parameters into which all such systems may be fit for comparison and clarity. These parameters include nodes, states, connections, parameters, interaction rules, learning algorithm and metadynamics. In table 2, we have catalogued

Table 2
Farmer's Rosetta Stone [23] for adaptive dynamic systems applied to microtubule automata (MTA).

	Single MT	MTA networks
nodes	cellular automata cells; MTA subunits	cellular automata networks; MTA-MAPA
states	α and β state patterns	α and β state patterns
connections	lattice geometry	MAPA and other connections
parameters	local thresholds	coding in MAPA connections
interaction rules	local dipole coupling forces	MAPA connections
learning algorithm	input/output learning; selection of the best in a population (selectionist learning)	input/output learning; selection of the best in a population (selectionist learning)
metadynamics	local threshold changes; metastable states	threshold change in MTA-MAPA loci and assembly/disassembly

both a single MT automaton and the adaptive MT automaton network according to Farmer's connectionist classifications.

Connectionist networks within single elements of larger connectionist networks suggest a self-similar "fractal-like" relationship. Information capacity of a high-resolution scale within a self-similar structure is far greater than the capacity at larger scales. Roboticist Hans Moravec [56] has attempted to compare the computing power of a computer and of the human brain. Considering the number of next states available per time in binary digits, a microcomputer has a capacity of about 10^6 bits per second. Moravec calculates brain "computing" power by assuming 40 billion neurons, each with perhaps one thousand synaptic connections which can change states hundreds of times per second, resulting in 4×10^{15} bits per second. We estimate approximately 10^{14} MT tubulin subunits in a human brain based on MTs spaced 100 nm apart and 40% of brain volume being neuronal cytoplasm. In our model of MT automata these 10^{14} brain MT subunits can change states 10^9 to 10^{11} times per second leading to a total brain capacity of 10^{23} to 10^{25} bits per second! Such a huge capacity could provide a high resolution to the grain of cognition, permit massive parallelism and redundancy, and account for cytoplasmic "housekeeping" chores not directly related to brain/mind function. Still, is such a huge capacity necessary to explain cognitive function? That question is presently unanswerable. However, we surmise that the highest level of brain/mind cognitive functions are collective emergent effects whose fine grain may exist within the cytoskeleton.

We conclude with possible applications of MTA molecular cognition. Deciphering real-time MTA codes may permit interfacing to emerging technologies such as spinoffs of scanning tunneling microscopy. These techniques should allow nanometer scale monitoring and programming of dynamic actions of individual protein subunits within assemblies such as MTs. Combining genetic engineering of MT tubulin subunits with real-time programming could lead to replicating MTA with complex, purposeful activities. Perhaps akin to programmable viruses, mobile MTA might be capable of intracellular repair with profound medical applications.

Acknowledgements

We appreciate the discussions we had with Kuni Kaneko on how to design the input and the output for the adaptive microtubular automaton network.

Appendix. The learning algorithm

The process of teaching the MTA net to map two different sets of input to two different output patterns consists of six steps.

(1) Primarily depending on the shape of the desired output pattern, the thresholds for both MTA are set manually. This step demands certain knowledge about what kind of dynamics is possible for each value of the threshold parameter (recall fig. 8), and is not yet endogenous in the algorithm.

(2) A single random MAPA connection is placed at D successive MTA nets. Each of these MTA nets are then simulated for $l - (13 + 1)$ generations, at which time the Hamming distance, H_d, between the actual output O_{21}, and the desired output pattern, O_{21}^*, is calculated. The MAPA connection pattern which resulted in the smallest H_d with the two first input patterns (I_{11}, I_{21}) is kept for the next steps. This new connection topology is denoted by C_1.

(3) With C_1 as the basic (mother) system, D new random MAPA connections are tested in D different (daughter) systems. Again the system with the smallest $H_d = |O_{21}^* - O_{21}|$ is chosen. This new connection topology is denoted by C_2. (3) is then repeated until an acceptable H_d is reached. The connection topology for the new system is denoted by C_j.

(4) I_{11} and I_{21} are now replaced with I_{12} and I_{22}, still conserving C_j.

(5) Additional MAPA to C_j as in step (3) which is repeated until an acceptable $H_d = |O_{22}^* - O_{22}|$ is reached for a new C_k.

(6) The map $(I_{11}, I_{21}) \rightarrow O_{21}^*$ is then tested with C_k. If $|O_{21}^* - O_{21}|$ still is acceptable, the sequence is finished satisfactorily. If not, the sequence is either started all over again at (1) or repeated from (5).

References

[1] A.C. Allison and J.F. Nunn, Effects of general anaesthetics on microtubules. A possible mechanism of anaesthesia, Lancet 2 (1968) 1326–1329.

[2] J. Alvarez and B.J. Ramirez, Axonal microtubules: their regulation by the electrical activity of the nerve, Neurosci. Lett. 15 (1979) 19–22.

[3] D.J. Amit, Modeling Brain Function, The World of Attractor Neural Networks (Cambridge Univ. Press, Cambridge, 1989).

[4] L.A. Amos and A. Klug, Arrangement of subunits in flagellar microtubules, J. Cell Sci. 14 (1974) 523–550.

[5] C. Aoki and P. Siekevitz, Plasticity in brain development, Sci. Am. (December 1988) 34–42.

[6] J. Atema, Microtubule theory of sensory transduction, J. Theor. Biol. 38 (1973) 181–190.

[7] H. Athenstaedt, Pyroelectric and piezoelectric properties of vertebrates, Ann. NY Acad. Sci. 238 (1974) 68–93.

[8] R. Audenaert, L. Heremans, K. Heremans and Y. Engleborghs, Secondary structure analysis of tubulin and microtubules with Raman spectroscopy, Biochim. Biophys. Acta 996 (1989) 110–115.

[9] M.P. Barnett, Molecular systems to process analog and digital data associatively, in: Proceedings of the 3rd Molecular Electronic Device Conference, ed. Forrest Carter, Naval Research Laboratory, Washington, DC, 1987.

[10] J.S. Becker, J.M. Oliver and R.D. Berlin, Fluorescence techniques for following interactions of microtubule subunits and membranes, Nature 254 (1975) 152–154.

[11] A.D. Bershadsky and J.M. Vasiliev, Cytoskeleton, in: Cellular Organelles, ed. P. Siekevitz (Plenum Press, New York, 1988).

[12] R.B. Burns, Spatial organization of the microtubule associated proteins of reassembled brain microtubules, J. Ultrastruct. Res. 65 (1978) 73–82.

[13] P.S. Churchland and T.J. Sejnowski, Perspectives on cognitive neuroscience, Science 242 (1988) 741–745.

[14] M. Conrad, On design principles for a molecular computer, Commun. ACM 28 (1985) 464–480.

[15] M. Conrad, W. Guttinger and M. Dal Cin, Lecture Notes in Biomathematics, Vol. 4. Physics and Mathematics of the Nervous System, Proceedings of a Summer School Organized by the International Centre for Theoretical Physics, Trieste, and the Institute for Information Sciences, University of Tubingen, 1973.

[16] J. Cronly-Dillon, D. Carden and C. Birks, The possible involvement of brain microtubules in memory fixation, J. Exp. Biol. 61 (1974) 443–454.

[17] M. De Brabander, A model for the microtubule organizing activity of the centrosomes and kinetochores in mammalian cells, Cell Biol. Intern. Rep. 6 (1982) 901–915.

[18] M. De Brabander and J. DeMey, eds., Microtubules and Microtubule Inhibitors, Proceedings of the 3rd International Symposium, Beerse, Belgium, 3–6 September 1985 (Elsevier, Amsterdam, 1985).

[19] M. De Brabander, G. Geuens, J. DeMey and M. Joniav, The organized assembly and function of the microtubule system throughout the cell cycle, in: Cell Movement and Neoplasia, ed. M. De Brabander (Pergamon Press, Oxford, 1986).

[20] A.M. De Callataÿ, Natural and Artificial Intelligence. Processor Systems Compared to the Human Brain (North-Holland, Amsterdam, 1986).

[21] N.L. Desmond and W.B. Levy, Anatomy of associative long-term synaptic modification, in: Long-Term Potentiation: From Biophysics to Behavior, eds. P.W. Landfield and S.A. Deadwyler (1988).

[22] P. Dustin, Microtubules, 2nd revised Ed. (Springer, Berlin, 1984) p. 442.

[23] D. Farmer, A Rosetta Stone for connectionism, Physica D 42 (1990) 153–187, these Proceedings.

[24] S.R. Farmer, G.S. Robinson, D. Mbangkollo, J.F. Bond, G.B. Knight, M.J. Fenton and E.M. Berkowitz, Differential expression of the B-tubulin multigene family during rat brain development, Ann. NY Acad. Sci. 466 (1986) 41–50.

[25] N.P. Franks and W.R. Lieb, Molecular mechanisms of general anesthesia, Nature 300 (1982) 487–493.

[26] W.J. Freeman, Mass Action in the Nervous System (Academic Press, New York, 1975).

[27] H. Fröhlich, Long range coherence and the actions of enzymes, Nature 228 (1970) 1093.

[28] H. Fröhlich, The extraordinary dielectric properties of biological materials and the action of enzymes, Proc. Natl. Acad. Sci. 72 (1975) 4211–4215.

[29] H. Fröhlich, Coherent excitations in active biological systems, in: Modern Bioelectrochemistry, eds. F. Gutmann and H. Keyzer (Plenum Press, New York, 1986) pp. 241–261.

[30] M.S. Gazzaniga, The Social Brain – Discovering the Networks of the Mind (Basic Books, New York, 1985).

[31] I. Ginzburg, A. Teichman and V.Z. LiHaver, Isolation and characterization of two rat alpha-tubulin isotopes, Ann. NY Acad. Sci. 466 (1986) 31–40.

[32] S. Grossberg, ed., The Adaptive Brain I. Cognition, Learning, Reinforcement and Rhythm (North-Holland, Amsterdam, 1987).

[33] W. Grundler and F. Keilmann, Sharp resonances in yeast growth prove nonthermal sensitivity to microwaves, Phys. Rev. Lett. 51 (1983) 1214–1216.

[34] F. Gutmann, Some aspects of charge transfer in biological systems, in: Modern Bioelectrochemistry, eds. F. Gutmann and H. Keyzer (Plenum Press, New York, 1986) pp. 177–197.

[35] S.R. Hameroff, Ultimate Computing: Biomolecular Consciousness and Nanotechnology (North-Holland, Amsterdam, 1987).

[36] S.R. Hameroff and R.C. Watt, Information processing in microtubules, J. Theor. Biol. 98 (1982) 549–561.

[37] S.R. Hameroff, S.A. Smith and R.C. Watt, Nonlinear electrodynamics in cytoskeletal protein lattices, in: Nonlinear Electrodynamics in Biological Systems, eds. W.R. Adey and A.F. Lawrence (Plenum Press, New York, 1984) pp. 567–583.

[38] S.R. Hameroff, S.A. Smith and R.C. Watt, Automaton model of dynamic organization in microtubules, Ann. NY Acad. Sci. 466 (1986) 949–952.

[39] S.R. Hameroff, S. Rasmussen and B. Mansson, Molecular automata in microtubules: basic computational logic for the living state?, in: Artificial Life, the Santa Fe Institute Studies in the Sciences of Complexity, Vol. VI, ed. C. Langton (Addison–Wesley, Reading, MA, 1989) pp. 521–553.

[40] D.O. Hebb, The Organization of Behavior (Wiley, New York, 1949).

[41] J.J. Hopfield, Neural networks and physical systems with emergent collective computational abilities, Proc. Natl. Acad. Sci. 79 (1982) 2554–2558.

[42] J. Jaynes, The Origin of Consciousness and the Breakdown of the Bicameral Mind (Alan Payne, Penguin Books, London, 1976).

[43] M. Karplus and J.A. McCammon, Protein ion channels, gates, receptors, in: Dynamics of Proteins: Elements and Function, Ann. Rev. Biochem., ed. J. King (Benjamin/Cummings, Menlo Park, 1983) pp. 263–300.

[44] M. Karplus and J.A. McCammon, Protein structural fluctuations during a period of 100 ps, Nature 277 (1979) 578.

[45] M. Kirschner and T. Mitchison, Beyond self assembly: from microtubules to morphogenesis, Cell 45 (1986) 329–342.

[46] T. Kobayashi, S. Tsukita, S. Tsukita, Y. Yamamoto and G. Matsumoto, Subaxolemmal cytoskeleton in squid giant axon, I. Biochemical analysis of microtubules, microfilaments, and their associate high-molecular-weight proteins, J. Cell. Biol. 102 (1986) 1699–1709.

[47] C.G. Langton, Computation at the edge of chaos, Physica D 42 (1990) 12–37, these Proceedings.

[48] R.J. Lasek, The dynamic ordering of neuronal cytoskeletons, Neurosci. Res. Prog. Bull. 19 (1981) 7–31.

[49] J.C. Lee, D.J. Field, H.J. George and J. Head, Biochemical and chemical properties of tubulin subspecies, Ann. NY Acad. Sci. 466 (1986) 111–128.

[50] M.R. Lindeburg, Engineer in training review manual professional publications, San Carlos, CA (1982).
[51] G. Lynch and M. Baudry, Brain spectrin, calpain and long-term changes in synaptic efficacy, Brain. Res. Bull. 18 (1987) 809–815.
[52] S. Mascarenhas, The electret effect in bone and biopolymers and the bound water problem, Ann. NY Acad. Sci. 238 (1974) 36–52.
[53] G. Matsumoto and H. Sakai, Microtubules inside the plasma membrane of squid giant axons and their possible physiological function, J. Membr. Biol. 50 (1979) 1–14.
[54] R. Mileusnic, S.P. Rose and P. Tillson, Passive avoidance learning results in region specific changes in concentration of, and incorporation into, colchicine binding proteins in the chick forebrain, Neur. Chem. 34 (1980) 1007–1015.
[55] M. Minsky, The Society of Mind (Simon and Schuster, New York, 1986).
[56] H. Moravec, Mind Children (University Press, San Francisco, 1987).
[57] S. Ochs, Axoplasmic Transport and Its Relation to Other Nerve Functions (Wiley-Interscience, New York, 1982).
[58] N. Packard, Adaption towards the edge of chaos, preprint (1988).
[59] K.H. Pribram, Languages of the Brain: Experimental Paradoxes and Principles (Brandon House, New York, 1971).
[60] B. Pullman and A. Pullman, Electronic delocalization and biochemical evolution, Nature 196 (1983) 1127–1142.
[61] G.R. Reeke and G.M. Edelman, Selective networks and recognition automata, in: Computer Culture: The Scientific, Intellectual, and Social Impact of the Computer, ed. H.R. Pagels, Ann. NY Acad. Sci. 426 (1984) 181–201.

[62] L.E. Roth and D.J. Pihlaja, Gradionation: hypothesis for positioning and patterning, J. Protozoology 24 (1977) 2–9.
[63] A.C. Scott, Neurophysics (Wiley, New York, 1977).
[64] S.A. Smith, R.C. Watt and S.R. Hameroff, Cellular automata in cytoskeletal lattices, Physica D 10 (1984) 168–174.
[65] D. Soifer, Factors regulating the presence of microtubules in cells, in: Dynamic Aspects of Microtubule Biology, ed. D. Soifer, Ann. NY. Acad. Sci. 466 (1986) 1–7.
[66] G.C. Somjen, Neurophysiology, The Essentials (Williams and Wilkins, Baltimore, 1983).
[67] H. Stebbings and C. Hunt, The nature of the clear zone around microtubules, Cell Tissues Res. 227 (1982) 609–617.
[68] W.E. Theurkauf and R.B Vallee, Extensive cAMP-dependent and cAMP-independent phosphorylation of microtubule associated protein 2, J. Biol. Chem. 258 (1983) 7883–7886.
[69] S.N. Timasheff, R. Melki, M.F. Carlier and D. Pantaloni, The geometric control of tubulin assemblies: cold depolymerization of microtubules into double rings, J. Cell Biol. 107 (1988) 243a.
[70] S. Tsukita, S. Tsukita, T. Kobayashi and G. Matsumoto, Subaxolemmal cytoskeleton in squid giant axon, II. Morphological identification of microtubules and microfilament-associate domains of axolemma, J. Cell Biol. 102 (1986) 1710–1725.
[71] P. Vassilev, M. Kanazirska and H.T. Tien, Intermembrane linkage mediated by tubulin, Biochem. Biophys. Res. Comm. 126 (1985) 559–565.

LIST OF CONTRIBUTORS

Aine, C.J.   411

Baird, B.   365
Banzhaf, W.   257

Caruana, R.A.   244
Churchland, P.M.   281
Compiani, M.   202

Eshelman, L.J.   244

Farmer, J.D.   153
Feldberg, R.   111
Flynn, E.R.   411
Forrest, S.   1, 213

George, J.S.   411
Greening, D.R.   293

Haken, H.   257
Hameroff, S.   428
Hanson, S.J.   265
Harnad, S.   335

Herring, Ch.   99
Hillis, W.D.   228
Hindsholm, M.   111
Hofstadter, D.R.   322
Hogg, T.   48
Holland, J.H.   188
Huberman, B.A.   38, 48

Ikegami, T.   235

Jensen, K.S.   428

Kaneko, K.   235
Kanter, I.   273
Karampurwala, H.   428
Kauffman, S.A.   135
Keeler, J.D.   396
Kephart, J.O.   48
Knudsen, C.   111

Langton, C.G.   12

Machlin, R.   85

Maxion, R.A.   66
Miller, J.H.   213
Mitchell, M.   322
Montanari, D.   202

Omohundro, S.M.   307

Palmore, J.   99

Rasmussen, S.   111, 428
Reeke Jr., G.N.   347

Schaffer, J.D.   244
Serra, R.   202
Siegel, R.M.   385
Sporns, O.   347
Stout, Q.F.   85

Vaidyanath, R.   428

Wilson, S.W.   249

ANALYTIC SUBJECT INDEX

abduction   281
abstract perception   322
adaptability   111
adaptation   66, 188
analogy   322
anomaly detection   66
artificial life   111
assembly/disassembly of microtubules   428
asynchronous algorithms   293
autocatalytic networks   153
automata   347

bifurcation theory   365
Boolean networks   135, 213
brain models   428
busy beaver problem   85

category learning   335
cellular automata   12, 111, 273, 428
cerebellum   396
chaos   385
classification   257
classifier systems   153, 202, 213
co-evolution   48, 135, 228
cognition   428
cognitive models   322, 335
collective phenomena   48
combinatorial optimization   293
complexity classes   12
complexity measure   111
computation   12
computational ecosystems   48
computational geometry   307
computer arithmetic   99
computer virus   111
concepts   322
concurrent computation   38
conditional branching   99
connectionism   153, 335
cooperative algorithms   38
cortex   385

cortical oscillations   365
credit assignment   249
critical slowing-down   12
cytoskeleton   428

deterministic chaos   99
diagnosis   66
distributed computing   293

emergence   322
emergent behavior   244
emergent computation   66, 111, 249, 307
emergent systems   85
energy function   257
evolution   235
evolvability   135
explanation   281

fitness landscapes   135
fractals   99

genetic algorithms   188, 244, 249

halting probability   85
halting problem   12, 85
host   235

IEEE Standard 754   99
immune networks   153

learning   202, 213, 249, 265
learning algorithms   307, 365, 428
local optima   228
lookahead   188
loose symbiosis   235

machine learning   188, 244
metadynamics   153
microtubules   428
motor control   396
mutation   235

natural selection   428
neural architecture   411
neural models   335
neural networks   153, 244, 257, 265, 273, 281, 307, 347, 365, 385, 396, 428
neuromagnetic measurements   411
neurons   385
nonlinear dynamical systems   202, 213
nonlinear dynamics   48, 385

open-ended evolution   111
optimization   228
origin of life   111

parallel processing   293
parallel-processing networks   135
parasites   228, 235
pattern recognition   244, 365
perception   281
perceptron   249
phase transitions   12
prediction   48, 396
prototypes   281

Red Queen hypothesis   228
representing normal behavior   66
robotics   307, 347

search   265
self-organization   111
simulated annealing   293
specialization   257
statistical physics   273
stochastic environments   202
strange attractors   385
symbiosis   235
symbol systems   335
synaptic noise   265
synaptic plasticity   428

TIT for TAT   235
transients   12
Turing machines   85, 347

universal law   38
unsupervised learning   257

vision   347, 385, 411
von Neumann machine   111